CLINICAL CHEMISTRY

A Fundamental Textbook

CLINICAL CHEMISTRY
A Fundamental Textbook

Donald F. Calbreath, PhD

Diplomate of the American Board of Clinical Chemistry
Associate Professor and Chair
Chemistry Department
Whitworth College
Spokane, Washington

Objectives and Review Questions Prepared by
Anna Ciulla, MCC, MT(ASCP)SC

Director, Medical Technology Program
University of Delaware
Newark, Delaware

W. B. SAUNDERS COMPANY

Harcourt Brace Jovanovich, Inc.

Philadelphia   London   Toronto   Montreal   Sydney   Tokyo

W. B. Saunders Company
Harcourt Brace Jovanovich, Inc.

The Curtis Center
Independence Square West
Philadelphia, PA 19106

Library of Congress Cataloging-in-Publication Data

Calbreath, Donald F.
    Clinical chemistry : A fundamental textbook / Donald F.
Calbreath.
        p.   cm.
    ISBN 0-7216-2621-1
    1. Clinical chemistry.   I. Title.
    [DNLM: 1. Chemistry, Clinical.   QY 90 C143c]
    RB40.C235   1992
    616.07′56—dc20
    DNLM/DLC   91-18611

Editor: Selma Ozmat
Developmental Editor: Martha Tanner
Designer: Bill Donnelly
Production Manager: Peter Faber
Manuscript Editor: Linda Davoli
Illustration Specialist: Peg Shaw
Indexer: Alexandra Nickerson
Cover Designer: Karen O'Keefe
Cover Photo: Dominique Sarra/THE IMAGE BANK

Clinical Chemistry: A Fundamental Textbook   ISBN 0-7216-2621-1

Printed in the United States of America.

Last digit is the print number: 9   8   7   6   5   4   3   2   1

To Sandy, with everlasting love.

PREFACE

• • • • • •

The field of clinical chemistry is complex and rapidly growing. Demands made on the laboratory staff also change rapidly. And yet, there are many fundamental concepts that must be comprehended beyond the flashing lights and clicks and hisses of our machinery.

This book is designed for the bench technician who must know the basic theory of testing, but also has some interest in the application of the data generated by the laboratory. This text focuses on the chemistry and the testing process. Principles of instrumentation are discussed from the perspective of broad concepts, leaving the operational details of a particular piece of equipment to the instruction manual provided by the manufacturer.

Enzymology and immunoassay techniques are closely intertwined in this book (Chapters 8–10). First, the fundamental concepts of enzymes and general approaches for measuring their activity (Chapter 8) are introduced. These basic principles are then applied to the development of the many enzyme-linked immunoassay techniques currently available (Chapter 9). The third chapter in the sequence (Chapter 10) utilizes both the concepts of enzymology and the applications of immunoassays to explore areas of clinical enzymology when both measurement of activity and determination of mass by immunoassay are currently of significance.

Learning objectives in each chapter indicate the direction of the chapter material and provide concrete ways to assess mastery. Learning activities are available at the end of each chapter to challenge the student in applying the principles discussed. An accompanying manual for the instructor provides additional resources for specialized material to enhance the instructional task. It also includes an Answer Key to the review questions that appear at the end of each chapter.

There is currently a great deal of debate about the level of expertise required by each segment of the laboratory community. The philosophy guiding this book is one of "stretching"—more is being demanded of the bench technician at all levels of training. New technology breeds new testing procedures. Staffing and salary realities frequently redesign job descriptions. Some of the traditional barriers to more complex responsibilities within the laboratory are falling. Each member of the technical staff (no matter what the educational background) must gain new proficiencies and cannot remain static. While focusing on current practical needs within the laboratory, this text also gently moves the technician into new territory and new functions as technology grows and changes.

DONALD F. CALBREATH, PH.D.

ACKNOWLEDGMENTS

• • • • • • • • • • • • •

This book is the product of the labors of many people and the support of caring family and friends. I can mention only some of the more significant individuals who have played a part in making this textbook possible.

Special thanks go to Larry Armstrong, who listened to a dream I had and started the process that has led to this book. Selma Ozmat, my editor (and many people before her), provided the push necessary when I became tired and somewhat frustrated. Her enthusiasm and support have been invaluable in bringing the book into production. My developmental editor, Martha Tanner, has been gentle with her corrections and suggestions. Linda Davoli, my copy editor, provided careful scrutiny of the text to ensure accuracy and clarity while offering encouragement through some very tight deadlines. Anna Ciulla, who wrote the objectives and review questions, has given helpful input.

My Whitworth College "family" has followed the progress of this book with interest and encouragement. The first question many students have had after returning from vacation is, "How is the book coming, Dr. C?" Other faculty and administration have been equally enthusiastic about its progress.

Special thanks go to my family for their ongoing encouragement and toleration of my early risings, tight deadlines, and times of distraction as I wrestled with a topic or a decision about some area of the text. Thank you, Gwendolyn and Caroline (and our lovely new granddaughter Marisa), for your patience. Special thanks to my wife Sandy, who has lived with this creative effort, has been patient and loving, and has provided the supportive atmosphere necessary for this book to be written.

CONTENTS

• • • • • • • •

CLINICAL CHEMISTRY
A Fundamental Textbook

Chapter One

THE SCOPE OF CLINICAL CHEMISTRY

UPON COMPLETION OF THIS CHAPTER, THE STUDENT WILL BE ABLE TO

1. Define clinical chemistry.
2. Discuss how the ability to measure small amounts of body fluids has contributed to health care.
3. Discuss how the development of the pH meter and the colorimeter has contributed to the growth of the field of clinical chemistry.
4. Discuss why the invention of the Technicon Autoanalyzer revolutionized chemistry testing.
5. Explain the role of automation in the clinical chemistry laboratory.
6. Discuss how laboratory testing has been facilitated by the coupling of computers with automated instruments.
7. Identify the techniques that have contributed to the laboratory's ability to accurately measure specific hormones, enzymes, and drugs in body fluids.
8. Explain how the following areas of interest in clinical chemistry may affect health care in the future:
   A. cholesterol and lipoprotein analyses
   B. biochemical markers for mental illness
   C. glucose and glucose products of long-term formation
9. Discuss the expected impact the following scientific areas will have on the future development of clinical chemistry:
   A. DNA probes
   B. forensic chemistry
   C. drug screening
   D. biosensors
   E. bedside testing
10. Name the two nonscientific factors that are expected to affect clinical chemistry testing and health care in the future.

INTRODUCTION

The field of clinical chemistry is broad, encompassing many scientific disciplines (Fig. 1–1). We might define clinical chemistry as the study of biochemical processes associated with health and disease and the measurement of constituents in body fluids or tissues to facilitate diagnosis of disease. In recent years, the concept has been expanded to include studies that monitor the effect of treatment, either by noting changes in specific parameters (for example, enzyme levels after a myocardial infarction) or by measuring the drug levels in blood or other body fluids to see that appropriate therapy is being administered. To the clinical chemist, any measurement of a biochemical parameter falls within the domain of the field.

Clinical chemistry is both an old and a new science. Diagnosis of diabetes mellitus by analysis of urine sugar dates back to the Middle Ages (the test involved tasting the sample to see how sweet it was). As scientific knowledge developed during the 19th century, it was quickly applied to issues related to health. Current technology has brought tremendous advances in the field of clinical chemistry. As we look to the future, we see an expansion of knowledge and techniques applied to the investigation and monitoring of disease states.

1.1

HISTORICAL BACKGROUND

The field of chemistry exploded in the early 1800s when Friedrich Wöhler synthesized a compound, urea, which had been thought to be produced only by living beings. This breakthrough led to exploration of all life phenomena, since the belief arose that life was really only a series of chemical reactions (we are slowly coming to see that this assumption is not valid). Measurement of biochemical changes developed slowly, because only crude techniques were available for quantitation. In spite of these limita-

tions, tremendous progress was made toward a basic understanding of the composition of "living" material.

By the end of the 19th century a significant amount of biochemical information was available: The compositions of starch and fats were known. A number of proteins from blood had been isolated and characterized. Cholesterol was shown to be present in gallstones. The chemical make-up of urine was elucidated, and the measurement of urine sugar in diabetics was carried out by 1815. Concurrently, carbohydrate metabolism was also being actively investigated, including studies on blood levels of carbohydrates in both healthy and ill individuals. The first clinical chemistry text was written in 1836.

1.2

IMPROVEMENTS IN INSTRUMENTATION

Progress continued into the 20th century, but was hindered greatly by the lack of methods allowing analysis of small amounts of material. Early studies required isolation and chemical identification of the component of interest, often including crystallization of the pure material and study of melting point, chemical composition, and other necessary parameters. Analyses were slow, tedious, and usually not very accurate.

Two developments contributed to the revolution in clinical chemistry (in fact, to all of chemistry) in the 1900s. One was the invention of the pH meter. First devised to measure the acidity of citrus fruits, this instrument was quickly adapted to a wide variety of applications. With this instrument we could control the pH of an enzyme assay and explore the effect pH had on the rate of reaction. Modifications of the basic instrument led to the development of specific measuring devices for blood pH, allowing identification and treatment of acid-base disorders. An important parameter in all biochemical reactions could now be monitored and carefully controlled.

The other significant invention was the colorimeter. Prior to the development of this device, the relative colors of solutions were determined by comparing them to standards through visual observation. This approach was tedious and inaccurate. It does not take long for the eye to become tired and for mistakes to be made. The electronic measurement of the amount of light absorbed by a solution greatly enhanced both accuracy of measurement and productivity in the laboratory. From this basic concept came the variety of automated instruments and computer-driven systems we have today.

Even with these improved means of measurement, the processing of laboratory specimens was

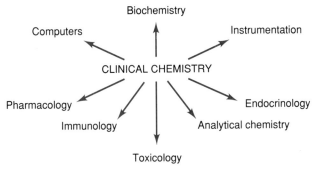

FIGURE 1–1 The scope of clinical chemistry.

still a tedious job. All pipeting was done by hand (and mouth; pipeting bulbs were not required and slowed the process). Rows of technicians would stand at the lab bench analyzing batches of samples, watching the clock on the wall as their timer, and getting progressively wearier and more inaccurate as the day went on. As the demands on the laboratory grew, the human capability for meeting those demands became less and less adequate.

One solution resulted from a clinical chemist's enjoyment of tinkering in his garage. Dr. Leonard Skeggs became concerned about the workload and the strain on his staff. In his spare time, he began to develop a device that would automatically pipet a sample, add reagents to it, mix and incubate the resulting solution, and measure the color change with a colorimeter. The initial model would run only one test, but it never got tired. Industry reception of the device was lukewarm; it took a small company to see the potential in this crude piece of equipment. But the Technicon Autoanalyzer in all its varied forms quickly became a mainstay of the clinical laboratory. The instrument has grown from a piece of equipment that could perform only one assay to the modern 20-channel computer-driven systems capable of producing well over a thousand test results in an hour. The tedious generation of data was at an end, and the human component of the laboratory could move on to more complicated and more interesting matters. In recent years, this approach to automation has been supplanted by a variety of other techniques, but it still represents an important contribution to the field.

Other instrumentation followed close behind. As the use of radioisotopes became more prevalent in the 1970s, scintillation counting devices grew in sophistication. No longer was it sufficient for an instrument simply to register the number of radioactive decay particles produced per unit time; computers provided automatic correction for background, plotting of standard curves in a variety of formats, and the calculation of results. The lengthy process of data reduction was now done automatically. This change eliminated a source of human error and again provided more time for other necessary activities.

Similar developments took place all over the clinical chemistry laboratory. Other automated chemistry systems were developed to meet demands other than the mass analysis of large numbers of samples. As nonisotopic immunoassay grew in popularity, specific equipment was constructed to measure the products of these reactions. The simple device built by Skeggs led to a significant shift in our concepts of how a laboratory should be operated. The coupling of automated instrumentation with computers now gives us the ability to process and monitor all the data produced, to assure accuracy of measurement, and to look for new relationships among previously isolated bits of information.

1.3

EXPLOSION IN BIOCHEMICAL KNOWLEDGE

Our skill in instrumentation provided us with new tools for probing more deeply the mysteries of the universe. Measurement at lower and lower levels of concentration became commonplace. With immunoassay techniques we could now study the amount of specific hormones in the body and explore the changes in the levels of these compounds in health and disease. New kinetic and electrophoretic techniques permitted rapid and accurate analysis of enzyme activity. This research opened the door to the study of isoenzymes, an important component of the diagnostic process in many disease states. Measurement of pharmacologic parameters furnishes information on the changing levels of drugs after administration. In addition, we now know that some drug metabolites may be therapeutically effective in their own right. Improvement in the techniques and database for the clinical laboratory sciences now makes the clinical chemistry laboratory an indispensable partner in the diagnosis and treatment process.

1.4

CURRENT AREAS OF INTEREST

Only a few of the areas of interest in the clinical chemistry laboratory can be highlighted in the limited space of this book. The accurate measurement of serum cholesterol has taken on new importance with the growing interest in healthy life-style. In addition, other predictors of heart disease (such as lipoproteins) are being intensively studied. The effects of diet and exercise, as well as genetic influences, are observed, and improvements in risk factors for heart disease are more readily monitored.

The study of biochemical markers for mental illness is still in its infancy, but some significant contributions are being made in the clinical chemistry laboratory. There appears to be a significant relationship between depression and plasma cortisol levels during the day and night. A number of biochemical constituents in platelets and cerebrospinal fluid (CSF) are being intensively studied as possible diagnostic aids in schizophrenia.

Measurement of long-term formation of products of impaired glucose metabolism may lead to improved prediction and monitoring of the damaging effects of diabetes mellitus. Development of devices for personal measuring of glucose and more effective administration of insulin should allow individual patients to be better able to manage their disease.

Increasing awareness of biochemical rhythms in our bodies is leading to more sophisticated means of documenting these regular and important changes within us. These biochemical rhythms regulate other processes, among which may be mental illness and problems of infertility and the regulation of reproduction.

Developments in the field of immunology are many, both in the diagnosis of diseases that produce altered levels of proteins in body fluids and in the use of antibodies as analytical agents. Arthritis and related disorders are better understood and more readily diagnosed with new laboratory tests. Allergy testing and treatment have improved greatly as a result of developments in clinical laboratory diagnosis.

1.5

THE FUTURE—SCIENTIFIC AND POLITICAL

Even with the explosion of knowledge that brought us to our present position, there are still many important roads to explore. Some significant trends in technology will influence analysis in the clinical chemistry laboratory of the future. DNA probes are increasingly anticipated to be a major tool for biochemical studies in the next several years. The ability to study genetic disease at the gene level permits rapid and accurate diagnosis of metabolic disorders and other inborn errors such as muscular dystrophy.

The field of forensic chemistry will grow, and the clinical laboratory may well be a part of that growth. DNA measurements can now be used for identification purposes to determine paternity and identify criminals. The technique is not trouble-free and has experienced some drawbacks when used in court, but it has shown itself to be valuable in many instances.

Both technologic and societal forces drive the increased market for drug screening. Advertisements now appear in magazines to encourage home use of drug-screening techniques. Substance abuse in the workplace has become a significant issue. Improved instrumentation (such as a small tabletop gas chromatograph/mass spectrometer) permits more access to drug screening at lower cost.

Sensors have become more varied and sophisticated in their construction and application. The formerly clumsy pH electrode system is now a miniaturized electrode. Ion-selective electrodes can measure specific constituents in body fluids with high accuracy. Biosensors that use "living" material as part of the sensing process are being developed.

Laboratory testing is moving closer to the patient. Blood glucose levels may be measured at the bedside or at home. Assessment of occult blood (in the screening for colon cancer) can now be done with a kit on an outpatient basis. Home pregnancy tests can detect pregnancy with a minimal effort and complete privacy, although with less than perfect reliability. The educational role of the laboratory staff is sure to grow as delivery of services shifts from a central laboratory to a variety of other settings.

Shifting population demographics are beginning to have a major impact on the delivery of health care. Improvements in health-care technology have resulted in a growing geriatric patient population. Instead of the acute (but relatively short-term) needs of younger patients, we are experiencing expansion of long-term care. This shift in patient demographics means that laboratory personnel must be more aware of the biochemical differences between older and younger patients. Laboratory values often must be assessed according to different criteria as a person ages. In some instances, we have only recently recognized this factor as an important part of the interpretive role of the laboratory.

Nutritional assessment has taken on new importance as a vital component of the healing process. Current testing focuses primarily on protein measurements and some determinations of mineral status. Proper nutrition is of great benefit, particularly to elderly patients, in decreasing the severity of an illness and improving the rate of recovery.

The future direction of clinical chemistry will be determined only in part by the scientific advances made and the new relationships and techniques we discover. Major change will be driven more by political and economic factors. The rising cost of health care has led to a close examination of what the billions of dollars we have spent have bought us. The need for certain laboratory services and the quality of those services are being carefully scrutinized, both inside and outside the field. Limits on the numbers of health-care providers will create new approaches to the delivery of services, including in the laboratory. The 1980s have seen some significant shifts in how laboratory services are provided and reimbursed; the long-term effect on policy is not yet clear. The technologists of today must be open to change, both in understanding the scientific developments that take place in the world around them and in dealing with the political realities and financial constraints already appearing.

SUMMARY

The field of clinical chemistry saw its real inception in the 19th century with the development of organic chemistry and biochemistry. Significant progress was made in measurement of biochemical constituents but was hindered by the lack of precise and reliable techniques for measuring low levels of materials in small volumes of body fluids. The development of instrumentation (particularly the pH meter and the colorimeter) opened the door to accurate quantitative techniques and reasonable time frames for analysis. With automation, the clinical chemistry laboratory was finally able to provide timely information on the ever-growing number of samples. Developments in modern biochemistry quickly found their way into the realm of clinical chemistry as the areas encompassed by the field expanded. Current forces working to shape the field are both scientific and political, with financial factors driving the direction and practice of clinical chemistry almost as much as the development of new scientific ideas.

FOR REVIEW

Directions: For each question, choose the best response.

1. The study of the biochemical processes associated with the analysis of body fluids to diagnose disease best describes
   A. analytical chemistry
   B. biochemistry
   C. clinical chemistry
   D. physical chemistry

2. Testing performed in a clinical chemistry laboratory was revolutionized by the invention of the
   A. analytical balance
   B. microscope
   C. refractometer
   D. colorimeter

3. The instrument credited with being the first automated analyzer to be used in the clinical chemistry laboratory is the
   A. autoanalyzer
   B. automatic clinical analyzer
   C. sequential multiple analyzer
   D. scintillation counting system

4. The fact that modern automated instruments require a small volume of sample from a patient to perform a large number of assays is advantageous since most chemistry analyses are performed on
   A. cerebrospinal fluid
   B. pleural fluid
   C. serum
   D. urine

5. Processing and monitoring of test data and patient demographics have been greatly facilitated by the development of
   A. automated instruments
   B. computers
   C. fluorometry
   D. spectrophotometry

6. The diagnosis of diseases that produce antibodies and the use of antibodies as analytical reagents have become possible through advances made in the area of
   A. cytology
   B. histology

C. immunology

D. pathology

7. Which of the following will affect health care in the future?

A. scientific developments

B. political factors

C. financial factors

D. both A and C

E. A, B, and C

BIBLIOGRAPHY

Annesley, T., and Patel, J., "Extending the clinical laboratory into forensic urine drug testing," *J. Clin. Immunoassay.* 12:205–211, 1989.

Barman, M. R., "Personnel licensure: A raging issue that won't go away," *Med. Lab. Observer.* January: 22–29, 1990.

Bernstein, L. H., "The lab's role in nutritional assessment of patients," *Med. Lab. Observer.* April: 25–27, 1988.

Bernstein, L. H., "New markers of nutritional status," *Am. Clin. Lab.* February: 20–23, 1989.

Caplan, Y. H., and Levine, B., "Laboratory testing in forensic postmortem cases," *Am. Clin. Lab.* October: 8–20, 1988.

Castleberry, B. M., and Kuby, A. K., "Who will staff the laboratory of the '90s?," *Med. Lab. Observer.* July: 59–66, 1989.

"Clinical chemistry's man on horseback—Leonard T. Skeggs, Jr.," *Chem. Eng. News.* August 10, 1970.

De Cresce, R. P., and Lifshitz, M. S., "The impact of the new technology," *Med. Lab. Observer.* July: 41–46, 1989.

Eisenstein, B. I., "The polymerase chain reaction. A new method of using molecular genetics for medical diagnosis," *N. Engl. J. Med.* 322:178–183, 1990.

Eisner, R., "Drug-of-abuse testing: Changing law, changing markets," *Diagn. Clin. Testing.* 27:20–23, 1989.

Hood, L., "Biotechnology and medicine of the future," *JAMA* 259:1837–1844, 1988.

Koepke, M. D., et al., "Legal aspects of self-testing," *Med. Lab. Observer.* November: 22–26, 1989.

Lefebvre, R. C., "The accuracy of those proliferating screening programs," *Med. Lab. Observer.* March: 51–60, 1990.

Lifshitz, M. S., and De Cresce, R. P., "Automation: Trends in instrumentation, robotics, computers," *Med. Lab. Observer.* July: 73–77, 1989.

Mani, N., "The historical background of clinical chemistry," *J. Clin. Chem. Clin. Biochem.* 19:311–322, 1981.

Moody, M. D., "DNA analysis in forensic science," *Bioscience.* 39:31–36, 1989.

Parsons, G., "Development of DNA probe-based commercial assays," *J. Clin. Immunoassay.* 11:152–160, 1988.

Rechnitz, G. A., "Biosensors," *Chem. Eng. News.* September 5: 24–36, 1988.

Rowe, J., "Health care of the elderly," *N. Engl. J. Med.* 312:827–835, 1985.

Schneider, B. H. et al., "Microelectrode probes for biomedical applications," *Am. Biotech. Lab.* February: 17–23, 1990.

Sisk, F. A., "Trends in regulation and reimbursement," *Med. Lab. Observer.* July: 49–55, 1989.

Statland, B. E., "How illness demographics will affect the lab," *Med. Lab. Observer.* July: 79–86, 1989.

CHAPTER OUTLINE

PRINCIPLES OF MEASUREMENT I: STATISTICS, REFERENCE VALUES, AND QUALITY ASSURANCE

UPON COMPLETION OF THIS CHAPTER, THE STUDENT WILL BE ABLE TO

1. Define the following terms:
   A. accuracy
   B. precision
   C. Gaussian distribution
   D. mean
   E. median
   F. mode
   G. standard deviation
   H. coefficient of variation
   I. 95% limits
   J. reference range
   K. quality control
2. Discuss the application of the Gaussian distribution and the 2-standard-deviation limit to evaluating quality control data.
3. Calculate the mean, standard deviation, and coefficient of variation for a given set of laboratory data.
4. Discuss how each of the following factors must be taken into consideration when developing reference values:
   A. age
   B. sex
   C. diet
   D. medications
   E. physical activity
   F. pregnancy
   G. personal habits
   H. body weight
   I. biologic rhythms
   J. assay methodology
5. Explain the function of using control materials when performing a chemical analysis.
6. Describe the differences between assayed and unassayed control sera in terms of:
   A. cost
   B. statistical database
   C. use
7. Discuss the importance of using quality control sera as part of a laboratory quality assurance program.
8. Explain the purpose of monitoring the mean, standard deviation, and coefficient of variation of quality control materials.
9. Evaluate the acceptability of control values using the Levy-Jennings plot and the Westgard rules as guides.
10. Explain the steps that should be followed when the values of control materials fall outside the 2-standard-deviation limit.
11. Differentiate between systematic and random errors in terms of possible causes and identification of the problem.
12. Explain the purpose of participating in a proficiency survey program.

INTRODUCTION

Clinical laboratories around the world generate a large amount of information every day. Much of that information is in the form of a value for a specific biochemical parameter. When we report a test result for serum cholesterol, two questions are always asked about the data:

1. How reliable is the result?
2. What does this value mean in relation to the physiologic health of the patient?

Understanding some basic statistics helps us answer both of these questions and provides confidence in the laboratory data we furnish the healthcare team.

2.1

THE GAUSSIAN DISTRIBUTION

If we analyze a specimen one time for a given constituent, we get one value, but we have no indication of the reliability of that number. We have no idea what value will be obtained for the same constituent in the same sample the next time the analysis is performed. The only way to acquire the necessary information concerning the reproducibility of the analysis is to perform the assay many times and evaluate the data using several statistical techniques.

When we perform a specific assay several times on the same sample, we find that all the results are not the same. If the method is reliable and the technician is meticulous, the values will all be reasonably close together, but they will not be identical. When we graph the data, plotting the assay values obtained on the *x* axis and the number of times each value was seen (frequency) on the *y* axis, a symmetrical bell-shaped curve is generated (Fig. 2–1). This graph is called a **Gaussian distribution,** after the mathematician Johann Karl F. Gauss, who developed many of these basic concepts.

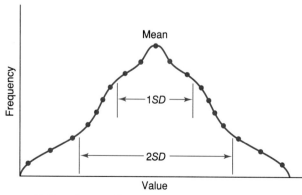

FIGURE 2–1 Gaussian distribution of values.

2.2

MEAN AND STANDARD DEVIATION

The peak of the curve (Fig. 2–1) indicates the mean value for the data set. The **mean** value is the average of all the values in the set if the numbers are distributed equally on both sides of the curve (a symmetrical curve). When the curve is nonsymmetrical (unequal distribution of values on the two sides), the definition for *mean* is still valid, but other descriptive terms also apply and will be defined later in this chapter.

We calculate the mean by adding all the individual values together and dividing by the number of values in the data set. If we have the numbers 12, 23, 56, 41, 10, 78, 52 as members of our data set, the sum of the individual values is 272. There are seven individual values in the data set, so the mean is calculated as follows:

$$\text{Mean} = \frac{\text{Sum}}{\text{Number of values}}$$
$$= \frac{272}{7}$$
$$= 38.86$$

The shape of this curve is also significant. Little spread on both sides of the mean indicates the values are all very close together. The term indicating the amount of difference among the individual data points is the **standard deviation** *(SD)*. This value can be calculated according to the following formula:

$$SD = \sqrt{\frac{\Sigma(x - \overline{x})^2}{n - 1}}$$

Most scientific calculators and statistical software for computers perform this calculation.

The practical meaning for the standard deviation lies in its giving an indication of how the values in a data set are distributed. Of all the values in the group of numbers, 68.7% of them will lie within 1 standard deviation from the mean; approximately 95.5% of the values fall within 2 standard deviations from the mean; and 99.9% fall within 3 standard deviations. Most statistical data will be reported as the mean value ± 1 standard deviation, unless some other range is specifically indicated.

Suppose we performed repeated analyses on a serum sample to determine the concentration of sodium. Our analyses gave a set of data with a mean of 137.6 mmol/L and a standard deviation of ±1.7 mmol/L. We would report the value as 137.6 ± 1.7 mmol/L for the sample. These figures indicate that 68.3% of the times we run an assay on the material we can expect to obtain a range of values between 135.9 and 139.3 mmol/L for the assay results (±1*SD*). More realistically, we would anticipate a

±2-standard-deviation range (95.5% of the assay results) of 134.4–141.0 mmol/L for our sodium results. In most circumstances, values that fall within a ±2-standard-deviation range are acceptable for assuring the reproducibility of an assay.

2.3

CONCEPT OF ACCURACY

As valuable as the Gaussian distribution is, it does not answer one important question: How accurate is the test result? **Accuracy** is an indication of how close the answer obtained lies to the "true" value. This parameter is not defined by statistics, but by comparison of the measured value to the "known" value obtained from a reference material. We might look at accuracy as showing how close we come to the bull's-eye of a target. The closer to the center of the target, the more accurate the result. We should note that a number of single values may not all be directly on target themselves, but the mean value of these single determinations may be accurate. Thus, it is important to replicate assays when establishing a standard value or a reference number for quality control data.

2.4

CONCEPT OF PRECISION

Although accuracy is not amenable to simple statistical expression, the concept of precision can be defined numerically. **Precision** indicates how close the single values are to one another and is expressed by the standard deviation. The smaller the standard deviation, the more precise the values are. If we have two methods for the analysis of our sodium pool, they may differ in their precision, as well as in their accuracy. One method may provide a standard deviation of ±1.3 mmol/L, whereas the other approach gives a standard deviation of ±2.8 mmol/L. The data indicate that the first method (smaller standard deviation) is more precise than the second.

Note that these figures say nothing about the relative accuracies of the two methods, only the degree to which the two sets of values agree with one another.

A term related to precision is **coefficient of variation** (*CV*). This parameter is expressed as a percentage and defined by the following equation:

$$CV(\%) = \frac{SD \times 100}{\text{Mean}}$$

Using our previous data for sodium analysis, we can determine the coefficient of variation for the data set:

$$CV = \frac{1.7 \times 100}{137.6}$$
$$= 1.24\%$$

The coefficient of variation can be monitored to assess consistency in precision of an assay.

Accuracy and precision are independent of one another. A set of data (Fig. 2–2) may be accurate as far as the mean value is concerned (set A), but the individual points may show a wide degree of variability and lack precision. Another set of values (set B) may all be very close to one another (good precision) and yet deviate markedly from the "true" value (poor accuracy). The goal of quality analysis is to achieve a high level of both precision and accuracy in the set of measurements (set C).

2.5

STATISTICS IN CLINICAL CHEMISTRY

The clinical chemistry laboratory relies on statistical measurement in several ways. Quality control procedures (which assure the reliable performance of instruments, materials, and personnel) are all based on the monitoring of the mean and standard deviation for any assay. Assessment of the presence of disease depends in many instances on the differentiation between a "normal" value for any parameter and the level seen in a given disease state. The method of analysis or the type of instrument is se-

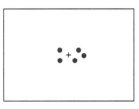

A. Good accuracy
Poor precision

B. Poor accuracy
Good precision

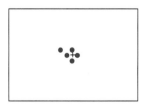

C. Good accuracy
Good precision

+ = Target value (reference value)
● = Values obtained from individual analyses

FIGURE 2–2 Relationship between accuracy and precision.

lected to assure reliable data and is determined in part by the accuracy and precision of the statistics of the system. Troubleshooting of procedures or equipment is assisted by proper statistical manipulation of data. As we see more computer monitoring of laboratory statistics, we become more aware of the important role played by these parameters in clinical chemistry.

2.6

CONCEPT OF "NORMAL"

We all have some intuitive sense of what is "normal." We know when we are in good health, and when ill we assess our degree of illness with reference to how we feel when well. "Oh, I'm just feeling a little run-down today" could signify a mild illness, but a comment such as "I won't make it through the day" or "I can hardly stand up" suggests more serious illness.

For the members of the health-care delivery team, these vague general comments are not enough. Expressing the amount of enzyme in serum after a heart attack as "pretty high" does not communicate sufficient information for diagnosis and treatment. Clinical laboratory measurements call for a degree of precision and accuracy far beyond the qualitative. Slight shifts in the values of some biochemical parameters may reflect significant changes in the clinical status. We need to be able to define clearly what values are consistent with health and what changes in analytical values constitute movement into a pathologic state.

The range of values for a given constituent in healthy individuals is referred to as the **reference range,** sometimes called the **normal range.** The term *reference range* (or *reference values*) better conveys the sense of an expected range of values without any implication of health or illness. This set of numbers simply indicates the range of values one would expect in a defined population with no apparent clinical problems.

2.7

THE GAUSSIAN DISTRIBUTION AND PATIENT VALUES

Application to Reference Range

The concepts of Gaussian distribution of data and standard deviation can be directly applied to our discussion of reference values. If we analyze fasting serum glucose in samples from 100 presumably healthy adults, the resulting set of values gives a Gaussian distribution when plotted. The mean serum glucose value represents the average of all the values in the population, and the standard deviation indicates the range of values. Note that the standard deviation in this case is a function of much more than the methodology. The range of values is much wider than that expected if the only variable were the care with which the assay was performed. Factors related to gender, age, and other variables contribute to the spread of results seen in any human population studied.

In our hypothetical situation, let us assume that the mean value for glucose in this particular population was 110 mg/dL and the standard deviation was ± 9 mg/dL. Therefore, 68.3% of the values fall between 101 mg/dL and 119 mg/dL (1SD). The 2SD range would be between 92 mg/dL and 128 mg/dL, encompassing 95.5% of the population investigated in this study.

The reference range is usually set at the mean $\pm$ 2 standard deviations, sometimes referred to as **95% limits.** In our example above, the reference range for this population would be 92–128 mg/dL. Note that this range encompasses only 95% of the presumably healthy group of people studied. The use of reference values carries with it the realization that for 5% of the healthy population (1 person in 20), the value falls outside the reference range. Reference range values cannot be used blindly as arbitrary cutoff points between "health" and "disease"; other factors need to be considered.

Overlap Between Healthy and Unhealthy Individuals

One other consideration further complicates the picture. Not every person with a specific disease shows an abnormal value for a given laboratory test. Using our example above and applying the data to a group of diabetic patients (in whom elevated serum glucose would be expected in the untreated disease), we find that a significant number of these patients may present with normal serum glucose values in the fasting state. Therefore, we see an overlap between the range of glucose values in our healthy group and in our diabetic group (Fig. 2–3). Frequently, testing under other conditions is required to discover the true clinical situation.

Precautions in Using Reference Values

There are several drawbacks to using reference values uncritically. Quite often the reference population studied was small (100 people or fewer). Blood sampling may not have been done under controlled conditions. Consideration may not have been given to all medications used by the subjects. Other factors, such as smoking or physical exercise, alter results and need to be taken into account. Reference values should be assessed carefully before they are used as a benchmark for diagnostic and monitoring purposes.

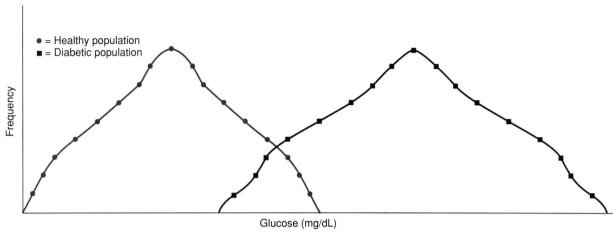

FIGURE 2-3   Overlap between normal and pathologic values.

2.8

NON-GAUSSIAN DISTRIBUTION

Although the Gaussian distribution plot may be valid for many data sets, in numerous situations the distribution of values does not result in this symmetrical curve. For example, in a study on enzyme values, we may see a curve similar to the one illustrated in Figure 2-4. There is a somewhat sharp rise to a cluster of values at the peak of the curve and then a gradual tailing off to higher values within the population. This pattern of data is referred to as a non-Gaussian, or **log-normal, distribution.**

With the log-normal distribution, we no longer talk of the mean value. Because distribution of values on both sides of the peak is not equal, we cannot mathematically take an average and have that rep-

resent the highest frequency of results. Instead we determine the **mode value,** the point where the most values lie. If our hypothetical glucose study gives this type of curve, we then examine the data closer. Suppose we find the following:

Value (mg/dL)	Frequency
90	5
95	20
100	37
105	26
110	19
115	11
120	6

From the table we can see that the value with the highest frequency is 100 mg/dL; this is our mode

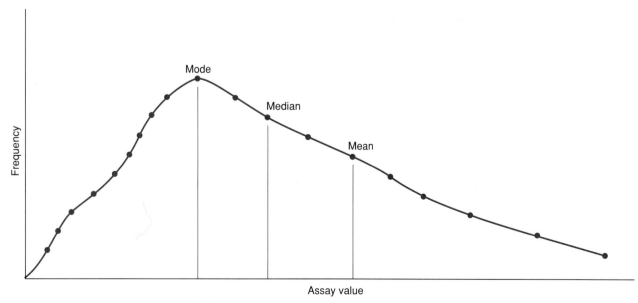

FIGURE 2-4   Non-Gaussian distribution of data.

value. The calculated mean, on the other hand, is higher (103.7 mg/dL), but does not represent the peak frequency of values.

Another parameter related somewhat to the mean and the mode is the median value. In our non-Gaussian distribution, the **median value** is the figure for which 50% of the values are higher and 50% of the values are lower. In our data, the median value lies somewhere between 100 and 105 mg/dL. Sixty-two of the data points are 100 mg/dL or less, and 61 are 105 mg/dL or more. With more detailed information on the subject data, we could calculate an exact figure for the median value. Note that the median value lies somewhere between the mode and the mean figures.

2.9

FACTORS AFFECTING REFERENCE VALUES

As we understand better what determines the "normalcy" of a set of laboratory data, a shift in terminology occurs, reflecting the growing complexity of the field. Not too many years ago, we spoke of the "normal range" and often gave the impression that there was only one such range for a given constituent. As our database increased, we began to see that each of various population groups had its own normal range. These sets of values overlapped to some extent but had enough differences to be noticeable. We now recognize that men and women may have different reference ranges for a specific biochemical parameter. These reference ranges may also differ depending on the age of the individual and a host of other factors, including the time of day. Our current understanding of reference ranges takes into account the nonpathologic factors that alter the level of a component of serum or other body fluids. There are no firm rules for how a specific parameter affects the reference range of any biochemical constituent. Each situation must be studied experimentally.

Age

There are marked differences between the normal values for several parameters of children and adults. Many enzymes are found in much higher quantities in the serum of children. The values change over the course of years in some instances. Blood levels of cholesterol and other lipids gradually rise with age. There is some debate about how much of this increase is healthy and at what point a medical problem exists. In the elderly, values for some serum proteins and enzymes may decrease from levels seen in younger adults. In part, this might reflect a less active life-style, or it may be a sign of shifts in diet. Immunoglobulin levels show significant changes within the first year after birth, indicating adaptation to life outside the womb and the loss of the pro-

Age (yr)	Mean Uric Acid Values (mg/dL)	
	Men	*Women*
20–29	6.2	4.8
30–39	6.4	4.4
40–49	6.6	4.6
50–59	5.7	5.3
60–69	5.8	5.3

Table 2–1
EFFECT OF AGE AND SEX ON URIC ACID VALUES IN URINE

Data from Goldberg, D. M., Handyside, A. J., and Winfield, D. A., "Influence of demographic factors on serum concentrations of seven chemical constituents in healthy human subjects," *Clin. Chem.* 19:395–402, 1973.

tective factors provided by the mother during gestation. Table 2–1 illustrates the changes in serum uric acid levels as a function of age.

Sex

Significant differences in reference values exist between men and women. Serum uric acid values for men tend to be higher than those for women (Table 2–1). Men also have levels for urea, magnesium, calcium, and iron that are greater than the values for women. Perhaps the most striking differences in reference values are those for the plasma levels of various hormones. Men have high testosterone levels, but the values for women are much lower. Estrogen levels show the reverse pattern, with women showing much higher values than men do. A number of other hormones also show sex-linked differences.

One interesting pattern of change that links both sex differences and changes in age involves the values for the enzyme alkaline phosphatase (Fig. 2–5). The serum levels of this enzyme for both boys and girls before puberty are higher than adult values,

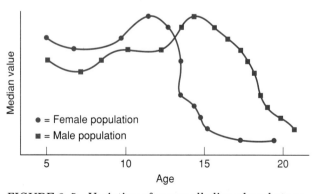

FIGURE 2–5 Variation of serum alkaline phosphatase values by age and sex. (Data adapted from Cherian, A. G., and Hill, J. G., "Age dependence of serum enzymatic activities (alkaline phosphatase, aspartate aminotransferase, and creatine kinase) in healthy children and adolescents." *Am. J. Clin. Pathol.* 70:783, 1978.)

with girls having somewhat more elevated values. At the onset of puberty, there is a marked increase in the serum level of this enzyme (owing to its role in bone formation). Because females usually undergo changes related to puberty earlier than males, we see some increase in serum levels during their early teens. The values tend to drop rather rapidly into the adult range after about age 15. Males, on the other hand, show a marked increase from about age 10 to age 16 or older. The decrease to adult values in young men may not take place until close to age 20.

Diet

One of the more controversial subjects today is the effect of blood levels of cholesterol on the development of heart disease. When we measure cholesterol levels in blood samples from a normal (non-diseased) population, we obtain a certain set of values. Over the years, this normal range has increased somewhat as our diet and our population changed. We have now learned that these normal individuals actually have cholesterol levels that are frequently higher than they should be. In this case, independent assessment of health status is needed to establish the reference range for cholesterol. These assessments have been extensively carried out to demonstrate the link between increased serum cholesterol and heart disease. In this case, it was not sufficient simply to measure the parameter of interest in apparently healthy persons and establish our reference range in the usual manner.

Often diet produces observable changes in blood levels of specific constituents, which represent long-term alterations that are not affected by fasting before the sample is collected. Carbohydrate and lipid metabolism are very sensitive to long-range dietary intake. Serum or plasma levels of components of these two classes of compounds fluctuate markedly as a result of dietary shifts. Quite often, vitamin levels and some trace elements (particularly iron) are elevated because of diet and not some clinical condition.

Medications

The effect of medications on body biochemistry is always a concern. Long lists of changes induced by pharmacologic agents are available from several other sources, but comprehensive coverage of this topic is not possible in the text. One class of compounds—the oral contraceptives—has been shown to produce changes in a wide variety of measurements (Table 2–2). Coagulation parameters are markedly affected by the presence of oral contraceptives. In addition, these medications alter thyroid function, leading to an elevated thyroid hormone level and a artifactual increase in the amount of thyroid-binding globulin seen. A complete medication history should be obtained for all patients in order

Table 2–2
EFFECT OF ORAL CONTRACEPTIVES ON LABORATORY VALUES IN SERUM

Laboratory Test	Effect Produced by Oral Contraceptives
Albumin	Moderate decrease
Antithrombin III	Marked decrease
Cholesterol	Slight/moderate increase
All coagulation factors	Slight/moderate increase
Folate	Marked decrease
Glucose tolerance	Slight/moderate decrease
Iron	Moderate increase
Iron-binding capacity	Marked increase
Sodium	Moderate increase
Thyroxine	Moderate increase
Thyroid-binding globulin	Moderate increase

Adapted from Miale, J. B., and Kent, J. W., "Effects of oral contraceptives on the results of laboratory tests," *Am. J. Obstet. Gynecol.* 120:264–272, 1974.

to interpret appropriately the meaning of abnormal results.

Physical Activity

The amount of physical activity, particularly activity occurring during the collection of timed samples, often profoundly affects laboratory results. During a glucose tolerance test, the normal rise and fall of blood glucose after a glucose load is administered becomes erratic if a person is physically active. Serum albumin levels show a small increase immediately after strenuous exercise. Hospitalized patients may demonstrate decreased serum albumin values, particularly if they are long-term patients. Creatine kinase in plasma is elevated after intense exercise and tends to be lower in persons who are not very active physically. Some lipoprotein values are altered by physical exercise.

Pregnancy

Pregnancy is a dynamic, ever-changing state. A large number of biochemical parameters are altered during this time (Table 2–3). The value for blood levels of several common parameters usually increases moderately; as the pregnancy progresses, increases in uric acid, cholesterol, cortisol, alkaline phosphatase, and lactate dehydrogenase can be detected. Slight decreases are observed in values for calcium and glucose. The alterations are more apparent at different stages in the pregnancy. Periodic monitoring of pregnant women being treated for other medical conditions allows for detection of trends that could prove hazardous to mother and child.

Personal Habits

Smoking and drinking have been shown to produce alterations of some parameters in otherwise

Table 2-3
EFFECT OF PREGNANCY ON
NORMAL SERUM VALUES

Constituent	Mean for Nonpregnant	Mean for Pregnant
Glucose	95 mg/dL	83 mg/dL
Uric acid	4.6 mg/dL	5.9 mg/dL
Cholesterol	212 mg/dL	255 mg/dL
Albumin	4.2 g/dL	3.1 g/dL
Alkaline phosphatase	43 IU/L	102 IU/L
Lactate dehydrogenase	109 IU/L	146 IU/L
Aspartate aminotransferase	33 IU/L	40 IU/L

Adapted from Elliot, J. R., and O'Kell, R. T., "Normal clinical chemistry values for pregnant women at birth," *Clin. Chem.* 17:156–157, 1971.

presumably healthy individuals. Smoking increases the level of fatty acids in blood, reflecting alteration in lipid metabolism. Values for serum carcinoembryonic antigen are approximately twice as high in smokers as in nonsmokers. The significance of the effect of smoking on the levels of this tumor marker is unclear at present. Moderate drinking has been shown to increase the levels of a particular lipoprotein thought to be beneficial in removing cholesterol from the body. Of course, many of the changes seen in patients who drink more heavily are the reflection of damage to the body produced by the intake of alcohol.

Body Weight

A growing body of literature relates body weight to changes in biochemical parameters. For both men and women, serum uric acid levels have been shown to be correlated somewhat with body weight; an increase in weight results in a rise in uric acid levels (Table 2–4). As body weight increases, there is a measurable rise in cortisol production. Urinary creatinine excretion increases as the lean body weight

becomes higher (owing to increase in muscle mass). As obesity becomes more pronounced, testosterone production in men decreases, whereas the levels of estradiol and estrone in women rise in obesity. Thyroid hormone values in serum decline in obesity. We must always keep the question open, however, about how much of this change is actually due to some underlying pathologic problem associated with obesity. A major cause of one type of diabetes mellitus is obesity; the most effective treatment in early stages of the disorder is weight reduction and changes in diet.

Biologic Rhythms

A number of biochemical rhythms occur on a daily, monthly, or yearly basis. Knowledge of these cyclic changes can aid in the interpretation of laboratory data in several instances. Probably the best known daily rise and fall of a biochemical parameter is the **circadian** ("approximately a day") cycle of serum cortisol (Fig. 2–6). Values for this steroid rise in the morning, reaching a peak at approximately 10 A.M. There is a gradual decline during the afternoon and evening, with the lower (trough) level occurring at around 11 P.M. The value then starts to rise again, and the process repeats itself. There are other similar changes taking place with some other hormones. Thyroid-stimulating hormone evidently has a short burst of production every night from roughly 11 P.M. to 1 A.M.

Monthly cycles are known for several hormones, usually associated with the menstrual cycle in women. These changes are not cyclic over the entire month, but occur in specific timed releases of a certain hormone over a period of a few days. Some of the changes may take place over 1 to 2 weeks.

Although longer cycles are very difficult to document, some evidence suggests that we experience yearly cyclic changes in our body chemistry. For example, vitamin D and its metabolites show some alterations over the course of a year. It is difficult at present to explain these shifts, but there may be

Table 2-4
EFFECT OF BODY WEIGHT ON SERUM
URIC ACID LEVELS

Body Weight (lb)	Men (mg/dL)	Women (mg/dL)
100–109	—	4.2
120–129	5.1	4.8
140–149	5.8	5.0
160–169	5.9	5.5
180–189	6.5	5.9
>199	6.8	—

Adapted from Goldberg, D. M., Handyside, A. J., and Winfield, D. A., "Influence of demographic factors on serum concentrations of seven chemical constituents in healthy human subjects," *Clin. Chem.* 19:395–402, 1973.

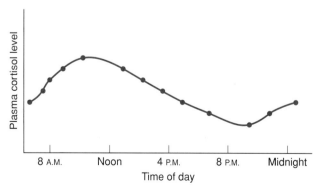

FIGURE 2-6   Variation of plasma cortisol values with time of day.

Table 2–5
REFERENCE VALUES FOR SERUM
ALKALINE PHOSPHATASE AS
A FUNCTION OF METHOD

Method	Adult Reference Values
Bodansky	1.5–4.0 Bodansky units
Bessey-Lowry-Brock	0.8–3.0 Bessey-Lowry-Brock units
King-Armstrong	3–13 King-Armstrong units
Kinetic	21–85 IU/L

Adapted from information appearing in *NEJM*. From Kaplan, M. M., "Alkaline phosphatase," *N. Eng. J. Med.* 286:200–202, 1972.

some relationship to the amount of sunlight available during different months. Other cycles are being explored, and discoveries about them may provide useful knowledge.

Assay Methodology

Although various assay methodologies are expected to measure the same chemical in a body fluid and give roughly the same results, the manner in which we report these results often leads to confusion. Nowhere is this problem more apparent than in the area of enzyme studies. Depending on the particular set of conditions employed, we can report a given enzyme level in five or more different units (Table 2–5). The values may be equivalent, but the casual observer has no way of knowing that the results all say essentially the same thing. Similar confusion occurs when results for a calcium assay are reported in milligrams per deciliter (mg/dL) or milliequivalents per liter (mEq/L). Very different numbers result from these two ways of stating the values, although the number of calcium ions present in the sample is the same in both cases. Care must be taken to see that all members of the health-care team understand the values reported and the appropriate reference range. This precaution is especially important at a time when there is a shift on the part of many laboratories to the International System (SI) of units. A number reported as micromoles per liter (μmol/L) is very different from one stating the parameter in milligrams per deciliter.

2.10

QUALITY CONTROL IN THE CLINICAL CHEMISTRY LABORATORY

There is much legitimate concern expressed about the accuracy of laboratory results. Patients want to have confidence in the diagnosis made, which is based (in part) on laboratory data. Physicians, nurses, and other health-care providers need reliable laboratory information for more accurate diagnosis and treatment decisions. Laboratory personnel are professional staff who desire to provide the best information available. An important part of any laboratory analytical program is quality control.

Definition of Quality Control

Quality control is the process of monitoring laboratory analyses to ensure accuracy of results. A good quality control program monitors test performance, helps identify problems with a specific assay system, and helps health-care staff assess the reliability of results. Comprehensive quality control programs are required by government and private accrediting agents, but these mandates should not be the major reason the laboratory assesses the reliability of the results produced. A good quality control program is part of the overall goal of quality assurance, which encompasses every aspect of the laboratory operation from initial patient identification and sample acquisition to the clear reporting of the final laboratory result.

Basic quality control involves the analysis of specific control fluids (serum, urine, CSF) at the same time patient samples are analyzed for the constituent of interest. The quality control materials are drawn from a pool that allows the same sample to be assayed at frequent intervals over a long period. These controls are marketed commercially or (more rarely) may be prepared by the laboratory itself. Aliquots of a large pool are prepared and freeze-dried to allow long-term stability. At the time of analysis, the aliquot is reconstituted (usually with distilled water) and analyzed along with patient samples. This process provides data on the same sample over an extended period. It is common for a specific control material to be analyzed several times a day for a period of a year or more. Data obtained from these assays are analyzed statistically, and trends in the values are monitored to assess testing reliability.

Quality Control Materials

The pool from which aliquots are prepared may be from either human or animal sources. A human serum pool is more expensive than a pool from bovine serum (the most common animal source), but is preferred by many laboratory workers. The properties of human pools more closely resemble those of the patient samples being analyzed. This similarity is particularly important in the area of enzyme assays, since a specific enzyme from animal sources may not react in exactly the same way as the enzyme from a human source does.

Most quality control pools (particularly serum pools) have two levels for any given constituent: one value is somewhere in the "normal range" for the analyte, and the other value lies somewhere in the range expected for common disease states involving that particular constituent. For example, a control material for serum albumin may have a value of 4.6 g/dL (within the 3.5 to 5.5 g/dL reference range for the analyte). The second-level control for albumin may have an albumin content of 2.0 g/dL (a result in the range expected for patients with altered albumin levels). The two-level control material allows assessment of test accuracy within both the normal and pathologic ranges for any given biochemical component. Some three-level programs are available but have not been shown to have any distinct advantage over two-level controls.

Controls may be purchased either assayed or unassayed; assayed controls are much more expensive than unassayed ones. An assayed control has the control value and $\pm 2SD$ range established for it through analyses performed by several laboratories. In most instances, values for specific methods and instruments are stated on the assay sheet. Assayed controls are most useful for tests that are run infrequently, since the statistical data base for the material has already been established.

Unassayed controls usually have some "target" values, with the actual results falling somewhere near the target values. These controls are appropriate for analyses that are performed frequently enough that a statistical data base can be quickly accumulated. When a new pool is introduced, the new materials are analyzed in parallel with other quality control materials using established values until an adequate number of runs has been completed on the new materials. Most commercial quality control programs employ unassayed materials with monthly group statistics provided by the company. This information allows an individual laboratory to compare its data with other facilities' data. The data may help identify a particular problem with an analysis or may indicate a better method for a procedure that is to be modified.

The Levy-Jennings Plot

Quality control data can be employed to identify two basic situations: (1) the assay is obviously malfunctioning at the present moment and none of the data obtained are reliable, or (2) a problem with the assay is developing and needs to be corrected before unreliable information is produced. The first situation does not require much comment; the test is inaccurate and no laboratory results should be reported until corrections are made. The second situation demonstrates the value of quality control programs in monitoring trends and dealing with a situation before it becomes a serious problem.

Trend analysis of quality control data can be performed using a **Levy-Jennings plot** (Fig. 2–7), sometimes referred to as a Shewhart plot. This type of data analysis involves a graph that can be considered as a Gaussian curve on its side. The value ob-

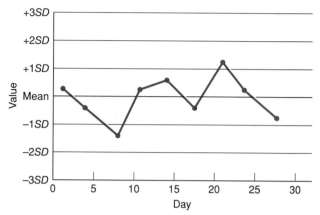

FIGURE 2–7   The Levy-Jennings plot.

tained for the control material is plotted on the y axis versus the date of the analysis on the x axis. Ranges for 1, 2, and 3SD are marked to delineate test limits. The resulting plot represents the trend of the values for the constituent of interest obtained over a period.

The data for a controlled test resemble that in Figure 2–7. Individual values fluctuate from one run to the next, usually falling with the $\pm 2SD$ range. Occasionally, a result may lie outside the $\pm 2SD$ range but be within the $\pm 3SD$ range. This is to be expected, since statistics tell us that 1 acceptable value in 20 falls within the wider $\pm 3SD$ distribution. Values that lie outside the $\pm 3SD$ range require corrective action, since they indicate there may be some problem with the specific test. Examples of quality control trends and the issues they highlight will be provided later in this section.

Westgard Rules for Quality Control

As useful as the Levy-Jennings plot is for monitoring test performance, the actual implementation of this approach is rather cumbersome. After the data are obtained, the individual points are plotted on a graph and analyzed visually. The process is time-consuming and somewhat subjective. Some trends may be overlooked that could point to a specific problem with the analysis. A set of criteria were therefore developed to improve quality monitoring, decrease subjectivity in data analysis, and provide some help in troubleshooting. These criteria have come to be known as the **Westgard rules,** named for the clinical chemist who developed them.

The control rules can be summarized as follows:

1. 1(2S) One control value exceeds $\pm 2SD$ from the mean.
2. 1(3S) One control value exceeds $\pm 3SD$ from the mean.
3. 2(2S) Two consecutive control values exceed

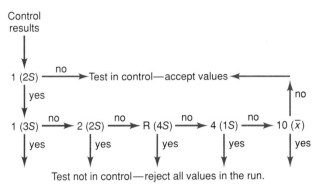

FIGURE 2-8   Use of the Westgard rules.

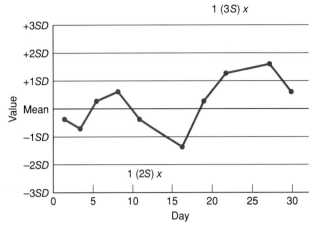

FIGURE 2-9   Violations of the 1(2S) and 1(3S) Westgard rules.

the same limit, either $+2SD$ or $-2SD$. The rule can apply to two different control materials in the same run, or to two successive analyses of the same control material.

4. $R(4S)$ The numerical difference between two control values within the same run exceeds $4SD$. One material could show a value greater than $-2SD$ while the other may provide a value in excess of $+2SD$.

5. $4(1S)$ Four consecutive control values exceed either $+1SD$ or $-1SD$. The values must all be in the same direction.

6. $10(\bar{x})$ Ten consecutive control results all lie on the same side of the mean.

As the control data are analyzed, the analytical run is rejected and all samples are reassayed if any of the above rules are violated. The rules are assessed sequentially (Fig. 2-8) to see if the quality control data fit the established criteria. This system is easily adapted to a computer, allowing automatic analysis of data and rapid response to problems in a given assay.

Quality Control and Troubleshooting

Two types of errors may be encountered during the analytical process: random and systematic. The random error is one with no trend or means of predicting it. Random errors include such situations as mislabeling a sample, pipeting errors, improper mixing of sample and reagent, voltage fluctuations not compensated for by the instrument circuitry, and temperature fluctuations. Although it is easy enough to identify the presence of a random error through quality control techniques, it is more difficult to pinpoint the specific problem associated with the incorrect measurement. Violations of the 1(2S) (Fig. 2-9), 1(3S) (Fig. 2-9), and $R(4S)$ (Fig. 2-10) Westgard rules are usually associated with random error. To assess the situation, the sample is reas-

sayed using the same reagents. If a random error occurred, the same mistake may not be made again, and the result will be within appropriate control limits.

A systematic error, on the other hand, will be seen as a trend in the data. Control values gradually rise (or fall) from the previously established limits. This type of error includes improper calibration, deterioration of reagents, sample instability, instrument drift, or changes in standard materials. All the Westgard rules that indicate trends identify systematic errors. Figure 2-11 illustrates violations of the 2(2S) and 4(1S) Westgard rules, and Figure 2-12 illustrates the trend seen when the $10(\bar{x})$ rule is violated.

If reassay does not correct the problem by bringing the control value within the $\pm 2SD$ range, further analysis of the data is necessary. A stepwise evaluation of the procedure needs to be carried out to determine where the problem lies. This examination could include preparing new control materials, restandardizing the assay, checking wavelength or other instrument settings, or making new reagents. The past history of the specific test may be helpful in deciding which steps to take first. Re-

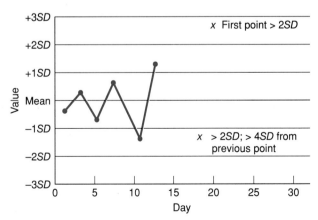

FIGURE 2-10   Violation of the $R(4S)$ Westgard rule.

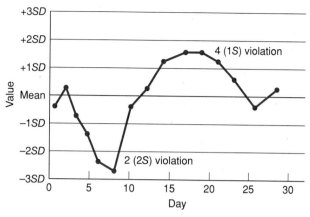

FIGURE 2–11 Violations of the 2(2S) and 4(1S) Westgard rules.

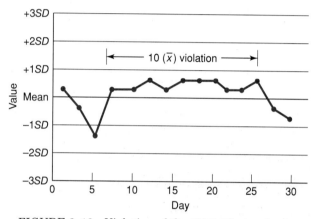

FIGURE 2–12 Violation of the 10($\bar{x}$) Westgard rule.

agents that are close to their expiration date should be discarded and remade. The laboratory records for calibration and the calibration schedule may point to a need for a new standard curve. The process should proceed in a logical fashion to identify and correct the problem. Adequate records should be kept of each step. Good documentation will permit easier correction of the problem in the future.

Proficiency Surveys

In addition to the in-house quality control program carried out by a laboratory, most facilities also participate in one or more proficiency survey programs. Accrediting agencies require outside proficiency testing as a part of the laboratory quality assurance program.

The laboratory receives a series of lyophilized samples in the mail from the survey agency. Information is provided on reconstitution of the samples and the assays to be performed. After the analyses have been carried out, the results are mailed back to the agency for comparison with those from other laboratories involved in the program. Some time later, individual and group data are returned to the laboratory with comments and recommendations.

Proficiency survey samples are to be treated like any other specimens, with no special handling or precautions taken. In reality, these samples are often analyzed apart from other samples under carefully controlled conditions. Since continued accreditation of the laboratory frequently depends (in part) on a successful result in the proficiency survey, it is understandable that these practices exist.

Although there is some value to a proficiency survey, there are also drawbacks. The data are returned too late to take appropriate corrective action if such action is needed. There is no way to assess the reason for a poor result weeks after the date of the analysis. However, continued poor showing in proficiency testing should alert the laboratory to ongoing difficulties in their testing program.

2.11

QUALITY ASSURANCE—MOVING BEYOND THE TEST

Assuring that the correct results are provided for the right patient involves much more than ascertaining that the assay was performed properly. A major emphasis in the clinical laboratory today is on maintaining the integrity of the entire process, from patient identification to accurate reporting and logging of the laboratory data. Quality assurance is concerned with the total process, not simply assay validation.

Patient Identification

The most accurate assay in the world is of no value if it is performed on a sample from the wrong patient. Before a venipuncture or other sample collection procedure is initiated, patient identity must be verified. If the patient is alert, ask for the name (not "are you Ms. Jones?"). Verification of identification by checking the arm band is mandatory. Many institutions are using bar-coded devices for sample and patient I.D., a step which also enables a higher level of automation in later stages of the analysis.

Sampling

Laboratory staff who collect blood samples should be thoroughly familiar with the circumstances that affect the handling of samples for each assay. For many blood tests, serum may be appropriate; for other analyses, a serum sample may create problems in the assay procedure or in the instrument. Constituent stability is often affected by the preservative or anticoagulant. Anticoagulants may seriously interfere with some assays. A manual or table listing the sampling requirements for each assay performed or sent out by the laboratory should be available.

Testing

Not only should the specific assay requested be performed in an accurate and timely manner, but the proper assay must be run. Careful checking of the laboratory request form is obligatory. Be sure that the proper specimen is available (a urine human chorionic gonadotropin [hCG] screen is not very useful when a serum hCG quantitation was requested). There will be instances when a pattern of testing on a patient deviates suddenly for some reason. The laboratory staff may need to contact others on the health-care team to see if an actual change has occurred, or if a test was accidentally not ordered. Test requests that are not clear should be confirmed and clarified before sampling is done and assays are performed.

Data Reporting

In some cases the proper sample was collected from the right patient, the assay was performed in a timely manner (with good quality control results), but the process was futile because the test results were not recorded properly (or at all). Errors in data reporting can vary widely. A handwritten report may be unreadable. Data entered into a computer may have figures transposed or the wrong numbers entered. A telephone message may be poorly heard and improperly recorded. Accurate transmission of data is a vital part of the laboratory testing process.

All results should be checked at least twice. The person performing the test should verify that all results have been recorded correctly and are legible. A laboratory supervisor or other professional responsible for technical operations needs to review all laboratory data, looking for inconsistencies and other problems. If you are telephoning to a unit, have the recipient repeat the report to make sure it was received properly. Every precaution must be taken to assure that the careful work done up to this point is not negated by poor communication.

SUMMARY

The reference range is defined as the set of values for a parameter that is expected in a healthy population of defined characteristics. Parameters usually employed to define such a group include age and sex. In any population with a Gaussian distribution of values, the data are distributed symmetrically around a central point. This point is the average of all values and is called the mean. The set of values that are ± 2 standard deviations from the mean defines the reference range for that parameter. Approximately 5% (1 person in 20) of healthy individuals have a value that falls outside the reference range. Many data sets are non-Gaussian in their distribution. The concepts of mode and median are employed to discuss the information contained in this type of data.

A variety of factors affect the distribution of values within a reference range. Values for a given constituent may be different for men and women, or the value may change with age. Factors such as personal habits, weight, pregnancy, medications, or time of day influence the reference range. Careful thought must be given to both the acquisition of data and the use of the information in the interpretation of a patient's value.

Quality control plays an important role in the proper performance of laboratory tests. Control materials may be assayed or unassayed, depending on the needs of the laboratory and the frequency of testing. Proficiency surveys also provide a means of evaluating the reliability of laboratory data. Two major (and interrelated) systems are employed for monitoring of quality control data: the Levy-Jennings plot and the Westgard rules. The Levy-Jennings plot provides a running graphical display of the control data; the Westgard rules monitor the same trends mathematically. Both random and systematic errors can be identified and corrected.

FOR REVIEW

Directions: For each question, choose the best response.

1. The extent to which measurements agree with the true value of the quantity being measured is known as
   A. acceptable limits
   B. accuracy
   C. precision
   D. reliability

2. The reproducibility of test measurements is referred to as
   A. accuracy
   B. precision
   C. quality control
   D. reliability

3. The middle value of a set of numbers that are arranged according to their magnitude is known as the
   A. arithmetic mean
   B. geometric mean
   C. median
   D. mode

4. In a Gaussian distribution, the $\pm 2SD$ range includes the following percentage of values:
   A. 31.6%
   B. 68.3%
   C. 95.5%
   D. 99.7%

5. Calculate the coefficient of variation when $\pm SD = \pm 7$ mg/dL and the mean $= 89$ mg/dL.
   A. 6.4%
   B. 7.9%
   C. 12.7%
   D. 15.7%

6. When establishing a reference range, it is most desirable to obtain specimens from at least the following number of individuals:
   A. 25
   B. 50
   C. 75
   D. 100

7. Which of the following should be taken into consideration when establishing a reference range?
   A. age
   B. sex
   C. sampling time
   D. both A and B
   E. A, B, and C

8. When data fluctuations are due to chance and results are seen to vary in either direction, the problem is referred to as
   A. the coefficient of variation
   B. experimental error
   C. random error
   D. systematic error

9. The process that encompasses all aspects of laboratory operation including

patient identification, specimen collection, equipment maintenance, and the reporting of patient results is
A. accuracy
B. reliability
C. quality assurance
D. quality control

10. The process that monitors each laboratory analysis, using material with known constituent concentrations, in order to ensure the accuracy of test results is
A. pooled control
B. quality assurance
C. quality control
D. accuracy monitoring

11. Which of the following may be used to analyze quality control data?
A. Levy-Jennings plot
B. Shewhart plot
C. Westgard rules
D. both A and C
E. A, B, and C

BIBLIOGRAPHY

Arkin, C. F., "Quality control in the new environment: Statistics," *Med. Lab. Observer.* December: 31–36, 1986.

Ash, K. O., "Research needed to set geriatric reference values," *Clin. Chem. News.* November:7–8, 1986.

Feinstein, A. R., *Clinical Biostatistics.* St. Louis: C. V. Mosby, 1977.

Fraser, C. G., "Desirable performance standards for clinical chemistry tests," *Adv. Clin. Chem.* 23:300–339, 1983.

Galen, R. S., "New math in the lab: Notions of normality," *Diagn. Med.* July/August:59–69, 1980.

Gambert, S. R. et al., "Interpretation of laboratory results in the elderly. 1. A clinician's guide to hematologic and hepatorenal function tests," *Postgrad. Med.* 72:147–152, 1982.

Grannis, G. F., and Caragher, T. E., "Quality control programs in clinical chemistry," *CRC Crit. Rev. Clin. Lab. Sci.* 7:326–364, 1977.

Johnson, T. R., "How growing up can alter lab values," *Diagn. Med.* June:13–18, 1982.

Lewis, A. E., *Biostatistics.* New York: Reinhold, 1966.

Munan, L. et al., "Associations with body weight of selected chemical constituents in blood: Epidemiological data," *Clin. Chem.* 24:772–777, 1978.

Munan, L. et al., "Population-based pediatric reference intervals," *Diagn. Med.* June:21–41, 1982.

Riegelman, R. K., "The limits of the laboratory: Nuances of normality," *Postgrad. Med.* 70:203–207, 1981.

Solberg, H. E., and Graesbeck, R., "Reference values," *Adv. Clin. Chem.* 27:2–29, 1989.

Sollberger, A., "Rhythmic changes in clinical laboratory values," *CRC Crit. Rev. Clin. Lab. Sci.* 9:247–285, 1975.

Swinscow, T. V. D., *Statistics at Square One.* Bath, England: Dawson and Goodall, 1982.

Tikkanen, M. W., and Wilson, G., "Statistical process control: A versatile tool of scientific management," *Spectroscopy.* 4:30–35, 1989.

Werner, M., and Marsh, W. L., "Normal values: Theoretical and practical aspects," *CRC Crit. Rev. Clin. Lab. Sci.* 9:81–100, 1975.

Westgard, J. O., "Better quality control through microcomputers," *Diagn. Med.* January/February: 60–74, 1982.

Westgard, J. O., et al., "A multi-rule Shewhart chart for quality control in clinical chemistry," *Clin. Chem.* 27:493–501, 1981.

Williams, G. W., and Schork, M. A., "Basic statistics for quality control in the clinical laboratory," *CRC Crit. Rev. Clin. Lab. Sci.* 17:171–199, 1982.

Winsten, S., "The ecology of normal values in clinical chemistry," *CRC Crit. Rev. Clin. Lab. Sci.* 10:319–330, 1976.

Wisser, H. and Breuer, H., "Circadian changes of clinical chemistry and endocrinological parameters," *J. Clin. Chem. Clin. Biochem.* 19:323–337, 1981.

CHAPTER OUTLINE

Chapter Three

PRINCIPLES OF MEASUREMENT II: CONCEPTS IN INSTRUMENTATION

• • • • • • • • • • • • • • • • • •

UPON COMPLETION OF THIS CHAPTER, THE STUDENT WILL BE ABLE TO

1. Identify the basic components of a spectrophotometer and a fluorometer.
2. Apply the Beer-Lambert law in calculating the concentration of an unknown specimen.
3. List three possible constituents of serum that may interfere with spectrophotometric analysis.
4. Explain how the examination of a standard curve can help reveal limitations in an assay.
5. Discuss the purposes for using a reagent blank or a sample blank as part of an assay system.
6. Explain the principle of fluorescence and how it affects the positioning of the emission wavelength selector and detector components of a fluorometer.
7. Discuss the effects that each of the following may have on fluorescent measurements:
   A. scattered light
   B. absorbance of sample
   C. detection limits
   D. quenching
   E. quality of reagents
   F. cleanliness of glassware
   G. pH
8. Explain the principle of flame emission photometry.
9. Explain the purpose of using an "internal standard" when performing analyses using flame emission photometry.
10. Discuss the effect of abnormal serum protein and lipid levels on electrolyte measurements when flame emission photometry is used rather than an ion-selective electrode system.
11. Describe the composition of the glass and reference electrodes which together make up the pH meter.
12. Explain the general principle of the ion-selective electrode.
13. Explain the principles of the carbon dioxide and oxygen electrodes.
14. Explain the principles of gas chromatography (GC), GC/mass spectrometry, and high-performance liquid chromatography (HPLC).
15. Identify the component parts of the GC and HPLC systems.
16. Name carrier gases commonly used in GC systems.
17. Explain the purposes of the inert supporting phase and the stationary liquid phase of the column in GC systems.
18. Define retention time.
19. Explain how the identification and quantitation of a compound may be achieved through peak analysis in GC.
20. Identify the types of detectors commonly used in GC and HPLC systems.
21. Explain the difference between normal-phase and reversed-phase partition systems as they relate to HPLC.
22. Discuss the clinical applications of GC, GC/mass spectrometry, and HPLC.

INTRODUCTION

The productivity of the modern clinical chemistry laboratory depends on instrumentation. The high output and accuracy of results obtained by modern analytical devices could not be achieved by the tedious manual separation and quantitation methods of classical analytical chemistry. From the simplest colorimeter to the most complex automated system, all instrumentation uses certain fundamental principles discussed in this chapter. With an understanding of basic concepts, one can master the purpose and function of any analytical instrument, in spite of the flashing lights and complex computer-driven components.

3.1

SPECTROPHOTOMETRY

Quantitation of a substance in solution by **spectrophotometry** is accomplished by measuring the amount of light absorbed by that solution after appropriate treatment. The more light absorbed, the higher the concentration of the material under study. An associated term, **colorimetry,** is often used to describe the same technique. The major difference in the two techniques is the sophistication of the instrument used. A colorimeter is a simple device and very limited in its flexibility. A spectrophotometer allows for the application of several different approaches that provide a full analysis of the materials being studied.

Theory of Light Waves

To better understand the theoretical basis of spectrophotometry, we need to review some fundamental properties of light. For our purposes we will consider light as a wave (although it can also be pictured as a particle in some circumstances). Each light wave possesses certain characteristic properties (Fig. 3–1). As a wave, the light ray will have a peak and a trough. The distance between two successive peaks is defined as the **wavelength** and

gives the light its characteristic color (if the wavelength is in the visible region). We define the wavelength as a distance and express it in nanometers (nm). One nanometer is equal to 10^{-9} m. An older term, which still appears in some literature, is *millimicron* (mμ); 1 mμ is equivalent in length to 1 nm. Light below about 300 nm is considered to be in the ultraviolet region. If the light has a wavelength between approximately 400 nm (blue) and 800 nm (red), it is in the visible region. When the wavelength is above approximately 1000 nm, we say the light is in the infrared region.

Amplitude, the other term used to define a light wave, is the distance between the peak and trough. The higher the amplitude, the more intense the light and the more light energy produced at that wavelength. Frequently we do not concern ourselves so much with amplitude because we work with a constant light source in which the amplitude of a given light ray is constant.

Beer-Lambert Law

The basic principle of spectrophotometry involves measuring the amount of light absorbed by a solution and relating that absorption to the solution's concentration.

If we shine light through a liquid material, part of the light energy is absorbed by the molecules in solution. This absorption of light depends on the structure of the molecule, primarily the types of covalent bonds present. By comparing the amount of light entering the solution (the incident light) with the amount of light passing through without being absorbed (the transmitted light), we can calculate the concentration of material in the solution. The difference between the amount of incident light and transmitted light can be expressed mathematically and is referred to as the **absorbance** of the solution.

The **Beer-Lambert law** (often referred to as Beer's law) expresses the relationship between concentration and absorbance:

$$A = abc$$

where A is measured absorbance; a is coefficient of absorptivity (specific for each compound); b is path length of the cuvet; and c is concentration of solution. Because we usually want to measure absorbance and calculate the concentration from that data, we rearrange the equation to give

$$c = \frac{A}{ab}$$

Since a and b are constants, there is a linear relationship between the absorbance of a material and the concentration of that material in a solution. As the concentration increases, the absorbance increases.

Some textbooks also discuss the relationship be-

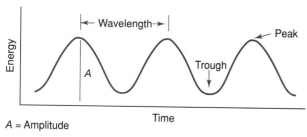

FIGURE 3–1 Properties of light waves.

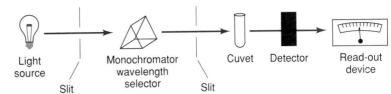

FIGURE 3–2   Block diagram of a spectro-photometer.

tween transmittance and concentration, dealing with the amount of light actually passing through the cell. Although this value is the parameter actually measured, the mathematics requires a logarithmic expression in order to obtain a straight line. We will employ the almost universal practice of absorbance measurements.

Operation of a Spectrophotometer

Any **spectrophotometer** has the following basic components (Fig. 3–2):

1. light source
2. wavelength selector
3. sample holder
4. detector system
5. data readout

A specific system may be very simple and involve manual operation or it may be coupled to a computer or microprocessor that carries out a number of data-acquisition and processing steps automatically. However, all spectrophotometers operate in the same fundamental way.

The light source provides the incident light for the system. In most cases we work in the visible range (approximately 400–800 nm) and can use a tungsten lamp. For work at wavelengths below about 300 nm, a different light source is needed, a deuterium lamp. The deuterium lamp requires a separate power supply and tends to be much more expensive and sensitive than a tungsten light source.

The light source emits light of a wide variety of wavelengths. Our interest is in using light of only a single wavelength (or at least a very narrow band of wavelengths) for our measurements. The proper wavelength is selected by a prism or grating system **(monochromator).** Light shines on the wavelength selector and is spread into a wide band of rays. By moving the selector, we can direct the specific desired wavelength through the sample. Less expensive instruments use a filter instead of a monochromator. The filter allows a wider range of wavelengths to pass through but is much less expensive. Some loss of sensitivity can be expected if a filter system is used for wavelength selection.

The sample holder is a glass or plastic container (either round or square) called a **cuvet,** which is designed to pass most of the incident light through without absorbing it. More sophisticated instruments use square cuvets to minimize the light scatter more likely to occur with a round cuvet. The liquid sample is poured into the cuvet, which is then placed in the sample holder so the light can pass through it to the detector. The composition of the cuvet is important for proper analysis. Although regular glass cuvets are transparent in the visible region and the near-UV, measurements below about 330 nm require special quartz cuvets that do not absorb UV light strongly. These cuvets are much more expensive than glass ones.

The detector is a phototube that responds to light striking it. When the transmitted light hits the tube, an electric signal is generated, going through the detector system to a readout (a moving needle on a dial or a digital display) to indicate the amount of light passing through the sample.

Absorbance Measurements

To use the spectrophotometer for quantitative purposes, we must first ascertain the wavelength of maximum absorbance for the compound of interest. Each chemical has a region where light is absorbed and other regions where little or no light is absorbed. For example, oxygenated hemoglobin absorbs strongly at approximately 412–415 nm and has weaker bands at 541 nm and 576–578 nm (Fig. 3–3). If we wish to quantitate the amount of hemoglobin in a solution, we determine the absorbance at 410 nm and calculate the concentration using techniques to be discussed later in this chapter. As long as the light absorption at that wavelength is due only to the hemoglobin present, we can use the technique with confidence.

Often a small amount of light is absorbed simply

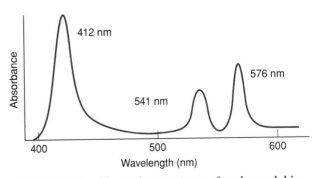

FIGURE 3–3   Absorption spectrum of oxyhemoglobin.

because the cuvet absorbs (or scatters) some of the light rays. Some of the absorbance reading may be due to other materials in the reaction mixture. To compensate for this unwanted light absorption, we measure a reagent blank in the same fashion we measure a sample. The light absorbed by the reagent blank can be subtracted from the measurement using the sample, and a "true" absorbance reading can be obtained. Many instruments do this blank correction automatically; simple systems require the operator to make a separate reading and then subtract the value. Measurement of the reagent blank can provide a good, inexpensive quality control procedure. As long as the absorbance value of the reagent blank is relatively constant, the reagents may be functioning properly. If the blank value changes markedly (usually seen as an increase in the blank absorbance), some deterioration of reagent has occurred, and new material needs to be prepared and employed in the assay.

We often find that the sample itself provides some nonspecific absorbance of light not related to the chemical reaction we are studying. A serum sample that has hemoglobin (due to hemolysis), one with high levels of bilirubin (giving a yellow color to the sample), or a sample that is lipemic (high levels of lipids, causing turbidity) absorbs some light owing to these interfering substances. A sample blank can compensate for at least part of this unwanted absorbance. The sample is added to a mixture containing all the components of the reaction except the reagent, which reacts to form the final colored product. The absorbance of this mixture can be measured and subtracted from the absorbance of the entire reaction system to obtain a more accurate final reading.

Calculation of Concentration from Absorbance Measurements

There are three major methods for calculation of the concentration of an unknown solution:

1. ratio of standard to unknown
2. standard curve
3. use of molar absorptivity value

The selection of a method depends on several factors. In some instances, the absorption value for the standard varies significantly from one day to the next. A standard curve then needs to be prepared every time a measurement is performed. If we know the linear range of the assay, we can use one standard within that range to compare with our unknown sample. If the standard curve is stable for an extended time, only one such curve needs to be run at some predetermined interval (once a week, once a month, every time new reagents are prepared), and unknown values can be determined from this curve. The use of molar absorptivity requires a knowledge

of the specific a value for the compound of interest. In the clinical laboratory, the only material commonly measured by this means is reduced nicotinamide adenine dinucleotide (NADH), a coenzyme in several enzyme reactions.

RATIO OF STANDARD TO UNKNOWN

The simplest type of concentration measurement involves determination of the absorbance values for a known concentration of the compound of interest (standard) and the measurement of that parameter in a patient sample (unknown). If we let A_s and A_u indicate the absorbance values of the standard and unknown, respectively, and c_s and c_u indicate the concentrations of standard and unknown, we can write the following equations:

$$A_s = abc_s$$

and

$$A_u = abc_u$$

These two equations can be stated as a ratio.

$$\frac{A_s}{A_u} = \frac{abc_s}{abc_u}$$

Since the a and b terms are the same in both equations, we have the following ratio:

$$\frac{A_s}{A_u} = \frac{c_s}{c_u}$$

and we can rearrange the equation to determine the concentration of the unknown sample.

$$c_u = \frac{A_u c_s}{A_s}$$

The following example illustrates the principle. We wish to determine the bilirubin concentration in a sample. The absorbance of our unknown at 450 nm is 0.428, and the absorbance of our 5.0-mg/dL standard is 0.372. If we put the figures into the previous equation, we get the following:

$$c_u = \frac{0.428 \times 5.0 \text{ mg/dL}}{0.372}$$
$$= 5.75 \text{ mg/dL bilirubin concentration}$$

USE OF A STANDARD CURVE

When we perform an assay using the ratio technique, we are making some assumptions that may not be valid. Without further study, we do not know the sensitivity of the assay, that is, the lower limit of detection. We also have no information on how high a concentration will still give data that obeys Beer's law. To define properly the limits of the assay, we need to measure the absorbance of a number of solutions of known concentration and examine the data. A calibration of this type is referred to as establishing a **standard curve.**

To set up a standard curve, we first determine the range of values we expect to obtain, both normal and abnormal. If we are measuring for plasma glucose, we anticipate the range of values in healthy individuals to be approximately 70–120 mg/dL. In addition, many clinical situations produce blood glucose levels that can be as high as 400–500 mg/dL. If a patient has low blood glucose, the results may be in the 30–40-mg/dL range. All these values might be expected in our patient population. Therefore, the standard curve needs to assess whether or not this range can be accurately measured with our method.

Solutions containing known concentrations of glucose are prepared in the range we wish to assess. These known materials are assayed in exactly the same way that a patient sample would be treated. After the absorbance of each solution is determined and corrected for the blank, a standard curve can be plotted with concentration on the x axis and the absorbance reading on the y axis. The result should be a straight line in accordance with Beer's law (Fig. 3–4).

Examination of the graph can quickly reveal any limitations in the assay. If the upper portion of the line (high concentration) deviates from the straight-line pattern obtained with lower levels of glucose, we have defined the upper limit of detection under the stated assay conditions. Because concentrations above this level do not follow Beer's law, we cannot measure them without employing some further process. Usually, these samples would be diluted and reassayed. The new value obtained from the standard curve would then be multiplied by the dilution factor to obtain the correct concentration.

The other useful piece of information that can be obtained from a standard curve is an indication of how low we can accurately measure. Again, we look for deviations from the straight line expected if Beer's law is obeyed. The concentrations of samples with absorbance values below this limit cannot be

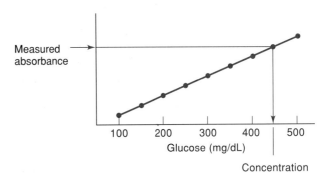

FIGURE 3–5   Determination of concentration with standard curve.

accurately assessed. These values are then reported as "less than (lower limit) mg/dL." There is no way to determine more accurately the concentration in a situation of this sort.

To use the standard curve, we simply measure the absorbance of our unknown sample after the assay reaction has been run. The absorbance value is located on the y axis of the curve and a horizontal line is run over to the standard curve until it intersects the calibration line (Fig. 3–5). Then a vertical line is dropped down to the x axis, where the concentration of the unknown can be read. The use of a standard curve eliminates the need to run known materials for standardization every time an assay is performed.

However, undue reliance on a standard curve can create problems. The curve needs to be rerun every time new reagents are prepared. Some assays use materials that deteriorate over time; the absorbance response one day may not be observed the next day or the next week. Studies need to be carried out to document the stability of the standard curve for each assay. Often, this information is provided by the manufacturer of the reagents. It is good laboratory practice to validate these product statements by some in-house checking, particularly if a new method is being employed or a method change is occurring. If these data are not available, the laboratory staff must perform the necessary studies to document the stability and reliability of the standard curve.

USE OF MOLAR ABSORPTIVITY

For a few assays (usually those dealing with the formation or oxidation of NADH), a direct calculation can be made using the absorbance data. For NADH, the molar absorption coefficient has been carefully determined as 6.22×10^3 L/cm-mol. This value can be used in the equation $A = abc$ to calculate the concentration of a specific material. However, the calculations turn out to be more complicated than might appear at first glance. We illustrate this type of measurement with the assay

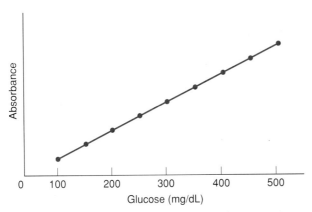

FIGURE 3–4   Glucose standard curve.

for ethanol in body fluids using an enzymatic reaction.

The reaction used to quantitate ethanol in blood involves enzymatic conversion of ethanol to acetaldehyde with reduction of NAD^+ to NADH:

$$ethanol + NAD^+ \rightarrow acetaldehyde + NADH + H^+$$

For every molecule of ethanol oxidized, one molecule of NAD is reduced to form NADH. If we measure the concentration of NADH, we can then indirectly determine the amount of ethanol originally present. NADH absorbs light strongly at 340 nm, but NAD^+ does not absorb significantly at this wavelength. When we add NAD^+ to a sample in buffer containing the enzyme alcohol dehydrogenase, we effect the conversion to the desired product. By measuring the change in absorbance at 340 nm after the reaction is complete, we have the necessary data to calculate the concentration of ethanol in the original sample. Appropriate checks must be carried out to determine linearity and limits of the reaction as in any other assay approach.

Our assay data give the following:

change in A_{340}: 0.421
cuvet path length: 1.0 cm
sample volume: 3.0 mL

Using the previously stated coefficient for NADH, we can write the following equations:

$$A = abc$$

or

$$c = \frac{A}{ab}$$
$$= \frac{0.421}{(6.22 \times 10^3 \text{ L/cm-mol})(1.0 \text{ cm})}$$

Canceling units and carrying out the calculations, we get:

$$c = 6.77 \times 10^{-5} \text{ mol/L concentration of ethanol}$$

If we wanted to find the exact amount of ethanol in that particular sample, we could take the calculations one step further:

$$\text{Sample volume} = 3.0 \text{ mL}$$
$$\text{Total amount} = \text{Concentration} \times \text{Volume of sample}$$
$$= (6.77 \times 10^{-5} \text{ mol/L})(3.0 \text{ mL})$$
$$\times (1 \text{ L/1000 mL})$$
$$= 2.03 \times 10^{-7} \text{ mol ethanol}$$

Factors Involved in Making Absorbance Measurements

BLANK CORRECTION

The proper use of reagent-blank and sample-blank data is one of the often neglected aspects of training in analytical techniques. Even when using very automated equipment, awareness of problems associated with spurious absorbance of light can quickly solve a number of problems.

A common technique in manual spectrophotometry is to "zero" the instrument with a reagent blank and then make readings on the samples. Although this approach accomplishes the automatic subtraction of reagent-blank absorbance from the total absorbance, some very useful quality control information is lost in the process. If the same lot of reagent is being used and the system is stable, we would expect a reasonably constant blank absorbance reading from one day to the next. In addition, we expect the blank absorbance reading to be fairly low in most cases. If we adopt the procedure of setting the absorbance to 0.000 with distilled water and then determining (and recording) the actual blank absorbance, we have a daily record of reagent stability. Any deterioration in reagents that leads to a change in the blank reading is easily detected. If an error is made in making up the blank, this problem is often detected by noticing the difference between the blank reading for the day and the trend already established for that set of reagents.

Sample blanks are somewhat more troublesome in that a separate blank is needed for each individual sample, instead of one reagent blank for the entire batch of assays. Frequently there is not a lot of extra sample available for blanking purposes. Not every patient requires a sample blank. The cost of running sample blanks on all patients must be carefully considered, particularly in a time when finances are the driving force in a number of decisions made in the laboratory. The use of a sample blank is dictated by the parameters of the assay and the condition of the sample. If a reaction product absorbs in the same region as bilirubin or hemoglobin, careful attention must be paid to those samples with hemolysis and/or high bilirubin levels. It is a wise practice to assess the exact contribution these various interfering parameters make to the end results of a specific assay before the test is implemented in the laboratory for routine patient use.

Useful information can often be obtained about the need for blanks by examining the standard curve for the assay. If the standard curve is linear (which should be the case for all common colorimetric methods), the line connecting the data points should pass through the origin of the graph (the $x = 0$, $y = 0$ interception point). Sometimes a linear relationship between absorbance and concentration is obtained, but the line intercepts the absorbance axis at some point greater than zero. The resulting graph tells us that a sample with "zero" concentration of the analyte of interest has a measurable absorbance value. If data of this type are obtained, the method must be examined for errors in the blanking system. If no reagent blank is being run, a proper one must be developed and employed in further assays. If some type of blank is a part of the assay and there is still a problem, the blank composition needs to be changed to compensate correctly for the nonspecific absorbance.

LOWER LIMIT OF DETECTION

When plotting a standard curve, our natural tendency is to draw the line through all the points and then down to the x, y intercept of the graph. When we do this, our unconscious assumption often is that we can then detect any concentration on the line all the way down to a zero level. In practice, this assumption is never valid. For maximum reliability in reporting results, we need to assess experimentally the lower limit of detection.

To determine exactly how low a given assay will measure a constituent accurately, we simply assay lower and lower concentrations of standards. Some judgment is needed about the exact concentrations for each standard. This decision is dictated by the expected range of values for the specific test and the clinical need for exact information at various levels. The lower limit of detection is seen when two solutions of different concentration give essentially the same absorbance reading, but the next highest concentration in the series follows a linear relationship between absorbance and concentration.

For our example, let us assume we established a standard curve for glucose using the following data:

Absorbance (540 nm)	Glucose (mg/dL)
0.740	500
0.444	300
0.148	100
0.074	50
0.045	30

If we plot the values, we see a linear relationship between glucose concentration and absorbance at the stated wavelength. To assess the actual lower limit of detection, we then prepare further dilutions of our standard and run the assay under the same conditions. The following data are obtained:

Absorbance (540 nm)	Glucose (mg/dL)
0.045	30
0.030	20
0.028	10

Notice that this extended standard curve is no longer completely linear. Although a concentration of 20 mg/dL gives an absorbance reading that still obeys Beer's law, the 10 mg/dL standard does not. Therefore, the lowest glucose level we can measure accurately is 20 mg/dL. Any results giving absor-

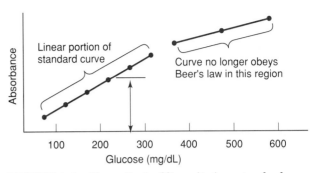

FIGURE 3-6   Upper limit of linearity in a standard curve.

bance values below this must be reported as "less than 20 mg/dL."

UPPER LIMIT OF DETECTION

Deviations from linearity at the upper end of the standard curve are easily detected. By running standards of higher concentrations, a point is reached where the relationship between absorbance and concentration is no longer linear. In most situations, the absorbance does not rise as rapidly as the concentration increase, so the actual level is underestimated. Figure 3-6 illustrates a situation in which the standard curve is not linear above about 300 mg/dL. This situation may occur when the absorbance of the solution is simply too strong for the instrument to measure accurately. In other cases, there is more analyte present than detecting reagent. Not all the analyte reacts with the assay system, and the concentration appears to be less than it actually is. The upper limit of detection must be established so that high levels of a constituent are not underreported.

Two approaches may be taken to deal with the sample with a high value. The report may be sent out as "greater than" whatever the upper limit for the standard curve might be. This option is taken in cases where the value is obviously so elevated that no further useful clinical information would be obtained by further reanalysis of the sample. In some situations, a retest might take much longer and the health-care team would want the initial information so that appropriate treatment might be initiated. Another option is dilution of the original sample followed by reassay. The value obtained on the diluted sample is multiplied by the dilution factor for the final result. It is not good laboratory practice simply to dilute the final assay mixture, because the constituent may be present in excess of the assay reagent. Care must be taken to indicate the exact nature of the dilution. A dilution of "1:5" could mean one volume of sample to a total of five volumes of mixture, or it could indicate one part sample plus five parts diluent. Procedures should state clearly how dilutions are to be handled.

3.2

FLUOROMETRY

An increasingly powerful tool in the analytical field is the use of fluorescence measurements for identification and quantitative measurement. In many instances, more sensitive assays have been developed with fluorometric techniques than were possible with spectrophotometry. This change in approach often allows the use of smaller samples and less reagent or permits the analysis to be carried out in less time. Frequently, problems with interferences seen in spectrophotometric measurement can be eliminated or decreased if a fluorescence technique is employed. However, other interferences that were not seen with colorimetric methodologies may arise.

Basic Concepts of Fluorescence

As with spectrophotometry, fluorescence measurement involves the interaction of light with a chemical compound. In this case, however, the compound emits light (usually of a longer wavelength) in response to the light striking it. This emitted light is detected by a phototube that determines the intensity. The amount of light the compound emits is roughly proportional to the concentration of the compound in most cases.

When light strikes a compound, it may be absorbed. Part of the energy provided by the light is used to excite one or more electrons, moving them out of their normal pathway. **Fluorescence** occurs when these electrons then give off light as they drop from the excited state back to their ground-state level within the molecule. The amount of light emitted is a function of the number of electrons excited, and the wavelength of light depends on the quantity of energy released as they drop back down from their previous excited state.

The related technique of **phosphorescence measurement** is sometimes employed for analysis. For all practical purposes, fluorescence and phosphorescence differ only in the exact route the electron takes in returning to the ground state. Applications, interferences, and techniques employing either fluorescence or phosphorescence involve the same basic principles.

Fluorescence Spectra

A major difference between absorption spectra and **fluorescence spectra** is that we actually measure two different spectra in fluorescence methods. The **excitation spectrum** gives us information on the proper wavelength of light in order to produce the maximum number of electrons raised to the excited state. Light of different wavelengths passes through the sample, and the amount of fluorescence is determined. The wavelength of excitation light that produces the maximum fluorescence is then used as the excitation wavelength. The second fluorescence spectrum, the **emission spectrum,** is then determined. Using the wavelength that stimulates the maximum number of electrons, the intensity of the fluorescence produced is determined at different wavelengths. The wavelength where the fluorescence is seen at its highest is called the **emission wavelength.** The proper combination of excitation and emission wavelengths must be determined for each individual compound under consideration.

Fluorescence Instrumentation

The basic components of a fluorometer are illustrated in Figure 3–7. Light passes through an entrance slit and then interacts with a wavelength selector (filter or grating) before striking a sample in a cuvet. The emitted light radiates out of the cuvet in all directions. A wavelength selector and detector systems are located at 90° to the path of the incident beam to avoid detection of the light coming into the sample. Emitted light from the sample passes through the wavelength selector and strikes the phototube of the detector. An electronic signal is sent to

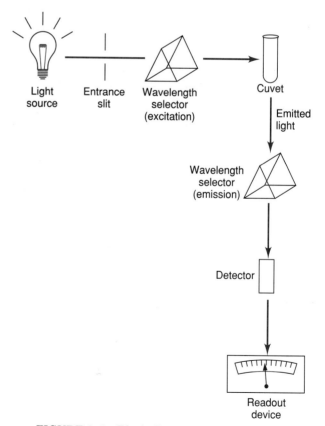

FIGURE 3–7 Block diagram of a fluorometer.

the readout system to indicate the intensity of the fluorescence detected.

LIGHT SOURCE

The most common light source for fluorometry is a xenon lamp, which furnishes a fairly intense light output over a wide range of wavelengths. Mercury lamps have been employed in some systems, but the mercury signal is a series of lines that do not always provide a wavelength in the desired region. The xenon lamp has a continuous spectrum in the UV and visible regions, with the most useful wavelength range being approximately 300–550 nm. At wavelengths above and below this range, the light intensity changes markedly with a small change in the wavelength. Lamps are to be handled carefully since fingerprints greatly reduce the amount of UV light output.

The source of excitation energy has always been somewhat problematic in fluorescence measurements. The usual light sources are much more expensive than those for absorption spectrophotometry and have shorter lives. More careful monitoring of fluorescence light sources is required since they are much more susceptible to fluctuations in output.

WAVELENGTH SELECTORS

For many routine fluorometric applications, a filter system is acceptable. In most cases, a combination of filters is employed, particularly for the emission wavelength selector. Light of wavelengths above and below the region of interest is blocked by these filters, allowing only the light of the desired wavelength range to strike the sample. On the emission side, a filter is employed to screen out unwanted higher wavelengths. Often a neutral density filter is included to lower the intensity of emitted light to a level within the capability of the detection system.

An instrument using a filter set-up is usually referred to as a **fluorometer.** More sophisticated systems employ monochromators for both the excitation and emission sides. These instruments are called **fluorescence spectrometers.** This type of arrangement is especially useful for research and development purposes to determine the actual excitation-emission spectrum of a compound. In addition, greater specificity and sensitivity are possible.

SAMPLE HOLDERS

In most situations, glass cuvets will be acceptable for fluorometric measurements. If UV work is required, some silica materials may be called for. Cuvets must be kept free from fingerprints, spots, and scratches. Disposable cuvets (frequently used in colorimetric analyses) may create problems caused by extraneous fluorescence from the material used to construct the cuvet. These items need to be assessed under the conditions employed for assay in order to detect either any native fluorescence from the cuvet material or any unwanted interactions between cuvet and reagents.

DETECTOR SYSTEMS

The fluorescence emitted by a sample strikes a photomultiplier tube, which converts light energy into an electric signal. Coupled to the photomultiplier tube are an amplifier and a readout device, either a dial or a digital system. Special phototubes must be used if the wavelength of light striking the phototube is in the 600-nm range or above. Since much of the background fluorescence reading comes from thermal activation, this factor can be lessened by cooling the tube.

Quantitation Using Fluorescence Measurements

The most common method of quantitation with fluorometry involves use of a standard curve or a single standard. It is very difficult (if not impossible) to determine the fluorometric equivalent of the molar absorptivity coefficient, so reference must be made to some standard measurement performed at the same time using the same conditions. Because the intensity of the lamp changes over time, the primary value of a standard curve is to assess the approximate range of linearity and limits of detection. For daily use, one or more standards need to be included in the assay sequence, with unknown values being determined by the ratio method.

Factors Involved in Making Fluorescence Measurements

SCATTERED LIGHT

The light beam measured by the fluorescence phototube is at right angles to the beam of incident light. This arrangement allows the system to eliminate any light produced by the original light source and to "see" only the fluorescence caused by that excitation beam. If the sample is turbid (lipemia, suspended particles), a portion of the incident light is directed out of the sample cuvet at angles other than the straight line of entry. Some of this scattered light can be detected by the phototube and registered as fluorescence. The value observed then is higher than the true value.

ABSORBANCE OF SAMPLE

In fluorescence measurements, the assumption is made that the incident light beam passes all the way through the solution. If this occurs, all the molecules within the beam path have the opportunity to become excited and produce fluorescence. The detector system is set up to measure energy from the entire width of the cuvet. A given sample could absorb enough light so that not all the incident beam penetrates the entire solution. Since the amount of fluorescence observed is not representative of the complete reaction system, a lower value is expected.

LIMITS OF DETECTION

As with spectrophotometry, it is necessary to determine experimentally the lower limit of detection for a fluorometric assay. The technique employed follows the principles outlined earlier in this chapter. The major concern is with random, nonspecific fluorescence, which suggests the presence of a certain level of the compound being measured when none was actually present. Preparation of blank materials and proper calibration of the instrument minimize this problem.

Concern for the proper upper limit of detection for a method is often overlooked since it is so easy to change ranges on the instrument to compensate for heavily fluorescent materials. However, the same considerations hold as was the case with spectrophotometry. Reagent depletion can lead to less than complete reaction. Not all the material under study would be converted to the fluorescent derivative, and the concentration would be underestimated. Linearity checks are as important in fluorescence work as they are with other techniques.

QUENCHING OF FLUORESCENCE

On occasion, the fluorescence obtained may be much lower than expected for a given compound. The decrease in response may be due to **quenching.** This phenomenon occurs when the excited state of the molecule (after the excitation light has interacted with the structure) loses some of that energy by interaction with another component of the reaction system. Part of the energy which might have been seen as emission of light is dissipated by transfer to another molecule and not detected by the phototube.

A similar phenomenon may occur if there is another component of the reaction mixture that itself absorbs light. A portion of the excitation light could be absorbed by the interfering material, with less energy left to excite the molecules of interest in the analysis. Conversely, a part of the emission light could be absorbed by other molecules, leaving a smaller than expected amount to reach the detector.

This phenomenon differs from quenching in that the light energy itself (either excitation or emission) is absorbed and not the energy of the excited molecule.

OTHER FACTORS

Reagents must be selected carefully to minimize native fluorescence due to contaminants. Solvents must be checked regularly and stored in glass containers since many plastic materials contain substances that leach into the solvent and produce highly fluorescent backgrounds.

Glassware used in fluorescence procedures should be carefully cleaned. Most detergents contain highly fluorescent materials, and failure to rinse carefully results in high blank values and erratic readings of samples due to contaminants that fluoresce.

Fluorescence is usually very pH-dependent. A small change in the H^+ concentration can lead to significant changes in the electron arrangements within the structure. These alterations may result in differences in both the amount of energy absorbed and the amount released. Excitation and emission wavelength maxima may be affected by pH shifts, leading to erroneous results in the analysis carried out under the conditions previously assumed to be correct.

Fluorometry or Spectrophotometry

The selection of fluorometry or spectrophotometry for a specific analysis depends on a number of factors. The obvious first question is: Does the material fluoresce strongly? If the substance we wish to analyze has strong native fluorescence (as does the antiarrhythmic drug quinidine) or can be fairly easily converted to a fluorescent product (catecholamines, for example), a fluorescent analysis has some advantages. In some instances, no spectrophotometric method is available because neither the starting material nor a reaction product has a strong absorption spectrum.

Fluorescence techniques can be much more sensitive than corresponding colorimetric ones. A spectrophotometric technique may be able to detect a colored compound at a concentration as low as 1 part in 10 million. Under the proper circumstances, fluorometry is able to extend that detection limit to 1 part in 10 billion, expanding the detection range by a factor of 1000.

Fluorescence often provides a higher degree of selectivity in an assay. A large number of compounds may absorb light and produce color; however, not all these compounds fluoresce, a fact that eliminates a number of interfering substances. In addition, because both excitation and emission wavelengths are optimized for the material being analyzed, this double wavelength restriction allows greater selectivity and specificity in a given analysis.

Fluorescence techniques do have some limitations not seen in spectrophotometric analyses. Problems of quenching and spurious fluorescence owing to contaminants create some difficulties in fluorescence analysis. Both fluorescence and colorimetric techniques are subject to effects brought about by changes in pH. A shift in pH may either alter the fluorescence characteristics of the system or shift the wavelength of maximum absorption.

3.3

FLAME EMISSION

Basic Concepts

Flame emission analysis shares some of the characteristics of fluorescence spectrometry. Energy is put into a system, producing electron transitions within an ion. As the excited electrons drop back to the ground state, light energy of a specific wavelength is emitted by the ion. This light can be detected and measured using specific optical systems.

One major difference between the two techniques involves the source of excitation energy. For fluorescence, light from a xenon tube provides the impetus for electron transitions. In flame emission studies, the excitation energy comes from heating the ion. By raising the temperature of the material sufficiently, electrons in atoms such as sodium and potassium can be excited so that detectable amounts of energy are released as they return to ground state. This energy input is provided by atomizing a solution containing the ion of interest and burning it in a flame.

In the routine clinical laboratory, flame analyses have been restricted primarily to sodium and potassium measurements. Lithium levels in serum are determined by flame photometry for manic-depressive patients who are on lithium therapy for the manic phase of the disorder. Although calcium and magnesium give detectable signals with this type of instrumentation, the sensitivity and reliability are not sufficient for these cations to be assayed using flame techniques.

Principles of Instrumentation

Flame emission spectroscopy requires a sample-handling system, a burner, and a light-detecting system. Principles of light measurement are essentially the same as for other spectrometric techniques, including the need for dispersion control, a monochromator, and a detector. Many of the problems associated with flame photometry concern the initial sample-handling stages of the device.

The sample is burned by diluting it and presenting it to the flame in the form of a spray. The sample enters the aspirator by suction and passes through an atomizer, which disperses the sample in a fine mist. This mist then interacts with the flame for excitation of the ions in solution. For most commonly used flame photometers, a mixture of propane and compressed air is used as the fuel source for the flame.

As the excited electrons in the sample drop back to a lower energy level, the light generated passes through a slit mechanism to eliminate stray radiation and collimate the beam to some extent. Interaction with a monochromator determines the wavelengths that impinge on the phototube in the detector system. The signal generated by the phototube is proportional to the concentration of the ion of interest in the sample. Many simpler pieces of equipment use a filter system for wavelength selection instead of a more complex apparatus, providing very satisfactory results.

Internal Standardization

Early flame photometers employed direct aspiration of sample with no compensation for background or variation in flame temperature or intensity. This practice created some obvious problems in accuracy and reproducibility. Currently, all sample, standard, and blank solutions are mixed with a diluent before entering the flame. The diluent contains a high concentration of lithium, employed as an internal standard. When lithium is burned, it produces light of a wavelength very different from those of other common analytes such as sodium and potassium. Light intensity at both wavelengths can be measured, and a ratio of intensities is determined. This ratio provides compensation for fluctuations in the flame or the aspiration rate, canceling out the effects of these variables. Modern instruments perform this internal standardization procedure automatically. Lithium in serum of patients who are being treated for manic-depressive illness is not present at a concentration high enough to provide interference.

Lipid and Protein Interference

A major source of error in flame measurements, discovered several years ago, has led to a striking decrease in the use of the flame emission photometer for electrolyte analysis in the routine clinical laboratory. All samples for flame photometry undergo a predilution step prior to flame analysis; dilution factors of 1:100 or more are commonly employed. These dilutions assume a plasma water volume of 100%. In many instances, when either the lipid concentration or the protein concentration (or both) is significantly elevated in the sample, the plasma water volume is considerably less than 100%. Dilution and analysis then produce erroneously low values for sodium and potassium.

Table 3-1

EFFECT OF LIPID CONTENT ON
ELECTROLYTE MEASUREMENT

Sodium (mmol/L)		Potassium (mmol/L)		Water Content (%)
Flame	ISE	Flame	ISE	
110	136	4.2	5.0	72.4
126	139	4.1	4.6	84.3
126	138	6.7	7.4	84.9
141	157	4.6	5.0	86.8

Data adapted from Ladenson, J. H., Apple, F. S., and Koch, D. D., "Misleading hyponatremia due to hyperlipemia: A Method-Dependent Error," *Ann. Intern. Med.* 95:707–708, 1981.

Measurement with ion-selective electrodes (to be discussed later in this chapter) eliminates, or at least minimizes, this problem. Since samples are analyzed undiluted and no assumptions are made about plasma water volume, the errors introduced by high lipid or protein concentrations are markedly decreased. Data from samples containing elevated amounts of lipids are illustrated in Table 3–1.

3.4

MEASUREMENT OF pH

Concept of pH

The ability to measure accurately the hydrogen ion concentration has had a significant impact on all areas of clinical chemistry. The success of colorimetric and fluorometric analyses is frequently dependent on maintaining proper hydrogen ion concentration. All enzyme assays require a specific and constant hydrogen ion concentration. Measurement of this parameter in whole blood allows accurate and reliable diagnosis of a number of disorders.

Because of the widespread use of this quantity, special means of expressing the hydrogen ion concentration have been developed. The concept of **pH** makes it easier to indicate the concentration of this ion in a simple, compact fashion. Instead of stating the hydrogen ion concentration of a solution as 1×10^{-5} M, we can simply say that the solution has a pH of 5.0 and communicate the same information.

We can define pH in the following way:

$$pH = -\log [H^+]$$

This equation indicates that the pH value for any solution is equal to the negative logarithm (to the base 10) of the hydrogen ion concentration. Another way of expressing this relationship is:

$$pH = \log \frac{1}{[H^+]}$$

As we will see in the examples to follow, when the hydrogen ion concentration decreases, the pH value increases.

Using our previous example, let us see how to calculate a pH value, given the hydrogen ion concentration. The $[H^+]$ is stated to be 10^{-5} M. Putting that value into our equation, we have

$$pH = \log \frac{1}{10^{-5}}$$

Since $1/10^{-5}$ is equal to 10^5, our equation now reads

$$pH = \log 10^5$$

The log of 10^5 is 5, so the pH of the solution is 5. If the concentration had been 10^{-3}, the pH would be 3.

Not all solutions are going to have such simple $[H^+]$ concentrations, however. What do we do with a solution that is 0.0045 M hydrogen ion concentration? First we need to express the concentration as an exponent, 4.5×10^{-3}. Fitting that number into our equation, we have

$$pH = \log \frac{1}{4.5 \times 10^{-3}}$$

or

$$pH = \log \left(\frac{1}{4.5} \times \frac{1}{10^{-3}} \right)$$

Rearrangement gives us

$$\begin{aligned} pH &= \log (0.2222 \times 10^3) \\ &= \log (2.222 \times 10^2) \\ &= \log 2.222 + \log 10^2 \end{aligned}$$

From the log tables (or our calculator) we get

$$\begin{aligned} pH &= 0.3467 + 2 \\ &= 2.3467 \text{ (or 2.35 for practical purposes)} \end{aligned}$$

We need to keep a technical point in mind. The pH value does not actually indicate the true hydrogen ion concentration, but rather the hydrogen ion activity, a somewhat different parameter. However, at lower concentrations the two terms come to represent essentially the same phenomenon.

The Hydrogen Electrode

Because of the widespread need to determine pH values quickly and accurately, instrumentation was developed for measurement of this parameter under a variety of conditions. Dr. Arnold Beckman (founder of Beckman Instruments, now one of the larger clinical instrumentation companies in the world) pioneered in the initial invention and improvement of the **pH meter.** The early models were crude and slow, but brought a quiet revolution to the chemistry world. Current instruments are small, rugged, and often contain microcomputers for ease in standardization.

The heart of the pH meter is the **hydrogen elec-**

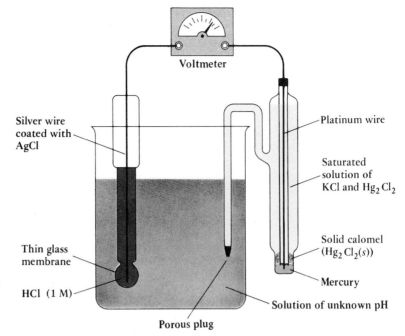

FIGURE 3-8   A pH meter consists of a glass electrode *(left)* and a calomel electrode *(right),* both of which dip into a solution of unknown hydronium ion concentration. (From Oxtoby, D., and Nachtrieb, N. H., *Principles of Modern Chemistry.* Philadelphia, Saunders College Publishing, 1990.)

trode. When this component interacts with hydrogen ions in solution, an electric potential is generated which can be detected by the pH meter acting as a type of voltmeter. By proper calibrations, the electric potential changes can be related to the [H$^+$] and thus to the pH of the solution.

We need to consider two electrode components of the pH meter: the glass electrode and the reference electrode (Fig. 3-8). These items may be separate parts of the measuring device, or they may be fabricated into one combination electrode. Although some aspects of technique and maintenance differ between separate and combination electrodes, the basic theoretical principles are the same.

Inside the **glass electrode** is a wire, usually composed of silver and coated with silver chloride. Surrounding the wire is a solution encased in the glass outer component of the electrode system. The glass electrode does not have any opening to the solution whose pH we wish to measure. A series of potential differences are established between the wire inside the electrode, the glass outer shell, and the solution in which the electrode is immersed. The voltage delivered to the meter by the glass electrode varies as a function of pH. These potential differences are translated into pH measurements by the instrument.

The **reference electrode** is a more complicated device and is somewhat open to the solution in which it is immersed. The basic components include the internal wire (either a silver/silver chloride system or platinum coated with calomel and other materials), a filling solution, and a permeable outer casing. A portion of the reference electrode has a small junction through which solution can flow to be in contact with the outside environment. This junction is usually formed from fritted glass or other

material that allows a slow release of solution from the interior of the electrode. The purpose of the reference electrode is to deliver a constant voltage to the meter.

Operation and Maintenance of pH Meters

The initial calibration procedure for a pH meter involves standardization with a known buffer (usually pH 7). Measurement of the potential generated by the glass electrode allows the instrument to be set so that this measured potential gives a meter reading corresponding to the stated pH value of the standard. This setting is accomplished by adjusting the millivolt output of the electrode so that it is equivalent to the millivolt value necessary to produce the desired meter reading.

The second adjustment allows the operator to compensate for whatever nonlinear response may take place at different pH values. The **slope** is the relationship between the change in measured millivolts and the change observed on the pH meter. The gain in the instrument is adjusted to produce the desired response so that a wider range of pH measurements can be made without frequent recalibration of the system.

A number of current pH meters contain microcomputers which automate much of the calibration process. It is imperative that the manufacturer's procedure be followed exactly. If you include a standard buffer of the wrong concentration in the sequence, the process halts. Calibration must then start again at the very beginning of the procedure.

Proper storage of electrodes is important for reliable operation of the system. Many combination electrodes are now stored in air with appropriate protective covering over the junction opening. Because of the wide variety of electrodes available, instructions of the manufacturer must be carefully followed.

One often neglected aspect of pH measurement is temperature. As the temperature changes, the $[H^+]$ changes. All pH measurements for a given system (both standards and solutions) should be made at the same temperature. If a buffer or other reagent is being used at a specific temperature, the pH of this material needs to be adjusted at that temperature (particularly if the working temperature for the assay is not room temperature).

3.5

ION-SELECTIVE ELECTRODES (ISE)

The success of the hydrogen electrode for the measurement of pH led to intensive investigation of other possible uses for these devices. Beginning with the same basic principles, electrodes specific for Na^+ and blood gases were developed. Expansion of the use of electrodes as analytical devices gained great impetus from the implementation of new techniques and the use of a variety of organic materials to provide selectivity in the electrode. In many clinical applications, **ion-selective electrodes (ISE)** are rapidly replacing other methods for the quantitative measurement of a variety of ions.

Electrode Principles

In many respects, the ion-selective electrode is simply an extension of the hydrogen electrode. A potential gradient is established between the measuring and reference electrodes in response to a given concentration of the ion of interest. This potential difference is detected by a meter similar to the pH meter. In many instances, instrumentation has been developed to determine pH or other ions in solution using the same piece of equipment. The measurement of any given ion is carried out using an electrode designed specifically for the detection of the component of interest.

The success of any specific electrode is based on its degree of selectivity for the desired ion. This selectivity is created in part by the use of porous organic polymers which allow (ideally) only one ion of a specific size to pass through. Other ions which might alter the potential gradient are excluded from the system because their ionic radius is different from that of the ion being selected. A wide variety of polymeric materials are employed in these electrodes, either as simple membranes to separate ions or as part of an ion-complexing, liquid-membrane system.

The formation of a gradient between the solution being analyzed and the interior of the electrode is facilitated by the trapping of ions by a complexing agent. Ideally, this material binds only the ion of interest, immobilizing it inside the electrode. This binding and trapping diminishes the tendency of the ion to move back to the outside of the electrode membrane, which would decrease the degree of charge separation and lower the potential difference. A variety of complexing agents are employed, depending on the specific ion being studied.

In many instances, an indirect measurement of the component of interest must be made. There may be no appreciable potential difference generated or selectivity of transfer across the membrane may be absent. In these situations, a chemical reaction is carried out to yield a product which can be measured using an ion-selective electrode. Urea (a nonionic compound) can be converted to CO_2 and NH_4^+ by the enzyme urease. Specific electrodes exist which allow the detection of the ammonium ion formed in the reaction. If the assay conditions are set so that all the urea present is converted to products, the concentration of ammonium ion can then be related to the concentration of the urea initially present in the sample.

Types of Measurements Possible

SPECIFIC IONS

Measurements for sodium and potassium in serum have employed ion-selective electrodes for a number of years. The sodium electrode is very similar to the hydrogen electrode, employing a different type of glass which is selectively permeable only to sodium. The assay for serum potassium is accomplished using a valinomycin derivative (a large cyclic organic structure) which traps the K^+ ion in its center (Fig. 3-9). Crown ethers are organic structures which have also been used successfully for potassium assays. Although a commercial system for lithium is not fully developed yet, research in crown ether chemistry has application in this instance also. Measurements of ionized calcium are carried out using agents which form complexes with the ion after it has diffused through a membrane.

Determination of CO_2 is accomplished by passing the gas through a selective membrane into a system where it can react with water to form H_2CO_3, which then dissociates to yield H^+ and HCO_3^-. The hydrogen ions produced in this reaction can then be detected by standard means.

Many of the ions of interest (particularly sodium and potassium) are measured in automated systems where large numbers of samples are processed at high throughput. In these instruments, the electrodes employed are flow-through devices in which the sample (and other necessary reagents) pass through an electrode coated with materials on the inside. As the sample flows through the system, it

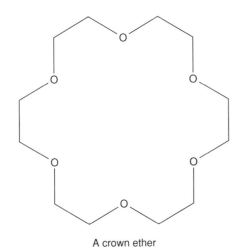

Valinomycin derivative

A crown ether

FIGURE 3-9   Selective ion-trapping agents in electrodes.

variety of techniques are being developed for glucose measurement which incorporate an enzymatic reaction producing hydrogen peroxide and measuring oxygen consumption or peroxide formation directly or indirectly.

The major drawbacks to widespread use of enzyme electrodes are slow response time and lack of long-term stability. An enzyme reaction needs to take place before the product can be detected by the electrode. These reactions may require up to a few minutes to occur, although proper control of conditions can decrease reaction time to as low as 30 s or so. Stability of the enzyme is a significant problem, since these proteins degrade easily and lose their activity. Even with these apparent shortcomings, a wide variety of analytical approaches are possible with enzyme-coupled, ion-selective electrodes. Commercial applications should develop rapidly after these technical drawbacks are overcome.

PROBLEMS ASSOCIATED WITH ION-SELECTIVE ELECTRODES

Although ion-selective electrodes perform well in aqueous solutions, the high concentrations of lipids and proteins found in serum or plasma create some problems in analysis of body fluids. The presence of these materials changes the actual water volume of the sample and can give rise to erroneous results. Interactions between these interferents and the chemical under analysis might affect the result, with bound material being unable to diffuse through the membrane and be detected by the electrode. Proteins (and lipids, to a lesser extent) can coat the electrode and decrease the amount of diffusion.

Calibration has not been standardized to any great extent. There is still more variability among instruments and manufacturers than is acceptable. Work by international standards commissions should aid in solving this problem.

Amperometric Measurements

In the potentiometric systems described above, the electrode is a passive component, responding to the changes in concentration taking place around it. **Amperometric methods** of analysis are being developed in which the electrode system participates in a reaction with the material being analyzed. The resulting oxidation/reduction process furnishes a flow of electrons which can be measured by the instrument and which is proportional to the amount of analyte present. Sensors for blood gas measurements are available which operate according to amperometric principles. Oxygen reacts with water to form hydroxide ions, consuming four electrons (e) per molecule in the process.

$$O_2 + 2H_2O + 4e \rightarrow 4OH^-$$

interacts with the electrode membrane, diffuses through the interface, and causes a potential change in the system.

MEASUREMENTS USING ENZYME ELECTRODES

Incorporation of an enzyme into the membrane has opened the door to a number of possibilities for specific analysis of biochemicals. The system which utilizes the enzyme urease to form ammonia from urea is one example of an **enzyme-coupled, ion-selective electrode.** Assays for oxalate have been developed using a decarboxylase enzyme and measuring CO_2 formed from the breakdown of oxalate. A

The net electron flow can be determined and the concentration of oxygen calculated. Some O_2 sensors for transcutaneous measurement of oxygen are commercially available.

Biosensors

The enzyme electrode described above might be considered as an early example of the **biosensor:** an electrode system using material from a living system as the detecting agent. This analytical approach is new, with little commercial application at present. However, research developments suggest biosensors may well provide versatile analyses at extremely low concentrations for a variety of clinically important biochemical molecules.

Many biosensors involve reasonably standard electrode technology coupled with the use of specific and sensitive detector systems isolated from living organisms. Enzymes are used in a variety of ways, either alone or in combination with antibodies, to catalyze reactions which produce a signal. This signal results in changes in potential at the electrode surface, permitting the quantitative analysis of some specific material. Receptors isolated from various organisms are also being employed in biosensor systems. Since a given receptor binds fairly specifically to a biochemical at extremely low concentrations (frequently in the range 10^{-6} to 10^{-9} M or less), the detecting capability of the system is greatly enhanced. Because of their specificity and high affinity, antibodies are finding wide application in this area.

The development of biosensor technology means that ion-selective electrode systems will be able to go well beyond the determination of a few inorganic ions and pH presently seen. The ability of the system to recognize specific organic and biochemical

molecules at physiological concentrations greatly expands opportunities for measurement. Although the response times of many electrodes are slow (2–5 min in some cases) and systems have stabilities of only one to two weeks, there is great potential for the development and application of biosensors in the clinical laboratory.

3.6

GAS CHROMATOGRAPHY (GC)

Principles of Chromatography

Ever since the Russian scientist, Mikhail Tswett, developed a new technique to separate plant pigments in 1906, **chromatography** has been a powerful tool for the identification and measurement of biochemical molecules. A wide variety of approaches have been explored for fractionation and detection of components, all based on one fundamental concept. A mixture of materials is placed on a support medium. As the mixture flows through the support (column or flat surface), molecules interact with functional groups on the support material. Each type of molecule binds to a somewhat different degree to the functional group. With appropriate selection of conditions, the compounds separate from one another, come off the support material, and pass through a detection system. Chromatography can be employed as a very selective separation device, allowing fractionation and quantitation of materials in complex mixtures.

Separation by **gas chromatography (GC)** utilizes a temperature-controlled column containing a support medium (Fig. 3–10). The sample is injected into the system and molecules interact with the material coating the inside of the column. A carrier gas

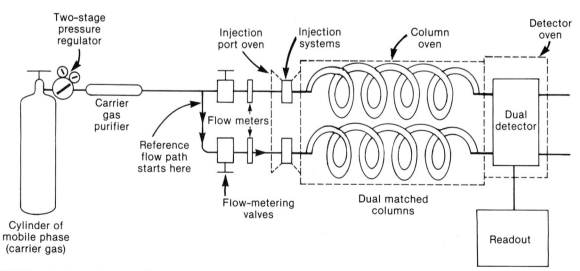

FIGURE 3–10 A modern gas chromatograph (not drawn to scale). (From Bender, G. T., *Principles of Chemical Instrumentation.* Philadelphia, W. B. Saunders Co., 1987.)

flows through the column to move the vaporized material down the system. Each compound exists in equilibrium between the solid (or liquid) phase bound to the column packing and the vapor phase produced by the temperature change around the column. As the temperature increases, more material is in the vapor phase to be swept through the column toward the detector. When each individual compound comes off the column, it interacts with a detection system, creating an electric signal whose intensity is proportional to the amount of material at the detector site.

Instrumentation for GC

The components of a gas chromatography system are

1. sample-injection chamber
2. column and oven for temperature control
3. detector
4. data recording/readout

SAMPLE INJECTION

For clinical work, samples are in liquid form, usually in solution. The solvent is organic and volatile, so it can be easily vaporized and removed from the system before any of the components of interest reach the detector. For maximum separation, sample-injection volumes are small, often 1 μL or less.

The sample-injection port is heated to a temperature sufficient to vaporize all the materials in the system. Since the sample must be volatile, it is often necessary to prepare a derivative of the native compound. Although the initial material may be incapable of being vaporized, the derivative will boil at a much lower temperature. Many derivatizing processes are available which require a simple mixing step followed by a solvent extraction before injection onto the chromatograph.

COLUMN FRACTIONATION

Gas chromatography columns come in a variety of sizes and packings. Column lengths can vary from a few feet to several hundred feet (for capillary columns). Composition of the column itself depends on the materials to be studied. For most biochemical analyses, glass columns are preferred. Even though metal columns are much less fragile, many biochemical molecules degrade through contact with the metal.

A carrier gas flows through the column to move the vapor phase along. In most applications, nitrogen is a satisfactory carrier gas; for others, helium is preferred. The gas must be very pure to avoid introduction of contaminants into the analysis. Gas com-

position and flow rates are determined by the type of column employed, the separation desired, and the specific detector in the system.

The heart of a gas chromatography column is its packing material, which provides the selectivity in the particular assay method. A solid support (usually a silica derivative) is coated with a high-boiling liquid in which the components to be separated dissolve as they pass through the column. The liquid portion is referred to as the **stationary phase.** A large number of stationary phases are available commercially which permit separation of most compounds of interest.

Capillary (or open tubular) columns have the coating applied directly to the inside of the tubing. The result is less obstruction to the flow of carrier gas and volatile components through the column since the solid phase is absent. Less pressure drop occurs, allowing much longer columns (sometimes up to 100 m in length) and greatly improved efficiency in separation.

The column is contained inside an oven which controls the temperature of the system. Some analyses can be done at constant temperature, but others require a gradual increase in temperature for optimum separation. These temperature changes can be programmed so that the desired increases in temperature and the necessary time intervals can all be dealt with automatically. Although it is possible to make some initial predictions about temperature conditions, all programming must be handled experimentally until the optimum conditions are achieved.

DETECTOR SYSTEMS

As the separated compounds elute off the column, they interact with a detection system to register their time of elution and relative concentrations. A number of detector systems are available, ranging from simple to complicated. Detectors are selected on the basis of the types of compounds to be analyzed and the sensitivity required in the assay. It is quite common for a single gas chromatograph to be equipped with two detectors (each with different capabilities) to encompass the wide range of analyses needed by the facility.

Thermal Conductivity Detector

The **thermal conductivity detector** is perhaps the least complex type of detector. The vapor from the eluting molecule strikes a heated wire, producing a change in the amount of electric current passing through it.

The thermal conductivity detector filaments are probably the Achilles heel of the system. They have a tendency to oxidize easily, which creates some instability in the baseline readings over time. Other limitations of this type of detector are its low sensitivity relative to other systems and the need to standardize individually for each component in the assay system if quantitative measurements are to be made.

Flame Ionization Detector

If a **flame ionization detector** is used, the sample is burned in a hydrogen flame after elution from the column. Ions are formed which interact with a wire collector through which electric current passes. There is a potential difference set up between the collector and the flame: The presence of ions alters that difference and is detected by the electronics. The flame ionization detector is the most widely used for clinical applications since it responds to all organic compounds, is sensitive, and is relatively trouble-free in operation.

There are two major limitations to the use of flame ionization detectors. First, this system is obviously a destructive one. Since the sample is burned in a flame, it cannot be recovered or further analyzed by another instrument (such as a mass spectrometer). If more study of the fractions is required, a separate run must be made under the same conditions, with the elution stream being directed to another instrument instead of passing through the flame ionization detector.

If the compound of interest contains elements other than carbon and hydrogen, sensitivity of detection decreases noticeably. Oxygen, nitrogen, sulfur, phosphorus, or halogens in the structure render the material less easy to detect by the flame ionization system. Since many biochemical molecules of interest contain these components, analytical sensitivity suffers somewhat.

Electron Capture Detector

The **electron capture** system is the most sensitive detector but is also very prone to operational malfunctions. This detector "sees" those compounds which have an affinity for electrons and thus is restricted (to a great extent) to the analysis of halogenated derivatives. In addition, the electron capture detector contains a radioactive source which requires special licensing. Material passes through an electrode system, with the radioisotope source connected to the cathode. The carrier gas (usually a mixture of nitrogen and other materials) is ionized by the radioisotope as it passes through the system, setting up a small constant potential across the two electrodes. As compounds capable of attracting electrons enter the system, the potential drops and a signal is sent to the detection system. Materials containing halogens or structures with conjugated carbonyls or nitrates make up the groups which produce the major response in the electron capture detector.

The primary limitation to the use of electron cap-

ture is the narrow range of compounds it can detect. In addition, calibration is difficult since the linear range for any given compound is quite small. The radioisotope also contributes to the problems of operating this type of detector. The radioactive source gradually becomes contaminated by deposition of the compounds being eluted, requiring replacement and recalibration.

Separation and Identification by GC

As compounds separate on the column and are detected at the end of their fractionation, the signal sent by the detecting system is translated into a chart showing the elution pattern of the mixture (Fig. 3–11). The initial peak on the trace is the solvent front, produced by the organic solvent employed in the assay. Since this material is of low molecular weight and quite volatile, it leaves the column before other components in the majority of analyses. Other peaks are displayed later at different time intervals. Many computerized systems can suppress the readout of the solvent front, allowing only the sample peaks to be displayed. The location of the peak provides an indication of the identity of the compound, and the peak height can be related to concentration.

The term **retention time** is used to describe how long a given compound requires to come off the column and be observed by the detector. Since the strip chart used for recording comes out of the recorder at a set speed, distance along the chart can be related to time. The retention time (usually expressed in minutes) is measured from the point of sample injection to the point of peak height for that given component. If the separation conditions (temperature, column, derivative) are changed, the retention time is altered.

Peak height is proportional to concentration of material, as is area under the peak. If the elution peak is sharp and has a small base, peak height is

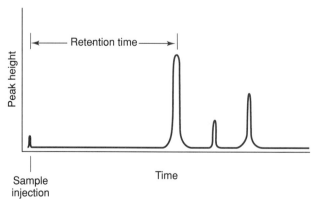

FIGURE 3–11 Gas chromatographic elution pattern.

the preferred parameter since it is easier to measure manually. Peaks with broad bases probably require determination of area under the peak. Modern instruments are usually computer-interfaced and have integrators as part of the data-reduction system, allowing automatic determination of the area under the peak. However, these data cannot be blindly accepted and must be reviewed for any anomalies in the trace which might suggest a problem.

Since we work with very small volumes in GC analysis, the raw peak height data cannot be relied on by itself if a quantitative estimation of a given compound is required. More helpful and reproducible information can be obtained by including an internal standard. This material is a compound which has a retention time different from any of the components for which we are analyzing. The internal standard is added to the assay system at a known concentration before analysis is started. After a separation is performed, the peak height ratio of standard to unknown is calculated and compared with standards for the compound of interest. Use of an internal standard allows us to compensate for any variability in sample treatment of difference in the amount of sample injected onto the column.

Applications to Clinical Chemistry

The major role for gas chromatography in the clinical chemistry laboratory has been for drug screening and drug analysis. Some assays, such as quantitative measurement of tricyclic antidepressants, use GC as the preferred method for identification and quantitation. For years, serum anticonvulsant levels were measured using GC, often with automated sample-loading devices to provide some degree of automation for laboratories with large workloads. More recently, immunoassay techniques have supplanted gas chromatography for some of these testing procedures.

Gas Chromatography/Mass Spectrometry

A recent innovation in the clinical applications of gas chromatography has been the coupling of GC to mass spectrometry. Organic molecules can be fragmented by a variety of means (electron bombardment being one simple technique). These organic ions then separate on the basis of charge and mass when they pass down a tube in a magnetic field. If the eluted material from a gas chromatograph then passes through a mass spectrometer, each component from the GC can then have a mass spectrum determined. The characteristic fragments for each molecule allow reliable identification of structure.

Two major applications for **gas chromatography/mass spectrometry** analysis are drug screening and steroid identification. In both instances, a complex mixture of materials must be analyzed to identify specific components of interest. For drug-screening work, the initial identification is performed by relatively nonspecific means which allow a rapid yes/no answer. Confirmatory testing to demonstrate the identity of a given drug may then be carried out by gas chromatography/mass spectrometry on samples which test positive by less specific means. Abuse of anabolic steroids by athletes has led to sophisticated techniques for identification of synthetic steroids present in the complex mixture of naturally occurring hormones in the body. Although in its infancy as a routine approach, gas chromatography/mass spectrometry may well play an important role in the clinical laboratory of the future.

3.7

HIGH-PERFORMANCE LIQUID CHROMATOGRAPHY (HPLC)

Even though gas chromatography greatly expanded the analytical horizons of the clinical laboratory, this technique suffered from one major shortcoming: any sample studied by this system had to be either volatile or capable of being converted to a volatile material for analysis. Although many organic molecules meet these conditions, a large number of biochemical components of clinical interest turn out to be rather fragile at high temperatures. Such molecules are often fragmented or totally destroyed rather than separated from other materials. Analysis of high-molecular-weight components (such as proteins and peptides) is not even feasible using GC. Obviously, another separation technique was needed that would permit rapid fractionation and identification of biomolecules without the destructive effects seen with gas chromatography. This need was filled in the last 15 years or so by the technique of **high-performance liquid chromatography (HPLC)**.

Basic Principles

The concept of **selective adsorption** of molecules onto a solid phase followed by removal with selected solvents has been known since the late 19th century. Prior to the development of HPLC, packed columns were used for fractionation of mixtures in a variety of biochemical research settings. A sample was placed at the top of a column, solvent washed through the column by gravity, and materials eluted from the column in an order determined by the packing material, the solvent, and the degree of binding of each solution component to the column material. The use of gravity produced slow flow rates through the column and long assay times. Development of specific equipment, along with improved column technology and new materials for packings, allowed HPLC to assume its present major role as a separation/analysis technique in the biochemical field.

To carry out an HPLC analysis (Fig. 3–12), a sample is dissolved into the appropriate solvent prior to injection into the system. Protein must be removed in most cases to avoid column damage. The material used to dissolve the sample is selected to be compatible with the elution solvent employed. A high-pressure pump moves the sample and elution solvent onto the column. Components of the mixture bind to the column packing with varying degrees of affinity, allowing the constituents to separate. As the solvent elutes fractions from the column, the material passes through a detection system where some property (light absorption, conductivity, fluorescence, or other parameter) is determined. The change in signal is sent to a data-processing system or recorder which gives a readout of elution time against peak intensity for each constituent of the mixture. An integrator allows quantitation of materials similarly to the gas chromatography method.

One advantage of HPLC over "conventional" chromatography is speed. In gravity-fed systems, fractionations often require hours to accomplish. With HPLC, a column analysis frequently can be accomplished within 5 to 10 min. Resolution of peaks is a problem in older methods due to the slow rate of passage through the column, with ample opportunity for each component to diffuse. With HPLC, the rapid movement of constituents minimizes sample diffusion and greatly improves peak sharpness.

Originally, the term *HPLC* was commonly thought to refer to high-*pressure* liquid chromatography, since pressures of 3000 psi or more were commonly required for proper operation. As technology has improved, lower and lower pressures are being required, and the term *high pressure* is no longer considered to be an accurate description of this approach to chromatography. Professionals in the field have come to adopt the concept of high-*performance* liquid chromatography as the proper designation for HPLC.

Instrumentation for HPLC

SAMPLE INJECTION MODULE

Samples for HPLC analysis are small, routinely less than 10 μL. The injection is made with a Hamilton syringe into a separate sample loop not connected with the solvent stream to the column. After injection, solvent will enter the loop and wash the sample onto the column.

PUMP

The pump plays a vital role in the success of an HPLC separation. The choice of constant-volume delivery or constant-pressure delivery needs to be made in conjunction with other factors. Constant-volume delivery minimizes problems with base-line readings through the detector. If less pulsing is seen in the solvent flow, the background noise will be lower. This factor can be particularly significant when measuring very low concentrations of a given material.

COLUMN

The key to selectivity in the HPLC methodology lies in proper choice of the column. A wide variety

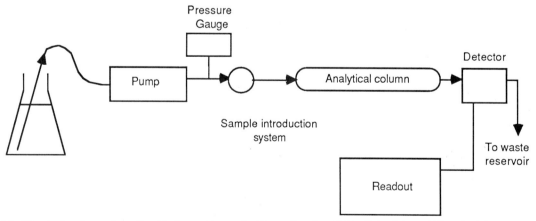

FIGURE 3–12   A simple modern liquid chromatograph. (From Bender, G.T., *Principles of Chemical Instrumentation.* Philadelphia, W. B. Saunders Co., 1987.)

of packings and supports are available for either general work or for very specialized applications. Although columns are expensive (they currently cost $200.00 or more), it is possible to obtain several hundred fractionations on a single column before it needs to be repacked.

Guard Columns

Prior to discussing types of columns and their functions, we should consider the use of guard columns. The major role of a guard column is protecting the analytical column from contamination and damage. By attaching a short column of the same packing into the system between the sample-injection valve and the column used for the actual separation, materials which create problems can be trapped and removed prior to sample analysis.

Guard columns serve a number of functions. Material in the sample which binds to the column packing, but does not elute, can be trapped by the guard column. Small particles in the solvent, which were not removed by prefiltration, or particulate contamination in the sample (protein or other) are removed by the guard column. The presence of a small column in line helps to minimize pressure surges on the main column, decreasing damage to this column. Since guard columns can easily be repacked in-house, they provide a relatively inexpensive means of protecting a rather costly analytical column.

Normal-Phase Columns

The first columns employed for HPLC work were referred to as **normal-phase columns** since they operated on the same general principles as the classic ion-exchange columns. The compounds of interest interact with a polar side chain on the column packing, either a hydroxyl group on the silica solid support itself or some functional group bonded to the silica particle. Typical bonded-phase groups are diols, cyano derivatives ($-C\equiv N$), or amino groups, either $-NH_2$ alone or a substituted amine. Although they are not ionic compounds, the various groups have electron pairs in their structures that can be polarized.

Highly polar compounds bind more tightly to normal-phase packings than nonpolar or less polar ones. As the polarity of the mobile phase (the solvent solution passing through the column) increases, binding to the column packing is lowered. The mobile phase is generally some mildly polar organic solvent such as an alcohol. Mixtures of alcohols, acetic acid, and water in varying proportions are common solvents used in the mobile phase for elution of materials.

Normal-phase packings are very useful for separation of isomers. In addition, the use of mild organic solvents allows a wide range of solubilities to be employed in the separation of biochemical molecules. Destruction of the silica solid support is minimized with these organic solvents since silica is much less soluble in organic material than in water-based elution systems.

Reversed-Phase Columns

A **reversed-phase** packing is one in which the silica stationary phase is coated with a nonpolar material, usually some hydrocarbon. One very popular organic material is octadecane, an 18-carbon compound. Materials are separated by the packing material on the basis of their ability to form nonpolar interactions with this C_{18} hydrocarbon. In these types of attractions, the length of the hydrocarbon chain and the presence of aromatic rings are important determinants of binding. The presence of polar groups decreases the binding of a given compound. Other nonpolar side chains include octane or benzene substituents.

Since the stationary phase is nonpolar, the removal of materials by solvent elution is based on polar interactions between solute and solvent. Solvents are usually water-based and may contain either organic or inorganic compounds. Often a mixture of water, a mildly polar organic solvent, and an inorganic salt are combined to facilitate removal of materials from a reversed-phase column.

The use of reversed-phase separation media allows water-based solvent elution, providing a wide variety of solvent combinations for maximum effectiveness. The presence of a predominantly aqueous medium provides stability for some biochemical molecules which might be degraded by harsher organic solvents. Many organic solvents absorb in the UV range or fluoresce, creating difficulties in detection which are markedly decreased in aqueous solutions.

Reversed-phase columns have some shortcomings. A limited pH range is available for fractionations (roughly pH 2–8 or less). When the silica particle is coated, only about half the particles react with the organic coating material, leaving a significant fraction of $-OH$ groups on the silica available for interaction with the molecules passing through the column.

DETECTOR

The ability to detect constituents eluted from an HPLC column appears to be limited only by the types of compounds we wish to study and the departmental budget. A wide variety of systems are currently available for monitoring and quantitating biochemical analytes. No longer are we limited to simple colorimetric detection. Fluorescence techniques, many applications of electrochemistry, and the coupling of HPLC to mass spectrometry and other very sophisticated systems provide a wide array of approaches for detection.

Absorbance measurement still remains the most versatile single approach to HPLC detection. Simple instruments make use of filters for wavelength selection, whereas more sophisticated systems have high-quality monochromators. A combination of UV and visible capabilities provides both sensitivity and flexibility in the vast majority of situations. Multiple-wavelength analysis allows identification of unknown compounds and permits detection of materials which coelute. Many commercial systems employ photodiode array spectrometers to scan each peak, obtaining a complete absorption spectrum within seconds. This data can be further processed by computer to provide detailed information about the characteristics of each component eluted from the column.

Fluorescence measurements play an important role in HPLC detection. Although subject to limitations previously discussed in this chapter, the use of fluorescence improves selectivity greatly. Sensitivity is much enhanced with fluorometric detection, sometimes as much as 100-fold. By proper choices of excitation and emission wavelengths, some coeluting materials can be identified. Laser applications with fluorometric detection have been studied but have a number of problems, mainly due to difficulties inherent in the laser.

Measurement of the refractive index provides a fairly universal means of detection without the need for formation of a compound which absorbs light or fluoresces. Any compound with a refractive index different from that of the solvent being used can be detected. However, use of refractive-index differences is not as sensitive a means of detection as other optical approaches.

Measurement of the ability of the solution to conduct an electric current has found some application in HPLC detection systems. This method is limited to ionic compounds and thus is not suitable for many biochemical applications.

Electrochemical detection has been shown to be a sensitive means of monitoring separations, particularly for catecholamines and related compounds. Passage of an electric current through the eluate produces an oxidation/reduction reaction for those compounds capable of undergoing electron transfer processes. Selectivity and sensitivity are comparable to absorbance and fluorescence techniques.

POSTCOLUMN DERIVATIZATION

Some biochemical materials have a native fluorescence or absorb light at specific wavelengths, which allows detection after column separation. For many compounds, a reaction forming a colored or fluorescent derivative can be carried out prior to sample injection on the column. However, there are a number of situations in which formation of a derivative prior to column application is not feasible. In these cases, it becomes necessary to form a **postcolumn derivative** before the material passes through a detector system.

A wide variety of techniques are available for generation of postcolumn derivatives. The simplest approach is to mix analyte and reacting species in a tubular reactor, either at room temperature, in a heated tube, or with light exposure to carry out a photochemical reaction. Great versatility in reaction chemistry is possible with minimal additional hardware, much of which is relatively inexpensive. Some interesting approaches involving chemiluminescent products have been developed using tubular reaction technology.

If the reaction time is longer than a few minutes, the material may be passed through a column or coil, which allows longer exposure to the derivatizing reaction and/or reaction conditions. Some coils employ immobilized enzymes to form products which can easily be detected. The use of enzymes provides specificity in the reaction, and the immobilization process gives extra stability to the enzyme itself. In addition, less enzyme is needed for the reaction.

Continuous-flow systems for automated analysis have been adapted to postcolumn derivative formation. The eluate stream is mixed with a reagent stream in a segmented-flow arrangement. Air bubbles are introduced into the final mixed stream to separate the sample. A number of ingenious systems have been reported using segmented streams coupled with solvents which do not mix with water for selective extraction of a particular analyte.

Applications to Clinical Chemistry

High-performance liquid chromatographic techniques have found wide application in clinical chemistry, particularly in the areas of therapeutic drug monitoring and drug screening. A large number of pharmacological studies are quickly and easily carried out using HPLC processes. Work has been done using HPLC for steroid analysis, although this methodology has not yet achieved routine clinical use. Some studies have been published which employ HPLC for fractionation of isoenzymes, primarily for research interests. The method of choice for analysis of catecholamines in a variety of clinical situations has been HPLC fractionation followed by electrochemical detection. This methodology has seen wide application in the study of the biochemical basis of mental illness.

SUMMARY

Quantitative measurement by absorption spectrometry is based on the Beer-Lambert law: Absorption of light by a material in solution is proportional to the concentration of that material. Instruments for this analytical technique consist of a light source, a monochromator, and some type of phototube detecting system. A variety of approaches are available for calculation of concentrations, including ratio of unknown to standard, use of a standard curve, or use of molar absorptivity coefficients. Careful attention must be given to establishing the linearity of an assay method, the lower limit of detection, and proper blanking procedures.

When a compound emits light after being excited by light of a shorter wavelength, fluorescence measurements are possible. In this type of apparatus, the detection system is at right angles to the light source. Factors which affect the efficiency of fluorescence quantitation include blanking, scattered light, absorbance by the sample, and quenching.

Flame emission spectrometry provides quantitation through detection of energy given off during electron transitions after excitation at high temperatures. Sodium, potassium, and lithium assays are the primary applications of flame photometric measurements. Variations in flame background emissions are compensated for with an internal standard. High concentrations of lipids or proteins in the sample can produce erroneously low values for electrolytes.

Determination of hydrogen ion concentration was the first application for electrodes as measuring devices. Measurement of pH is accomplished by detecting potential differences between reference and sample electrodes. The concept has been extended greatly to the assay of a wide variety of constituents by the use of ion-selective electrodes. Biosensors have been developed which allow quantitation of nonionic materials in solution.

Separation and quantitation of biochemical components in a mixture is often easily accomplished with gas chromatography. A liquid sample is injected into a sample port where it is vaporized and passed through a column. Interactions between the stationary and mobile phases allow fractionation to take place. Several detecting systems are available for monitoring components as they elute from the column.

For samples which might be destroyed at high temperatures and for those materials which are not volatile, high-performance liquid chromatography provides a versatile approach to fractionation. The sample is injected into a valve, washed onto the column by the eluting solvent, and partitioned between the stationary and mobile phases. As the eluting material passes through the column, materials fractionate on the basis of relative affinities. Detecting systems include measurement of light absorbance, fluorescence, or index of refraction. Electrochemical approaches have application for catecholamines and related types of compounds. Postcolumn derivatization techniques have been developed for those compounds which do not have any native light absorption or fluorescent properties.

FOR REVIEW

Directions: For each question, choose the best response.

1. The more light absorbed, the higher the concentration of analyte in this technique of measuring the amount of light absorbed by a solution.
   A. atomic absorption
   B. fluorometry

C. nephelometry

D. spectrophotometry

2. The visible region of the electromagnetic spectrum lies between
   A. 200 and 700 nm
   B. 340 and 850 nm
   C. 400 and 800 nm
   D. 500 and 900 nm

3. A monochromator is a device that
   A. measures the intensity of light energy falling on a detector
   B. determines the critical angle of refraction of dispersed light
   C. disperses polychromatic light into its separate wavelengths
   D. measures the intensity of light which falls on a diffraction grating

4. The component of a spectrophotometer that is responsible for detecting transmitted light and converting this light energy to electrical energy is the
   A. detector
   B. galvanometer
   C. monochromator
   D. readout device

5. When measuring the concentration of a solution, the Beer-Lambert law takes all of the following parameters into consideration *except*
   A. wavelength of assay
   B. absorptivity coefficient
   C. measured absorbance
   D. path length of cuvet

6. All of the following methods may be used to calculate the concentration of an unknown solution from absorbance measurements *except*
   A. ratio of standard to unknown
   B. spectral transmittance curve
   C. standard curve
   D. use of molar absorptivity value

7. Which of the following parameters may be determined from a standard curve?
   A. sensitivity of assay at low concentrations
   B. necessity of using individual sample blanks
   C. upper concentration limit of detection
   D. both A and B
   E. both A and C

8. Which of the following activities would be the most useful in identifying reagent deterioration?
   A. constructing a standard curve
   B. measuring the absorbance of the standards
   C. measuring the reagent blank absorbance
   D. measuring the sample blank absorbance

9. Using a sample blank will help correct for absorbance interferences caused by which of the following substances?
   A. bilirubin
   B. cholesterol
   C. hemoglobin
   D. both A and C
   E. both B and C

10. A fluorescent substance absorbs light of one wavelength and emits light of a
    A. longer wavelength and lower energy
    B. shorter wavelength and lower energy

C. longer wavelength and higher energy

D. shorter wavelength and higher energy

11. Which of the following components of a fluorescence spectrometer would be positioned at a right angle relative to the light source?

A. primary monochromator and cuvet

B. secondary monochromator and cuvet

C. secondary monochromator and detector

D. primary and secondary monochromators

12. The internal components of a fluorometer would include

A. one filter

B. two filters

C. two monochromators

D. two detectors

13. The process by which the fluorescence of an analyte is reduced due to the excited molecule losing some of its energy by interacting with other substances in solution is known as

A. ionization

B. quenching

C. phosphorescence

D. self-absorption

14. In flame emission photometry the concentration of the analyte is directly proportional to the amount of light

A. emitted by the flame

B. emitted by the analyte

C. absorbed by the flame

D. absorbed by the analyte

15. In flame emission photometry the internal standard is used primarily to

A. give direct readings in concentration units

B. eliminate the need for precision in specimen dilution

C. facilitate the separation of sodium and potassium signals

D. correct for variations in the flame and aspiration rate

16. In flame emission photometry, a photon of light of a specific wavelength is emitted when

A. an electron is raised to a higher energy orbital

B. an electron absorbs ultraviolet radiation

C. an excited electron returns to its ground state

D. the bonds in a molecule vibrate

17. The statement that most correctly describes the measurement of pH with a glass electrode is

A. hydrogen ions pass through a special type of permeable glass

B. the internal and external half-cells have the same potential

C. a special type of glass sensitive to hydrogen ions is used

D. the temperature knob on the pH meter compensates for the influence of temperature changes

18. When measuring pH, the external reference electrode is usually a

A. calomel electrode

B. glass electrode

C. hydrogen electrode

D. sodium electrode

19. Which of the following may be associated with ion-selective electrodes (ISE)?

A. responds to ions in aqueous solution

B. responds to one type of ion

C. the sensing membrane is ion-selective

    D. both A and B

    E. all of the above

20. When measuring CO_2, the dissolved CO_2 gas in the specimen reacts within the ISE electrode to form a product which is measured as
    A. bicarbonate ion
    B. carbonic acid
    C. hydrogen ion
    D. silver chloride

21. An amperometric system, where electrons are consumed as part of the reaction and the net electron flow is measured, best describes
    A. a carbon dioxide electrode
    B. a sodium electrode
    C. an oxygen electrode
    D. a potassium electrode

22. The packing of a column in gas chromatography consists of
    A. inert solid support material
    B. a liquid stationary phase
    C. a gas mobile phase
    D. both A and B
    E. both A and C

23. In gas chromatography a compound may be identified by comparing its _____ to that of a known standard.
    A. solvent front
    B. retention time
    C. peak height
    D. area under the peak

24. The selective adsorption of molecules by a solid phase which is followed by their removal by a selected solvent system best describes
    A. GC
    B. GC/mass spectrometry
    C. HPLC
    D. TLC

25. The chromatography system that employs the use of spectrophotometric, fluorescent, and electrochemical techniques for its detector system best describes
    A. GC
    B. GC/mass spectrometry
    C. HPLC
    D. TLC

BIBLIOGRAPHY

Abbott, S. R., "Practical aspects of normal-phase chromatography," *J. Chromatog. Sci.* 18:540–550, 1980.

Bachman, W. J., and Stewart, J. T., "HPLC postcolumn derivatization techniques," *LC-GC.* 7:38–50, 1989.

Borman, S., "Biosensors: Potentiometric and amperometric," *Anal. Chem.* 59:1091A–1098A, 1987.

Braun, R. D., *Introduction to Instrumental Analysis.* New York: McGraw-Hill, 1987.

Conroe, K., and Bidlingmeyer, B., "Overview of the growth of HPLC applications," *Am. Lab.* 82–87, October, 1988.

Czaban, J. D., "Electrochemical sensors in clinical chemistry: Yesterday, today, tomorrow," *Anal. Chem.* 57:345A–356A, 1985.

Dolan, J. W., et al., "HPLC method development and column reproducibility," *Am. Lab.* 43–47, August, 1987.

Ettre, L. S., "Open-tubular columns: Evolution, present status, and future," *Anal. Chem.* 57:1419A–1438A, 1985.

Fisher, J. E., "Measurement of pH," *Am. Clin. Prod. Rev.* August: 22–27, 1984.

Freeman, D. H., "Liquid chromatography in 1982," *Science* 218:235–241, 1982.

Frew, J. E., and Hill, H. A. O., "Electrochemical biosensors," *Anal. Chem.* 59:933A–944A, 1987.

Froehlich, P., "Fluorescence of organic compounds," *Instrumentation Research.* March:98–103, 1985.

Halasz, I., "Columns for reversed phase liquid chromatography," *Anal. Chem.* 52:1393A–1403A, 1980.

Ichida, A., "Chromatographic resolution on chiral stationary phases," *Am. Lab.* 100–103, 1988.

Meyerhoff, M. E., and Opdycke, W. N., "Ion-selective electrodes," *Adv. Clin. Chem.* 25:1–47, 1986.

Monroe, D., "Bioselective electrodes, parts I and II," *Am. Clin. Prod. Rev.* January:24–34, March:46–57, 1985.

Oesch, U., et al., "Ion-selective membrane electrodes for clinical use," *Clin. Chem.* 32:1448–1459, 1986.

Pickering, M. V., "Assembling an HPLC postcolumn system: Practical considerations," *LC-GC.* 6:994–997, 1988.

Rechnitz, G. A., "Biosensors," *Chem. Eng. News.* 24–36, September 5, 1988.

Rothstein, F., and Fisher, J. E., "pH measurement: The meter," *Am. Clin. Prod. Rev.* August:26–32, 1985.

Stadalius, M. A., et al., "Reversed-phase HPLC of basic samples," *LC-GC.* 6:494–500, 1988.

Tarbet, B. J., et al., "The chemistry of capillary column technology," *LC-GC.* 6:232–248, 1988.

Turner, A. P. F., and Swain, A., "Commercial perspectives for diagnostics using biosensor technologies," *Am. Biotech. Lab.* 10–18, 1988.

van den Driest, P. J., et al., "Octadecyl bonded phases on silica gel supports," *LC-GC.* 6:124–132, 1988.

Yeung, E. S., "Chromatographic detectors: Current status and future prospects," *LC-GC.* 7:118–128, 1989.

Yeung, E. S., and Synovec, R. E., "Detectors for liquid chromatography," *Anal. Chem.* 58:1237A–1256A, 1986.

CHAPTER OUTLINE

Chapter Four

PRINCIPLES OF MEASUREMENT III: AUTOMATION

● ●

1. Describe three factors that have supported the development of automated laboratory equipment.
2. Name three factors that have facilitated our ability to automate laboratory instruments and to handle large quantities of data.
3. Explain the process of "continuous flow" analysis.
4. Compare and contrast segmented stream continuous flow analysis with flow injection analysis.
5. Describe the process used in continuous flow analysis to separate a protein-bound analyte.
6. Explain the basic concept of centrifugal analysis.
7. Describe the function of the rotor in centrifugal analysis.
8. Explain how reagent blanks are handled in each of the following systems:
   A. manual photometric
   B. continuous flow analysis
   C. centrifugal analysis
9. Differentiate between an end-point measurement and a rate measurement.
10. Explain how each of the following may be considered a benefit of automation:
    A. reproducibility
    B. speed
    C. reduction in personnel time
11. Discuss the significance of the bar-code on the aca reagent test pack.
12. Describe the function of each layer in the Kodak dry slide technology:
    A. spreading   D. indicator
    B. scavenger   E. support
    C. reagent
13. Compare and contrast light measurements made using spectrophotometry with reflectance spectroscopy.
14. Describe how the Kodak system compensates for hemoglobin or bilirubin that may be present in a sample.
15. Explain the purpose of dual-wavelength analysis.
16. Define the terms batch mode and random-access.
17. Discuss the use of a fluorocarbon to coat the inside of sampler and reagent transfer probes in Technicon's newer equipment.
18. Discuss the pros and cons associated with performing laboratory tests in physician office laboratories.

INTRODUCTION

The mechanization of laboratory operations can take many forms. The pipet, perhaps one of the simplest pieces of equipment in the laboratory, has undergone striking changes within the last 30 years. The traditional glass pipet, operated by mouth suction, is virtually unchanged from devices of a century ago. However, safety precautions led to the introduction of a suction device (rubber bulb or more complicated system) to eliminate the need for mouth pipeting. Further automation of this basic process came about with the development of the repetitive pipeting system which allows liquid measurement and transfer with the push of a button. Systems now exist which permit the pipeting of several samples simultaneously. From a simple shaft of fashioned glass to the vastly more accurate and precise pipet systems of today, we see laboratory automation at its least complex.

Usually we do not think of the pipet as an example of laboratory automation. Our minds turn rather to the major instruments in the laboratory which process hundreds (sometimes thousands) of samples each day, allowing one person to produce the same amount of laboratory data as an entire laboratory staff could provide three decades ago. The classic picture of a row of laboratory workers arranged along a bench, wielding pipets, making manual spectrophotometer readings, and hand-writing the results is now obsolete. The laboratory of today is more likely to be populated with a sometimes confusing variety of highly complex instruments which automatically measure samples, initiate chemical reactions, determine optical properties, generate laboratory results, create reports, and perform a variety of self-analysis and troubleshooting checks to assess their own performance.

4.1

RATIONALE FOR LABORATORY AUTOMATION

Why automate a procedure? The reasons are many, ranging from the erratic nature of human performance to cost considerations. Perhaps the initial reason for the development of automated methods of analysis was a recognition that human performance varies over the course of time. The physical energy and mental alertness present at 8:00 A.M. certainly diminishes by lunchtime. As the workday draws to a close, fatigue increases and the acuity of the individual drops off. As people tire, they have greater difficulty in maintaining accuracy in both the preciseness of physical motions and in their ability to perform mental tasks. Instruments, on the other hand, can pipet with the same high degree of reliability at midnight as they did at noon.

An associated problem lies in the repetitiveness of many laboratory tasks. The manual pipeting of several hundred serum samples does not provide the emotional or mental challenge a person needs to remain alert. Boredom sets in, and the number of errors increases. An instrument cannot become bored and carries out the same task over and over again as long as it does not experience any mechanical malfunction.

A third reason for automation becomes particularly pertinent when we look at the growing shortage of trained laboratory personnel. Use of automated analytical equipment releases the laboratory staff for the performance of other more complex duties. Previously, where 10 laboratory workers were required for the routine analysis of enzymes, electrolytes, and some basic constituents, the same tasks can now be fulfilled with one or two individuals who operate a complex chemistry analyzer. Although the initial cost of the equipment may far exceed the personnel salary budget for the year, the expense spread over a longer time proves to be a very cost-effective means of providing efficient laboratory services.

4.2

FACTORS CONTRIBUTING TO ADVANCES IN AUTOMATION

Developments in automation in the clinical laboratory are intertwined with advances in technology in a variety of fields. Three major areas contribute greatly to our ability to automate laboratory analysis and data handling:

1. robotics
2. computers and microprocessors
3. sensors for monitoring reactions

As progress in any of these areas is made, the improvements are quickly incorporated into automated clinical analyzers.

Robotics

One of the early challenges in laboratory automation was to develop a system for introducing a patient sample into a reaction chamber for analysis. The early probes were effective but were crude and cumbersome compared with modern sampling technologies. In addition, the sample-processing stage was still very much a manual operation. Systems are currently available (or under intensive development) which process the patient specimen without human intervention almost from the time the sample is collected. The use of bar codes permits sample identification to be automated to a great extent, minimizing the possibility of sample mix-up. Once the sample has reached the analytical instrument, a variety of means are available for introducing the sample into the reaction mixture for analysis.

Computers

No automated system today can function without one or more computers to carry out a variety of tasks. Process control is becoming an increasingly important function of computers. Not only is the computer responsible for instruction (what steps to carry out in what order), present-day monitoring capabilities allow the computer to provide a measure of control to the operation of the system. By continually assessing the status of each component, the computer can provide warnings of equipment malfunction or reagent depletion. In many instances, the operating condition can be adjusted automatically to provide optimum performance for the analyses.

Data handling has long been a strength of computers. The absorbance reading generated by a chemical reaction, the change in voltage observed by an electrode—these signals can be converted to a readout of concentration by computer processing. The data obtained from different tests can be collated for a specific patient and a report generated. All the elevated (or decreased) patient values for a test can be collected and examined statistically. Quality control can be monitored with data readouts, trend analysis, and ongoing computer information regarding the operation of the instrument and the reliability of the results. Information can then be fed into another computer for billing purposes, stored in the patient medical record, or processed in a variety of other ways.

Sensors

Advances in sensor technology have greatly enhanced our ability to automate chemical analyses. Development of a wide variety of sensors expanded the types of analyses possible, allowing the laboratory to move far beyond the simple spectrophotometric procedures originally employed. As sensor response time decreased, more samples could be processed in a shorter period, making automation of the particular procedure more feasible. Improvements in sensor reliability meant that the instrument could be operated for longer times without the need for sensor replacement or maintenance. Some sensors can now be built on a microchip, creating an even closer interface between sensor and computer, while decreasing (or eliminating) the need for chemical reagents as part of the analytical process.

4.3

CONTINUOUS-FLOW ANALYSIS: SEGMENTED STREAM

Overview

Perhaps the earliest approach to automated chemical analysis was the continuous-flow system developed in the late 1950s by Leonard Skeggs. The term **continuous-flow** describes the way in which the chemical and samples travel through the instrument. A unique characteristic of this type of analysis was the use of air bubbles in the sample and reagent streams. Air is injected into each stream as a series of small bubbles which travel along with the reaction system. The air bubbles minimize diffusion of reagents and mixing between samples, preserving the integrity of each individual reaction.

Initial systems could only analyze for one constituent, although large batches of samples could be processed. To measure another analyte, some components of the instrument had to be reconfigured for the new procedure. Current instruments can analyze 20 or more different constituents in a single sample in less than 1 min. Although the details are different, the basic processes of continuous-flow are the same, whether the analyzer be single- or multichannel.

A Typical Application

To facilitate our understanding of the processes involved in continuous-flow analysis, we will examine a typical procedure and see how the chemical reactions are translated into a series of steps using the components of the automated system. The assay under investigation measures serum calcium. The following reactions are employed:

1. Ca bound to protein + HCl → protein + unbound Ca
2. Ca + cresolphathalein complexone → colored product
3. colored product + diethylamine → intensified color

Measurement of absorbance at 580 nm is used to quantitate the amount of color and indicate the level of Ca present. The entire process in a continuous-flow system is outlined in Figure 4–1.

Much of the calcium in serum is attached loosely to proteins, mainly serum albumin. Before this calcium can react with a color reagent, it must first be dissociated from the carrier proteins. Dilute HCl is added to the serum sample to facilitate this dissociation. Passage of the HCl-serum mixture through a mixing coil allows the reaction to go to completion.

After the protein and calcium are dissociated,

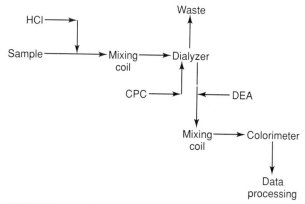

FIGURE 4-1 Continuous-flow analysis of serum calcium. CPC = cresolphthalein complexone. DEA = diethylamine.

there must be a way to physically separate these two components. The next step in the process is dialysis (Fig. 4-2). The protein-calcium mixture enters one side of the dialysis cell. Since calcium is a small molecule, it passes through the membranes of the cell to enter the recipient stream. Proteins are too large to penetrate the pores of the dialysis material and remain on the donor stream side to flow into a waste receptacle.

Following dialysis, the calcium present reacts immediately with cresolphthalein complexone to form a colored complex. Addition of diethylamine produces a basic pH, which results in a more intense color. Complete mixing and reaction take place in the double coils after both reagents have been added. The stream then passes through the colorimeter system where the bubbles are removed (Fig.

4-3) and the absorbance of each sample is determined.

4.4

CONTINUOUS-FLOW ANALYSIS: FLOW INJECTION

Within 20 years of the inception of continuous-flow analysis, a radically different concept was developed. Instead of a stream segmented by air bubbles to promote sample integrity, direct injection of sample into very small diameter tubing was utilized for reactions and measurements. This new technique came to be called **flow-injection analysis.**

In flow-injection analysis, the reagent stream is pumped through small tubing (usually about 0.5 mm in diameter). The sample slug is introduced directly into that stream, allowing intimate mixing within a few seconds. Sample volumes can be very small (50 μL or less). Reactions times are quite short, often less than 1 min. As the mixture passes through a detector, absorbance is measured and the concentration determined. The reaction does not need to be completed before being monitored since the exact timing in the system allows all reactions to be measured under exactly the same conditions.

The major reason for air bubbles (used in segmented stream systems) is eliminated due to the design of the flow tubing. The extremely narrow diameter, coupled with direct injection of sample and short reaction times, minimizes the extent of lateral diffusion. In addition, the coiling in these small tubes generates a force which helps cut down diffusion.

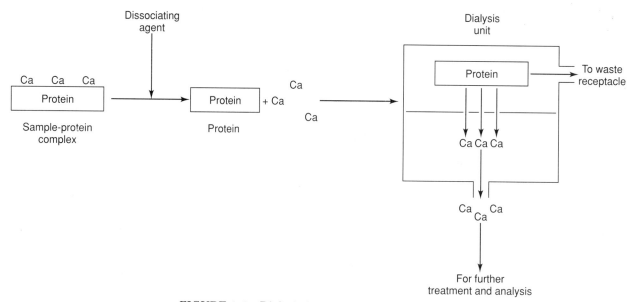

FIGURE 4-2 Dialysis in continuous-flow analysis.

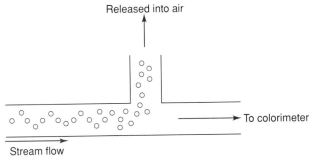

Released into air

To colorimeter

Stream flow

FIGURE 4-3 Debubbler.

One advantage of flow-injection analysis can be seen in the timing of reactions. Because there are no air bubbles, the integrity of a portion of the stream can be maintained. With precise pumping, the stream can be stopped for a time to allow a longer reaction time if needed. Instead of passing the mixture through a long reaction coil (which permits more sample diffusion to take place), the flow through the system is interrupted for a defined time. If the flow is stopped when the sample has entered the cuvet for absorbance measurements, a series of timed measurements can be made and the reaction kinetics determined. This approach is particularly valuable when studying enzyme levels.

Flow-injection analysis is in its infancy in the clinical laboratory, yet a number of common assays have been adapted to this type of analysis. Measurement of glucose, albumin, total protein, electrolytes, calcium, urea, phosphate, and a variety of other constituents have been reported in the literature. At least one company has flow-injection systems commercially available (in addition to the segmented stream equipment they also sell). Some experts predict that flow-injection will ultimately replace segmented stream techniques completely in the clinical chemistry laboratory.

4.5

CENTRIFUGAL ANALYSIS

History and Basic Concepts

The centrifugal analyzer was born in the mid-1960s as a result of discussions on alternative methods for automation. Although continuous-flow analysis had shown the value of automation, there was some dissatisfaction with its flexibility and ease of operation. Research at the Oak Ridge National Laboratory, directed by Dr. Norman Anderson, resulted in the development of an innovative approach to automated analysis called the **centrifugal analyzer.** Commercialization rapidly followed, with models available from several companies by the early 1970s.

A miniaturized version was used in some space exploration programs, allowing scientists to perform chemistry analyses in outer space, where the lack of gravity would allow fluids to float away if left to themselves.

The basic concept is amazingly simple in retrospect. Sample and reagent are placed in separate chambers in a rotor. When centrifugal force is applied, the rotor spins and causes the two components to flow into a reaction chamber where they combine to form a product. The product solution then moves to a cuvet portion on the outer rim of the rotor. As each individual cuvet in the rotor passes over a light source, a beam of light shines through that cuvet. Light absorbance for each sample is detected by a phototube, the resulting signal is stored and processed in a computer, and analytical results are quickly determined and printed out for a permanent record.

Sample and Reagent Loading

Samples and reagents can be added in one of two ways, depending on the system employed. Many older instruments have an auxiliary piece of equipment for placing materials into the rotor. In separate operations, the samples to be analyzed and the necessary reagents are pipeted into their respective chambers. The rotor is then manually placed in the instrument and the assay carried out. In more modern models, the sample and reagent loading are both accomplished in the instrument itself as a part of the overall program for analysis. Test requests can be individually entered for each sample, with the loading and assay steps being carried out automatically. Signals are given when the supply of reagents, samples, or rotors requires replenishing.

The Rotor: Heart of the System

Mixing, reaction, and analysis of product all take place within the confines of the small plastic rotor of the centrifugal analyzer (Fig. 4–4). In some instruments, the **rotor** is a disposable item to be discarded after an analytical run has been made. Other systems incorporate a permanent rotor which is washed and dried automatically between assays.

The inner chamber (closest to the center of the rotor) is the sample compartment. The range of sample types is limited only by the sensitivity of the specific analysis and the level of that analyte detectable by the method being employed. Serum, plasma, CSF, urine can all be assayed using a centrifugal analyzer. The small volumes employed for analysis (frequently less than 10 μL for a single assay) are a distinct advantage, particularly when multiple analyses must be performed on a single sample.

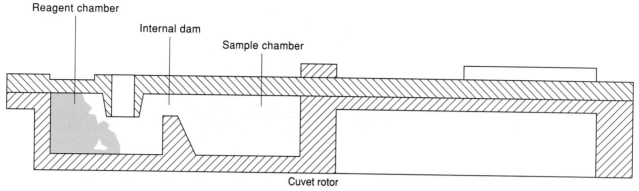

FIGURE 4-4   Rotor for centrifugal analysis. (Courtesy of Instrumentation Laboratory.)

The reagent chamber is located on the outer edge of the rotor. Separating the two chambers is a plastic wall, which is slanted toward the reagent portion. This slant facilitates transfer of sample to reagent area while the rotor is turning, but still permits separation during the stationary stage of the assay.

The outer edge of the rotor also serves as the cuvet for the analysis (Fig. 4–5). Light is directed through the rotor from above, passing through the reaction solution. The light not absorbed is detected by a phototube under the rotor, permitting measurement of absorbance. If a fluorometric technique is being performed, the light beam is directed edge-on into the cuvet and the 90° emitted light is then detected by the same phototube. Light-scattering assays (measurement of turbidity) are accomplished in the same way.

The rotor spins in a predetermined manner to

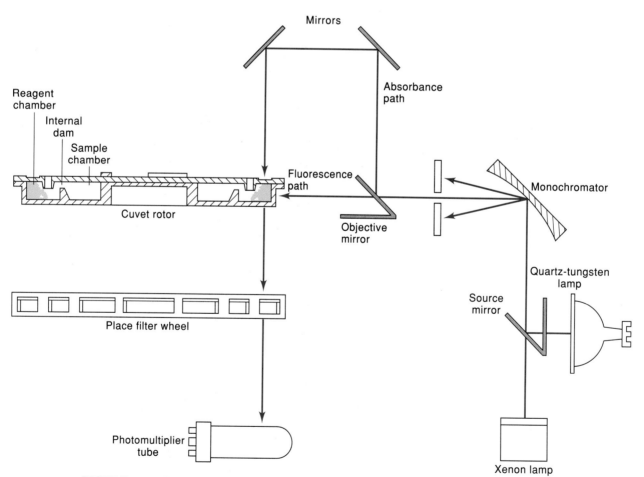

FIGURE 4-5   Optics for a centrifugal analyzer. (Courtesy of Instrumentation Laboratory.)

allow mixing to take place and light readings to be made. An initial rapid spin transfers the sample to the reagent chamber for mixing. The spin is then slowed for incubation and to allow the readings to be made. By having a preset rate of spin, the computer system can make readings at timed intervals and determine the optical properties (either absorbance or fluorescence) of materials in each individual reaction chamber.

Blank Measurements

In manual photometric measurements, a separate reagent blank could be prepared and measured easily. Continuous-flow analysis requires a separate channel for blank measurements, but blank correction is automatic and requires no specific manipulation by the operator. In centrifugal analysis, the preparation of a separate rotor for blank measurements takes extra time and materials, rendering this approach impractical. However, the rapid reading capability of the system can be exploited to provide a reliable blank measurement. When the sample is transferred to the reagent chamber, no product is formed within the first few milliseconds after mixing. The light measurement made at this time, therefore, represents the sum of absorbances (or fluorescences) of the components of the mixture. Both reagent absorbance and any native absorbance of the sample can be detected at the start of the run and canceled out by the computer when the final readings and calculations are made. A separate blank reading is thus avoided, resulting in greater productivity.

An alternative approach can be taken if all the samples on the rotor are being analyzed for the same component. One rotor cell can be designated as the blank and contains only the color reagents, not the sample. Any absorbance detected in this cell is subtracted from other readings before final values are calculated. Although this method compensates for reagent blanks, it does not compensate for any light absorbed by the serum samples (with a high bilirubin, for example).

Rate Measurements

For many analyses, an end-point measurement of optical properties is sufficient. Sample and reagent are mixed, a colored product forms, and the absorbance is measured after the reaction is complete. The results are determined after blank correction has been carried out and a report is generated. However, a number of assays are based on rate measurements—the amount of product formed in a given time. This rate is then related in some manner to the concentration of constituent or (in the case of an enzyme) is reported directly as the index of concentration. When rate measurements are made, a blank reading is usually not required since the results are calculated on the basis of changes in absorbance from one reading to another. In this situation, the system already compensates for the blank.

Rate measurements are easily accomplished with the centrifugal analyzer. If, for example, the rotor is spinning at a precise 1400 rpm (revolutions per minute), the computer can "read" each cell at specified intervals. These data are stored until the analysis is completed. Following statistical analysis to check for linearity, the final result is calculated. In a given assay, a reading may be made initially, to determine the blank value, followed by readings at 10-s intervals for 5 min. Each data point is stored as part of a set for that analysis for that patient. A least squares or other regression calculation is made, the linear portion of the data is selected, and results reported. Rate measurements are one of the strong points of the centrifugal analyzer.

Possible Analyses

SCOPE OF THE SYSTEM

A wide variety of quantitative measurements are possible with a centrifugal analyzer. Methods are commercially available for essentially all common analytes, using prepackaged reagents. A number of assays involving enzyme immunoassay techniques have been successfully adapted to these instruments.

Initially, centrifugal analyzers were employed in a **batch mode**—each rotor was used to measure a single component and all assays run on that rotor were for the same constituent. With the advent of small, powerful computer systems as part of the instrumentation, it is possible to employ the centrifugal analyzer as a more flexible **random-access** instrument to a variety of analytes in a single sample. The random-access capabilities will be described later in this chapter.

MEASUREMENT OF ENZYME ACTIVITY

The power of centrifugal analysis is most strikingly seen in its ability to assay enzymes rapidly and accurately. When we assess the amount of enzyme present in a body fluid, we look at the reaction catalyzed by the enzyme. Determination of the amount of product formed per unit time allows us to calculate the enzyme activity. It is therefore very important to be able to make rapid sequential measurements on each cuvet in the rotor.

To illustrate the use of a centrifugal analyzer for

enzyme studies, consider the reaction catalyzed by the enzyme alkaline phosphatase

$$p\text{-nitrophenylphosphate} \xrightarrow[\text{phosphatase}]{\text{alkaline}}$$
(colorless)

$$p\text{-nitrophenol} + \text{phosphate}$$
(yellow)

The enzyme hydrolyzes a phosphate group from the substrate p-nitrophenylphosphate. The product, p-nitrophenol, is yellow. The reaction can be monitored by measuring the increase in absorbance at 405 nm as the product forms. The more rapidly the absorbance increases, the more enzyme is present.

Serum and buffer are pipeted into the inner well and substrate (p-nitrophenylphosphate) is added to the outer well. The rotor is incubated at 30°C, and then the reaction is initiated by rapid spinning to mix sample and substrate in the outer chamber. After 30 s (to allow for mixing and for the enzyme reaction to begin), absorbance readings are recorded every 15 s for the next 2.5 min (a total of 10 data points). Each recorded reading is actually the sum of several readings made while the rotor is spinning. A least squares fit of the data is made to determine the best straight line, from which the rate measurement is then determined.

The power of this method is its ability to make several readings within a short time (15-s intervals in this case). Although a trained technician could also make readings in this interval for a single sample, the individual certainly could not carry out this operation for 40 or more samples simultaneously.

KINETIC ASSAYS

Application of enzyme rate measurements opens another set of possibilities for exploiting the unique capabilities of the centrifugal analyzer. Many body fluid constituents can be assayed by using an enzyme system to convert the material to a product with distinct optical properties. If the enzyme concentration is held constant, the rate of the reaction depends on the amount of the compound to be acted on by the enzyme. If more of the compound is available, the enzyme rate is higher. By measuring the rate of formation of product in the presence of a known amount of enzyme and comparing this rate with that observed for a standard amount of the analyte, the material in a body fluid can be quantitated. Time is saved since the reaction does not need to go to completion; the initial rate of reaction indicates the quantity present.

We can illustrate this principle by examining an assay for serum lactic acid. The enzyme reaction is as follows:

$$\text{lactic acid} + \text{NAD}^+ \xrightarrow[\text{dehydrogenase}]{\text{lactate}}$$
(no absorbance at 340 nm)

$$\text{pyruvate} + \text{NADH}$$
(absorbs at 340 nm)

Sample containing lactic acid is added to the inner sample chamber and the enzyme, lactate dehydrogenase (with buffer), is placed in the outer reagent chamber. Incubation at 35°C is followed by spinning to mix sample and enzyme. Both the substrates (lactic acid and NAD^+) have no appreciable absorbance at 340 nm, although one of the products (NADH) absorbs strongly at this wavelength. Timed measurements of the increase in absorbance at 340 nm are made for several minutes, and the amount of lactic acid originally present in the sample is calculated. The time required for total conversion would be perhaps 15 min or more, depending on concentrations and conditions. By use of a **kinetic method,** a shorter assay is possible and blank correction is not necessary.

4.6

REAGENT PACKAGE TECHNOLOGY: DUPONT aca

The development of the DuPont aca series of instruments represented a conceptual breakthrough in laboratory analysis. With the continuous-flow analyzers, samples were processed for a batch of tests and not for single items. All the assays in the profile were performed, whether or not they had been requested. Analysis on the aca, on the other hand, permits test selectivity while retaining the other benefits of automation—reproducibility, speed, reduction in personnel time, and other factors. A given patient sample could have 1 test performed or 10, depending upon the specific requests. There was no need to batch patient samples for efficiency of operation; the instrument could be used effectively for intermittent requests throughout the day or night as well as heavy runs where several samples needed to be processed with multiple analyses run on each sample.

The heart of the system (first released in 1970) is the **reagent test pack** (Fig. 4–6). This small plastic pouch contains all the reagents necessary for a single analysis. Each reagent is contained in a small bubble at the top of the pack. When needed, the bubble is popped, releasing the reagent into the main compartment of the test pack where it can be mixed with diluent, buffer, other reagents, and sample. The packs can be loaded onto the instrument behind the sample pack for a given patient and each

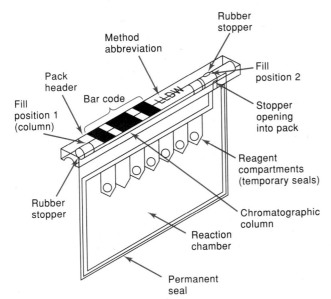

FIGURE 4-6   aca test pack. (Courtesy of DuPont Instruments, Inc., E. I. DuPont de Nemours and Company.)

analysis carried out in sequence. In early systems, the sample pack had a handwritten patient I.D. which the instrument "read" when reports were generated. Reagent packs were bar-coded to identify the specific test carried out by that set of reagents. Current models include bar codes for both patient I.D. and reagent identification.

To initiate testing, a sample injector removes the necessary amount of serum or other body fluid from the sample-pack well. This material is then injected into a specific test pack; the volume is determined through reading the test-pack bar code by the instrument. Buffer is added to the system and the reagents are mixed. The reaction is timed by its travel along the pack transport system (Fig. 4–7). If incubation at a specified temperature is required (for an enzyme reaction or other test where temperature control is critical), a heating block encloses the pack for the necessary time. Before optical processing takes place, the test pack enters a system which presses a portion of the cell, creating an optically clear bubble, which serves as a cuvet for spectrophotometric measurement. The necessary absorbance

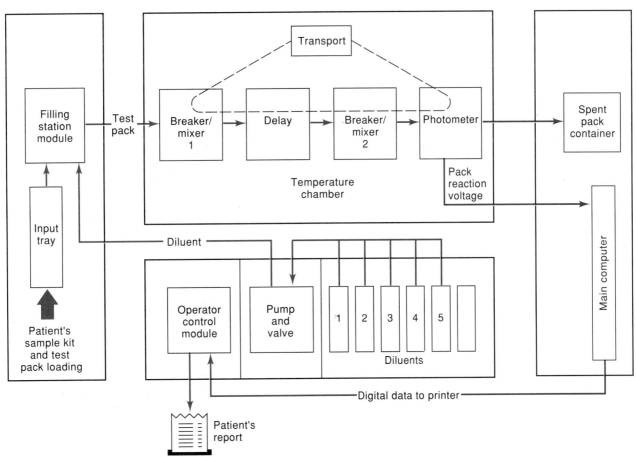

FIGURE 4-7   Transport path for aca test pack. (Courtesy of DuPont Instruments, Inc., E. I. DuPont de Nemours and Company.)

readings are taken (wavelength is set according to the bar code of the test pack), and the computer calculates the appropriate results.

More recently, an ion-selective electrode system has been added for electrolyte measurements. This module is a stationary flow-through system which operates in parallel with the existing pack processes. The ISE provides for analyses of sodium, potassium, chloride, and CO_2.

Compared with some other systems, throughput on the aca is rather slow, usually under 200 tests per hour. This decrease in speed is offset by the rapid availability of the system for emergency analyses at any time. In addition, new tests can be added to the instrument with a minimum of effort. A few minutes spent programming the new parameters into the computer is all that is required. Through a combination of in-house research and licensing of technology from other companies, the aca can provide a wide variety of assays using modern analytical techniques.

4.7

DRY-SLIDE TECHNOLOGY: EASTMAN KODAK

An innovative approach for chemical analysis, which was developed by Kodak, exploits their major areas of expertise—layering of chemicals on a film and reagent preparation. The result was a slide which contained all the materials necessary for a single analysis. No reagents were needed to prepare the slide for use. The **dry-slide technology** allows longer shelf life and provides a high degree of reproducibility.

Composition of a Dry Slide

Figure 4–8 illustrates the construction of a typical slide employed in a chemical analysis. The slide is usually composed of four layers:

1. spreading layer
2. reagent layer
3. indicator layer
4. support layer

Other special layers are occasionally employed and will be described later.

The spreading layer is the point of contact with the sample (Fig. 4–9). In current instrumentation, a small sample (something in the general range of 10 μL or less) is automatically pipeted onto the outer surface of the slide, making contact with the spreading layer. The porous network in the spreading layer causes the sample to spread over a defined area of this section of the slide, equally distributing the

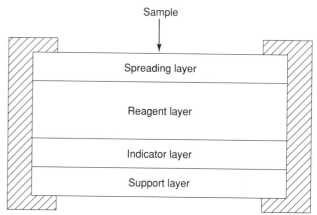

FIGURE 4–8   Dry-slide technology. (Courtesy of Eastman Kodak, Inc.)

sample over the layer. In addition, the capillary network serves as a sieve; large molecules, such as proteins, are not able to pass through the network and are effectively removed from the system. This filtration process serves the same purpose as manual precipitation and centrifugation of protein or the dialysis which takes place in a continuous-flow system. Some spreading layers are designed to exclude lipids, such as cholesterol and triglycerides (in addition to proteins), to diminish interference in selected tests.

The reagent layer contains all the chemicals necessary for the specific reaction. This layer perhaps represents the major challenge to slide manufacturing. In a complex assay, several reagents are successively layered onto the separating barrier in exactly the right amounts and in the correct sequence for the assay to proceed properly. The slide for the analysis of glucose, for example, contains two enzymes, color reagents for reaction with the product of the enzyme reactions, and a buffer to maintain the proper pH.

On occasion, a scavenger layer is included between the spreading layer and the reagent layer (Fig. 4–10). This special section of the slide removes materials present in the sample which might interfere with the reaction taking place in the reagent layer. The slide for uric acid has a scavenger layer consisting of the enzyme ascorbate oxidase. Since vitamin C (ascorbic acid) interferes with the enzymatic reaction used to measure uric acid concentrations, vitamin C must be removed prior to the color reaction taking place. Ascorbate oxidase converts the vitamin C to a noninterfering product as the materials diffuse through the scavenger layer into the reagent layer.

The indicator layer is another optional layer, not present on all slides. This layer contains a dye or some other type of indicator which reacts with the product of reactions taking place in the reagent layer, forming a colored complex. Many of the com-

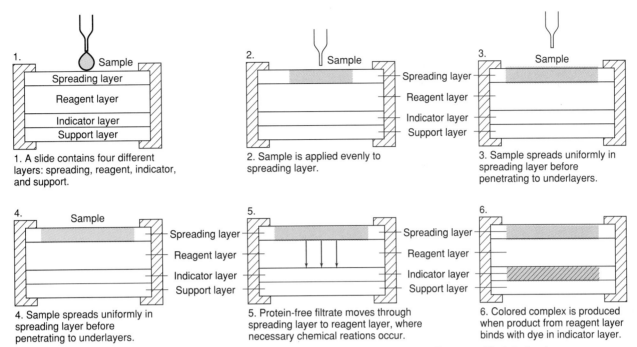

FIGURE 4–9   Analysis using dry-slide technology. (Courtesy Eastman Kodak, Inc.)

mon assays produce the colored product in the re-agent layer and do not require a separate indicator layer.

The support layer at the bottom of the slide is composed of clear plastic and is the layer on which the entire slide is constructed. The plastic is still transparent when the colored complex is being measured.

Reflectance Measurements

Measurement of the amount of color formed in a reaction requires a somewhat different approach to spectrophotometry. In traditional systems, light is shone through the cuvet (whether a tube, a flow cell, or a rotor cell) and strikes a phototube, producing a signal proportional to the amount of light absorbed by the sample. This direct approach is not possible with the dry-slide system since the spreading layer is opaque.

The opacity of the top spreading layer is utilized in **reflectance spectroscopy** (Fig. 4–11). In this method, light of a selected wavelength shines

Ektachem clinical chemistry slide (URIC)

Sample

Spreading layer	
Scavenger layer	Ascorbate oxidase
Reagent layer	Uricase peroxidase indicator dye buffer at pH 8.7
Indicator and support layers	

1. Spreading layer—ascorbic acid present in sample.
2. Scavenger layer—ascorbate oxidase binds with ascorbic acid.
3. Reagent layer—ascorbic acid interference has been removed.

FIGURE 4–10   Dry-slide technology with scavenger layer. (Courtesy of Eastman Kodak, Inc.)

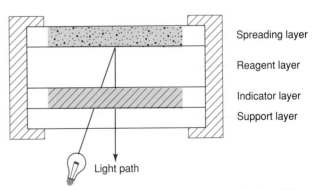

Spreading layer

Reagent layer

Indicator layer

Support layer

Light path

FIGURE 4–11   Reflectance spectroscopy with dry-slide assay. (Courtesy of Eastman Kodak, Inc.)

through the bottom of the cell. The light beam reflects off the underside of the spreading layer and passes through the reagent and indicator layers to a photodetector. Since the clear support layer is essentially transparent, any absorbance which takes place must be due to the colored material formed by reactions in the slide.

The light source is positioned at an angle to the underside of the slide. A filter system is employed to screen out light of unwanted wavelengths, allowing selectivity of the light entering the slide area. The spreading layer reflects the light back in a different direction, which eliminates most (or all) the possible interference due to stray light.

With this system, there is no need for sample blanks. Materials which are highly colored (due to hemoglobin or bilirubin, for example) are trapped in the spreading layer and removed from the reaction before entering the reagent area. The interfering compounds are not present in the area through which the light passes and hence do not absorb any additional light.

Instrumentation for Dry-Slide Technology

Systems employing dry-slide technology run the gamut from simple desk-top instruments for physicians' office laboratories to large, fully automated, programmable machines involved in mass production of laboratory analyses. All employ the same basic slide technology, but differ in the degree of automation of sampling and initiation of the testing procedure.

The small office system does require more hands-on labor than the larger instruments. The technician selects the specific slide, pipets the sample onto it, and places the slide into the machine. A bar code on the slide identifies the test, allowing the instrument to select the proper light filter, incubation period, and other parameters necessary for proper performance of the test. A data printout is available within a few minutes.

The fully automated, hospital-based system is completely programmable. Samples and slides are loaded into their respective compartments, tests are requested for each patient sample using the computer, and the machine does the rest. Testing parameters, test sequences, and patient data are all automatically organized through the computer programming.

Types of Tests Possible with Dry-Slide Technology

The dry-slide systems are capable of a variety of approaches in terms of the timing of tests and the monitoring of reactions. Although the simpler office systems may not have all these capabilities, the larger instruments can exploit the full range of assay options.

Most of the common tests are **end-point analyses:** The reaction is carried out under the specified conditions and the absorbance of the product is determined at a set time. Knowledge of the assay timing allows programming of the computer to read the absorbance after the reaction is complete. The common colorimetric tests for components such as albumin, total protein, calcium, and other constituents all employ end-point analysis.

Dual-wavelength analysis is an approach adopted by several instruments. By measuring absorbance at two different wavelengths, some specific blank corrections can be made or selectivity in measuring the amount of a certain component is possible. Dry-slide technology employs the dual-wavelength technique to quantitate two bilirubin fractions which have different molar absorptivities at selected wavelengths.

Blank-corrected measurement is occasionally utilized. In the creatinine assay, the product measured is ammonia produced by an enzymatic reaction on the creatinine. Since there is some endogenous ammonia present in samples, a second slide is employed to quantitate this ammonia. The amount of endogenous ammonia is then subtracted from the ammonia detected by the creatinine slide, allowing a more accurate measurement of serum creatinine.

Enzyme assays can be performed using either a **two-point** or a multipoint measurement. In either case, measurements are made on the slide at fixed times after the reaction is initiated. If **multipoint measurement** is used, the instrument will make 54 individual readings while the reaction is taking place. The change in absorbance is an indication of the amount of enzyme activity, calculated by the computer on the basis of the readings. The multipoint readings offer the additional advantage of checking the linearity of the reaction, permitting detection of samples with high enzyme activity.

Potentiometric analyses are also possible, permitting measurement of sodium, potassium, chloride, and carbonate. These assays will be covered in more detail in Chapter 18, which deals with automated analysis of electrolytes.

4.8

RANDOM-ACCESS ANALYZERS

Concept of Random Access

Although early automated systems provided the ease and convenience of having the instrument perform the requested assays, they were seriously limited in terms of test selectivity. The continuous-flow equipment pumped each sample through every channel; all the assays were performed whether or not they had been requested on a specific patient. Centrifugal analyzers were initially limited to one assay per rotor. All the tests for calcium were performed, then all the tests for glucose or other requested component. There was little flexibility in these systems. Tests had to be batched for efficiency and there was almost no opportunity for changes in the testing sequence once the work pattern for the day had been established.

Two major factors forced a change in this mode of operation. There was increasing awareness of the need for test selectivity. Not all patients required a 20-channel sequential multiple analyzer, computerized (SMAC) analysis when only a few selected tests would provide the necessary diagnostic information. Economics also provided a significant driving force for change. With increasing emphasis by government and private insurers on cost-cutting, there was greater scrutiny of test requests. Overtesting was increasingly challenged, with insurance companies simply not paying for excessive test orders. Diagnostic companies responded to these challenges (and were, in fact, responsible for helping to raise some of the initial questions) by producing instruments which provided a high degree of flexibility in patient testing while maintaining some controls on cost.

The concept of **random-access** is simple in theory, but it provides some interesting scientific and engineering challenges. The instrument should analyze patient samples for only those constituents specifically ordered. The system should be flexible enough to accommodate any reasonable combination of requests for a single patient. Stat samples must be provided for so emergency assays can be carried out in a reasonable time by momentarily interrupting the normal sequence of patient analyses. After the stat request has been filled, the system should automatically return to normal operation without the need for reprogramming or reorganization of samples and test patterns. The instrument must be capable of incorporating new tests into the analytical scheme by addition of the appropriate reagents and simple reprogramming for the new test parameters. The new generation of random-access analyzers meet these conditions, furnishing both the speed and flexibility required in today's laboratory.

It is impossible to describe all available instrumentation in the short amount of space available here, so we restrict ourselves to several representative techologies which demonstrate the versatility of approaches taken to the development of random-access technology.

Some Representative Instrumentation

TECHNICON

Moving from the segmented stream, batch-processing approach long characterized by Technicon to a random-access system required a significant shift in outlook. This transition was successfully accomplished through development of a sampling system which essentially eliminated interaction between samples and carry-over between reagents without the use of the traditional bubble system.

The key to success of the Technicon system is a fluorocarbon which coats the inside of both sampler and reagent transfer probes. This patented material eliminates contact between samples (or reagent) and the inside wall of the probe. As a sampling cycle begins, a small air bubble is aspirated into the probe, followed by the necessary volume of sample. The fluorocarbon coats the sample completely, forming a "slug" which is then dispensed using air pressure. This slug is pipeted directly into a reagent cuvet on a reaction tray which holds 100 sites for assay. The reagent probe has already pipeted the necessary chemicals into the cuvet. Because of its density and its total lack of interaction with the aqueous material of both sample and reagents, the fluorocarbon sinks to the bottom of the cuvet out of the light path for the colorimeter. The encapsulation process completely surrounds both sample and reagents with fluorocarbon. No contact with the interior wall of either probe can take place, so there is no carry-over to provide contamination.

After addition of reagents and sample, the cuvet is mixed by motion of the reaction tray. Depending on the assay being run, a blank reading may then be taken, followed by one or more measurements of absorbance during the course of a reaction or at its completion. The light source for the colorimeter is located outside the radius of the reaction tray. Light is passed through a filter wheel before entering the individual cuvet. The detector for the light passing through the solution is located on the interior of the reaction wheel. All data are computer processed and printed out.

Three different types of optical measurements are possible. A rate reaction (for enzymes or immunoassay procedures) involves the reading of several absorbance values over a specified period. The rate of the reaction is calculated from the change in absorbance as a function of time. A modified rate reaction can be measured beginning with a reagent-blank reading followed by two absorbance readings at a specified interval. End-point assays involve a reagent-blank measurement and a second absorbance reading after the reaction is complete. These different options permit the use of a variety of assay technologies, providing a high degree of flexibility and innovation.

In an attempt to get the best of both worlds, Technicon has coupled the random-access instrumentation to their continuous-flow system. Tests are assigned to one or the other of these two instruments by computer for the most efficient, cost-effective analysis. Back-up is available in either mode for the other instrument, as is convertibility of tests from one system to the other. This approach furnishes some interesting possibilities for the high-volume production laboratory.

BECKMAN

The Synchron series of analyzers (originally marketed as Astra systems by Beckman) began as a glucose blood urea nitrogen (BUN) electrolyte profile instrument. Samples could be analyzed for any single parameter or combinations of tests as needed. The technology was developed to a great extent by Beckman. A unique method used oxygen-consumption rate to determine glucose levels, and a conductometric analysis for ammonia formed the basis for urea assay. Ion-selective electrodes measure the electrolytes. Since that time, the system has been expanded to allow a variety of other tests, including several common enzyme assays. The instrument is currently capable of 24 different chemistry analyses (Fig. 4–12). The possibility of programming user-developed methodologies adds to the versatility of the instrument.

Bar codes provide the basis for reagent identification and automatic inventory. Automatic sampling is available at 16-s intervals. Samples are loaded in small sector trays, each of which will hold 10 samples. Automatic transfer to the sample turntable is computer directed. Samples and reagents are added to glass cuvets on a carousel. Reactions are monitored photometrically and data stored for further processing. A separate ion-selective electrode module provides assays for sodium, potassium, chloride, and CO_2.

DUPONT

Utilizing the test packet approach, which they developed and introduced in 1970, the DuPont aca system allows a high degree of flexibility in patient testing. This flexibility, however, comes at the cost

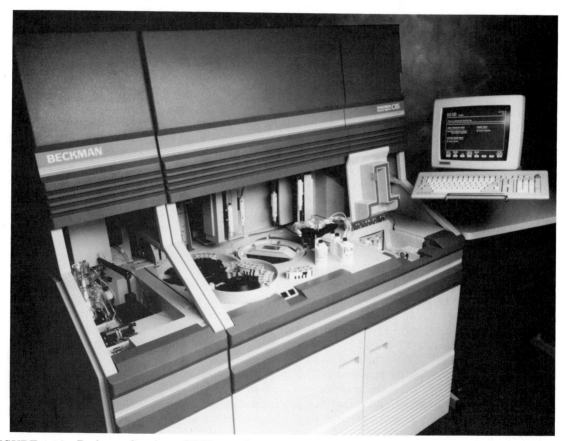

FIGURE 4–12 Beckman Synchron CX(R)5 random-access analyzer. (Courtesy of Beckman Instruments, Inc.)

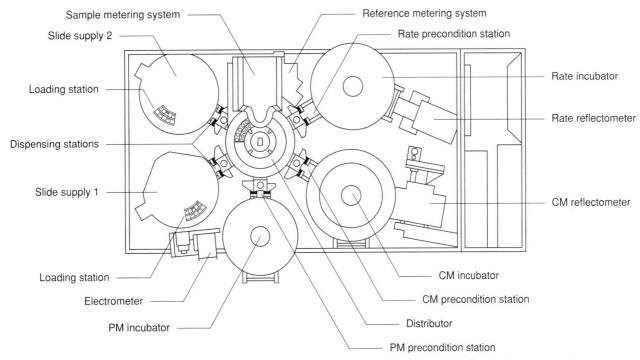

FIGURE 4-13   Kodak Ektachem analyzer. (Courtesy of Eastman Kodak, Inc.)

of a greater amount of personnel interaction with the instrument than is the case with other technologies. Since sample and test packets are manually loaded onto the system, an emergency request can easily be accommodated by simply inserting the sample/test-packet combination ahead of other tests already loaded on the instrument. With the bar-code capability of the aca, test requests are read automatically with no other programming needed. A major drawback of the aca is the low throughput of only some 100 tests or fewer per hour.

Some of these drawbacks can be alleviated with the Dimension system, a more automated version of the aca (Fig. 4-13). Bar-coded reagents are loaded on the instrument for each assay. Samples are pipeted into individual reaction cuvets which are formed from a polymer film and can be discarded at the end of the assay. Reagents from the appropriate reagent cartridge are added to the sample in the cuvet, the reaction is run, and products monitored photometrically. Data are stored in the computer and processed.

KODAK

The Ektachem series of analyzers from Kodak utilize the slide technology described earlier in this chapter. Although the small instruments for physicians' offices require direct technician interaction at each step of the analysis, the large automated sys-

tems can be programmed to carry out a complex series of processes. An overview of the components of the Ektachem 700 series is provided in Figure 4-13.

Two slide-supply chambers dispense specific slides as needed for the requested tests. Bar-code readers detect the code for a specific test, and that slide is passed to the sample-metering station using the slide rotor. At this point, 10 μL of patient sample is added to the slide. The rotor then moves the slide to the appropriate testing station (rate incubator, colorimetric incubator, or potentiometric incubator). In the rate incubator, the slide is maintained at a particular temperature for the enzyme assay. Reflectance readings are taken at intervals determined by the enzyme being assayed. The colorimetric incubator holds the slide until the reaction is complete, then a reflectance reading is made. Electrode insertion and potentiometric readings take place in the potentiometric incubator. All data are transmitted to the computer for processing and reporting. The slide is automatically disposed of at the end of the particular analysis. The larger instruments are capable of some 750 individual tests per hour.

INSTRUMENTATION LABORATORY

The Monarch from Instrumentation Laboratory is an outgrowth of technology developed for the Multistat centrifugal analyzers. The instrument is capable of colorimetric, fluorometric, potentiometric (on some models), and turbidimetric analyses. The chemistries employed are analogous to those utilized in a centrifugal analyzer since the basic spinning rotor technology is still in place.

Test requests can be programmed either on a first-come, first-served basis or for optimum use of time, rotors, and reagents. Rotors are kept in a feed stack at a specified temperature (to save time in any incubation step). When needed, a rotor moves to a loader table to receive reagents. Both samples and reagents are stored in a separate section of the instrument, with space for 44 samples and 20 barcoded reagent boats. After loading, the rotor moves to the analyzer table for spinning to perform the specific test. When the assay is complete, the rotor moves either to the discard portion (for disposal) or to a park table for reuse (if all the slots in the rotor have not already been employed for analyses). Optical readings go to the computer for processing and data readout. Potentiometric analyses are performed using a separate ion-selective electrode module which is not a part of the rotor system. The Monarch has the rated capability of approximately 600 tests per hour.

4.9

AUTOMATION IN THE OFFICE LABORATORY

In recent years, there has been a significant movement of testing away from the hospital laboratory and into the out-patient arena. This shift has been prompted to a large extent by reimbursement practices from Medicare and private insurers. To respond to these changes, many hospital laboratories have developed out-patient services through delivery of laboratory testing to physicians at their offices. This practice has been in place for many years, providing good coordination of testing for the patient, whether out-patient or hospitalized. A more significant trend, the implications of which are not yet well understood, has been an increase in the volume of laboratory testing by physicians in their offices or clinics.

Whatever the reason for the shift in testing site, instrument companies have been quick to respond to this new effort. Small, easy-to-operate instruments have been developed specifically for the out-patient testing market. Although they generally do not have all the automated features of the larger pieces of equipment used in hospital laboratories, the programming and reagent-dispensing approaches are the same as the more sophisticated counterparts. Manual steps are kept to a minimum and usually involve only dispensing the sample. With packaged reagents (the aca and Kodak slide systems being preeminent examples), the lab operator has little or no direct reagent handling. Possible errors are minimized and the speed of the assay is enhanced. Test results are available within a reasonably short time after the sample is collected, allowing access to the results while the patient is still at the office. Significant time is saved by the patient since another trip is not necessary to obtain many laboratory results.

Several troublesome issues arise as these products and practices proliferate. Frequently, untrained personnel are employed to operate the instruments (one article described the "office nurse or secretary" pipeting the sample into a specific assay system). Quality control is frequently minimal or nonexistent. Although more conscientious office facilities form liaisons with nearby hospital laboratories in order to maintain quality performance, many offices do not take the time to deal with these issues. The federal government is taking an increasingly close look at the practices of the office laboratories. Although no certification standards currently exist, it is possible that legislation will soon change that situation. Trained laboratory technologists can play an important role in upgrading the standards and performance of office laboratories through a variety of collaborative efforts. Some larger clinics are employing qualified laboratory staff to oversee their operations.

SUMMARY

A variety of factors have contributed to an explosion in automated approaches to clinical chemistry. Advances in robotics, computers, and other forms of technology have been rapidly incorporated into automated systems for the clinical laboratory.

One initial entrant into the field was continuous-flow analysis. Materials are segmented by bubbles and flow through an apparatus. Separation of large molecules from small ones is possible by dialysis, reaction temperatures are controlled by heating baths, and the time of reactions is regulated by the length of the reaction coil. A wide variety of analytes could be easily measured with this methodology.

Centrifugal analysis has also played a major role in automated systems. Sample and reagent are mixed in a reaction chamber while the disk is spinning. Transfer of the product solution to the cuvet area of the rotating disk allows colorimetric or fluorometric analyses on either an end-point or kinetic basis.

Individual reagent-pack technology was developed by DuPont (with the aca) and Eastman Kodak (using dry-slide analysis). The aca system involves a package containing all materials. Sample and buffer are added, mixing occurs, and reactions are read using a portion of the packet as a cuvet. In the Kodak slide system, a series of layers separates unwanted interferents, contains reaction materials, and allows reflectance spectrophotometry for product quantitation. Ion-selective electrodes are incorporated into the methodologies for both systems.

The demand for increasingly efficient testing and the need for cost containment have resulted in development of a wide variety of random-access analyzers. These instruments, often adapting existing automated chemistry technology, allow variety in the analyzing sequence and the capability of performing only requested tests. Stat needs can quickly be accommodated without having to reconfigure the system after the emergency determination is performed.

Automation has reached into the physician's office laboratory. Benefits include more accurate testing and quicker results. Drawbacks relate to deficiencies in operator training and the frequent lack of an ongoing quality control program.

FOR REVIEW

Directions: For each question, choose the best response.

1. One sample sequentially following another through the system so that different analytical functions are being carried on simultaneously on more than one sample best describes
   A. automatic clinical analysis
   B. centrifugal analysis
   C. continuous-flow analysis
   D. dry-slide analysis

2. The process used in continuous-flow analysis to separate out protein from the specimen is
   A. chemical precipitation
   B. column chromatography
   C. ion-exchange chromatography
   D. dialysis

3. Direct injection of a sample into very small diameter tubing, thus minimizing lateral diffusion best describes
   A. automatic clinical analysis
   B. centrifugal analysis

Much of the material in this chapter has graciously been provided by the manufacturers of the instruments described. We thank them for their many contributions to this effort.

C. continuous-flow analysis

D. flow-injection analysis

4. Which analyzer requires that sample and reagent be pipeted into separate chambers in a rotor prior to the chemical analysis being performed?

    A. centrifugal

    B. continuous flow

    C. DuPont aca

    D. Kodak dry slide

5. In a chemical reaction the amount of product formed is measured at specific intervals during a specified period and then related to the concentration of the analyte in the unknown. This type of measurement is known as

    A. colorimetric

    B. end-point

    C. rate

    D. ultraviolet

6. The DuPont aca is instructed about the test method by means of a

    A. technician programming the instrument using the appropriate buttons

    B. magnetic-tape computer system

    C. test card with punched holes at the top for test identification

    D. bar code located on the header of the reagent pack

7. All of the following terms may be used to describe the layers of the Kodak dry slide *except*

    A. spreading

    B. scavenger

    C. indicator

    D. dialysis

8. Which system utilizes reflectance spectrophotometry for measuring the color intensity of reactions?

    A. centrifugal analyzer

    B. continuous-flow analyzer

    C. DuPont aca

    D. Kodak dry-slide analyzer

9. The purpose of dual-wavelength analysis is to

    A. read the absorbance of two different tests simultaneously

    B. correct for sample pipeting errors

    C. read the absorbance of the reagent blank prior to sample addition

    D. make specific blank corrections to minimize the effects of turbidity and hemolysis in specimens

10. An instrument that can analyze patient samples for only those tests specifically ordered and can analyze stat samples by interrupting the normal sequence of patient analyses is referred to as a

    A. batch analyzer

    B. discrete analyzer

    C. multitest analyzer

    D. random-access analyzer

BIBLIOGRAPHY

Burtis, C. A., "Advanced technology and its impact on the clinical laboratory," *Clin. Chem.* 33:352–357, 1987.

Burtis, C. A., and Painter, P. A., "Analytical systems for the clinical laboratory, parts 1 and 2," *Amer. Clin. Lab.* February: 14–19, March: 43–49, 1989.

Dobbins, Jr., J. T., and Martin, J. M., "Flow injection analysis: Twelve years old and growing," *Spectroscopy* 1:20–29, 1986.

Mayer, T. K., and Kubasik, N. P., "Dry-film chemical analysis: A brief history," *Lab. Manage.* April: 43–50, 1986.

Rocks, B., and Riley, C., "Flow-injection analysis: A new approach to quantitative measurements in clinical chemistry," *Clin. Chem.* 28:409–421, 1982.

Sharp, R. L., Whitfield, R. G., and Fox, L. E., "Robotics in the laboratory: A generic approach," *Anal. Chem.* 60:1056A–1062A, 1988.

Snyder, L. et al, "Automated chemical analysis: Update on continuous-flow approach," *Anal. Chem.* 48:942A–956A, 1976.

Walter, B., "Dry reagent chemistries in clinical analysis," *Anal. Chem.* 55:498A–511A, 1983.

Chapter Five

PROTEIN CHEMISTRY I: FUNDAMENTALS OF PROTEIN STRUCTURE AND ANALYSIS

● ● ● ● ● ● ● ● ● ● ● ● ● ● ● ● ● ●

UPON COMPLETION OF THIS CHAPTER, THE STUDENT WILL BE ABLE TO

1. Define each of the following terms:
   A. protein
   B. amino acid   F. zwitterion
   C. peptide bond   G. isoelectric point
   D. carboxyl group   H. net charge
   E. amino group   I. polypeptide

2. Describe how each of the following relates to protein configuration:
   A. primary structure   C. tertiary structure
   B. secondary structure   D. quaternary structure

3. Describe the functions of each of the following classes of proteins:
   A. enzymes   D. antibodies
   B. structural proteins   E. transport proteins
   C. contractile proteins   F. peptide hormones

4. Explain the process of protein digestion from the stomach through the small intestine.

5. Discuss the excretion and retention of proteins by the glomerulus.

6. State the principle of the biuret method used for the quantitation of serum total protein.

7. Describe how each of the following may interfere with the biuret assay of serum total protein:
   A. lipids   C. hemoglobin
   B. bilirubin   D. ammonia

8. List the reference ranges for each of the following:
   A. serum total protein
   B. serum albumin
   C. urine total protein (random and 24 h)
   D. cerebrospinal fluid protein

9. Discuss how various disease states may cause an increase or a decrease in serum total protein.

10. State the principles of the bromcresol green (BCG) and 2-(4'-hydroxyazobenzene)-benzoic acid (HABA) methods used for determining serum albumin.

11. Describe how each of the following may interfere with the BCG and HABA assays for the quantitation of serum albumin:
    A. hemolysis   D. salicylate
    B. bilirubin   E. heparin
    C. lipemia

12. Discuss the clinical significance of detecting abnormal total protein and albumin levels in the following body fluids:
    A. serum   D. synovial fluid
    B. urine   E. pleural fluid
    C. cerebrospinal fluid (CSF)

13. State the principles of the sulfosalicylic acid turbidimetric assay and the dipstick method for urine protein determination.

14. Describe causes of false positive results in the qualitative turbidimetric analysis of urine protein.
15. Contrast the specificity of the turbidimetric method with the dipstick method for urine protein determination.
16. Discuss the clinical significance of detecting Bence Jones protein.
17. State the methods used and the specimen requirements for performing quantitative urine protein assays.
18. State the principles of the following methods used for quantitating protein in CSF:
    A. sulfosalicylic acid
    B. trichloroacetic acid
    C. Coomassie blue dye
19. Describe causes of false positive results when quantitating CSF protein.
20. Explain the difference between pleural fluid classified as a transudate and the same classified as an exudate.
21. Identify the protein that is measured in amniotic fluid to screen for neural tube defects.

INTRODUCTION

Proteins are perhaps the most versatile molecules in our bodies. These structures furnish the support which holds us together, carry a large number of molecules through the bloodstream, move materials in and out of cells, provide the means of converting one molecule to another, and make it possible for the body to resist bacterial, viral, and chemical attacks from outside.

It is estimated that between 10,000 and 50,000 different proteins exist in our bodies. We know very little about most of these proteins; only about 1 to 2% have been studied to any degree. One of the most exciting areas of biochemical research today is the study of protein structure and function.

5.1

AMINO ACIDS

Proteins are large molecules, often referred to as **macromolecules.** No matter how different their structures and properties, all proteins are composed of the same fundamental building blocks—**amino acids.** It is the composition and arrangement of the amino acids in the protein which establishes the size and shape of each of these macromolecules. Our ability to assay for the presence of protein is also due, to a great extent, to the properties of these amino acids.

Structure of Amino Acids

All amino acids have the same basic components. Each is composed of an amino group $-NH_2$, a car-

boxylic acid group $-COOH$, and a carbon skeleton. We can illustrate a "generic" amino acid as follows:

$$\begin{array}{c} NH_2 \\ | \\ R-C-COOH \\ | \\ H \end{array}$$

The R can stand for a number of groups ranging from a simple $-H$ to a much more complex phenolic derivative. It is the structure of this R portion which gives each amino acid its particular unique characteristics. Some of the specific amino acids are illustrated in Figure 5-1. Although there are several

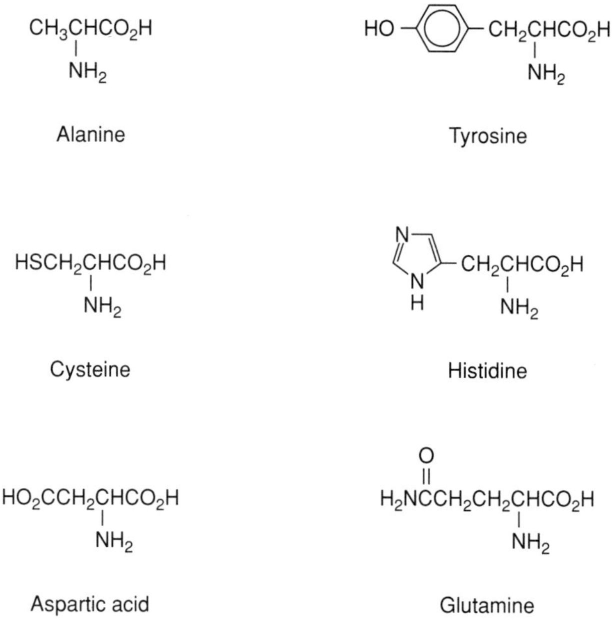

FIGURE 5-1   Representative amino acids.

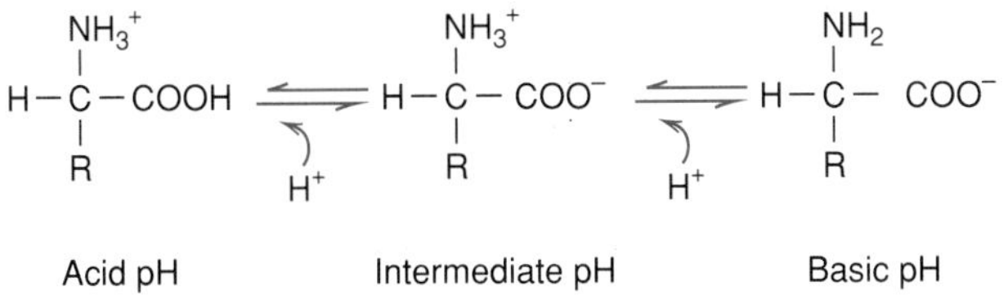

FIGURE 5-2   Effect of pH on amino acid charge.

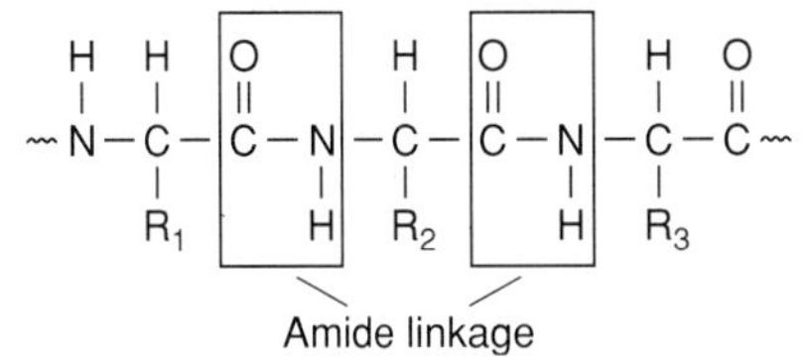

FIGURE 5-3   Amide linkage and primary structure.

thousand proteins in the body, they are all composed of the same 20+ amino acids.

Electric Charge on Amino Acids

It is a little misleading to write the structure of an amino acid as an electrically neutral compound because each amino acid really has a charge on it. Whether the charge is positive or negative depends on both the pH of the solution and the type of R group making up the specific amino acid. In basic solution, the net charge is negative. A proton dissociates from the carboxyl group present, forming COO^-. In acid solution, there is an excess of protons (H^+) which can attach to the $-NH_2$ group to form $-NH_3^+$. Other types of substituents found in amino acids also show changes in charge due to shifts in pH (Fig. 5-2). These charges affect how each protein moves in an electric field. This property allows us to separate proteins and study the changes in the amounts of these molecules in various pathological states.

At intermediate pH values (specific for each amino acid), both plus and minus charges may exist on the same molecule. This form of the amino acid, referred to as a **zwitterion,** has two differing charges. However, the *net* charge on the molecule may be zero. The pH value at which the sum of a molecule's charges equals zero is called the **isoelectric point.**

5.2

PROTEIN STRUCTURE

Three different components contribute to defining protein structure: primary, secondary, and tertiary structures. Sometimes there is a fourth component, known as quaternary structure. Each contribution to protein structure will be defined separately. These aspects of structure determine the shape of a given protein molecule and affect the function of that protein. In some instances, knowledge of the structure gives us insights into how to analyze for the presence and amount of a given protein. This is especially true for **hemoglobin** (the protein involved in the transport of oxygen in blood) and the **enzymes,** the biochemical catalysts responsible for almost all the chemical reactions which take place in the body.

The Peptide Bond

The amino acids in all proteins are linked together by a specific bond called the **peptide bond** (also referred to as an **amide linkage**). The structure of the peptide bond is illustrated below:

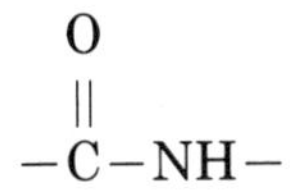

$$
\overset{\text{O}}{\overset{\|}{-\text{C}-\text{NH}-}}
$$

The peptide bond is formed when the carboxyl group of one amino acid joins to the amino group of a second, forming a molecule of water as a by-product. In cells, this reaction is carried out as part of the very complex process of protein synthesis. If the carboxyl group of amino acid I has joined to the amino group of amino acid II (as shown in Fig. 5-3), the carboxyl group of amino acid II can then react with the amino group of amino acid III to add another component to the growing protein chain. In this fashion the protein is created, one amino acid at a time.

Primary Structure

The amino acid sequence of any given protein is referred to as its **primary structure.** Each protein has its amino acid components arranged in a specific sequence unique for that protein. Two **polypeptides** (long sequences of amino acids) may have the same amino acid composition (Fig. 5-4), but have these amino acids in different orders. Therefore, although the amino acid composition of the two proteins may be the same, the sequence of the amino acids is different.

An important characteristic of the primary structure is that the amino acid sequence determines the overall shape of the protein—how it folds, bends, and twists. The way the amino acids are arranged determines whether the protein lies flat like a sheet or coils up in a ball.

Protein one: -alanine-leucine-glycine-tyrosine-

Protein two: -alanine-glycine-tyrosine-leucine-

FIGURE 5-4   Amino acid sequence and protein structure.

Alpha helix   Beta pleated sheet   Random

FIGURE 5–5   Secondary structure of proteins.

Table 5–1
SUMMARY OF PROTEIN STRUCTURE

1. Primary: amino acid sequence
2. Secondary: alpha helix, beta pleated sheet, random
3. Tertiary: three-dimensional configuration
4. Quaternary: combination of subunits

Secondary Structure

The coiling and folding of the protein chain is referred to as the **secondary structure** of the molecule. This refers not to the overall three-dimensional shape of the molecule, but to how the protein chain twists and bends along its length. There are three major types of shapes the secondary structure may assume:

1. alpha helix
2. beta pleated sheet
3. random

These shapes are illustrated in Figure 5–5.

The **alpha helix** is a coil, resembling a spring. As the coil turns, the twists are weakly linked together by hydrogen bonds. These interactions give stability to the protein molecule. The **beta pleated sheet** is a flat, corrugated structure which also contributes to molecular stability. If there is no apparent pattern, the term **random** is applied.

Tertiary Structure

The amino acid sequence (primary structure) not only determines the coiling and folding of the polypeptide chain (secondary structure), but also establishes the overall three-dimensional shape of the molecule, or **tertiary structure.** A protein may have a globular tertiary structure—an irregular, somewhat ball-shaped form. Many enzymes have this configuration. Some proteins twist into long thin rods called **fibrous proteins.** Other, less well defined shapes are possible. In each case, the shape of the molecule is established by the amino acid sequence unique to that protein. Knowledge of protein tertiary structure allows us to develop specific methods of separation using column chromatography or filtration techniques.

Quaternary Structure

When two or more polypeptide chains associate closely together to make up a multichain complex, the resulting molecule has **quaternary structure.** Not all proteins have this property, but it is demonstrated by several important protein molecules in the body. Hemoglobin has quaternary structure because four protein chains (two alpha and two beta chains) assemble together to constitute the intact hemoglobin molecule. The separate chains (called **subunits**) do not individually perform the function made possible when the four chains join together. No chemical bonds form to hold these chains in place, and, therefore, they can be easily separated from one another and recombined to form the intact, functional molecule again. Many enzymes possess quaternary structure, giving them unique and interesting properties, particularly in the regulation of enzyme behavior. Table 5–1 summarizes the aspects of protein structure discussed.

5.3

FUNCTIONS OF PROTEINS IN THE BODY

Types of Proteins

Proteins perform a variety of useful tasks in the body. We can identify at least six different classes of proteins based on function (Table 5–2). Each protein has a specific role in the organism. It is often important to know the concentration of a given protein in a body fluid so we can learn if that protein is available in sufficient quantity to perform its given function.

Table 5–2
FUNCTIONS OF PROTEINS

1. Enzymes: catalyze biochemical reactions over 1500 known
2. Structural: provide cellular or body support
   collagen (bone, skin)
   keratin (hair, nails)
3. Contractile: contract muscles
4. Antibodies: neutralize foreign materials
5. Transport: carry molecules
   transferrin (iron)
   HDL, LDL (cholesterol, other lipids)
6. Peptide hormones: regulate metabolism
   insulin (glucose metabolism)
   endorphins (pain perception)

PROTEIN CHEMISTRY I: FUNDAMENTALS OF PROTEIN STRUCTURE AND ANALYSIS **79**

Enzymes

One of the most complicated types of proteins, in terms of both structure and function, is the enzyme. Currently, over 1500 enzymes are known to be present in a human being. We know something of the roles these enzymes play, but there are still a large number of unanswered questions about each particular protein.

Enzymes are biochemical catalysts—they cause reactions in the body to take place more rapidly and with less need for outside sources of energy. The molecule taking part in the reaction binds to the enzyme and is transformed rapidly into another compound by that enzyme. Since enzymes play an important role in normal body biochemistry and are extremely useful in providing important diagnostic information, we will discuss these complex and fascinating proteins in two later chapters of the book.

Structural Proteins

An important role for proteins is to provide structural support for the body, a tissue, or a cell. **Structural proteins** are usually long, fibrous molecules such as collagen and keratin. **Collagen** is found primarily in bone and provides the support for calcium to bind and form the solid structure for bone. Another structural protein is **keratin,** mainly seen in hair and nails. Again, it is the fibrous quality of the protein which makes it a good structural material. The red cell membrane and other tissues have unique structural proteins which help keep the cell walls intact. Although analysis of structural proteins has found little application in clinical chemistry to date, new findings may expand the use of quantitative measurement of these biomolecules.

Contractile Proteins

Proteins involved in the contraction and relaxation of muscles are known as **contractile proteins.** These molecules are also usually long, fibrous materials. Often the subunits of a contractile protein are assembled so that they slide back and forth as the muscle fiber (or other tissue) shortens or lengthens. Although they carry out important functions, measurement of contractile proteins in body fluids has not been a part of modern clinical chemistry.

Antibodies

A number of developments have taken place in antibody biochemistry within the last 20 years. As evidence of the importance of this field, a Nobel Prize was awarded in 1987 to researchers who studied how these molecules were synthesized. Antibodies form in response to the presence of "foreign" substances in blood or tissues. When the body is stimulated by these materials, antibodies are produced and bind to the unwanted chemical or microorganism to either destroy or neutralize it.

Our knowledge of antibody structure and function has opened the door to a greater understanding of a number of disease states. Since we can quantitate specific antibodies very accurately, we can often obtain critical diagnostic information. In addition, a large number of modern biochemical assays use antibodies as analytical reagents because of their specificity.

Transport Proteins

Transport proteins serve a vital function in carrying materials from one part of the body to another through the circulation. Other transport proteins are involved in the movement of chemicals through the cell membrane. These molecules play important roles in helping regulate the amount of specific biochemicals in living systems and in contributing to control of the rate of biochemical processes.

A large number of materials in the body do not dissolve in water. Since whole blood is approximately 70% water, these components must bind to proteins to move through the circulatory system to the sites where they are needed. Some transport proteins carry a wide variety of substances. **Albumin** serves as a transport medium for bilirubin, calcium (to some extent), fatty acids, and a number of drugs. Other transport proteins are more selective about the compounds they will pick up and deliver. Two such proteins are responsible for the transport of iron (**ferritin** and **transferrin**), but each has a somewhat different role in regulating iron use. Many hormones have specific transport proteins which move these molecules through the circulation. For example, the steroid hormone cortisol is carried by a protein known as **transcortin** (or **cortisol-binding globulin, CBG**), and the thyroid hormones circulate bound to thyroid-binding globulin. Vitamin B_{12} has three proteins responsible for its transport. Perhaps the best-known transport proteins are **HDL (high-density lipoprotein)** and **LDL (low-density lipoprotein),** which carry cholesterol and other lipids in the bloodstream. The levels of these two proteins are important in predicting the degree of risk for coronary heart disease.

Peptide Hormones

Hormones are molecules formed by the endocrine glands and released into the bloodstream to be carried to another part of the body. At the new location, the hormone binds to a specific site on the cell and produces certain biochemical changes. Each hormone generates a unique set of responses to its production and cellular uptake. Therefore, hormones are very important in controlling metabolism.

Although not all hormones are proteins, several hormones have the basic protein structure. **Insulin** is perhaps the best-known protein hormone. This molecule consists of two peptide chains joined together by sulfur bridges (called **disulfide bonds**). Glucose uptake and use by the cell are regulated primarily by the level of insulin in the circulation. Some of the hormones that are involved in the development of pregnancy and that regulate time of delivery of the baby are proteins. Several peptide hormones **(endorphins)** are believed to be responsible for how we perceive pain. These endorphins may also play a role in the development of some aspects of mental illness.

5.4

PROTEIN INTAKE AND METABOLISM

Sources of Dietary Protein

The proteins found in cells and body fluids are not the same intact molecules consumed in the diet. Proteins and amino acids found in food serve only to supply the necessary components used by the organism to build the proteins required by the body. Obviously, an adequate intake is necessary to provide all the amino acids needed in the right ratios. Fad diets that may be "high in protein," the use of protein or amino acid supplements, and other alternatives to eating a well-balanced menu serve no useful nutritional purpose.

Digestion of Protein

Protein digestion begins in the stomach and is completed in the small intestine. The presence of a highly acid environment (pH 2–3) helps to unfold the proteins, making them more susceptible to enzyme attack. **Pepsin** (an enzyme in the stomach) begins the process of protein digestion by cleaving specific amide linkages to form a variety of smaller peptides. These smaller units then promote release of hormones and enzymes by the pancreas and small intestine.

Several inactive enzyme precursors are released into the small intestine from the pancreas. These **proenzymes** are first cleaved by other **proteolytic** (protein-splitting) enzymes, removing a small portion of the polypeptide chain and forming the active enzyme. The activated enzyme can then begin to attack proteins in the small intestine at specific sites; each enzyme cleaves the dietary proteins at a different location in the peptide chain. Trypsin, chymo-

trypsin, elastase, carboxypeptidase, and other enzymes are responsible for the breakdown of proteins into small peptides. These smaller molecules are then hydrolyzed to single amino acids by aminopeptidases and related enzymes.

Intestinal Absorption of Amino Acids

Amino acids enter the mucosal cells of the small intestine and are then transported to the liver. Different enzyme systems transport specific groups of amino acids into the mucosal cells. At least three different transport groups are known: one for neutral amino acids, one for basic amino acids, and one for movement of proline, hydroxyproline, and several other compounds. After transport into the mucosal cell, the amino acids passively diffuse from the cell into the bloodstream for eventual utilization by the liver or other tissues.

Formation of New Proteins

The process of protein synthesis is a complicated one, regulated by chromosomes contained in the nucleus of each cell. Although we will not go into detail about the mechanism by which proteins are formed, we can say that the amino acid sequence for each protein and the amount of protein produced are determined by the genetic code each individual possesses. When the cell receives appropriate biochemical signals, the machinery for protein synthesis is activated and production of the needed proteins takes place. Many of these proteins are used within the cell to perform metabolic tasks; other proteins are released into the circulation to function in various roles.

Metabolism of Protein

Once proteins are manufactured by the cells, they do not stay in the body indefinitely. As each cell ages, it either divides into two cells or breaks down. In both cases some proteins are lost into the circulation. There are also enzymes within the cell which catabolize other proteins, reducing them once again to small peptides and individual amino acids (which are then recycled for new protein synthesis).

If a protein is released into the circulation or CSF as a result of cellular turnover, a similar fate awaits it. A number of proteolytic enzymes are present in both fluids, gradually consuming the proteins and reducing the circulating levels of the macromolecules. This explains why a protein which increases in concentration in blood does not remain indefinitely at a high level, but decreases in concentration over a period of hours or days. The breakdown process is especially important for regulating the levels of hormones and other proteins whose concentrations in the circulation produce significant biochemical changes.

Excretion of Protein

Most proteins are not excreted to any appreciable extent in urine. The filtering apparatus of the kidney (the glomerulus) has a cutoff of about 60,000 mol wt (molecular weight). Compounds larger than the cutoff size are not filtered and excreted; compounds with lower molecular weights generally pass through the glomerulus and into the urine. A few proteins can be lost from the body in this manner. The enzyme amylase has a molecular weight of approximately 40,000–50,000. This molecule is small enough to be filtered at the glomerulus and found in the urine. Certain immunoglobulin fragments are produced in high amounts in some types of bone cancer. These proteins have low molecular weights and are readily excreted by the kidney. Detection of protein in the urine can have great clinical significance.

Protein Concentrations and Disease States

In most circumstances, the level of any given protein in blood, CSF, or other body fluid increases in various disease states. This elevation of protein concentration may be the result of massive cell breakdown releasing protein into the fluid. In other instances, the disease state itself may lead to production of an increased amount of a protein or proteins manufactured by a given tissue. However, there are situations in which the amount of protein present may decrease from the normal level due to the disease process. This decline in concentration may be due to impaired synthesis of the protein or enhanced loss of protein in the urine. An important factor in setting up an assay for a given protein is knowing how the level of that material changes in disease states. Does the level increase or decrease? How much change takes place? These questions must be answered in order to select the appropriate methodology for analyzing a given protein.

5.5

PRINCIPLES OF PROTEIN ANALYSIS: GENERAL CONSIDERATIONS

Qualitative or Quantitative

Protein analysis encompasses a wide variety of techniques. Each method must be selected on the basis of the information needed. Quite often the only information needed is qualitative—is there too much protein present? In this case we are not concerned about the exact level, but need only a "ballpark" indication. Urine protein measurements are usually qualitative, so the methods are designed to give a yes/no answer. On occasion, a quantitative urine protein determination may be of value. In the majority of situations, there is also no particular need to measure a specific protein in urine, so sophisticated assay approaches are not needed for routine urine protein analysis. For serum and other body fluids, however, the protein concentration is normally fairly high. Often, information about the concentration of a specific protein is needed, even though the protein may be present in low amounts. In these instances, our need for information is such that quantitative measurement provides the only useful data.

Total Protein or Specific Assay

When we consider quantitation of proteins in serum, plasma, or CSF, we must ask several questions about the information we wish to obtain. In some instances we may need only a value for the total amount of protein present. A relatively nonspecific chemical test provides this information quickly. However, we may wish to know the level of a specific protein. In this case, more selective analytical procedures must be developed.

The analysis may rely on a particular chemical change. This approach has been widely used for the assay of serum albumin. On the other hand, the measurement could involve the use of an antibody which reacts specifically with the protein under consideration. These powerful techniques have found extensive use in recent years. Often we wish to learn the amounts of several proteins in a sample. The technique of electrophoresis allows us to separate different proteins in a body fluid or tissue and study patterns of protein distribution.

5.6

QUANTITATION OF SERUM TOTAL PROTEIN

General Considerations

When discussing serum protein analysis, we are dealing with a complex body fluid containing high concentrations of a wide variety of proteins. Although urine protein concentrations do not normally exceed 15–20 mg/dL, the level of serum total protein may reach 8.5 g/dL—a 200-fold increase in concentration. Moreover, the protein composition of serum is much more complex than that of urine. Urine protein is made up predominantly of low-molecular-weight globulin components and albumin, with the globulins being present at approximately twice the concentration of albumin. In contrast, over half the serum protein is albumin. Some of the other specific proteins in serum are lipoproteins, transferrin, a number of immunoglobulins, and low concentrations of a variety of enzymes. Because this mixture is so complex, separation of the various protein fractions by electrophoresis yields much more useful information than most nonspecific chemical tests. We will consider some of the widely used assays in this chapter, leaving a detailed analysis of enzymes and a few specialized proteins to other areas of this book.

Although measurement of serum total protein rarely gives much clinically useful information by itself, the data provide a baseline for comparing the relative amounts of the several protein fractions. Quite often, the serum levels of total protein and albumin will furnish a quick overview of the protein metabolic status of the patient.

The Biuret Assay for Total Protein

If we are interested in only the total amount of protein in serum and not in any specific fraction or protein, the assay we select must be one which measures a property common to all proteins. In our earlier discussion of protein structure, we noted that all proteins are amino acids joined together by an amide linkage (or peptide bond). One of the properties of this bond is its ability to form a complex with copper ions, which can be measured spectrophotometrically. This copper/protein complex is the basis for the quantitative measurement of serum total protein.

The reagent most widely used for quantitation of serum total protein is called the **biuret reagent.** Biuret is formed when urea (an end product of metabolism) is heated. Two molecules of urea form one

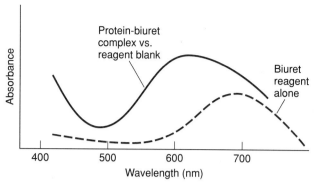

FIGURE 5-7   Absorbance spectra of biuret reagent and Cu complex.

molecule of biuret. One characteristic of the biuret structure is the presence of an amide-type linkage. When copper reacts with this linkage, a specific complex is formed (Fig. 5-6). A similar complex is formed when copper reacts with the amide bond in proteins. Biuret is not formed when we assay for protein; biuret is used only to describe the type of reaction which takes place in the assay.

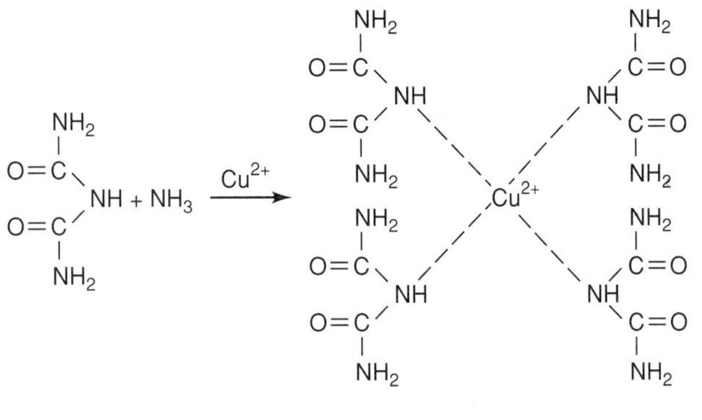

The biuret reagent is prepared by dissolving copper sulfate in water; a tartrate salt is added to keep the copper in solution. Often, potassium iodide is added to stabilize the Cu^{2+}. Otherwise, reduction to Cu^{1+} would take place; the cuprous form of the ion does not react with the amide bond. When mixed with a protein solution, the copper forms a complex with the amide linkage. Although the copper solution itself is pale blue, the purple complex absorbs light at a wavelength different from that absorbed by the biuret reagent alone (Fig. 5-7). This assay forms the basis of almost every manual and automated procedure used today for quantitation of total protein in serum.

Interferences with the Biuret Method

If the serum sample is turbid (due to elevated lipids), has a high bilirubin level, or has significant amounts of hemoglobin present (from hemolysis), the total protein value will be falsely elevated. Although there are a variety of clinical situations which result in lowered total protein levels, there are no significant drug interferences which produce falsely decreased values. High blood-ammonia levels may give rise to falsely low total protein because of a competing reaction between the ammonia and the copper ion. Some medications (particularly steroids) produce metabolic changes leading to either decreased or increased total protein values. These are real changes, however, and not artifacts of the analytical procedure.

FIGURE 5-6   The biuret reaction.

5.7

CLINICAL SIGNIFICANCE OF SERUM TOTAL PROTEIN VALUES

When we look at serum total protein results, we need to keep in mind that the value we obtain represents a general overview of how proteins are being produced, utilized, and excreted by the body. A wide variety of factors affect this process, including damage to tissues resulting in decreased protein synthesis, abnormal processes leading to increased protein production, and alterations of protein excretion. The serum total protein value may reflect one or more of these processes. Therefore, we see only a crude and hazy picture of the dynamic changes occurring. An abnormality may be shown to exist, but little detail is obtained until more specific analyses are performed.

Increases in Serum Total Protein

The adult reference range for serum total protein is 6.5–8.5 g/dL. Serum total protein elevations (Table 5–3) are generally due either to loss of fluid from the body (dehydration) or to an increase in the globulin component. Serum albumin is almost never selectively increased as a result of a pathological process, so it does not contribute to increases in total protein. Elevations in total protein from dehydration or from severe exercise are due to depletion of body fluids, leading to relative increases in the concentrations (but not the total amount) of all protein fractions. These elevations can be reversed within a short time by replacement of water. In some cases of infection, the elevation may be due to increase in the gamma globulin fraction as antibodies are formed to deal with the attacking microorganisms. In various types of cancer, the increase in total protein often reflects synthesis of abnormal proteins by the tumor (examples include carcinoembryonic antigen and alpha fetoprotein).

Decreases in Serum Total Protein

Serum total protein values lower than normal (Table 5–3) may reflect protein loss, increased utilization, decreased intake, or diminished rates of protein synthesis. Patients with cancer of the stomach or intestinal tract experience significant gastrointestinal loss of protein. In some instances of liver disease protein synthesis decreases. Generalized malnutrition or thiamin deficiency results in diminished levels of total protein due to impaired dietary intake and decreased synthesis (caused by lowered amino acid availability). Glomerulonephritis is a disorder characterized by an inability of the kidney to retain large-molecular-weight proteins. Loss of these proteins in the urine produces a decrease in serum total protein.

5.8

MEASUREMENT OF SERUM ALBUMIN

Although the quantitative measurement of most specific proteins will be discussed in later chapters, we do want to examine aspects of the assay for serum albumin. The combined determination of albumin and total protein in serum can give a good index of the overall status of protein metabolism in the body. Since the albumin concentration constitutes over half the serum total protein, changes in the level of this component can provide significant information.

Dye-Binding Methods for Serum Albumin

Because the level of albumin in serum is high (approximately 3.5–5.0 g/dL), this protein can be easily

Table 5–3
TOTAL PROTEIN AND ALBUMIN VALUES: CLINICAL SIGNIFICANCE

Parameter	Reference Range	Increased Values	Decreased Values
Total protein	6.5–8.5 g/dL	Dehydration Severe exercise Infections Cancer	Gastrointestinal cancers Liver disease Malnutrition Low thiamin Glomerulonephritis
Albumin	3.5–5.0 g/dL (women approx. 0.5 g/dL lower than men)	Dehydration Sunstroke Exercise Multiple sclerosis Hypothyroidism	Pregnancy Malnutrition Malabsorption Liver disease Kidney disease Burns

quantitated by a simple colorimetric assay. Most techniques employ some sort of dye which more or less specifically binds to albumin but is unreactive with other protein fractions. Although a wide variety of dyes have been explored, two seem to be most suited for the assay of serum albumin. Bromcresol green (BCG) and 2-(4'-hydroxyazobenzene)-benzoic acid (HABA) have found wide use in both manual and automated procedures.

Both dyes use the same analytical principle. Binding of dye to albumin results in a shift in the wavelength of maximum absorbance and in an increase in absorbance of the dye-albumin complex relative to the absorbance of the dye in the absence of albumin. Reactivities of the two dyes with albumin are similar. There are some reports that BCG overestimates albumin when compared with electrophoretic methods.

Differences in Results Using HABA and BCG

An examination of the spectra of the dyes alone and when complexed with albumin suggests that BCG may possess some advantages over HABA in terms of sensitivity and in minimizing problems of interferences (Fig. 5-8). Both dyes absorb strongly below 500 nm; BCG has an additional small peak at about 625 nm. When albumin is added to either system, a complex forms and the absorbance maximum of the solution changes. Both dyes have an increase in absorbance in the region around 500 nm. However, there is a striking increase in the absorbance maximum at 630 nm for BCG. These data suggest two advantages in using BCG as the indicator for serum albumin quantitation. The test using BCG is more sensitive; the same concentration of albumin gives a much greater absorbance change at 630 nm than it would at 500 nm. In addition, the absorbance at 630 nm is less likely to be affected by absorbance due to bilirubin or hemoglobin in the sample.

Significant differences between the two dyes become apparent when we look at substances which

interfere with the assay (Table 5-4). Hemolysis and lipemia cause false elevations of albumin with both assays but appear to have much more effect on HABA techniques than when BCG is used. Elevated bilirubin levels produce lowered albumin levels with the two systems (due to binding of bilirubin to albumin, displacing some of the dye), although the effect is much more pronounced with HABA. High blood levels of salicylate produce a slight decrease in values obtained with HABA but do not seem to affect the BCG method. Elevated heparin concentrations produce apparent decreases in albumin when measured by the BCG approach, apparently because of a precipitate which forms between the dye and the anticoagulant. Addition of detergent to the assay system overcomes this problem. Heparin has been reported to increase the observed albumin value in the HABA assay. Enhancement of dye binding to fibrinogen (a coagulation protein found in plasma, but not in serum) and to globulins appears to be the reason for this artifactual increase.

Other approaches to the quantitation of serum albumin include quantitation by electrophoretic fractionation of proteins and the formation of a specific albumin/antibody complex, which can be quantitated by several means. These methods will be described in detail later.

Table 5-4
INTERFERENCES WITH ALBUMIN ASSAY USING BCG AND HABA

Interfering Substance	Effect on Albumin Value Using	
	HABA	BCG
Hemolysis	Increase	Slight increase
Bilirubin	Decrease	Decrease
Salicylate	Slight decrease	No apparent effect
Heparin	Increase	Decrease

5.9

CLINICAL SIGNIFICANCE OF SERUM ALBUMIN VALUES

Increases in Serum Albumin

Elevated levels of serum albumin are infrequently seen (Table 5-3). In some cases of dehydration or significant fluid loss, we might expect to see slight increases in the value for albumin (again, due to a concentration effect, not an increase in albumin production). This increase can occasionally be seen in patients with sunstroke or persons who have been exercising strenuously. About 10% of patients with multiple sclerosis show increases in serum albumin (CSF globulin changes are much more diagnostic for this disease). There have been a few reported in-

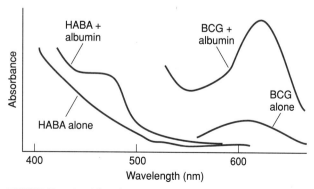

FIGURE 5-8 Absorbance spectra of BCG and HABA with albumin.

stances of increased serum albumin in patients with hypothyroidism.

Decreases in Serum Albumin

Serum albumin levels below the normal range occur in a wide variety of disorders (Table 5-3). Most of the clinical situations resulting in decreased serum total protein also produce lowered serum albumin values. Among these clinical problems are those associated with malnutrition and malabsorption, where protein either is not consumed in the diet or is lost through the gastrointestinal tract. Patients with a number of disorders affecting the liver frequently show a decrease in serum albumin levels because the liver is the primary organ for synthesizing this protein. Albumin levels are also decreased in a wide variety of cancer patients. Kidney diseases, such as glomerulonephritis or nephrotic syndrome, produce lowered serum levels of albumin because of excessive loss of this protein in the urine. Burn patients lose large amounts of albumin through the capillary system of the affected area of the body, resulting in low albumin values. During the last two trimesters of pregnancy, albumin levels are somewhat diminished.

5.10

QUANTITATION OF SERUM GLOBULINS

Nothing has been mentioned in our discussion about the quantitation of serum globulin levels. Although methods are described in the older clinical chemistry literature and some reference laboratories make the procedure available, there is no clinical utility to the measurement of total serum globulins. Any useful information regarding elevations or decreases in this parameter are more properly obtained from the assay of the specific protein in question and not from a nonspecific methodology.

5.11

ANALYSIS OF URINE PROTEIN

General Considerations

Most protein analysis of urine samples is performed to look for possible kidney damage or the excretion of abnormal proteins. If the kidneys are leaking protein, we expect to see an increase in urine protein levels. Analysis of urine protein (and other constituents) is generally carried out during a routine examination on an out-patient basis or during an initial laboratory work-up when a patient is admitted to the hospital. The specimen is a random

Table 5-5
TURBIDIMETRIC ASSESSMENT OF URINE PROTEIN

Result Reported	Observations Made	Protein Range (mg/dL)
Negative	Clear sample, no turbidity	Below 20
Trace	Very faint precipitate	20–100
1+	Small degree of turbidity	100–1000
2+	Moderate turbidity	1000–2500
3+	Heavy turbidity	2500–4500
4+	Heavy flocculation (clumping)	>4500

one, since there is no need to quantitate the output of protein over a given period.

Turbidimetric Analysis

Two major approaches are used for qualitative protein determination of a random urine specimen. In the turbidimetric assay, an acid (usually sulfosalicylic acid) is added to the urine specimen. The low pH causes denaturation of the proteins, and they form a precipitate. The concentration of protein is usually low enough that the precipitate remains suspended in the liquid (at least for a while). The amount of precipitate (or turbidity) observed is an indication of the level of protein present.

The turbidity is graded on a scale which indicates increasing amounts of total protein. The sample is viewed against a dark background and the turbidity assessed according to the criteria listed in Table 5-5. Urine samples with normal levels of protein give a "negative" or "trace" reaction. A sample containing a large amount of protein gives a greater amount of precipitate. The reaction is reasonably nonspecific, with most proteins giving much the same amount of precipitate.

Interferences with the Turbidimetric Method

A number of situations will give rise to false positive results (Table 5-6). High bacterial levels in

Table 5-6
INTERFERENCES WITH TURBIDIMETRIC PROTEIN ASSAY

False Positive	False Negative
Bacteriuria	none known
Cephalosporin	
X-ray contrast media	
Sulfa drugs	
Penicillin	
Tolbutamide	

urine produce turbidity. Cephalosporin drugs form precipitates. If radiopaque media are used for x-ray studies, precipitates form in the urine. Turbidity is observed with several sulfa drugs and penicillin. Diabetic patients on tolbutamide may show false positive results due to urine turbidity produced by this drug. There are no known false negative interferences.

Qualitative Dipstick Method

The second qualitative method is commonly employed as part of a multitest dipstick screen for routine urinalysis. Commercially prepared sticks are treated with a dye which changes color in the presence of protein. When the protein binds to the dye, a pH shift results in an alteration in the color of the dye. The degree of color change is related to the concentration of protein in the specimen. Color-coded charts on the dipstick container allow a graded assessment of the dye color and the protein content. The only known interference is with alkaline urines. These samples produce false positive values and overestimate the amount of protein present.

Specificity of Qualitative Tests

Testing urine samples with both the turbidimetric and dipstick methods reveals major discrepancies with the results of some samples. Although the acid precipitation method is fairly nonspecific, the dye method gives a much stronger reaction with albumin than with other proteins which may be present. Albumin is the major protein in urine in many diseases involving simple kidney damage and generalized loss of protein. However, there are certain disorders in which significant amounts of globulins (rather than albumin) are excreted in the urine. The dipstick method would seriously underestimate the amount of protein present in these situations, giving misleading diagnostic information. A routine urinalysis should include both the dipstick assay and an acid precipitation for urine proteins to provide reliable data for clinical purposes.

Bence Jones Proteins

In the past, a specific assay for Bence Jones proteins (immunoglobulin fragments) was considered important for patients with possible multiple myeloma (a type of cancer affecting bone marrow). These proteins possess the unique property of precipitating in acid solution and redissolving upon heating of the mixture. This test has largely been abandoned because of the high percentage of false positive results obtained, although there are still many references in the literature to this class of pro-

teins. A more useful alternative is the screening of the urine sample by protein electrophoresis.

Quantitative Urine Protein Determination

A qualitative assay for urine protein provides sufficient information in most clinical situations, with protein levels greater than 20–25 mg/dL indicating an abnormal situation. In some instances, quantitation of urine protein output as a function of time has been found useful. The generally preferred specimen is a 24-h collection, refrigerated to avoid bacterial contamination. The validity of the collection time can be determined by measuring the urinary creatinine output with the same specimen; more will be said about this approach later.

Quantitation of urine protein is accomplished with a modification of the turbidimetric method used for qualitative analysis. A standard curve is prepared employing an accurately measured serum sample which has been diluted to the appropriate range. Precipitation of protein can be accomplished with either sulfosalicylic or trichloroacetic acid. Turbidity is assessed using a spectrophotometer, the increase in absorbance readings being correlated with an increase in turbidity. By comparing the absorbance of the patient sample with the standard curve, the concentration of total protein can easily be obtained. Total 24-h output is then calculated using the urine volume. Results have an accuracy of about ±15%. This test is not designed for high precision, but provides useful information when employed appropriately.

Clinical Significance of Urine Protein Values

Unlike serum, the protein level in urine is either normal or increased; low levels have no clinical significance. Increased total protein in urine (proteinuria) may result from a number of causes (Table 5-7). Any situation involving kidney damage produces loss of protein from the body, leading to increased protein excretion by the kidney into the urine. A

Table 5-7
CLINICAL SIGNIFICANCE OF URINE PROTEIN VALUES

Proteinuria	Albuminuria
Kidney damage	Kidney damage
Toxic compounds	Cancer
High blood pressure	Infectious diseases
Congestive heart failure	Complications of pregnancy
Multiple myeloma	

large number of toxic compounds (lead, arsenic, mercury, carbon monoxide) promote kidney damage and urine protein loss. Situations involving high blood pressure or congestive heart failure lead to increased urinary protein. The clinical condition generates extra pressure on the kidney, forcing protein out of the bloodstream into the urine. Some infections produce proteinuria. Patients with multiple myeloma produce abnormal immunoglobulins and fragments of immunoglobulins. These fragments are small enough to pass through the glomerulus of the kidney and enter the urine in large quantities.

Often the protein loss in kidney damage is primarily due to excretion of albumin with elevated levels seen in urine. Excessive loss of albumin in urine (albuminuria) is also seen in patients with a wide variety of cancers. Women who experience complications in pregnancy quite frequently show albuminuria. Urine albumin is increased in patients with a number of infectious diseases, such as typhoid fever, diphtheria, and malaria.

5.12

QUANTITATION OF CSF TOTAL PROTEIN

Assay Limitations

Measurement of CSF total protein presents many of the same challenges and limitations seen in urine protein analysis. The total CSF protein level is low, normally about 45 mg/dL or less. The predominant fraction is albumin (over 60% of the total) with low levels of most globulin fractions. In clinical situations we may expect to see elevations of all protein fractions or only selective increases of certain immunoglobulins (particularly IgG in cases of multiple sclerosis). The assay technique selected must be versatile and reliable enough to encompass all these possibilities without a compromise in accuracy and speed.

Turbidimetric Assay

Because of the low concentration of protein in CSF, the biuret method cannot be used. As is the case with urine protein, one method for CSF total protein quantitation is turbidimetric. The acids selected for precipitation are generally sulfosalicylic or trichloroacetic. Precipitation of albumin and globulins seems to be more equivalent with trichloroacetic acid, although less turbidity is seen than if sulfosalicylic acid is used. Whatever the choice of precipitant, care must be taken to run the assay at a reasonably constant temperature. Changes of only a few degrees affect the ratio of albumin/globulin precipitation. The tubes must not be allowed to

stand for more than a few minutes after adding the acid. If the precipitate settles out at the bottom of the tube, it does not give an accurate reading when resuspended.

All the materials which provide false positive results with urine total protein measured by turbidimetry (Table 5-6) also give false positive results with this assay for CSF total protein.

Quantitation with Coomassie Blue

In recent years, other approaches to the quantitation of CSF total protein have been developed to avoid some of the problems associated with the turbidimetric assay. One of the most promising alternatives has been the use of Coomassie blue dye. This reagent has long been employed as a stain for detecting low concentrations of proteins after electrophoresis. Binding the dye to protonated amine groups on the side chains of the CSF proteins produces a blue color. The unbound dye and the dye/protein complex have different absorption properties; measurement of light absorbed at two different wavelengths allows quantitation. The procedure is simple: the dye is mixed with a CSF sample, color is allowed to develop and the absorbance is measured on a spectrophotometer. The time after addition before a reading is made is not critical; the color remains stable for over an hour. As is also true of the turbidimetric method, there is some difference in the reaction with albumin and with globulins, so protein quantitation requires a standard with an albumin/globulin ratio similar to that of CSF. A mixture of 70% albumin and 30% globulins should provide satisfactory results for standardization.

The presence of hemolysis gives false elevations in the CSF total protein value obtained by the dye-binding method. CSF protein values determined by the Coomassie blue procedure are some 9% higher than the assay results obtained using a sulfosalicylic acid turbidimetric method.

Clinical Significance of CSF Protein Values

Normal values for CSF total protein are approximately 15–45 mg/dL. As with urine, the only clinically significant alterations in CSF protein are those resulting in elevated values. A wide variety of infectious diseases produce elevations, often fairly modest ones. One of the characteristic signs of bacterial meningitis is an increase in CSF total protein. Many viral disorders result in elevations of CSF protein. On examination by electrophoresis, these changes appear to be complex and nonspecific. An important diagnostic sign of multiple sclerosis is an elevation in CSF total protein, with the increase seen primarily in the globulin fraction. Because of the importance of CSF studies in multiple sclerosis, we will deal with this subject more thoroughly in Chapter 7. Albumin is usually not quantitated separately when CSF total protein is measured. This fraction is best measured using electrophoresis or an immunochemistry method.

5.13

PROTEIN MEASUREMENTS OF OTHER BODY FLUIDS

Total protein measurements of other body fluids may be made. Synovial fluid (from joints) might have elevated protein levels in cases of gout, systemic lupus erythematosus, and some types of arthritis. Pleural fluid (from lung) protein is elevated in pulmonary tuberculosis and pulmonary embolism. Fluid low in protein (less than approximately 3 g/dL) is termed a **transudate** and reflects changes in the ability of the filtering membrane to retain protein. If the concentration of protein in pleural fluid is above 3 g/dL, the material is referred to as an **exudate** and is produced as a result of infection or malignancy. The measurement of alpha fetoprotein in amniotic fluid has received a great deal of attention in recent years as a screen for neural tube defects. Because of the legal and ethical implications surrounding the diagnosis and treatment of this disorder, testing for alpha fetoprotein is being performed only in special licensed laboratories.

SUMMARY

Amino acids are the basic building blocks of proteins. Each amino acid has a carboxyl group and an amino group. These compounds are positively charged at acid pH values and negatively charged at basic pH values. The zwitterion form (where the total net charge on the molecule is zero) occurs at intermediate pH values.

The basic unit of structure for all proteins is the peptide bond, an amide linkage joining two successive amino acids. Each protein possesses primary structure (the order in which amino acids are arranged in the protein chain), secondary structure (local coiling or bending), and tertiary structure (the overall three-dimensional structure of the protein). If the protein consists of two or more polypeptide chains, it has quaternary structure (the arrangement of the chains in relation to one another).

Six major categories of proteins exist. Enzymes catalyze biochemical reactions. Structural proteins provide support for the body or for specific tissues. Contractile proteins are involved with the processes of muscle contraction. Antibodies protect the organism by responding to the presence of foreign material in the body and neutralizing it. Transport proteins facilitate movements of materials throughout the circulation, particularly substances that are not very water soluble. Some hormones are peptides, serving as biochemical signals for regulation of metabolic processes.

Some digestion of protein takes place in the acid environment of the stomach. Most protein breakdown occurs in the small intestine as specific proteolytic proenzymes are activated and then cleave the protein into constituent amino acids. Groups of amino acids are then transported into intestinal mucosal cells to diffuse into the circulation. After being delivered to the liver and other tissues, these amino acids are involved in the synthesis of new protein. Protein turnover in cells occurs continually, with the amino acids generated being recycled. If the protein is of low molecular weight (below about 60,000 mol wt), it can be filtered at the glomerulus and excreted in the urine. Large proteins do not undergo excretion in this manner.

Protein analysis can be either qualitative or quantitative, depending on the specific protein and the information needed. For serum total protein quantitation, a reaction with a copper salt (called the biuret reaction) provides adequate quantitation. Albumin in serum is assayed by dye-binding methods. The absorption maximum and intensity of absorption of the dye changes after it complexes with albumin.

The reference range for serum total protein in adults is 6.5–8.5 g/dL. Increases in total protein are seen in patients who are dehydrated, those who have performed intense exercise, and in individuals who have infections or cancer. Low levels are observed in cases of malnutrition, gastrointestinal cancer, liver disease, and some patients with glomerulonephritis. The reference range for serum albumin is 3.5–5.0 g/dL, with women having approximately 0.5 g/dL less albumin than men. Increases in serum albumin are uncommon, being seen primarily in sunstroke, dehydration, following severe exercise, and in multiple sclerosis or hypothyroidism. Decreases in serum albumin are often observed during pregnancy, following severe burns, and in patients with liver or kidney disease or malabsorption.

Urine protein can be measured semiquantitatively by turbidimetry or with a dipstick dye-binding method. Quantitation of urine total protein is accomplished through turbidimetric measurement. The presence of bacteria, sulfa drugs, penicillin, cephalosporin, tolbutamide, or x-ray contrast media in the urine produces false positive increases in the measurement. Increases in urine total protein may be seen in patients with high blood pressure, congestive heart failure, kidney damage, or multiple myeloma, or increases may be seen following exposure to certain toxic compounds. Elevated albumin in urine may result from kidney damage, from cancer, from infections, or as a complication of pregnancy.

CSF total protein may be measured using turbidimetry or by dye-binding with Coomassie brilliant blue. Normal values for CSF total protein are approximately 15–45 mg/dL. Elevations are seen in both bacterial and viral infections and in multiple sclerosis.

FOR REVIEW

Directions: For each question, choose the best response.

1. The primary structure of a protein is amino acids
   A. in a helical formation
   B. linked by covalent bonds
   C. linked by peptide bonds
   D. linked by S—S bonds

2. The term *zwitterion* refers to an amino acid
   A. that is positively charged
   B. that is negatively charged
   C. that has both positive and negative charges on the molecule at the same time
   D. that has a net charge of zero

3. The pH value at which the sum of the electric charges on a protein equals zero is referred to as the
   A. balanced point
   B. equalization point
   C. isoelectric point
   D. zwitterion point

4. Which of the following functions may be ascribed to proteins?
   A. act as catalysts
   B. form antibodies
   C. provide structural support for tissues

    D. transport other compounds in the blood

    E. all of the above

5. Proteins are absorbed from the small intestine into the circulation as

    A. intact proteins

    B. polypeptides

    C. single amino acids

    D. chains of 100 amino acids

6. When using the biuret method for quantitating serum total protein, the intensity of the color produced is dependent on the

    A. acidity of the reaction products

    B. molecular weight of the protein

    C. nitrogen content of the protein

    D. number of peptide bonds that react

7. A dye-binding technique commonly employed for the quantitation of serum albumin is

    A. biuret

    B. bromcresol green (BCG)

    C. Coomassie blue

    D. sulfosalicylic acid

8. A turbidimetric method used for the quantitation of total protein in urine and cerebrospinal fluid specimens is

    A. biuret

    B. 2-(4′-hydroxyazobenzene)-benzoic acid (HABA)

    C. Coomassie blue

    D. sulfosalicylic acid

9. All of the following when present in a specimen may cause a false elevation of serum total protein when the biuret method is employed *except*

    A. ammonia

    B. bilirubin

    C. hemoglobin

    D. lipids

10. Decreased serum albumin levels may be associated with

    A. malnutrition

    B. liver disease

    C. kidney disease

    D. both B and C

    E. A, B, and C

11. A protein that precipitates in acid solution but redissolves upon heating best describes

    A. albumin

    B. Bence Jones

    C. haptoglobin

    D. transferrin

12. The term *transudate* may be associated with

    A. a protein concentration greater than 3 g/dL

    B. pleural fluid

    C. a change in the filtering ability of cell membranes

    D. both B and C

    E. A, B, and C

13. Elevated total protein levels in cerebrospinal fluid may be associated with

    A. bacterial infections

    B. viral infections

    C. multiple sclerosis

    D. both A and B

    E. A, B, and C

14. Which of the following proteins seldom exhibits an increased level in the blood?
    A. alpha$_1$ globulin
    B. alpha$_2$ globulin
    C. beta globulin
    D. albumin

15. Which of the following colorimetric methods may be used to quantitate total protein in cerebrospinal fluid?
    A. bromcresol green
    B. Coomassie blue
    C. sulfosalicylic acid
    D. both B and C
    E. A, B, and C

16. Which of the following may be associated with alpha fetoprotein?
    A. measured in amniotic fluid
    B. assay performed routinely by all clinical labs
    C. used to detect neural tube defects
    D. both A and C
    E. A, B, and C

BIBLIOGRAPHY

American Association for Clinical Chemistry, "Effects of disease on clinical laboratory tests," *Clin. Chem.* 26(4), supplement, March 1980.

Carone, F. A. et al., "Renal tubular transport and catabolism of proteins and peptides," *Kidney Intern.* 16:271–278, 1979.

Cotran, R. S., and Rennke, H. G., "Anionic sites and the mechanism of proteinuria," *New Eng. J. Med.* 309:1050–1052, 1983.

Freeman, H. J. et al., "Protein digestion and absorption in man: Normal mechanisms and protein-energy malnutrition," *Amer. J. Med.* 67:1030–1036, 1979.

Goldberg, D. M., "Hepatic protein synthesis in health and disease, with special reference to microsomal enzyme induction," *Clin. Biochem.* 13:216–226, 1980.

Gyure, W. L., "Comparison of several methods for semiquantitative determination of urinary protein," *Clin. Chem.* 23:876–879, 1977.

Henry, R. J. et al. (ed.), *Clinical Chemistry: Principles and Technics* (2nd ed.). Hagerstown, MD: Harper and Row, 1974.

Johnson, J., and Lott, J. A., "Standardization of the Coomassie blue method for cerebrospinal fluid proteins," *Clin. Chem.* 24:1931–1933, 1978.

Klein, B., "Standardization of serum protein analyses," *Ann. Clin. Lab. Sci.* 8:249–253, 1978.

Matthews, D. M. et al., "Protein absorption—Then and now," *Gastroenterol.* 73:1267–1279, 1977.

McElderry, L. A. et al., "Six methods for urinary protein compared," *Clin. Chem.* 28:356–360, 1982.

Pesce, A. J., "Methods used for the analysis of proteins in the urine," *Nephron* 13:93–104, 1974.

Peters, Jr., T., "Serum albumin: Recent progress in the understanding of its structure and biosynthesis," *Clin. Chem.* 23:5–12, 1977.

Rosenfeld, L. *Origins of Clinical Chemistry. The Evolution of Protein Analysis.* New York: Academic Press, 1982.

Rothschild, M. A. et al., "Albumin metabolism," *Gastroenterol.* 64:324–337, 1973.

Stryer, L. *Biochemistry* (3rd ed.). New York: W. H. Freeman and Co., 1988.

Tietz, N. W. (ed.), *Textbook of Clinical Chemistry.* Philadelphia: W. B. Saunders, 1986.

Chapter Six

PROTEIN CHEMISTRY II: ELECTROPHORESIS TECHNIQUES

UPON COMPLETION OF THIS
CHAPTER, THE STUDENT WILL BE
ABLE TO

1. Define the following terms:
   A. cation
   B. anion
   C. cathode
   D. anode
   E. electrophoresis
   F. zone electrophoresis
   G. polyacrylamide gel electrophoresis
   H. isoelectric focusing
   I. isoelectric point
   J. ampholyte
   K. electroendosmosis

2. Discuss how the following factors influence the electrophoretic rate of migration:
   A. net charge of the molecule
   B. size and shape of the molecule
   C. electric field strength
   D. nature of the support medium

3. Name the three basic components of an electrophoresis system.

4. Explain the general procedure used for performing electrophoresis.

5. State the purpose of the support medium.

6. Discuss the advantages and disadvantages associated with using each of the following types of support media:
   A. paper
   B. starch gel
   C. cellulose acetate
   D. agarose
   E. polyacrylamide gel

7. List four techniques that may be employed to aid in the identification of protein fractions following electrophoretic separation.

8. Name two colored dyes traditionally used to identify protein fractions and two stains used to identify lipoprotein patterns.

9. Explain why Coomassie blue and silver stains are used to detect urine and CSF protein fractions.

10. Explain the use of densitometry for quantitating electrophoretic fractions.

11. Compare and contrast zone electrophoresis, polyacrylamide gel electrophoresis, isoelectric focusing, and two-dimensional electrophoresis in relationship to
    A. principle of system
    B. type of support medium employed
    C. types of proteins separated
    D. clinical usefulness

12. Discuss the influence of the following parameters on protein electrophoresis:
    A. buffer
    B. ionic strength
    C. voltage
    D. current
    E. heat

13. Discuss the phenomenon of electroendosmosis and its effect on protein fraction separation.

14. Name the five basic bands seen on agarose for a normal serum following protein electrophoresis (use the anode as a reference point).

15. Identify the serum protein fraction that migrates the greatest distance on agarose electrophoresis (assume cathodic application of the sample).
16. Name the specific proteins that migrate with each of the globulin fractions and their functions.
17. Describe how bisalbuminemia looks on agarose following electrophoresis.
18. Describe the serum protein electrophoresis patterns characteristic of the following disorders:
    A. acute inflammation   D. acute cirrhosis
    B. chronic inflammation   E. renal protein loss
    C. hepatitis
19. State the migration patterns (use the anode as a reference point) of the normal and abnormal hemoglobin fractions when hemoglobin electrophoresis is performed at pH 8.6 on either cellulose acetate or agarose and at pH 6.2 on citrate agar.
20. Identify the type of hemoglobin present, the amino acid and/or chain variation, and the clinical manifestations associated with
    A. sickle-cell disease   C. alpha thalassemia
    B. sickle-cell trait   D. beta thalassemia

INTRODUCTION

From the clumsy moving-boundary apparatus invented by Arne Tiselius in 1926, to the sophistication of current support media, the technique of electrophoresis has made significant strides in the last 50 years. Where we once were satisfied with five poorly resolved bands, we now obtain so many protein fractions a computer is required to keep track of them.

The ability to separate mixtures of proteins rapidly and reliably has opened the door to significant advances in clinical chemistry. Specific protein markers for a number of diseases can now be identified, abnormal proteins can be detected, and changes in the distribution of proteins give valuable clues to the underlying pathological situation.

6.1

FUNDAMENTALS OF PROTEIN ELECTROPHORESIS

Movement of Charged Particles

In an electric field, any charged particle (either positive or negative) will migrate to one pole or the other. If a particle has a positive charge (**cation**), it will move toward the negative pole (**cathode**). Conversely, a negatively charged particle (**anion**) will migrate in the direction of the positive pole (**anode**). The rate of migration depends on several factors, including the amount of charge on the molecule, the

size and shape of the molecule, the strength of the electric field, and the support medium employed in the electrophoresis procedure.

Protein Structure and Electrophoretic Mobility

As was discussed previously, proteins are composed of amino acids. Certain amino acids have side chains which may possess an electric charge at a given pH. Each protein has a specific net charge on the molecule depending on the number of these charged groups present. Two representative examples are given in Figure 6–1. We see that protein one has four groups that are negatively charged, but protein two has only three such groups. In addition, protein two contains a side chain with a positive charge. The net charge of protein 1 is −4, whereas the net charge on protein two is −2 (−3 + 1). We would then expect protein one to move twice as far in an electric field as protein two (all other factors being equal).

Electric charge is not the only influence on protein migration, although it is a major contributor. In most techniques, molecular shape plays a role in determining how far a given protein moves. We cannot make general statements about this parameter, because the various techniques are affected in different ways by the protein shape (whether it is globular, fibrous, or something in between). Some electrophoretic methods use this property of a protein as much as the charge on the molecule in order to achieve separation of various fractions.

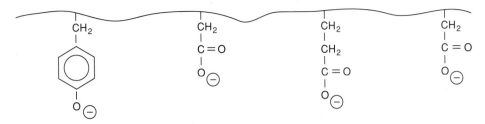

Protein one. Relative charge = −4

Protein two. Relative charge = −2

FIGURE 6-1   Protein structure and electric charge.

Components of Electrophoresis Systems

There are three basic components to any electrophoresis system (Fig. 6–2):

1. support medium
2. electrophoresis chamber
3. power supply

In addition, some technique (either chemical or immunochemical) is needed to detect the various fractions separated by the procedure. Every specific system is simply some variation on the above theme.

The purpose of the support medium is to hold the sample and provide a path for the migration of the proteins. The medium may be a passive one, with little or no physical interaction with the protein other than support. In other systems, the support

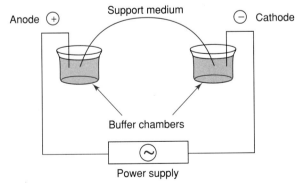

FIGURE 6-2   A general electrophoresis system.

medium plays an important role in the separation process, either by serving as a sieve (separation on the basis of size) or by affecting the charge in the system (separation on basis of net electric charge). For some support media (such as cellulose acetate), a presoak in buffer is required prior to sample application. Agarose and other gel media do not require a buffer pretreatment, since buffer is incorporated into the gel during the initial preparation.

The electrophoretic separation takes place on the support medium. The material must not be chemically reactive to any significant degree, either to the biochemicals in the sample or to the dyes used to locate the proteins after separation. After a sample is applied to the support material, both ends of the material are placed in contact with buffer solution to allow passage of the electric current. Unless specifically indicated, the medium also has the same buffer incorporated into it.

The sample on the support medium is then placed in the electrophoresis chamber. There are generally partitions in the chamber for buffer. The ends of the medium are placed in buffer in separate buffer partitions; the support material then serves as a bridge between the two areas of buffer. Wiring of the chamber allows passage of electric current into one buffer chamber, where the electric flow then goes across the support material and into the second buffer chamber. Most systems are equipped with safety devices so that removal of a piece of the apparatus disconnects the flow of electricity, minimizing the danger of electric shock to the operator.

The power supply converts alternating line current into direct current for the operation of the system. This piece of equipment often plays a step-down role also. Voltages across the apparatus are frequently lower than those supplied by the line current. In addition, most power supplies provide constant voltage regulation so that the electric field remains very stable throughout the analysis. Many electrophoresis systems today are sold as a unit, with both chamber and power supply designed specifically for the support medium marketed as part of the overall system.

Detection of Proteins After Separation

A key ingredient to the success of any electrophoretic technique is its ability to detect the separated proteins on the support medium (Table 6-1). Traditional approaches to this problem have involved the use of nonspecific protein stains such as amido black and S Ponceau. These dyes bind to essentially all proteins; the color of the dye is seen wherever a protein is located. If enzymes are being studied, a reaction which relates to the function of that enzyme may be used to highlight sites of interest on the support medium. This approach allows location and identification of specific proteins without the "noise" generated by other materials which we do not want to consider at the time. Proteins which are not enzymes can be located with the use of specific antibodies (see Chapter 7) in a process called **immunofixation,** followed by staining to visualize the protein bands. Again, only the protein or proteins which we wish to locate are noted. More recently, silver staining has become a very sensitive method for the detection of extremely low levels of proteins. This detection system is particularly useful for urine and cerebrospinal fluid (CSF) protein electrophoresis. Each technique will be discussed further in the context of specific electrophoretic approaches in this and subsequent chapters.

Quantitation of Protein Fractions by Densitometry

A variety of approaches have been taken over the years to assess the level of a given protein (or proteins) after electrophoresis and staining. For some purposes, a simple visual examination is sufficient. The presence or absence of certain fractions is noted and no other data are required. This type of report is rarely useful in clinical situations. Cutting the electrophoresis strip and eluting each fraction into solution for further chemical analysis provides reasonably accurate results, but is very time-consuming. The method of choice in the clinical laboratory is **densitometry,** the measurement of the density of light passing through the fractions.

Most densitometric systems are based on colorimetry. The protein fractions are stained with a material which produces visible bands of color on the support medium. The strip is placed in a holder and is slowly moved through a beam of light (Fig. 6-3). In most instances, filters are used to permit only light of a certain wavelength range to strike the strip. As the light interacts with the dye bound to the specific protein fraction, a certain amount of light is absorbed, depending on the concentration of the protein and the amount of dye bound to it. Absorption of light is roughly proportional to concentration of dye (and, therefore, to concentration of protein fraction).

The readout from the system is a trace (Fig. 6-4) which relates the amount of light absorption to the location of protein fractions. The data can be further analyzed electronically to provide a percentage distribution of each fraction on the electrophoresis strip.

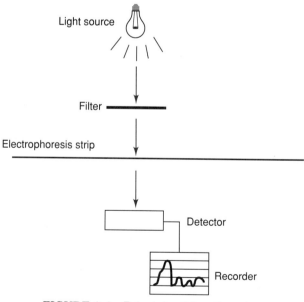

FIGURE 6-3 Principle of densitometry.

Table 6-1
IDENTIFICATION OF PROTEIN FRACTIONS AFTER ELECTROPHORESIS

Technique	Specificity	Comments
Colored dyes	All proteins	Nonspecific Sensitivity low
Enzyme reactions	Enzyme of interest	Specific Sensitivity varies
Immunofixation	Specific proteins	Specific Sensitivity varies
Silver staining	All proteins	Excellent for low-level detection

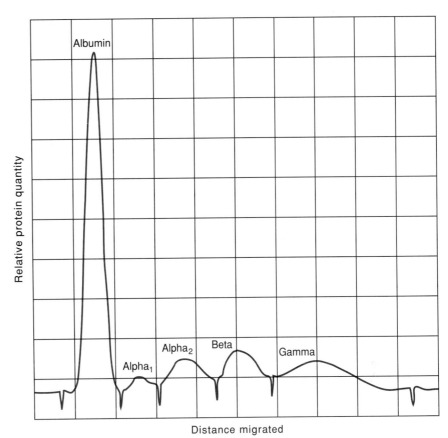

FIGURE 6-4   Densitometric scanning of protein electrophoresis. (From Fauchier, P., and Catalan, F., *Interpretive Guide to Electrophoresis: Important Normal and Abnormal Patterns.* 2nd ed. Beaumont, TX: Helena Laboratories, 1988, p. 3. Reprinted by permission of the publisher.)

6.2

ELECTROPHORETIC SUPPORT MEDIA AND TECHNIQUES

Paper

Perhaps the earliest support medium used for clinical studies was paper, introduced in 1950. This **zone electrophoresis** technique (so named because discrete zones of proteins were isolated) gave the first clear picture of serum proteins in health and disease. Of only historical interest now, the use of paper as a support medium was plagued by a number of problems. The paper was fragile and could be easily damaged. Since the material was quite porous, there was significant variability in the degree of separation and in the reproducibility of results. Staining of protein fractions was quite variable owing to inconsistencies in the composition of the paper. The mechanical properties of the medium were such that albumin would be seriously underestimated, whereas other fractions were overestimated. However, the history of the technique and the results achieved with this very limited material make for fascinating reading by all serious students of science.

Starch Gel

One attempt to overcome the inadequacies of paper as a support medium led to the development of **starch gel electrophoresis.** By using a starch base, more uniformity in support medium from lot to lot could be achieved. Because rather thick slabs were used (in contrast to the very thin paper), larger samples could be employed. This factor was especially important if preparative separations were being done. However, the advantages of greater uniformity and larger sample size did not compensate for the fragility of the material, the inability to store results permanently, and the need to prepare the gels in-house. Starch gel electrophoresis is still employed in some research applications but never found a role in the clinical laboratory.

Cellulose Acetate

Because of the drawbacks associated with paper electrophoresis, a search was launched for a more satisfactory support medium. In 1957, workers in England developed the technique of **cellulose acetate electrophoresis.** This approach soon became the standard method for protein electrophoresis until it was supplanted by superior techniques in the 1970s. The first commercial apparatus for cellulose acetate electrophoresis was marketed in the early 1960s.

Although the basic techniques remain the same, the cellulose acetate support medium has some significant differences from paper. As acquired from many companies, the strip consists of a clear plastic backing (not all commercial preparations have this type of backing) with a coating of cellulose acetate particles attached to it. Since the cellulose acetate is manufactured, the size and structure of the particles can be carefully controlled. Protein is applied as a narrow band onto the strip (giving better resolution than the earlier spot approach). After electrophoresis at pH 8.6 (separation times are usually 30–45 min), the fractions are stained to locate protein. One advantage of cellulose acetate over paper is that the strip can then be cleared of cellulose acetate, leaving the developed protein bands on a transparent plastic background. A permanent record of each electrophoresis can then be maintained.

The other major advantages over paper are the greater resolution and faster separation made possible with cellulose actate. In part because of the development of improved apparatus, temperature and other factors can be more closely controlled, allowing for faster electrophoresis procedures. Because of the uniformity of the particles, the bands are sharper and better separated (resolution is improved). In addition, albumin does not adsorb to the cellulose acetate to any great extent (as it would with paper), so the tailing seen earlier with albumin is eliminated. This decrease in adsorption also means that both albumin and other fractions are more accurately determined.

Proteins on cellulose acetate can be detected by a variety of techniques. Some involve treating the film first with dilute acetic acid to precipitate ("fix") the proteins. This technique decreases protein diffusion and spreading while the material is in the staining solution. A dye in dilute acid solution is then added to visualize the various proteins on the film. Alternatively, the fixing and staining steps are combined into one with the use of an appropriate acid solution of the dye. S Ponceau in 5% trichloroacetic acid has been widely used, giving a red color to the protein bands. In recent years, the use of amido black (naphthol blue black) has gained popularity. The dark blue color of amido black allows more sensitive visualization of protein fractions. After staining with dye, the strips are again soaked in a dilute acid solution to remove dye which has not reacted with protein. The resulting pattern can be scanned with a densitometer to determine relative amounts of the proteins seen on the strip.

However, cellulose acetate has some major drawbacks as a support medium. When dried, this material becomes extremely brittle, making it hard to work with and store. Some of the chemicals used to prepare the cellulose acetate strips inhibit developing reactions, particularly those for the visualization of enzymes. If a fluorometric development technique is employed, the background fluorescence of the cellulose acetate must be compensated for. The clearing reactions in some instances remove the developed colored bands (again, this is a major problem with some routinely studied enzymes of clinical interest). So the search for a more satisfactory material continued.

Agarose

The first reported use of agar as an electrophoresis support medium was in the 1920s. Separation of isotopes was attempted in agar-filled tubes which were 9 ft long, with each procedure running for several days. Techniques have improved markedly since that time, and a derivative of agar (**agarose**) has enjoyed immense popularity as an electrophoresis support medium.

Agar is a mixture of polysaccharides derived from seaweed; agarose is the predominant component of that mixture. A major advantage of agarose as a support medium is its electric neutrality. There is minimal electric charge on the agarose polymer itself, so little or no interaction with proteins is expected. Since the material is relatively porous, separation of proteins takes place strictly on the basis of electric charge. It is possible to purify the agarose fraction, which results in greater uniformity of materials and better reproducibility of electrophoretic patterns.

The procedure with agarose is much the same as with cellulose acetate. Samples for serum protein fractionation require an application volume of less than 1 μL. Urine and CSF samples require preconcentration and possibly a larger sample size. If isoenzyme fractionation is desired, samples greater than 1 L are frequently required. The buffer employed is a dilute barbital buffer at pH 8.6; at this pH, all the charge on most protein molecules is negative. Staining and destaining can be easily accom-

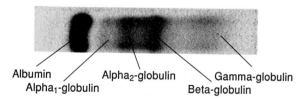

Albumin   Alpha$_2$-globulin   Gamma-globulin
Alpha$_1$-globulin   Beta-globulin

FIGURE 6–5   Serum protein pattern on agarose gel. (From Carstens, K. S., Sepulveda-Pacheco, A. M., and Romfh, P. C., *Introduction to High Resolution Protein Electrophoresis and Associated Techniques.* Beaumont, TX: Helena Laboratories, 1986, p. 2. Reprinted by permission of the publisher.)

FIGURE 6-6 High-resolution protein electrophoresis. (From Carstens, K. S., Sepulveda-Pacheco, A. M., and Romfh, P. C., *Introduction to High Resolution Protein Electrophoresis and Associated Techniques.* Beaumont, TX: Helena Laboratories, 1986, p. 2. Reprinted by permission of the publisher.)

plished in dilute acetic acid, followed by drying of the film at 70°C. The protein stain of choice for agarose is amido black. If CSF samples are run, Coomassie blue provides good staining for the resulting fractions. There are a number of commercial set-ups available, with a given manufacturer providing the entire package: apparatus, films, drying oven, chemicals, and densitometer for determining the relative amounts of protein on the strip. Figure 6-5 illustrates the pattern obtained for a serum protein electrophoresis using standard techniques employing agarose. Electrophoresis on cellulose acetate provides the same type of pattern.

Modifications of the buffer and the apparatus now permit the detection of 12 to 15 different protein fractions (Fig. 6-6). With the addition of calcium lactate to the buffer (still at pH 8.6) and the addition of a cooling system, higher voltages can now be used for the separation without damage to the gel or the proteins. This **high-resolution protein electrophoresis** is increasingly becoming the accepted approach to routine protein fractionation in the clinical laboratory.

Polyacrylamide Gel Electrophoresis

The previously described techniques permitted the separation of protein based solely on the differing electric charges of the molecules. Although protein fractionation using paper and cellulose acetate as support media was influenced to a small extent by the size of the molecule, the charge on the molecule was the major factor in determining electrophoretic mobility. Separation on agarose also was not affected to any great extent by protein size. In contrast, **polyacrylamide gel electrophoresis (PAGE)** uses protein size as a major factor in the separation process. Protein fractionation in PAGE is based both on the molecular shape and the net charge of the proteins in the mixture. This technique yields more detailed patterns of proteins than does agarose gel electrophoresis.

The key to separation on the basis of molecular size lies in the composition of the gel used as the support medium. This gel is composed of an acrylamide polymer, long-chain molecules which crosslink to a certain extent. As the proteins in the sample migrate through this polymer, driven by the electric current, some of them are held up more than others by the polymer matrix. Large molecules find it more difficult to move through this system, but smaller molecules are not retarded as much by the sieving effect of the gel. For a given total molecular charge, the smaller molecule moves faster than the larger one through the gel. So we have a combination of effects. The greater the molecular *charge,* the faster the protein moves. Conversely, the larger the molecular *size,* the slower its movement through the gel. The net result is a more detailed separation of protein molecules and a more complex pattern resulting from the separation.

At one time PAGE was widely used for the separation of lipoproteins, hemoglobin, and some isoenzymes. Because of the complexity of the technique (particularly in the gel preparation steps), this approach has been largely supplanted by agarose for routine work. However, there are still a number of specialized applications where PAGE has proven to be a valuable approach to protein fractionation.

Isoelectric Focusing

Whereas other methods for protein separation have been based on some aspect of the net charge of the molecule, **isoelectric focusing** exploits a different parameter associated with the protein charge: the isoelectric point. Each protein has a number of potentially charged groups on the molecule, both positive and negative. As we discussed earlier, the net charge on the protein molecule is determined (in part) by the pH of the solution.

For each protein molecule (as is also the situation for each amino acid), there is a pH where the net charge on the molecule is zero; this value is called the **isoelectric point** for that protein. At pH values greater than the isoelectric point, the net charge on the protein is negative. At lower pH values, the protein has a positive net charge. But at the isoelectric point, the molecule has no net charge. The protein does not become electrically neutral at the isoelectric point, but the various local charges at different points on the protein cancel each other, giving a net charge of zero for the entire molecule.

"Routine" protein electrophoresis employs a buffer incorporated into the support medium. This buffer is usually at a basic pH (often pH 8.6), giving a net negative charge to the protein. The buffer conditions are constant across the gel (or other support medium), so the charge on the protein remains relatively constant throughout the separation procedure. In isoelectric focusing, a different approach is used. Special molecules are incorporated into the system to stabilize the pH at a given point in the separating medium. These molecules (called **carrier ampholytes**) impart specific pH values to the region where they are located. An **ampholyte** is usually a structure with 300–600 mol. wt. Varying numbers of carboxyl and amino side chains are incorporated into the molecule to provide the desired charge on the ampholyte at a given pH. When a gel is prepared which contains carrier ampholytes, the pH in one area of the gel is not the same as it is in another region. It is possible to generate pH gradients as small as 0.001 pH unit within a gel with the proper selection of ampholytes.

To perform isoelectric focusing, protein and ampholytes are mixed together and applied to the gel. The ampholytes distribute themselves across the gel according to their pH values. As the protein moves across the gradient, it slows down as it enters a region close to its isoelectric point. When the molecule moves into a region where its net charge becomes zero, migration stops. If it diffuses into another area, the net charge on the molecule changes and the protein migrates back to its isoelectric point under the influence of the electric field. The resulting pattern is a series of very fine bands of protein, each one separated from the others on the basis of its unique isoelectric point.

There are a number of clinical applications available for isoelectric focusing. Major uses appear to be in the study of immunoglobulins, either in the characterization of the normal molecules or in the fractionation of these proteins in clinical situations. This technique has proved to be of great value in examining the various abnormal proteins found in CSF in patients with multiple sclerosis. Applications have also been developed for the study of isoenzyme distribution. Alkaline phosphatase isoenzymes in various body fluids have been examined with this technique. Isoelectric focusing will become increasingly valuable as a laboratory diagnostic tool.

Two-Dimensional Protein Fractionation

Routine separation of proteins on agarose can provide information about five major fractions of interest. High-resolution electrophoresis at least doubles that capability. If we move to polyacrylamide gel or isoelectric focusing, the increase in separating power shows us even more protein fractions. However, many body fluids contain several thousand unique proteins. Although a single technique cannot detect all these different entities, the combination of two different methodologies provides the capability to examine most of these thousands of proteins in a single analysis.

Two-dimensional protein electrophoresis is a synthesis and modification of two existing methodologies: isoelectric focusing and polyacrylamide gel electrophoresis. Proteins are first separated on the basis of their isoelectric points in a tube gel. The re-

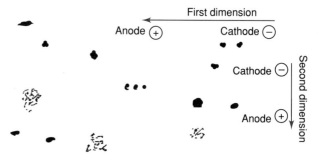

FIGURE 6–7   Two-dimensional protein fractionation.

sulting fractionation is then transferred to a slab for further treatment by polyacrylamide gel techniques. Patterns are usually detected with silver staining. A theoretical resolution of over 10,000 proteins is possible with this analytical approach.

The resulting pattern (Fig. 6–7) is a complex one. Many hundreds of different fractions are seen and the usual techniques for examination are not adequate or accurate. Sophisticated computer-driven scanners have been developed to examine the patterns, store the data, and do pattern comparison to quantitate materials and to detect the presence or absence of specific proteins. Although two-dimensional electrophoresis is currently a research tool, it may be of great value in studying some clinical situations.

6.3

PROBLEMS ASSOCIATED WITH PROTEIN DETECTION AND MEASUREMENT

Buffer Concentration and Ionic Strength

Mobility of proteins depends greatly on buffer concentration and the total charge composition of the system. Generally, a decrease in buffer concentration produces an increase in mobility of protein fractions, with a broadening of each band. If we increase the buffer concentration, the bands do not move as far in a given time but are much narrower. The ionic strength is related to buffer concentration and to the types and amounts of ionic charges present. Not only buffer concentration, but also buffer type, is important. Two buffers may have the same molar concentration but very different ionic strengths owing to the specific composition of each buffer. The higher the ionic strength, the slower the movement of protein fractions.

Electroendosmosis

We commented previously on the differences among paper, cellulose acetate, and agarose with regard to the electric neutrality (or lack thereof) of the support medium. To achieve optimum separation of protein fractions, we must minimize **electroendosmosis,** a phenomenon relating to induced charges in the medium. When the electrophoresis material comes into contact with water, a small amount of negative charge may be induced in the molecules making up the medium. The extent of this behavior is not always predictable. Interaction of charged protein molecules with the charged groups on the medium alters the mobility of the protein to some extent. This change in migration rate may be reflected in a displacement of the band (it either moves ahead of or behind its usual location). More likely, we see band distortion (a wavy fraction) or spreading, with the band occupying a larger area and not appearing as a sharp, clearly defined fraction.

This generation of charged particles on the solid support matrix also affects the buffer ionic strength. Since the charges on the solid phase are fixed (they cannot move), movement of charge must be brought about in part by movement of buffer ions through the system. As the buffer ions move, the ionic strength changes, resulting in altered mobility of protein fractions and distortions in the resulting bands.

Heat and Voltage Effects

It is intuitively obvious that if we apply more voltage to the electrophoresis system, we move the proteins more rapidly. However, there are some limitations to increasing mobility (and shortening the time of the run) by carrying out the fractionation at higher voltages. If we increase the current, we also increase the temperature. Several problems begin to appear as the temperature rises. Proteins are labile molecules, easily denatured by higher temperatures (usually 50°C or higher). The mobilities of the fractions are also enhanced, leading to each band being in a different (and unexpected) location than the one seen at a lower temperature. Evaporation of solvent increases at a higher temperature, causing an increase in ionic strength of the buffer. This increase is particularly noticeable along the edges of the electrophoresis strip, where distortion and trailing of the band are very likely to occur. In high-resolution systems, a cooling device is frequently incorporated into the apparatus to regulate temperature closely so that reproducibility and optimal pattern development may be enhanced.

Staining and Scanning Problems

The proteins separated on electrophoresis are heterogeneous in nature. There are varying concentrations of the different proteins and different structures resulting in different dye-binding capabilities.

Some proteins require specific developing agents, and others can be detected on the strip or tube without the need for a visualizing reagent. Conditions for staining and observing protein bands must be carefully selected to achieve the desired purpose.

Stain concentration and reaction time need to be selected to allow development of all the bands present. Whereas a protein in low concentration may react sufficiently in a given time, another protein present at a much higher level may have inadequate development owing to insufficient reagent or reaction time. The background contributed by the support medium is important when proteins are to be quantitated by densitometric scanning. Scratches, spots, or local discolorations falsely alter the levels of certain proteins. Comparison of hemoglobin fractions can result in a significant underestimation of the minor components in the presence of the very large proportion of normal hemoglobin A_1. All methods need to be examined carefully to ensure that artifacts are not introduced, leading to erroneous values for protein quantitation.

6.4

SERUM PROTEIN ELECTROPHORESIS

Patterns Obtained with Various Techniques

Where we once saw a few protein fractions in serum, we can now identify many hundreds of these molecules. Some similarities and differences in patterns can be seen using the various electrophoretic techniques. Protein fractionation on agarose using classic methods yields five bands, with albumin migrating the greatest distance. Although high resolution produces additional bands, the overall fractionation pattern of albumin migrating most rapidly and the gamma globulins showing the least mobility still holds. Once we move beyond these approaches into PAGE, isoelectric focusing, and two-dimensional techniques, no clear, consistent pattern is available. The separations are very technique dependent, and all protein mobilities must be described in specific terms for that system. References to protein fractions are generally based on mobilities and properties as defined by the classic agarose ("5-band") approach.

Protein Fractionation on Agarose

We have already seen (see Fig. 6–5) the pattern of protein fractionation obtained by agarose gel electrophoresis. With the exception of the albumin fraction, all the other bands are mixtures of several proteins.

The albumin fraction may occasionally contain a separate prealbumin component. Under most circumstances, this small component is not visible. The prealbumin fraction contains two proteins which bind thyroxine and alpha$_2$ macroglobulin. If a drug has attached itself tightly to albumin, some of this derivative may be in the prealbumin fraction also.

The albumin zone consists essentially of one protein. There may be some variation in this fraction owing to genetic effects on protein structure or because of binding to albumin by other molecules. The albumin component is the largest single contributor to the total protein level in the body. Some genetic variants of albumin result in either two bands in the albumin region or a band which is broader than usual. This phenomenon is known as **bisalbuminemia.**

In the **alpha$_1$-globulin** zone, the major protein is alpha$_1$ antitrypsin, responsible for inhibiting the activity of the proteolytic enzyme trypsin (and some other related enzymes). Some lipoproteins migrate in this zone (usually closer to albumin). Other minor proteins are often seen in the alpha$_1$-globulin fraction or between this component and the alpha$_2$ globulins.

The **alpha$_2$-globulin** fraction has a number of physiologically significant proteins. Ceruloplasmin (copper transport), lipoproteins, haptoglobin (responsible for binding free hemoglobin), and alpha$_2$ macroglobulin are found in this zone. The role of alpha$_2$ macroglobulin is largely unknown, with some evidence suggesting that it may bind trypsin and insulin.

Beta globulins are another heterogeneous mixture of proteins. Lipoproteins, transferrin (binding and transport of iron), and hemopexin (transport free heme) are some of the important components of this fraction. The complement system (a key part of the immune response) is located in the beta-globulin fraction.

Most **immunoglobulins** (IgG, IgA, IgM, IgD, and IgE) are found in the **gamma-globulin** fraction (although some may be located in the beta-globulin portion). Since the various antibodies are located here, change in the concentration of the gamma-globulin fraction is often the first indicator of a problem in the production of these important proteins.

Table 6-2
SIGNIFICANT PROTEINS IN ELECTROPHORETIC FRACTIONS

Fraction	Specific Proteins
Prealbumin	Prealbumin
Albumin	Albumin
Alpha$_1$ globulin	Alpha$_1$ antitrypsin
	Lipoproteins
Alpha$_2$ globulin	Ceruloplasmin
	Haptoglobin
	Lipoproteins
	Alpha$_2$ macroglobulin
Beta globulin	Lipoproteins
	Transferrin
	Hemopexin
	Complement system
Gamma globulin	Immunoglobulins

Table 6-2 summarizes the locations of several important proteins seen in an electrophoresis pattern for serum protein.

6.5

CLINICAL APPLICATIONS OF SERUM PROTEIN ELECTROPHORESIS

General Principles

Protein electrophoresis patterns can provide quick and useful information regarding altered levels of protein groups within the circulation. Some disorders have characteristic patterns which are readily recognized (Table 6-3). In most situations, however, the role of the electrophoresis pattern in providing specific clinical information is limited. Many patterns are somewhat nonspecific; more than one disease state produces the same electrophoretic distribution of proteins. Although a qualitative change in the level of a specific protein may be observed, the electrophoretic data are not accurate enough to deal with the exact amount of the change, but can provide some measure of specificity. Interpretation of protein electrophoresis patterns in the absence of detailed clinical information about the

Table 6-3
SERUM PROTEIN ELECTROPHORESIS CHANGES IN COMMON DISORDERS

Clinical State	Albumin	Alpha$_1$	Alpha$_2$	Beta	Gamma
Acute inflammation	N, D	I	I	N	N, D
Chronic inflammation	N, D	I	I	N, I	I
Hepatitis	D	D	D	D	D
Acute cirrhosis	D	N	D	beta/gamma bridge	
Renal protein loss	D	N	I	N	N, D

N = Normal; D = Decreased; I = Increased.

patient is fraught with hazards and should be avoided. Because of the complexity of data and possible interpretations, only a brief overview of diagnostic possibilities will be presented here.

Inflammation

A number of biochemical changes occur during the process of inflammation, stimulated by infection and fever. In the acute phase, the body responds to the stress by increasing the concentrations of several proteins involved in dealing with infectious agents. The serum protein electrophoresis (Fig. 6-8) shows elevations in the alpha$_1$- and alpha$_2$-globulin fractions. The concentrations of both albumin and the gamma globulins are either normal or decreased.

Chronic inflammation arises from a variety of sources, including chronic infections, allergies, collagen disease and related disorders, malignancies, and autoimmune problems. The major difference in the serum electrophoresis pattern of acute and chronic inflammation is the increase in the gamma-globulin fraction (Fig. 6-9). The beta-globulin region may be normal or elevated, and the albumin fraction is either normal or decreased. Alpha$_1$ and alpha$_2$ globulins are both elevated in the chronic inflammatory state.

Hepatic Disorders

In liver disease of whatever cause, a major result of the hepatic damage is a decrease in protein synthesis. Many of the serum proteins (particularly al-

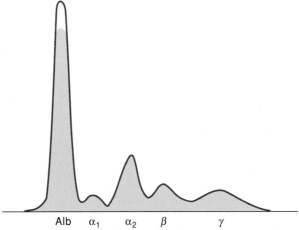

FIGURE 6-9   Serum protein electrophoresis—chronic inflammation. Note marked increase in alpha globulin. (From Fauchier, P., and Catalan, F., *Interpretive Guide to Electrophoresis: Important Normal and Abnormal Patterns.* 2nd ed. Beaumont, TX: Helena Laboratories, 1988, p. 5. Reprinted by permission of the publisher.)

bumin) are manufactured by the liver. When problems arise with this organ, low levels of albumin, lipoproteins, and other fractions occur (Fig. 6-10). Conversely, there are frequently increases in the gamma globulins owing to some infection. Patients with hepatitis manifest elevations of IgG, IgA, and possibly IgM. The characteristic pattern in cirrhosis is a single broad band in place of the distinct separation between the beta- and gamma-globulin fractions. This bridging effect is due to enhanced production of IgA.

Renal Protein Loss

In suspected loss of protein in the urine, both the serum and the urine protein electrophoresis patterns can provide useful information (Fig. 6-11). Low-molecular-weight materials (including albumin, transferrin, alpha$_1$ antitrypsin, and alpha$_1$ acid glycoprotein) are lost in the urine, but other proteins may be excreted if there is tubular damage. Albumin is frequently low in the serum electrophoresis pattern because of its loss into the urine. Some of the serum globulin fractions appear elevated because the albumin is low. Urine protein patterns show an elevation of albumin and possibly transferrin.

There are cases in which immunoglobulin synthesis is impaired and the intact immunoglobulin cannot be produced. The low-molecular-weight fragments are then seen in urine (and serum) in high concentrations. These monoclonal gammopathies are very diagnostic and exhibit themselves as discrete bands mainly in the gamma-globulin region. The use of electrophoresis in the diagnosis of multiple myeloma will be considered in Chapter 7.

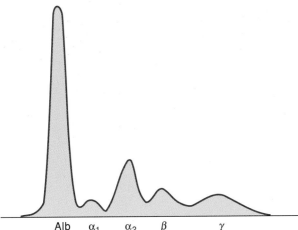

FIGURE 6-8   Serum protein electrophoresis—acute inflammation. Note the increases in alpha$_1$ and alpha$_2$ globulin. (From Fauchier, P., and Catalan, F., *Interpretive Guide to Electrophoresis: Important Normal and Abnormal Patterns.* 2nd ed. Beaumont, TX: Helena Laboratories, 1988, p. 4. Reprinted by permission of the publisher.)

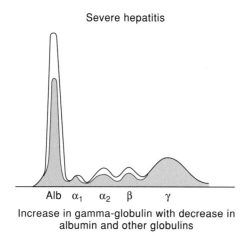

Severe hepatitis

Increase in gamma-globulin with decrease in albumin and other globulins

A

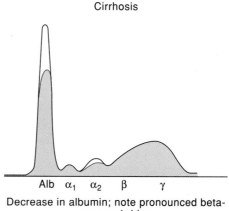

Cirrhosis

Decrease in albumin; note pronounced beta-gamma bridge

B

FIGURE 6-10   Serum protein electrophoresis—hepatic damage. A. Increase in gamma globulin with the decrease in albumin and other globulins. B. Decrease in albumin. Note pronounced beta/gamma bridge. (From Fauchier, P., and Catalan, F., *Interpretive Guide to Electrophoresis: Important Normal and Abnormal Patterns.* 2nd ed. Beaumont, TX: Helena Laboratories, 1988, p. 6. Reprinted by permission of the publisher.)

Lipoproteins

Fractionation of lipid-carrying proteins by electrophoresis enjoyed great popularity in the 1970s. By staining with a lipid-reacting material (Fat Red 7B or oil red O) instead of amido black or other protein reagent, the various lipoproteins could be easily detected. Relationships were developed between the presence and amount of each lipoprotein and the cholesterol or triglyceride levels in blood. Lipoprotein electrophoresis has declined in usefulness with the development of specific assays for the various lipid transport proteins in serum. Now that we can quantitate high-density or low-density lipoproteins directly, the less specific estimation by lipoprotein profiling is no longer extensively used.

Immunoglobulin Abnormalities

Serum protein electrophoresis cannot be used to quantitate the levels of the various immunoglobulins. We can roughly determine the overall immunoglobulin situation by examining the gamma-globulin fraction, but more specific techniques are required to quantitate any particular immunoglobulin. The major role of electrophoresis in the study of immunoglobulin abnormalities is to bring to light various abnormal fractions seen in serum or CSF. These atypical proteins are easily observed by electrophoresis, triggering further specific tests to identify the details of the underlying pathological state more clearly.

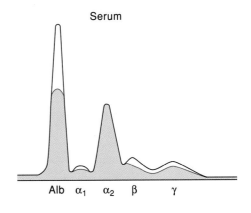

Serum

A

B

FIGURE 6-11   Serum and urine protein electrophoresis. Renal protein loss. A. (From Fauchier, P., and Catalan, F., *Interpretive Guide to Electrophoresis: Important Normal and Abnormal Patterns.* 2nd ed. Beaumont, TX: Helena Laboratories, 1988, p. 6. Reprinted by permission of the publisher.) B. (From Carstens, K. S., Sepulveda-Pacheco, A. M., and Romfh, P. C., *Introduction to High Resolution Protein Electrophoresis and Associated Techniques.* Beaumont, TX: Helena Laboratories, 1986, p. 32. Reprinted by permission of the publisher.)

Limitations of Electrophoretic Data

Before we developed the capability of assaying for specific proteins in serum or other body fluids, the major test which allowed any sort of assessment of detailed protein status was electrophoresis. We could gain useful information about classes of proteins, but in most cases were not able to quantitate any one protein with accuracy. In more recent years, a wide variety of procedures, equipment, and materials have become available for specific protein analysis. Protein electrophoresis no longer enjoys the premier status it once did. Electrophoretic patterns are now used as screening tests, to suggest further evaluation of a fraction or fractions by selective and specific analyses. There is still a great deal of debate about the relative merits of electrophoresis and specific protein assays, but more and more evidence (as well as changes in laboratory utilization and reimbursement patterns) points to the value of determining the quantity of certain proteins in any given disease state.

6.6

URINE PROTEIN ELECTROPHORESIS

Normal Studies

Urine from a healthy individual is usually characterized by a low level of total protein and an electrophoretic pattern (on agarose) that is diffuse and nondescript. A small amount of albumin may be observed, along with small amounts of other low-molecular-weight components. Quite often the urine sample needs to be concentrated several-fold before any distinct bands can be detected.

Clinical Utility

Although the analysis of urine protein by two-dimensional electrophoresis has revealed the presence of over 250 specific protein fractions, the use of urine protein fractionation in the routine clinical chemistry laboratory is limited. Other than the identification of immunoglobulin light chains in the urine of patients with elevated total protein who show diagnostic signs of multiple myeloma, urine protein electrophoresis is of marginal value in clinical diagnosis. Urine protein electrophoresis can indicate whether proteinuria is due primarily to albumin loss or to increased excretion of other proteins. Comparison of the urine electrophoresis pattern with a serum pattern from the patient may indicate something about the pattern of protein loss. Measurement of albumin excretion in urine is becoming increasingly important in monitoring diabetics, but is

better accomplished through a specific immunochemical assay.

6.7

CEREBROSPINAL FLUID PROTEIN ELECTROPHORESIS

Normal Studies

Spinal fluid normally has a total protein level of up to 45 mg/dL. This low amount indicates that concentration of the fluid is necessary before clearly discernable fractions can be seen on protein electrophoresis. Cerebrospinal fluid patterns are very much like serum, with the predominant fraction being albumin (Fig. 6–12). Other fractions are usually rather blurred and hard to quantitate with a densitometer. The major difference between serum and CSF is the presence of a very clear prealbumin fraction in samples from spinal fluid, amounting to some 4–5% of the total protein. Protein electrophoresis patterns of CSF can be distorted if there is difficulty in obtaining a clear specimen. Proteins from the blood obtained with the spinal fluid can seriously affect the results.

Clinical Utility

Diagnostic information from CSF protein electrophoresis is limited. The major purpose of performing CSF protein fractionation is to obtain information regarding the production of abnormal immunoglobulin bands seen in multiple sclerosis (discussed in Chapter 7). Occasionally, abnormal electrophoresis patterns in CSF may be seen secondary to changes in serum protein fractions. The

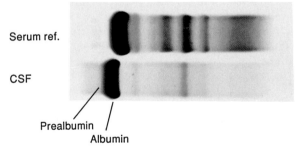

FIGURE 6–12   CSF protein electrophoresis. T.P. 44 mg/dL. Concentration factor 80x. (From Carstens, K. S., Sepulveda-Pacheco, A. M., and Romfh, P. C., *Introduction to High Resolution Protein Electrophoresis and Associated Techniques.* Beaumont, TX: Helena Laboratories, 1986, p. 7. Reprinted by permission of the publisher.)

clinical utility of these data are not clearly known at present.

6.8

HEMOGLOBIN ELECTROPHORESIS AND HEMOGLOBINOPATHIES

Hemoglobin Structure and Function

The protein **hemoglobin** (a major constituent of the erythrocyte) is responsible for the transport of oxygen through the circulation. Hemoglobin is composed of four polypeptide chains: two alpha chains and two beta chains. The four peptide subunits interact with one another to form the intact hemoglobin molecule. Formation of alpha and beta chains is regulated by separate genes; each type of chain is synthesized independently of the other type. Each chain also contains a porphyrin ring with an iron atom in its center. The iron is the site of binding for oxygen.

In the lung, oxygen attaches to the hemoglobin at the porphyrin site. Each hemoglobin molecule has the capacity to bind four oxygen molecules, one at each porphyrin. As the oxygen attaches to the porphyrin, the three-dimensional shape of the protein changes. The more oxygen which attaches, the more the protein conformation is altered. When the oxygen leaves the hemoglobin molecule, the shape returns to its original configuration.

There are three fractions of hemoglobin (often abbreviated as Hb) found in the normal red cell. Hemoglobin A is the major component (approximately 95% of the total), composed of two alpha and two beta chains. Hemoglobin A_2 consists of two alpha and two delta chains, and comprises up to 3.5% of the total hemoglobin in the red cell. The final 2% is hemoglobin F (two alpha and two gamma chains). Hemoglobin F is often referred to as **fetal hemoglobin,** since it is found in high concentrations in the red cells of newborns.

A wide variety of genetic disorders are known in which the amino acid sequence of either the alpha chain or the beta chain is altered owing to an error in the nucleic acid coding process for protein synthesis. These alterations frequently result in a decreased ability of the hemoglobin molecule to function properly. Oxygen transport may be impaired, red cells may break down sooner than normal, and cellular damage may be the end result of these abnormal processes.

Hemoglobin Electrophoresis

If the altered amino acid sequence produces a difference in the total charge on the hemoglobin mole-

cule, the abnormal hemoglobin can be separated from normal components by electrophoresis. The initial step in the process involves removal of hemoglobin from the erythrocyte. Blood is collected with an anticoagulant (several are suitable) and centrifuged to separate red cells from plasma. After the supernatant plasma is discarded, the cells are washed once with 0.155 M NaCl and centrifuged again. Distilled water (1:9 ratio of cells to water) is then added to the cellular precipitate to lyse the red cells. Centrifugation removes the stroma (red cell wall), leaving the hemolysate ready for electrophoresis.

The protocol for hemoglobin electrophoresis is similar to that for serum protein separations. Fractionation of hemoglobin takes place at pH 8.6 on either cellulose acetate or agarose. Either S Ponceau or amido black may be used for staining after the separation is complete. Since fractionation at pH 8.6 does not always reveal every abnormal fraction, a second separation at pH 6.2 on citrate agar is strongly recommended to obtain further data.

Hemoglobin Electrophoretic Patterns

At pH 8.6, Hb A migrates the furthest toward the anode (Fig. 6-13). Hb F is found somewhat behind Hb A, but Hb A_2 does not migrate very far from the origin. Several abnormal hemoglobin fractions (including Hb S) are found between Hb A_2 and Hb F. Hb C, Hb D, and Hb E travel the same distance as Hb S and cannot be fractionated at this pH.

Fractionation on citrate agar at pH 6.2 (Fig. 6-14) allows separation of hemoglobin fractions S, C, and D (among others). Although Hb A still moves toward the anode at pH 6.2, both Hb C and Hb S travel in the opposite direction. Hemoglobin S does not move far from the point of application, but Hb

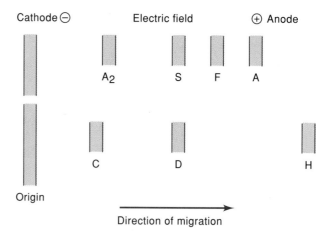

FIGURE 6-13 Hemoglobin electrophoresis at pH 8.6.

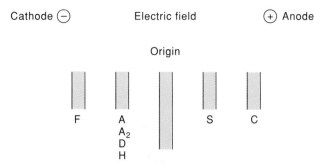

Cathode ⊖   Electric field   ⊕ Anode

Origin

F   A   S   C
     A₂
     D
     H

FIGURE 6–14   Hemoglobin electrophoresis (citrate agar at pH 6.2).

C moves further toward the cathode. Hemoglobin D comigrates with Hb A in this system, and Hb F moves the furthest toward the anode.

Sickle-Cell Disorder

Hemoglobin S is an abnormal hemoglobin in which the sixth amino acid in the beta chain is valine instead of glutamic acid. The presence of Hb S in the erythrocyte produces **sickle-cell disorder.** Although oxygenated Hb S is reasonably soluble, this hemoglobin precipitates from solution when oxygen is removed from the molecule. In the deoxygenated form, the shape of the molecule changes, allowing adjacent beta chains to interact and form long, insoluble polymers inside the red cell. These polymers precipitate from solution and distort the shape of the erythrocyte, producing the characteristic sickle shape of the cell.

Two forms of the disorder exist: sickle-cell trait and sickle-cell disease. The individual with **sickle-cell trait** is heterozygous for the disorder, having inherited the Hb S gene from one parent and the normal Hb A gene from the other. The hemoglobin electrophoresis pattern for this situation shows approximately 35% Hb S, with the remainder the normal Hb A. **Sickle-cell disease,** on the other hand, involves inheritance of Hb S genes from both parents (homozygous disorder). These patients show no Hb A in their electrophoresis studies, only Hb S. Frequently, higher-than-normal levels of Hb F are also seen in patients with sickle-cell disease.

These pathological states are found predominantly in Africans and people of African descent. Approximately 8% of the African American population have sickle-cell trait, whereas sickle-cell disease is found in less than 0.3% of this population. In some African countries, the incidence of the trait may be as high as 35% or more. The disorder can also be seen in small segments of some Mediterranean non-African populations. Several Latin American populations also demonstrate a high incidence of sickle-cell disorders.

Individuals with the trait may not manifest any significant medical problems over the course of their lives. Occasionally, a patient with sickle-cell trait may experience some problems in a low-oxygen environment (such as in an aircraft or other high-altitude situations), but this is an infrequent occurrence.

However, patients with sickle-cell disease exhibit a number of clinical difficulties. When oxygen content is low (particularly in the capillaries), sickling tends to occur. These "crises" lead to extreme pain in the joints, interruption of blood supply, thrombosis, and infarctions in bone, lungs, or brain. Death often is a result of an extreme sickle-cell crisis.

Thalassemia

In contrast to sickle-cell disorders, where an abnormal protein is synthesized, the **thalassemias** comprise a series of hemoglobin disorders in which one of the normal subunits is synthesized only in extremely low concentrations, or is not produced at all. In the alpha thalassemias, the formation of the alpha chain is impaired. Defects in beta-chain synthesis make up the group of beta-thalassemia syndromes. Excess production of the unaffected subunit may lead to formation of hemoglobin tetramers such as Hb H (four beta subunits) or Bart's Hb (four gamma subunits). The overall clinical effect of these disorders is a greatly diminished formation of red cells, with all the complications of severe anemia.

Screening for thalassemia involves a variety of hematology tests, in addition to hemoglobin electrophoresis. Hemoglobin F and A₂ analyses are crucial since these two fractions are frequently quite elevated in thalassemia. Electrophoresis may show an abnormal hemoglobin (Hb S is seen frequently), suggesting a mixed disorder. Studies of the individual globin chains provide useful information, but are usually carried out by a reference laboratory.

Other Disorders of Hemoglobin Production

Over 300 different hemoglobin mutations are presently known. In most instances, the amino acid substitution does not result in any deleterious effect. Genetic abnormalities resulting in synthesis of Hb C and Hb D are of clinical significance, since these disorders do have an adverse affect on the health of the individual. The clinical problems in any given hemoglobinopathy may be associated with stability of the hemoglobin molecule, the amount of hemoglobin available, or problems with formation of sufficient numbers of erythrocytes.

SUMMARY

Protein electrophoresis takes advantage of the fact that charged particles migrate in an electric field. Since proteins each have a net charge, they can be separated from one another using appropriate techniques. A simple electrophoresis set-up consists of a support/separation medium, a buffer, and a power supply. Following the protein fractionation, the different proteins can be detected using a variety of stains, including amido black, Coomassie blue (very useful for urine and CSF protein studies), or silver staining (extremely useful for low protein concentrations). Densitometric scanning allows quantitation of the different protein fractions.

A wide variety of separation approaches are available. The most commonly employed media are cellulose acetate and agarose. Proteins separate on the basis of molecular charge in both systems, but protein fractionation on the basis of size as well as charge is employed in polyacrylamide gel electrophoresis. Isoelectric focusing takes advantage of the isoelectric point of the proteins. Two-dimensional protein electrophoresis combines isoelectric focusing with subsequent further fractionation by size exclusion with polyacrylamide gel.

Changes in buffer concentration and ionic strength affect the migration patterns of proteins in a variety of ways. Although a higher voltage yields more rapid movement of proteins and better fractionation, the heat generated causes pattern distortion and destruction of protein fractions. All proteins do not stain the same way with a given dye, creating some difficulties with accurate quantitation of proteins and reproducibility of results.

The serum protein electrophoresis pattern using agarose or cellulose acetate observed in healthy individuals consists of a strong albumin peak and fairly well defined small peaks for the $alpha_1$, $alpha_2$, and beta globulins. The gamma-globulin fraction is a broad, diffuse band. Inflammation produces decreases in albumin and elevations of $alpha_1$ and $alpha_2$ globulins. If the inflammation is acute, gamma globulins may be normal or decreased; this fraction is usually elevated in chronic inflammation. In patients with liver disease, a lowering of the albumin fraction is frequently seen, accompanied by an increase in the gamma-globulin fraction. Occasionally a single band is observed in the beta/gamma region instead of the two distinct fractions normally present.

The normal urine protein electrophoresis pattern consists of albumin with very diffuse and low levels of the various globulin fractions. Urinary protein loss is characterized by elevations in albumin as well as one or more of the globulin fractions. Cerebrospinal fluid contains primarily albumin, with low levels of the various globulin fractions. In addition, a small prealbumin fraction may be observed. Increases in the gamma-globulin fraction may reflect multiple sclerosis or an infection.

Hemoglobin, a protein found in erythrocytes, is responsible for oxygen transport. This protein consists of four subunits (two alpha chains and two beta chains). Hemoglobin electrophoresis has proven invaluable in the detection of a variety of hemoglobinopathies. The normal pattern is primarily Hb A, with small amounts of Hb A_2 and Hb F. In sickle-cell disorders, Hb S appears, migrating behind Hb A. With sickle-cell trait, approximately 30–35% of the total hemoglobin present is Hb S, whereas in sickle-cell disease, 100% of the hemoglobin present is Hb S. The thalassemia syndromes are characterized by decreased production of either the alpha or beta subunit of hemoglobin. Increased amounts of Hb A_2 and Hb F may be seen in addition to an abnormal hemoglobin.

FOR REVIEW

Directions: For each question, choose the best response.

1. In an electric field, a molecule that is negatively charged migrates to the
   A. anode
   B. negative pole
   C. positive pole
   D. both A and B
   E. both A and C

2. Which of the following factors influences the rate of migration of a molecule in an electric field?
   A. amount of charge on the molecule
   B. size and shape of the molecule
   C. support medium employed
   D. both A and B
   E. A, B, and C

3. Electrophoresis performed on a solid support medium is known as
   A. curtain electrophoresis
   B. free electrophoresis
   C. moving-boundary electrophoresis
   D. zone electrophoresis

4. Which of the following is commonly used in the clinical lab as a support medium for electrophoresis?
   A. cellulose acetate
   B. agarose
   C. paper
   D. both A and B
   E. A, B, and C

5. Which dye may be used to stain serum protein fractions following electrophoresis?
   A. amido black
   B. S Ponceau
   C. fat red
   D. both A and B
   E. A, B, and C

6. The method of choice for quantitating protein fractions following electrophoresis is
   A. densitometry
   B. fluorometry
   C. nephelometry
   D. spectrophotometry

7. In polyacrylamide gel electrophoresis the separation process is primarily affected by protein
   A. charge
   B. size
   C. shape
   D. amino acid sequence

8. Which of the following may be associated with isoelectric focusing?
   A. functions at pH 8.6
   B. requires the use of ampholytes
   C. individual protein migration ceases at the isoelectric point
   D. both B and C
   E. A, B, and C

9. In zone electrophoresis when a buffer at pH 8.6 is used for the separation of serum proteins, the proteins will have a net
   A. positive charge and migrate toward the anode
   B. positive charge and migrate toward the cathode
   C. negative charge and migrate toward the anode
   D. negative charge and migrate toward the cathode

10. A congenital disorder characterized by a split in the albumin band when serum is subjected to electrophoresis is known as
    A. analbuminemia
    B. anodic albuminemia
    C. bisalbuminemia
    D. prealbuminemia

11. The immunoglobulins generally migrate electrophoretically as
    A. $alpha_1$ globulins
    B. $alpha_2$ globulins
    C. beta globulins
    D. gamma globulins

12. Ceruloplasmin and haptoglobin migrate electrophoretically as
    A. $alpha_1$ globulins
    B. $alpha_2$ globulins
    C. beta globulins
    D. gamma globulins

13. In cirrhosis a predominant characteristic observed on an electrophoretic serum protein pattern is a(n)
    A. increase in the $alpha_2$-globulin fraction
    B. bridging effect between the beta- and gamma-globulin fractions
    C. monoclonal band in the gamma-globulin region
    D. polyclonal band in the gamma-globulin region

14. In a healthy individual, the protein fraction present in the serum in the greatest concentration is
    A. $alpha_1$ globulin
    B. beta globulin
    C. gamma globulin
    D. albumin

15. Which disorder is generally characterized by an increase in the $alpha_1$ and $alpha_2$ globulins?
    A. acute cirrhosis
    B. acute inflammation
    C. hepatitis
    D. both A and B
    E. A, B, and C

16. When performing serum protein electrophoresis at pH 8.6, which fraction migrates the farthest toward the anode?
    A. $alpha_1$ globulin
    B. $alpha_2$ globulin
    C. gamma globulin
    D. albumin

17. When performing hemoglobin electrophoresis at pH 8.6, which fraction migrates the farthest toward the anode?
    A. Hb A
    B. Hb A_2
    C. Hb C
    D. Hb F

18. Hemoglobin electrophoresis fractionation on citrate agar at pH 6.2 is most useful in differentiating
    A. Hb A from Hb A_2
    B. Hb A from Hb D
    C. Hb C from Hb S
    D. Hb C from Hb A_2

BIBLIOGRAPHY

Allen, R. C., "Applications of polyacrylamide gel electrophoresis and polyacrylamide gel isoelectric focusing in clinical chemistry," *J. Chromatog.* 146:1–32, 1978.

Allen, R. C. et al., *Gel Electrophoresis and Isoelectric Focusing of Proteins.* Berlin New York: Walter de Gruyter, 1984.

Anderson, N. G., and Anderson, N. L., "The human protein index: A prospect for laboratory medicine," *Lab. Manage.,* June, 1982.

Fosslein, E. et al., "Two-dimensional electrophoresis: Recent applications and prospects," *Lab. Manage.,* October: 49–56, 1985.

Gordon, A. H., *Electrophoresis of Proteins in Polyacrylamide and Starch Gels.* Amsterdam: North-Holland Publishing Co., 1969.

International Committee for Standardization in Hematology, "Recommendations of a system for identifying abnormal hemoglobins," *Blood* 52:1065–1067, 1978.

Jeppsson, J.-O. et al., "Agarose gel electrophoresis," *Clin. Chem.* 25:629–638, 1979.

Jorgenson, J. W., "Electrophoresis," *Anal. Chem.* 58:743A–760A, 1986.

Katzmann, J. A., and Tracy, R. P., "Discovering marker proteins: A combination technology," *Lab. Manage.,* January, 1987.

Killingsworth, L. M. et al., "Protein analysis: The closer you look, the more you see," *Diag. Med.,* January/February: 47–59, 1980.

Lubin, B. H. et al., "Sickle-cell disease and the thalassemias: Diagnostic assays," *Lab. Manage.,* August: 38–47, 1980.

Macy, J., and Burke, M., "Hemoglobin separation by electrophoresis and the use of acid hemoglobin confirmation of variants," *Amer. Clin. Prod. Rev.,* June: 16–24, 1986.

Normansell, D. E., and Savory, J., "Diagnosis of congenital dysproteinemias," *Ann. Clin. Lab. Sci.* 12:154–157, 1982.

Orkin, S. H., and Kazazian, H. H., "The mutation and polymorphism of the human beta-globin gene and its surrounding DNA," *Ann. Rev. Genet.* 18:131–171, 1984.

Pribor, H. C., "Artificial intelligence in laboratory medicine. II: The electrophoresis profile," *Lab. Manage.,* August: 13–19, 1982.

Righetti, P. G., "Immobilized pH gradients," *Amer. Biotech. Lab.,* February: 22–27, 1988.

Rosenfeld, L., *Origins of Clinical Chemistry. The Evolution of Protein Analysis.* New York: Academic Press, 1982.

Schneider, R. G. et al., "Laboratory identification of hemoglobinopathies," *Lab. Manage.,* August: 29–43, 1981.

Simpson, C. F., and Whittaker, M. (ed.), *Electrophoretic Techniques.* New York: Academic Press, 1983.

Sun, T., "The present status of electrophoresis," *Lab. Manage.,* June: 43–49, 1979.

Sun, T. et al., "Clinical application of a high-resolution electrophoretic system. A review of electrophoretic patterns in disease," *Ann. Clin. Lab. Sci.* 8:219–227, 1978.

Tracy, R. P., and Young, D. S., "Two-dimensional gel electrophoresis: Methods and potential applications in the clinical laboratory," *J. Clin. Lab. Automation* 3:235–243, 1983.

Zak, B. et al., "Associated problems of protein electrophoresis, staining, and densitometry," *Ann. Clin. Lab. Sci.* 8:385–395, 1978.

Chapter Seven

PROTEIN CHEMISTRY III: IMMUNOCHEMICAL TECHNIQUES FOR THE ANALYSIS OF SPECIFIC PROTEINS

• • • • • • • • • • • • • • • • • •

UPON COMPLETION OF THIS CHAPTER, THE STUDENT WILL BE ABLE TO

1. Define the following terms:
   A. antigen   F. hapten
   B. antibody   G. polyclonal antibody
   C. heavy chain   H. monoclonal antibody
   D. light chain   I. monoclonal gammopathy
   E. autoantibody   J. oligoclonal bands
2. Name the five major immunoglobulin classes.
3. Identify the group of proteins that when separated by electrophoresis migrate in the gamma-globulin region.
4. Identify the portions of the heavy and light chains that determine the specificity of the antibody.
5. Describe how immunodiffusion techniques make use of the interaction between antigen and antibody.
6. Explain how the amount of precipitate formed is dependent on antigen and antibody concentrations and relate your response to antibody excess, equivalence, and antigen excess.
7. Describe the principles of the following immunochemical techniques:
   A. double diffusion   D. immunofixation
   B. radial immunodiffusion   E. rocket immunoelectrophoresis
   C. immunoelectrophoresis
8. Explain the principles of nephelometry and turbidimetry.
9. Contrast the positioning of the detector in relationship to the incident light beam in nephelometric and turbidimetric systems.
10. Explain how rate nephelometry differs from end-point nephelometry.
11. State the major function, reference range, and changes observed in disease states associated with each of the following serum proteins:
    A. prealbumin   E. transferrin
    B. alpha$_1$ antitrypsin   F. C3 complement
    C. alpha$_2$ macroglobulin   G. immunoglobulins
    D. haptoglobin
12. Describe the clinical symptoms and the laboratory findings that characterize multiple myeloma and multiple sclerosis.
13. Describe the chemical composition of Bence Jones proteins.
14. Differentiate between normal serum and urine protein electrophoretic patterns and those generally demonstrated in multiple myeloma.
15. Contrast a normal CSF protein electrophoretic pattern with that frequently demonstrated in multiple sclerosis.

INTRODUCTION

Our consideration of protein assays to this point has been restricted to quantitative analysis of total protein content in a body fluid, measurement of serum albumin, and fractionation of the proteins in serum, urine, and cerebrospinal fluid. Electrophoresis provides extremely useful information about the overall composition of the protein mixture, but does not allow sufficiently accurate quantitation of many proteins present in low concentrations. As the relationships between changes in the concentrations of specific proteins and disease states become more apparent, we see the need for enhanced ability to measure these proteins accurately and rapidly. The growing field of immunochemistry provides the tools for these analyses and has opened the door to the measurement of a wide variety of biochemical constituents. Applications of antibodies to clinical chemistry measurements have greatly expanded over the last 15 years, becoming one of the most rapidly developing areas of this discipline.

7.1

ANTIGENS AND ANTIBODIES

A diabetic individual injects the daily dosage of the protein hormone insulin to control blood sugar. Because the insulin is isolated from pork, the amino acid sequence is slightly different from that of human insulin. This difference in primary structure is sufficient to cause the formation of other proteins, which bind to the administered insulin (and to whatever human insulin may be present) and block its effect. These antiinsulin antibodies are formed in response to the presence of materials which the body does not recognize as being native to it and binds to the "foreign" material in an attempt to inactivate it.

Infection with hepatitis B virus leads to the generation of specific proteins which attach themselves to the virus (or certain structures within the virus) and help destroy this disease-causing entity. Measurement of the presence of these antibodies in serum serves as a diagnostic tool to confirm the presence of this disease.

The above are but two examples of the many proteins which can be formed in response to a specific stimulus. These proteins, called **antibodies,** have unique structural characteristics and functions which make them useful both in dealing with disease and in analytical processes for detecting the presence of other compounds in body fluids.

Definition of Antibody

Earlier we referred to a group of proteins known as immunoglobulins. When separated by electrophoresis from other protein fractions in serum, these immunoglobulins migrate predominantly in the gamma-globulin region. In samples from healthy individuals, the various classes of immunoglobulins do not separate clearly into distinct bands, but show a broad diffuse region of protein. The five major immunoglobulin classes (IgG, IgA, IgM, IgD, and IgE) constitute the proteins known as antibodies ("Ig" is an abbreviation for immunoglobulin).

The antibody proteins are formed in response to specific stimuli. Certain cells within the body produce antibody proteins of a given structure when an exogenous molecule (one not normally found within the organism) is present and is detected by the system. The process of antibody formation is a complicated one and is still not clearly understood. The Nobel Prize in medicine was awarded in 1987 to researchers who are actively studying how antibodies are generated.

One characteristic of antibodies is their specificity. This property indicates the selectivity of the protein molecule. Most (if not all) antibody proteins are synthesized to interact with only a certain molecule or molecules. The three-dimensional structure of this molecule determines the structure of the antibody which binds to it. Whether the antibody is formed through natural processes within the organism or through laboratory manipulation, the protein attaches itself to only one certain molecule. The myriad other molecules within the cell are not attacked by the antibody and do not bind to it. Because of this specificity, we really do not know exactly how many different antibody proteins there are. The body is capable of synthesizing millions of proteins with different structures and different specificities.

Structural Aspects of Antibodies

When we look at the structure of an antibody molecule, we can more easily understand how the many proteins in this category can achieve the high degree of specificity shown by each immunoglobulin. Figure 7–1 illustrates the overall structure of an IgG molecule, the predominant class of antibodies. IgD and IgE antibodies have the same general structure, differing only in the sequence of amino acids making up the various polypeptide chains. IgM (and frequently IgA) is composed of complexes constituting several of the fundamental protein units shown in the figure.

All antibodies have the same basic Y structure. The immunoglobulin is made up of two identical sections, each one composing one-half of the Y. The two sections are linked by disulfide bridges to form the intact molecule. Each half of the protein consists of a **heavy** chain (the longer polypeptide) and a **light** chain, with the heavy and light chain in each instance also joined by disulfide bridges. The leg of

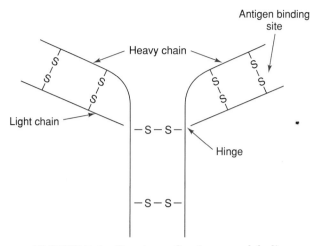

FIGURE 7-1 Structure of an immunoglobulin.

the Y is made up of two polypeptide chains held together by disulfide bridges. The amino acid sequence of each polypeptide is the same, terminating in the C-terminal amino acid for the protein. For any given antibody, this portion of the polypeptide chain is always composed of the same amino acid sequence. This segment is thus referred to as the **constant region.**

The **hinge region** provides the ability of the immunoglobulin to flex and adjust itself to the molecule to which it is binding. This region is also susceptible to attack by various enzymes, a property which can be useful when studying the structure of a given immunoglobulin.

When we examine the structure of a heavy chain, we see that there is also a constant region in part of the arm of the Y, connected to the leg by the hinge region. Beyond this constant region, ending with the N-terminal amino acid of the polypeptide chain, is the **variable region.** The amino acid sequence of this portion of the protein determines the exact specificity of the antibody molecule.

The light-chain portion of the immunoglobulin molecule also contains a constant region and a variable region. Note that the constant regions of the heavy chain and the light chain do not have the same structure. Neither are the variable regions of these two chains composed alike. It is this variety of structural possibilities which allows for synthesis by the organism of the large number of immunoglobulins which have been characterized to date.

Definition of Antigen

An **antigen** is any material which stimulates the formation of an antibody. Antigens are sometimes also referred to as **immunogens.** Under most circumstances, the antigen is a compound which is not normally present within the organism; it is considered to be "foreign." When the antigen is detected in an organism, the complicated process of antibody

synthesis is initiated. In response to this external stimulus, a specific antibody (or antibodies) is formed, binds to, and inactivates the invading material.

Although the antigen is usually not native to the organism, there are situations where antibodies are formed against a material which is present under normal circumstances. These autoantibodies can destroy cells and aggravate the particular disease state. Little is known currently about why these autoantibodies form, but detection of these molecules is important in the diagnosis of diseases such as myasthenia gravis and some forms of arthritis, and in the monitoring of the later stages of diabetes mellitus.

Not all foreign material automatically causes the formation of antibodies. Small-molecular-weight compounds known as **haptens** must first be attached to a protein of some sort to provide a large enough target for the stimulus of antibody production. By coupling a hapten to a protein, the molecular weight becomes large enough for antibody synthesis to occur. The presence of the hapten in the complex results in formation of an antibody with some specificity directed toward that hapten. Many antibodies used in assay procedures to detect drugs in body fluids are generated with the use of a hapten/protein complex as the stimulus.

7.2

ANTIGEN/ANTIBODY INTERACTIONS

Basic Concepts

In its simplest form, we can visualize the antigen/antibody interaction as the noncovalent binding between the two molecules. There are no formal bonds created where electrons are shared between an atom of antigen and another atom of antibody. The interactions are restricted to electrostatic (ionic) attachments, hydrogen bonds, and the weak van der Waals forces between nonpolar segments of the two molecules. As a result, a complex forms and then dissociates in an equilibrium between complexed and free materials.

The strength of the attachment varies from one antigen/antibody complex to the next. As indicated earlier, there are no covalent bonds formed. Therefore, antigen and antibody are able to come together and then break apart with relative ease. The property of loose attachment is exploited when we use antibodies in the quantitation of various molecules.

The specific site of attachment of antigen is to the variable portion of the arms of the immunoglobulin (Fig. 7–2). In both the light chain and heavy chain of each arm this variable region is at the extremity of the protein molecule, easily accessible to reaction

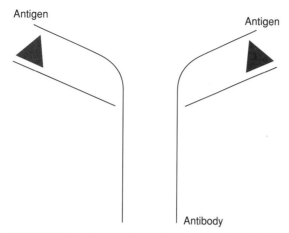

FIGURE 7-2   Interaction of antigen and antibody.

with the antigen. Connection to the antigen requires the participation of the two segments (heavy chain and light chain) in a concerted attack on the antigen molecule.

In the test tube and in the living system, however, the situation becomes somewhat more complicated. Here we are not looking only at a single antigen/antibody combination, but at interactions of a large number of molecules of both types. Every individual antibody has two arms, each of which is capable of attaching to separate antigen molecules. The result is not the simple one antigen/one antibody complex illustrated in Figure 7-2 above, but an interlocking chain of antigen/antibody complexes (Fig. 7-3). These macro complexes often precipitate from solution, a property that can be used as a means of measuring the amount of an antigen in biological fluids. Specific techniques will be considered later in this chapter.

Specificity

The specificity of an antigen/antibody reaction relates to how selective the binding is between the two entities in the complex. Ideally, an antigen binds to only one certain protein; no other compounds have any interaction at all with that antibody. This situation is one of perfect specificity. In the real world,

however, life becomes much more complicated. A particular antigen may bind to more than one specific antibody. Conversely, a given antibody molecule could attach itself to any of several compounds which are structurally similar. In either situation, we are dealing with a lowered degree of specificity in the reaction.

To examine the issue of specificity further, let us look first at the antigen molecule. Any given antigen, especially if it is a large one, may have more than one antigenic site. Different portions of the molecule stimulate production of antibody protein for that particular site. A structurally complex antigen (such as the hypothetical example in Fig. 7-4) may cause the formation of antibody proteins at more than one site on the antigen molecule. The result is the presence of a number of different proteins which bind the antigen at different locations on the molecule and with different binding strengths. As a result, specificity is decreased.

The other factor which contributes to less-than-perfect specificity involves the process of antibody synthesis. The various cells in the body have the capability of synthesizing antibodies to a given antigen. Each type of cell produces a protein of somewhat different structure. The constant region of an antibody is only constant when that antibody to a certain antigen is synthesized by a particular type of cell. Other types of cells produce antibodies with essentially the same specificity, but with a different primary protein sequence in the constant region. The net effect is the formation of several different proteins with somewhat similar (but differing) structures and somewhat different (but similar) selectivity for a given antigen.

The above process of antibody synthesis by several types of cells leads to the formation of **polyclonal antibodies.** To enhance the specificity of antibody production (and better control the quality of the final product), a single type of cell is isolated and used for antibody production. Since these antibody molecules are the product of only one gene, they are referred to as **monoclonal antibodies.** A monoclonal antibody has a much higher degree of specificity, since all the protein molecules in the batch have essentially the same structure. Monoclo-

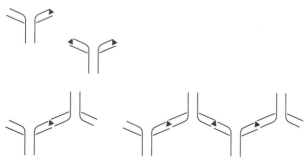

FIGURE 7-3   Antigen/antibody complexes.

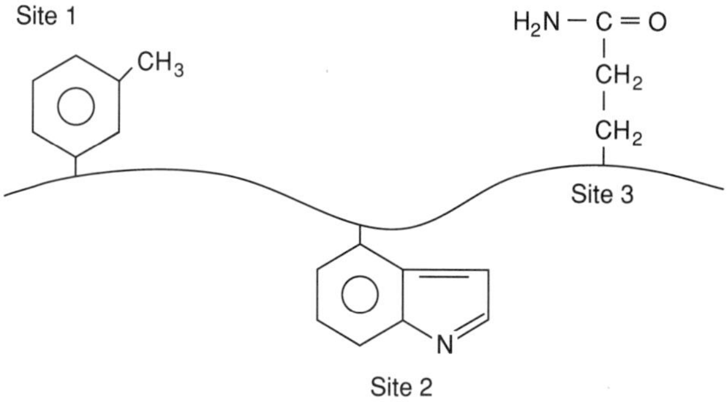

FIGURE 7-4   Hypothetical antigen with multiple binding sites.

nal antibodies are widely used in assays involving antigen/antibody complex formation because of this enhanced selectivity. Other uses for these molecules include tumor location in the body with a radioactively labeled antibody. Current research on some aspects of cancer treatment involves the use of monoclonal antibodies to locate tumors and carry dosages of radioisotope or other materials to the tumor in an attempt to destroy the malignant cells with minimal effect on other tissues.

Relation Between Bound and Free Antigen

We mentioned earlier that the complex between antigen and antibody was one which could dissociate fairly easily in most instances. Each different antigen/antibody complex has its own distinct tendency to bind and unbind. The degree of binding is referred to as the **affinity** of antigen to antibody. A complex with a high affinity shows more binding at a given concentration than one with low affinity.

However, there is usually (if not always) some **free** (unbound) antigen present in the reaction mixture as well as some free antibody. We can write the following equilibrium equation:

antigen (free)
   + antibody (free) $\rightleftharpoons$ antigen/antibody complex

If there is high affinity (and a low dissociation constant), the majority of material is in the form of the antigen/antibody complex. Conversely, compounds with lower affinity (and a higher dissociation constant) shift the equilibrium toward the left-hand side of the equation. In this case, there would be a greater percentage of free antigen and antibody with less of the complex being formed. Knowledge of the affinity and dissociation properties of a particular complex is important when designing a specific assay which employs antibody techniques.

Use in Quantitative Analysis

Measurement of some aspect of antigen/antibody complex formation gives us powerful tools for analyzing constituents of body fluids and tissues. If we use excess antibody and quantitate the antigen/antibody complex formed, the amount of complex we measure is proportional to the amount of antigen present. On the other hand, we can determine changes in the ratio of free antigen and that amount bound to antibody. This information also allows a quantitative estimation of the desired parameter. A wide variety of approaches have been developed based on these fundamental principles. We will discuss some of the more commonly used techniques in the field of immunoassay (the use of immunochemical techniques for the assay of constituents). In this chapter, we restrict our consideration only to protein

measurements. More versatile immunoassay techniques will be discussed in Chapter 9.

7.3

IMMUNODIFFUSION AND IMMUNOELECTROPHORESIS

Types of Immunodiffusion Measurements

It has been known for a long time that antibodies against large molecules (particularly proteins) could be easily formed. Early use of these antibodies mainly involved a reaction which simply indicates the presence or absence of a certain protein in a sample. More recently, these techniques have been greatly refined and expanded to allow the quantitation of specific proteins in blood, urine, cerebrospinal fluid (CSF), and other types of samples.

All immunodiffusion techniques (both qualitative and quantitative) are based on the same phenomenon: when the protein antigen and the antibody interact, a large-molecular-weight complex is formed which is somewhat insoluble in water (the exact degree of solubility depends on the particular proteins involved). The relationships among antigen concentration, antibody concentration, and degree of precipitation are seen in Figure 7–5.

We see from the graph that the amount of precipitate increases as the concentration of antigen increases. The greatest formation of precipitate occurs at a point where the concentrations of antigen and antibody are "equivalent" (not necessarily the same molar concentration, but at levels which shift the equilibrium most toward the formation of the complex). At levels below this equivalence point (called **antibody excess**), less precipitate is formed. If the concentration of antigen is increased further above the equivalence point, precipitation again decreases. This area is called **antigen excess** and is probably the result of several antigen proteins binding to a single antibody molecule, leading to the formation of smaller, more soluble complexes (Fig. 7–6).

Two types of immunodiffusion approaches have been used, depending on whether qualitative or quantitative results are desired. For a simple yes/no

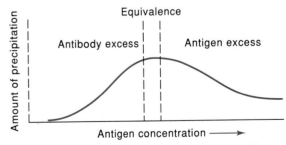

FIGURE 7–5 Precipitation of antigen by antibody.

A. Antigen excess B. Eqivalence C. Antibody excess

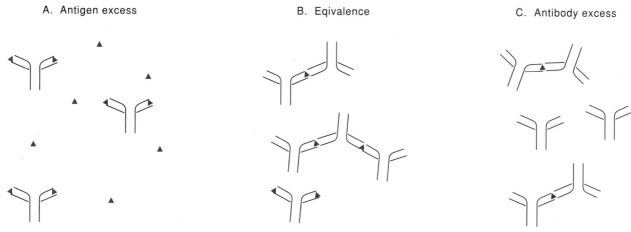

FIGURE 7-6   Complex size and degree of precipitation.

(qualitative) answer, the technique of double diffusion is sufficient. If a quantitative result is needed, **radial immunodiffusion (RID)** is the procedure to be used.

Double Diffusion (Qualitative)

For qualitative **double diffusion,** a plate containing agar gel is employed. A small center hole is punched in the agar and several other small holes are punched around the center one. The sample is placed in this center well. Solutions containing antibodies to known proteins are placed in the outer circle of wells (each solution contains only one antibody). As the protein in the center well and the antibodies in the outer wells diffuse, they come in con-

tact with one another. If an antibody reacts with the protein in the assay sample, a line (or arc) of precipitate forms at the equivalency point (Fig. 7-7). The appearance of a precipitate indicates the occurrence of an antigen/antibody reaction, confirming the presence of the protein specific for that given antibody. Multiple unknown samples can be screened for the presence of a single protein by placing the unknown materials in the outer wells with the known antibody in the center well. The technique of double diffusion is primarily of value in research studies and has little routine clinical application.

Radial Immunodiffusion (RID)

Quantitation of a given protein is accomplished with the technique of **RID.** An agar plate is employed, similar to that used in double diffusion, with a number of small sample wells punched into the agar. The major difference is that the antibody to the protein being quantitated is incorporated into the agar itself. When the solution containing the protein to be measured is placed in a well, it begins to diffuse outward in a circular fashion. As the protein moves through the agar, it reacts with the antibody. In areas of either antigen or antibody excess,

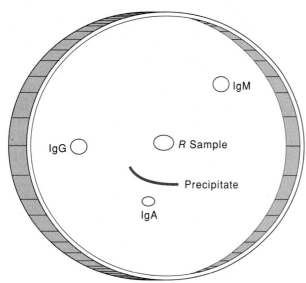

FIGURE 7-7   Double-diffusion assay for detection of proteins. Sample in middle well is positive for IgA only. Immunoglobulins IgG and IgM in outer wells.

Radial immunodiffusion plate

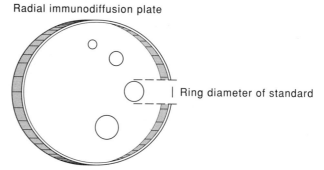

| Ring diameter of standard

FIGURE 7-8   Quantitation of protein by radial immunodiffusion.

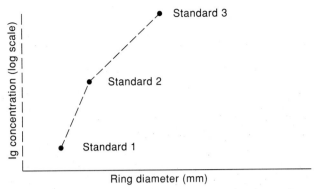

FIGURE 7-9   Radial immunodiffusion standard curve.

A. Protein electrophoresis

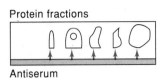

B. Antibody diffusion

C. Precipitation arc formed

FIGURE 7-10   Process of immunoelectrophoresis.

a weak precipitate forms. In the zone of equivalence for antigen and antibody, a strong ring of precipitate is seen (Fig. 7-8). The diameter of this ring is measured and used for calculations. Standard solutions containing known amounts of the protein are also placed in wells, with the diameters of the resultant rings being measured. By knowing the diameters resulting from known concentrations, a standard curve can be plotted and used to calculate the concentration of protein in the unknown sample. Note that the standard curve (Fig. 7-9) is not a simple linear relationship. We will not go through the mathematics involved in calculating antigen/antibody binding, but will simply state that logarithmic relationships are quite common in this type of interaction.

Two approaches to RID are commonly employed. The Fahey technique involves measuring the diameters of all the circles at a set time after the initiation of the diffusion process. Frequently, measurements are made 24 h after addition of samples to the RID plate, in part as a matter of convenience. The Mancini technique involves measurement of the diameters after diffusion has ceased. This technique often requires 2–3 days before results are available. For most situations, the Fahey technique is preferable since results are available much sooner. The Mancini method, on the other hand, provides a more reliable estimation of low levels of antigen.

Immunoelectrophoresis

We saw earlier how the technique of electrophoresis could be used to obtain valuable information regarding the protein composition of a body fluid. One drawback to the approach as it was described was the lack of specificity. The reagents employed to detect proteins were quite nonspecific, taking advantage only of those generic properties which all proteins possess. The addition of immunochemical reactions to the process of electrophoretic separation allows an added measure of specificity which neither approach can provide by itself.

In **immunoelectrophoresis,** the separation of protein fractions is carried out in the usual fashion. Agarose or (less commonly) cellulose acetate is used as a support medium. After protein fractionation, the strip does not undergo the usual staining procedure to visualize proteins. Instead, a solution containing a mixture of antibodies (polyvalent antiserum) is placed in two troughs, one on either side of the path of protein migration (Fig. 7-10). Since the protein fractions have not yet been chemically fixed, they are free to diffuse from their initial migration site. The antibody mixture also diffuses into the area where the proteins are located. As protein antigen and antibody react, precipitate is formed. These **precipitin lines** sometimes can be observed without further enhancement. More frequently, unreacted protein (which is not part of an insoluble complex) is washed from the film. The remaining antigen/antibody complexes are developed with some common protein stains, and the arcs of protein deposits are visualized (Fig. 7-11). The resulting patterns are complex and require a great deal of ex-

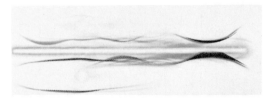

FIGURE 7-11   Immunoelectrophoresis of human serum. (From Carstens, K. S., Sepulveda-Pacheco, A. M., and Romfh, P. C., *Introduction to High Resolution Protein Electrophoresis and Associated Techniques.* Beaumont, TX: Helena Laboratories, 1986, p. 9. Reprinted by permission of the publisher.)

perience to interpret properly. Greater specificity is possible using immunoelectrophoresis since other possibly interfering proteins may have been separated before the immunochemical reaction took place. If desired, more specific antisera (reacting with perhaps only a single protein) may be used. In this case, we may see only a single arc of precipitation owing to the presence of a certain protein.

Immunofixation

Although immunoelectrophoresis provides a sensitive (and aesthetically pleasing) approach to the qualitative detection of proteins, it is obvious that the patterns obtained are often complex and difficult to interpret. With the advent of high-resolution protein electrophoresis, an alternative to classic immunoelectrophoresis became possible. **Immunofixation** provides a simple, yet powerful, alternative for the identification of proteins in body fluids.

The first step in immunofixation involves separation of protein fractions by high-resolution electrophoresis on agarose. This technique allows separation into some 12–15 distinct bands. As in immunoelectrophoresis, the protein bands are not stained after fractionation; instead specific antisera are layered directly over the protein areas and allowed to react with any protein present. Instead of arcs, we see simply a band of protein/antibody complex. After washing with a material such as sodium chloride (to remove unreacted protein), the film is stained with amido black or Coomassie blue to reveal the location of each protein.

A typical analysis is illustrated in Figure 7–12. The same serum sample is applied in each of the wells. Following electrophoresis, each track is treated with a different antiserum, and formation of a precipitin band confirms the presence of a protein of that particular type. In our example, the patient serum contained abnormal proteins of the IgG and IgM types. Free lambda light chains are found, associated with IgG, and the free kappa light chains

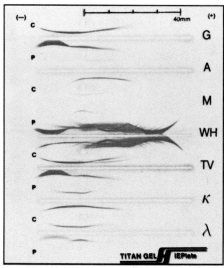

FIGURE 7–13   Immunoelectrophoresis of sample in Figure 7–12. Gammopathy with IgG lambda and IgM kappa monoclonal bands. (From Fauchier, P., and Catalan, F., *Interpretive Guide to Electrophoresis: Important Normal and Abnormal Patterns.* 2nd ed. Beaumont, TX: Helena Laboratories, 1988, p. 8. Reprinted by permission of the publisher.)

are associated with IgM. The pattern is less complex and much easier to interpret than one obtained with classic immunoelectrophoretic techniques (Fig. 7–13).

Rocket Immunoelectrophoresis

One of the drawbacks to the use of radial immunodiffusion for the quantitation of proteins is the great amount of time required to obtain any data. Even if the Fahey technique is employed, a minimum of 24 h is usually required before results are available. **Rocket immunoelectrophoresis** methods were developed to help overcome this time limitation, as well as to provide a more sensitive and accurate approach to protein quantitation.

The central component of this analytical method is an agarose film which has the antibody incorporated into the agarose. The agarose is buffered to permit protein migration when an electric field is applied. Because most proteins migrate readily, the pH is adjusted to the point where the antibody is essentially neutral (its isoelectric point) so it does not move when electrophoresis is carried out.

At one end of the agarose film is a row of sample wells. Samples containing the protein antigen of interest are placed in the wells. When the electric field is applied, the antigen protein moves out into the agarose and reacts with the antibody present in the gel. The main purpose of the field is to speed up

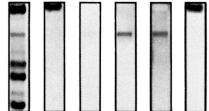

FIGURE 7–12   Immunofixation pattern of serum proteins. (From Fauchier, P., and Catalan, F., *Interpretive Guide to Electrophoresis: Important Normal and Abnormal Patterns.* 2nd ed. Beaumont, TX: Helena Laboratories, 1988, p. 8. Reprinted by permission of the publisher.)

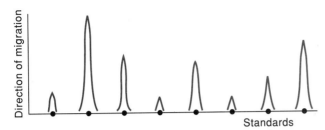

FIGURE 7–14　Rocket immunoelectrophoresis. ● indicates sites of sample application.

the movement of antigen so it comes into contact with antibody more readily.

As the antigen migrates in the electric field, some of it reacts with antibody to form the insoluble complex. Since this complex has essentially no mobility under the conditions of the assay, it remains in place where it is formed. The remaining antigen continues to migrate across the gel, reacting with antibody as it progresses. Eventually, all the antigen reacts with antibody and there is no further migration observed. The plate is then washed to remove unreacted proteins and stained in the usual manner to visualize the peaks (or "rockets") formed during the run (Fig. 7–14). Most assays require 1–2 h to perform.

Quantitative results are obtained by measuring the distance from the origin to the tip of the peak; this peak height is proportional to concentration. Standard solutions are run in some of the wells to provide a standard curve for the assay.

The height of the peak is a function only of antigen concentration. The more antigen present, the more antigen/antibody complex forms. Since the antigen concentration is decreasing as the sample moves across the field, eventually all the antigen is consumed. The antigen/antibody complex does not migrate and remains stationary even after all the antigen has reacted.

One limitation to the use of rocket immunoelectrophoresis is that immunoglobulins cannot be quantitated with this procedure. Under the conditions of the assay, these proteins do not migrate significantly in an electric field. If the immunoglobulin is first treated chemically, analysis is possible, but some technical problems arise.

Comparison of Techniques

Selection of a method for immunochemical measurement of a protein depends on many factors. Double diffusion, immunoelectrophoresis, and immunofixation are used only for a qualitative yes/no answer. Of the three techniques, immunofixation is perhaps the most rapid and sensitive. Materials are commercially available from several companies for immunofixation analyses. If the amount of available antibody is a factor (as it is with rare or unusual

proteins), double diffusion requires the least amount of sample and antibody.

For quantitative measurement of proteins, analysis using radial immunodiffusion is probably the best approach. The technique is simple, little special equipment is needed, and the cost is reasonable. Although rocket immunoelectrophoresis allows the simultaneous assay of a large number of samples, more elaborate equipment must be purchased. In addition, the cost of the antibody-impregnated electrophoresis media is relatively high. Rocket immunoelectrophoresis does not permit the quantitation of immunoglobulins, which can be carried out easily with the RID technique.

7.4

TURBIDIMETRIC AND NEPHELOMETRIC TECHNIQUES

Definition of Terms

Up to this point, we have restricted our discussion to the measurement of proteins using a solid phase as a matrix for the reaction. Recent developments in the immunoassay field have explored various facets of solution immunochemistry. The precipitate which forms in an agarose gel and can be quantitated will also form in solution. The amount of precipitate is proportional to the concentration of the antigen under consideration. Two techniques are used to quantitate this precipitate: turbidimetry and nephelometry.

When a solution of antigen and another solution of antibody are combined, the result is a cloudy suspension. The degree of cloudiness is an indication of the amount of antigen/antibody complex being formed. If we shine a beam of light through this mixture, less light passes through than if we had a clear solution. In a manner analogous to absorption spectrophotometry, the amount of light blocked by the precipitate is proportional to the concentration of material being analyzed. Measurement of this "absorption" of light is called **turbidimetry.** We quantitatively assess how turbid the solution is by determining how much light is prevented from reaching the phototube by the suspension which is formed. Turbidimetric measurements can be made using any spectrophotometer. The assay for urine total protein described earlier is an example of a turbidimetric analysis.

In contrast, a **nephelometric assay** involves measurement of the light scattered by the particles in the suspension. The detector is placed at some angle other than directly in the line of the light beam which comes into the sample. Quite frequently the detector is at 90° to the incident light beam (Fig. 7–15), although much smaller angles have been shown to be more effective in some situations. As light enters the tube or cell containing the

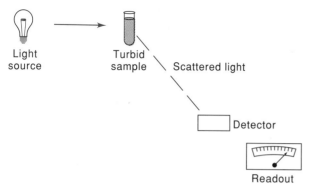

FIGURE 7-15   Instrumentation for nephelometry.

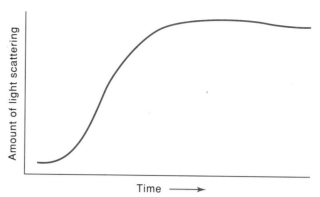

FIGURE 7-16   Time course of turbidity increase.

antigen/antibody suspension, some of the individual light rays strike a particle and are deflected off at an angle. The more suspension present (the higher the concentration of antigen), the more light is scattered. In nephelometric assays, we have a direct relationship between amount of light scattering and analyte concentration: the higher the degree of light scattering, the higher the concentration of material being quantitated.

Basic Techniques

The methodology for a turbidimetric assay has been covered to some extent in the discussion of total protein measurement in urine. To carry out the assay, some substance is added which precipitates the material of interest. For our present purposes, we are interested in the immunochemical precipitation of a protein antigen by an antibody. Conditions should be adjusted to obtain optimum particle size (too large a particle results in the suspension settling out of solution). Time of measurement of turbidity must be carefully standardized, since the reaction rarely reaches equilibrium in a reasonable time. Instrumentation is simple; any spectrophotometer serves adequately since the same parameter (absorbance of incident light) is being estimated.

Although the methodology and instrumentation for a turbidimetric assay are simple, improved accuracy and sensitivity are usually obtained if the assay is adapted to nephelometric measurement. The major changes required are in instrumentation and approach to data collection.

To measure light scattering in nephelometry, the incident light source, sample, and photodetecting system must not all be in the same line. Many instruments utilize a detector set at 90° from the light source and sample. A great deal of nephelometry has been done with fluorometers, since the fundamental optical design of this class of instruments fits the required configuration. In addition, the shorter wavelengths of light emitted by the fluorometer usually provide increased sensitivity. Centrifugal analyzers also lend themselves well to nephelometric

measurements. At least one company has designed such an analyzer, which is capable of nephelometric measurements using the fluorescent light source and detecting the 90° light scattering. Other approaches include the measurement of **forward light scattering,** in which the detector is set off at an angle greater than 90° from the light source and sample. For large molecules (particularly in the quantitation of immunoglobulins), this alternative significantly improves the results obtained. In some instruments, a laser is used as a source of coherent light of high intensity, allowing greater sensitivity.

When a nephelometric assay is developed, it is usually based on the antibody-excess approach. When antigen is added to a solution containing an excess amount of antibody, a complex forms after a few seconds. The turbidity of the solution increases rapidly, then stabilizes for some 5–15 min (Fig. 7-16). At antibody excess, the particles formed are not large aggregates, but small antigen/antibody units. These smaller complexes are less likely to precipitate from solution and provide more stable suspensions.

In most instances, the formation of these aggregates is reasonably independent of temperature, allowing greater flexibility in reaction conditions. Control of pH regulates the amount of turbidity to some extent. Increased amounts of neutral salts (Cl^-, Br^-, and others) decrease the rate of complex formation and result in smaller particles. Addition of polymers such as dextran or polyethylene glycol promotes more rapid formation of the protein aggregate and increases the size of the complex, scattering the light more. Suspension turbidity can be increased two to three times with the use of these polymers.

Rate Nephelometry

One of the assumptions of the nephelometric approach is that a condition of antibody excess exists in the reaction system. In reality, we do not know in advance what the concentration of a given material is in a body fluid. In a spectrophotometric assay, we get around this limitation in our knowledge by defining an upper limit to the assay. Because of the behavior of antigen/antibody systems, this upper limit cannot be clearly and easily defined in a nephelometric assay. More sophisticated techniques must be used to overcome this obstacle to accurate measurement.

There is an increase in light scattering as antigen concentration increases when antibody is present in excess amounts. Whatever antigen is added to the system immediately forms a complex with antibody, with no free antigen remaining. In the region of antibody excess, cross-linking can occur between one antigen molecule and several antibody molecules. At the **equivalence point,** antigen and antibody are present in similar amounts and the amount of the complex between the two is at its greatest, as reflected by the peak in the turbidity observed. If we further increase the concentration of antigen, we see a decrease in turbidity as a result of each antigen molecule (now in excess) binding only one antibody molecule. The resulting particles are now smaller in size and scatter the light less.

The data described above point out the difficulty in determining the actual amount of antigen present in a reaction mixture. Is the amount of turbidity or light scattering due to antigen reacting in an area of antibody excess or because we have insufficient antibody to form appropriate complexes with the excess antigen present? The same instrument reading may reflect two entirely different situations which are quite difficult to distinguish from one another using our usual spectrophotometric principles.

To overcome this ambiguity, the technique of rate nephelometry was developed. Instead of measuring the end point of the reaction (the final amount of light scattering), **rate nephelometry** determines how rapidly the light scattering is changing—the rate of change of suspension formation indicating the rate of formation of antigen/antibody complexes. Commercial instrumentation has been developed to employ this approach to the nephelometric quantitation of materials.

When we add antigen to a solution containing a given amount of antibody, a complex between the two begins to form and turbidity results. With modern equipment we can measure the rate of formation of the turbidity (Fig. 7–17) and relate this parameter to the concentration of antigen. At conditions of antibody excess, the rate of formation of the complex (and the rate of increase in turbidity) is directly proportional to the concentration of antigen. Thus, the greater the amount of added antigen (the higher the concentration of a given protein in a sample), the greater the rate of formation of the antigen/antibody complex and the greater the rate of light scattering.

How do we deal with the presence of excess antigen in the system? Not only is the rate of turbidity

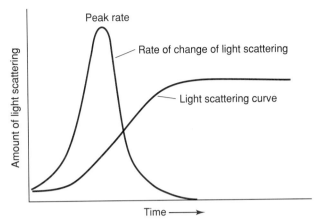

FIGURE 7–17   Principle of rate nephelometry.

related to antigen concentration, but the time of the peak rate is also a function of concentration of antigen. At low antigen concentration, the time required to reach that peak rate is rather long. As the antigen concentration increases, the time needed to achieve the peak rate decreases. Once the equivalence point is past and we enter a situation of antigen excess, the time for peak rate increases again. All these times are electronically monitored in a computerized system which automatically determines the peak rate after addition of antigen and the time required to reach that peak rate. With the use of materials of known antigen concentration, standardization can easily be carried out on a regular basis.

Nephelometric Inhibition Techniques

The analytical approaches described above are directly applicable to the quantitation of many proteins in serum, plasma, or other body fluids. If we wish to assay for the presence of small molecules, modifications are necessary before an antigen/antibody methodology can be employed. Small molecules do not generate antibodies by themselves, but first must be chemically attached to a large protein molecule. When this hapten/protein complex is injected into an animal, the appropriate antibodies to that hapten can be generated.

A common approach to nephelometric analysis of small molecules involves a modification of the usual assay approach called **nephelometric inhibition.** One component of this type of assay is the antibody to the hapten of interest, obtained as described in the previous paragraph. Another component, which must be prepared ahead of time, is called the **developer antigen.** This molecule is composed of a large protein (such as bovine fibrinogen) to which a number of hapten molecules are attached. When this developer antigen reacts with specific antibody, the ex-

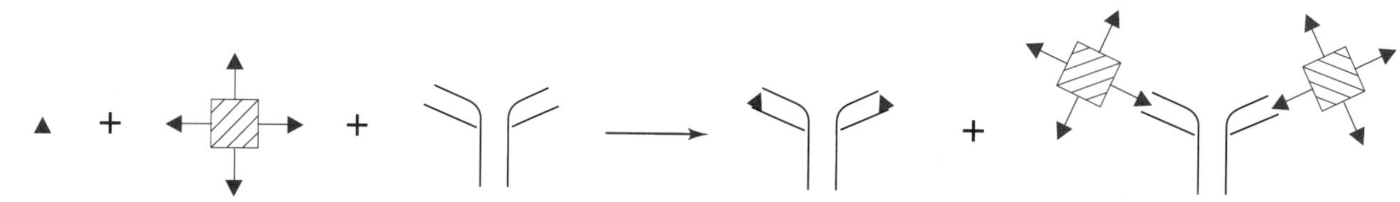

FIGURE 7-18 Nephelometric inhibition immunoassay.

pected complex forms and we can quantitate the amount using either regular or rate nephelometric techniques.

When we wish to quantitate the amount of a small molecule in serum, we mix the sample with specific antibody and the appropriate developer antigen (Fig. 7-18). The patient antigen and developer antigen compete for binding sites on the antibody. With no patient antigen present, antibody and developer antigen react to form large amounts of suspension. As the amount of patient antigen increases, it reacts with more and more antibody, leaving less antibody available for reaction with the developer antigen. With less antibody available, the amount of precipitate is also less. The higher the concentration of patient antigen, the lower the amount of developer antigen/antibody reaction and the lower the amount of turbidity in the reaction mixture.

Clinical Applications

Rate nephelometry is perhaps the method of choice for the quantitation of a large number of proteins in body fluids. Reagents are commercially available from several sources and at least one company has instrumentation specifically designed for this technique. In more recent years, rate nephelometric assays for small molecules (primarily drugs) have been developed. These techniques find widespread application in therapeutic drug monitoring. Some work is being done in the area of hormone quantitation, but progress is hindered to some extent by the extremely low levels of the materials being measured.

Nephelometry and Other Analytical Approaches

Rate nephelometry has generally been shown to be much more sensitive than the corresponding endpoint techniques. Levels of biochemical analytes in the 10-100 nmol (nanomole) range can be routinely measured. When compared with other approaches for the measurement of protein components in body fluids, rate nephelometry has been shown to be much more precise and reliable as well as being capable of detecting much lower levels of the analyte

protein. In addition, assays can be carried out in a matter of minutes instead of the 24-48 h required by immunodiffusion techniques or the several hours needed for electrophoresis.

7.5

CLINICAL SIGNIFICANCE OF SPECIFIC SERUM PROTEINS

As techniques for the rapid and accurate quantitation of specific serum or plasma proteins have become more sophisticated, we are now able to relate changes in particular protein levels to a variety of

Table 7-1
SPECIFIC PROTEIN CHANGES IN DISEASE STATES

Protein and Reference Range	Changes Observed
Prealbumin 10–40 mg/dL	Decreased: inflammatory response, hepatic damage, results of estrogen use
Alpha$_1$-antitrypsin 200–400 mg/dL	Increased: inflammatory response, results of estrogen use Decreased: hereditary defect
Alpha$_2$-macroglobulin 150–350 mg/dL (males) 175–420 mg/dL (females)	Increased: renal or gastrointestinal protein loss, diabetes mellitus Decreased: Disseminated intravascular coagulation, peptic ulcer
Haptoglobin 100–300 mg/dL	Increased: inflammatory response, biliary obstruction, steroid therapy Decreased: hemolysis, low erythrocyte production, liver disease
Transferrin 200–400 mg/dL	Increased: iron deficiency anemia, results of estrogen therapy Decreased: inflammatory response, burns, liver disease
C3 Complement 80–160 mg/dL	Increased: inflammation, biliary obstruction Decreased: liver or kidney disease, occurs after surgery or infections

disease states. A voluminous literature discusses protein measurements in detail. The following is a brief summary of some of the more important plasma proteins (Table 7–1). Reference ranges are taken from various literature sources and are only approximate. The ranges vary somewhat, depending on the population studied and the methodology employed.

Prealbumin

The major function of prealbumin is the transport of thyroxine. The normal range is 10–40 mg/dL. Prealbumin increases in patients receiving high-dose corticosteroid therapy. Low levels are seen in the acute inflammatory response and in situations where estrogen levels are high (mainly due to administration for estrogen replacement). Prealbumin appears to be a very sensitive indicator of liver damage, decreasing markedly in cases of hepatic problems.

Alpha₁ Antitrypsin

This protein is one of many protease inhibitors, acting to inhibit the proteolytic enzyme trypsin. Normal concentrations of alpha$_1$ antitrypsin are 200–400 mg/dL, with increases occurring in the acute inflammatory response (probably the most sensitive indicator) and after estrogen and androgen therapy. A hereditary defect in the production of this protein leads to a deficiency associated with emphysema and related pulmonary problems (possibly owing to increased lung destruction by trypsin).

Alpha₂ Macroglobulin

The normal levels of this protease inhibitor are approximately 150–350 mg/dL for males and 175–420 mg/dL for females. Values for infants and small children are higher. Elevations are seen in situations involving renal or gastrointestinal protein loss and in diabetes mellitus (also possibly due to renal protein loss in later stages of the disease). Patients receiving estrogens show increased levels of alpha$_2$ macroglobulin, whereas decreased amounts are seen in disseminated intravascular coagulation and peptic ulcer disease.

Haptoglobin

Hemoglobin released from disintegrating red cells is bound by haptoglobin. There are three types of this protein (genetically determined) and the normal range is type-dependent; values may run from 100–300 mg/dL in adults. Increased levels of haptoglobin are seen in the acute inflammatory response, biliary obstruction, or after some forms of steroid therapy. Decreases appear with hemolysis or impaired production of red blood cells. Low levels can also be seen in some cases of liver disease or in patients with a hereditary defect in haptoglobin synthesis.

Transferrin

Transferrin is the major transport protein for iron in the circulation, and its normal levels are 200–400 mg/dL. Increased amounts of transferrin are seen in iron deficiency anemia and after estrogen therapy. Low levels are observed in the acute inflammatory response, burns, and severe liver disease. The role of transferrin in iron metabolism and methods for analysis will be discussed in detail later in this text.

C3 Complement

The C3 complement is part of a complex series of serum proteins which interact to promote some of the functions of the immune system. The reference range is 80–160 mg/dL. Elevated C3 levels are seen in low-grade inflammation and biliary obstruction. C3 levels drop in cases of liver or kidney disease, after surgery, and in some infections.

Immunoglobulins

There are five major classes of immunoglobulins, three of which are primarily involved with combatting infections (Table 7–2). The classes are designated by Ig (for immunoglobulin) followed by a specific letter to indicate the immunoglobulin class. IgG is the immunoglobulin present in highest concentration (800–1800 mg/dL), followed by IgA (90–450 mg/dL) and IgM (60–250 mg/dL). These three classes can be quantitated using either radial immunodiffusion or specific nephelometric techniques.

Table 7–2
IMMUNOGLOBULIN CONCENTRATIONS AND DISEASE STATES

Immunoglobulin and Reference Range	Changes Observed
IgG 800–1800 mg/dL	Increased: infections, hepatocellular disease
IgA 90–450 mg/dL	Increased: infections, hepatocellular disease
IgM 60–250 mg/dL	Increased: infections

Little is known about the role of IgD (normal range 2–6 mg/dL), and it is rarely measured. IgE is associated primarily with allergic reactions. An IgE panel can be performed using radioimmunoassay techniques (discussed in Chapter 9) to assess the specific cause of an allergy.

The three classes associated with infections are all present in low amounts in newborns and infants, with values rising to those of normal adults within the first year of life. All three fractions are elevated after an infection. Both IgG and IgA increase after hepatocellular disease. The presence of increased amounts of IgM in newborns strongly suggests perinatal infection. The presence of a myeloma is indicated by an increase in the specific Ig fraction with a noticeable decrease in the other two fractions. For example, an IgG myeloma produces an elevation of IgG with low amounts of IgA and IgM being detected. All three fractions are low in light-chain disease.

7.6

MULTIPLE MYELOMA

There are very few disease states where protein analysis plays a more important role than multiple myeloma. Electrophoresis patterns showing the presence of atypical protein fractions are virtually diagnostic of this disease. Although the data are not yet conclusive, further studies to identify the exact abnormality present give some indication of survival and clues for the most appropriate therapy.

Clinical Description

Multiple myeloma is characterized by the presence of a number of soft tumors, found primarily in bone but occasionally elsewhere in the body. These tumors are made up of plasma cells, a normal constituent of the body responsible for the production of antibodies. When the plasma cells become malignant, they change some of their biochemical characteristics. Normal production of immunoglobulins is interrupted and the production of atypical proteins increases. The tumors spread, causing bone deterioration and other related complications. Most deaths from this disease result from infections. The body has a greatly lowered resistance to microorganisms owing to the impaired ability to produce antibodies. Renal insufficiency also contributes to overall mortality, since the abnormal protein metabolism in this disease creates some serious problems with kidney function.

Most patients who develop multiple myeloma are over the age of 40. Approximately 60% of those affected are men. At present there is no known cure for the disease. Few patients survive as long as five years after multiple myeloma develops.

Laboratory Findings

There seems to be a characteristic laboratory profile for patients with multiple myeloma. Hematological examination shows a striking increase in plasma cells in the blood (normally 5% or less are seen). Total protein is usually elevated, although albumin is generally unchanged. Serum creatinine is elevated in a high percentage of these patients, reflecting kidney damage. Serum calcium is frequently increased, possibly owing to the presence of increased protein and as a result of the breakdown of bone. Almost 80% of these patients show bone damage when examined by x-ray studies.

One of the most striking characteristics of this disease is the frequent occurrence of Bence Jones proteins in the urine (seen in approximately 50% of the cases). The Bence Jones proteins are immunoglobulin light chains (of either the kappa or lambda classes), resulting from impaired immunoglobulin production. These proteins were first studied intensively by Dr. Henry Bence Jones in the mid-1800s. He noted an interesting property of the material: the proteins in urine of a patient with multiple myeloma precipitated when acid was added, but redissolved if the sample was then heated. This qualitative test was used for many years as a screening procedure for samples with a high urine protein level. Unfortunately, the test has a high false negative and false positive rate. Today, the Bence Jones proteins are better identified by electrophoretic and immunochemical techniques.

Electrophoretic Patterns

Both the serum and urine protein electrophoretic patterns of a patient with multiple myeloma show distinctive patterns. In a normal serum protein pattern, the gamma-globulin region is seen as a broad diffuse band near the origin, reflecting the presence of a large number of different immunoglobulins. In contrast, the serum protein studies from a patient with multiple myeloma demonstrate the presence of

FIGURE 7–19   Serum protein electrophoresis—multiple myeloma. (From Fauchier, P., and Catalan, F., *Interpretive Guide to Electrophoresis: Important Normal and Abnormal Patterns.* 2nd ed. Beaumont, TX: Helena Laboratories, 1988, p. 33. Reprinted by permission of the publisher.)

a sharp peak in the gamma-globulin region superimposed over the lighter broad band normally seen (Fig. 7-19). This type of pattern is referred to as a **monoclonal gammopathy** and reflects the alteration of immunoglobulin synthesis found in the disease. Approximately 75% or more of the patients with multiple myeloma demonstrate this serum protein pattern.

An examination of the urine protein fractionation reveals similar findings. Normally, the urine protein pattern would be rather light and diffuse, with the only distinct band being albumin. In the multiple myeloma patient, there is a globulin "spike" in roughly 75% of the cases. This massive loss of protein probably gives rise to some of the kidney problems associated with the disorder.

Laboratory Results and Biochemical Defect

The abnormal protein fractions observed in both serum and urine are reflections of a disorder in immunoglobulin synthesis. Each protein chain is synthesized separately; the four subunits are then assembled to make the intact antibody. This synthesis and assembly is carried out in the plasma cell. In multiple myeloma there is a defect of some sort in the behavior of the plasma cell, giving rise to impaired manufacture of the whole immunoglobulin. The plasma cells produced are immature and not under normal biochemical control. As a result, instead of a variety of immunoglobulins being produced, an excess of certain subunits is manufactured. These subunits do not assemble into the complete antibody molecule and are released free into the circulation. When we carry out electrophoretic studies, these atypical proteins are what we see in the gamma-globulin region.

Monitoring the Disease State

After the initial diagnosis was made, some researchers have carried out further studies to determine the exact type of abnormal protein being produced. Light-chain studies can be done using either quantitative or qualitative methods to get more information as the structure of the protein fragments. Immunofixation electrophoresis with the use of antisera to specific light chains yields further information regarding specific atypical immunoglobulin synthesis. Light chains may be quantitated by rate nephelometry or immunodiffusion techniques. Although these studies provide interesting information about the details of the disease state, they do not give any useful data relative to treatment plan or mortality rates. Light-chain typing is not routinely carried out as a part of the work-up for multiple myeloma.

7.7

MULTIPLE SCLEROSIS

Description of the Disease

Multiple sclerosis is a frustrating disease to deal with clinically. The diagnostic criteria are ambiguous, and no therapy arrests the course of the disorder. The basic defect appears to be in the breakdown of the covering of the nerve fibers, leading to a number of neurological complications. The myelin sheath which surrounds most of the nerve tissue serves to insulate the fiber, allowing the nerve impulse to be transmitted. When this sheath breaks down, a type of "short circuit" occurs, impairing nerve function and decreasing body control. The cause of this myelin degeneration is not known at present.

The disorder has its highest incidence among young adults (late 20s and early 30s). There is an interesting and unexplained relationship between geography and the disorder; more cases are seen in areas closer to the poles than in areas near the equator. Occasional times of remission are common, but the general downward course of the disease does not reverse.

Clinical symptoms are varied. One of the most common early signs is the "pins and needles" feeling in the extremities. Gradual numbness and loss of function then develop. Muscular weakness and lack of coordination are common, with eventual total impairment, particularly of the lower part of the body. Blurring of vision is commonly seen later in the course of multiple sclerosis. The overall trend is the gradual loss of normal neurological function.

Laboratory Findings

The primary material for laboratory testing in multiple sclerosis is cerebrospinal fluid. A number of protein studies have been carried out on CSF to ascertain the basic biochemical defect in this disease. Myelin basic protein (the major protein in the myelin sheath) has been quantitated and is seen to be elevated in some instances, but does not give a clear diagnostic sign. In addition, the assay is hampered by problems of antibody specificity, so it is not routinely done. Quantitation of albumin and IgG in CSF gives important information regarding the presence or absence of multiple sclerosis. Measurement is usually carried out using radial immunodiffusion techniques, although quantitation by nephelometric approaches is becoming more widespread. Determination of the CSF IgG/albumin ratio by radial immunodiffusion or using nephelometric techniques allows differentiation of the abnormal production of immunoglobulins seen in multiple sclerosis from that in other disorders where both albumin and IgG are elevated (such as chronic inflam-

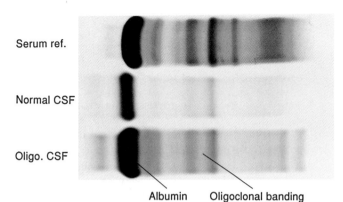

Serum ref.

Normal CSF

Oligo. CSF

Albumin   Oligoclonal banding

FIGURE 7–20
CSF protein electrophoresis—multiple sclerosis. (From Carstens, K. S., Sepulveda-Pacheco, A. M., and Romfh, P. C., *Introduction to High Resolution Protein Electrophoresis and Associated Techniques.* Beaumont, TX: Helena Laboratories, 1986, p. 33. Reprinted by permission.)

matory diseases). Abnormal IgG/albumin ratios (greater than 0.27) are seen in over 70% of patients with multiple sclerosis.

Electrophoretic Studies

CSF protein levels normally range up to 40 mg/dL or so. Even in multiple sclerosis, total CSF protein is not markedly elevated. Therefore, any electrophoretic studies require a preliminary concentration step before the material is applied to the film. After concentration, the sample may be run using the same materials and techniques available for serum proteins. In recent years, the use of high-resolution techniques for electrophoresis has become popular for CSF studies.

The characteristic pattern for normal CSF is one with a discernable albumin fraction and a diffuse pattern for the remainder of the fractions; little is seen in the way of distinct bands. In contrast, pronounced banding in the gamma-globulin region is observed for CSF samples from patients with multiple sclerosis (Fig. 7–20). These multiple bands (called **oligoclonal bands**) are frequently seen in this disorder. The pattern looked for is the presence of two or more bands; a single atypical band does not fit the definition of oligoclonal. Over 90% of patients with multiple sclerosis demonstrate the presence of oligoclonal bands at some time during the course of the disease. Unfortunately, these bands can be seen in other patients who have meningitis or viral encephalitis, so they cannot be considered completely diagnostic for multiple sclerosis. The measurement of the IgG/albumin ratio confirms the diagnosis. Both electrophoresis and quantitative studies of CSF proteins should be carried out for the laboratory diagnosis of multiple sclerosis.

SUMMARY

Antibodies are specialized proteins which bind to foreign materials and neutralize their effect on the organism. An antigen is any material which binds to an antibody. Although broad categories of antibodies can be defined, each protein specifically binds one or more antigens of a particular structure. In any mixture of antigen and antibody, an equilibrium exists among the amount of free (unbound) antigen, free antibody, and antigen/antibody complex. This property of antigens and antibodies can be employed to quantitate a wide variety of proteins and other materials in body fluids.

Immunodiffusion and immunoelectrophoresis have long been used to identify the presence of specific proteins and to quantitate proteins. Double-diffusion techniques are sensitive, but only provide qualitative information. Radial immunodiffusion furnishes quantitative results, but often requires 24 h or more for results to be available. Immunoelectrophoresis and immunofixation permit confirmatory identification of specific proteins in a body fluid after electrophoretic separation has taken place. Rocket immunoelectrophoresis allows quantitation of many proteins (except immunoglobulins) in a fairly short time.

Both turbidimetric and nephelometric approaches have proven to be very useful in quantitation of proteins and other materials. In a turbidimetric assay, the antigen/antibody complex forms a suspension which blocks the passage of light shining through the material. By measuring the absorbance of light by the solution, a relationship between light passed through the material and quantity of antigen can be established. Nephelometric techniques measure the amount of light scattered at an angle owing to the presence of suspended material (antigen/antibody complex). By measuring the rate at which the suspension forms, rate nephelometry allows more rapid assay of biochemical components.

Quantitation of several proteins can provide valuable clinical information for a variety of diseases. Measurements of prealbumin, alpha₁ antitrypsin, haptoglobin, transferrin, and immunoglobulins are technically feasible and clinically useful in several situations.

Multiple myeloma is a disorder in which incomplete immunoglobulins are formed. The light chains produced in this disease can be detected in serum and urine using protein electrophoresis and immunofixation techniques. Effectiveness of treatment can be followed by observing changes in the number of light chains released into the system.

Multiple sclerosis is a neurological disease characterized by loss of myelin from nerve fibers. Laboratory diagnosis includes the detection of oligoclonal bands on CSF protein electrophoresis and an altered CSF IgG/albumin ratio.

FOR REVIEW

Directions: For each question, choose the best response.

1. A protein that is formed in response to a specific stimulus and demonstrates a unique structure best describes a(n)
   A. antigen
   B. antibody
   C. light chain
   D. heavy chain

2. Antibody specificity is determined by the amino acid sequence located within the
   A. variable region of the light chain
   B. variable region of the heavy chain
   C. constant region of the light chain
   D. constant region of the heavy chain
   E. both A and B

3. Any material that stimulates the formation of immunoglobulins best describes a(n)
   A. antigen
   B. antibody
   C. autoantibody
   D. hapten

4. Which of the following may be associated with monoclonal antibodies?
   A. produced from single cell type
   B. product of only one gene
   C. low degree of specificity
   D. both A and B
   E. A, B, and C

5. The general principle of all immunodiffusion techniques is based on the phenomenon that antigen/antibody complexes are
   A. soluble in water
   B. insoluble in water
   C. soluble in organic solvents
   D. insoluble in organic solvents

6. Which of the following may be associated with radial immunodiffusion?
   A. qualitative technique
   B. uses ring diameter measurements
   C. antibody incorporated into the agar
   D. both A and C
   E. both B and C

7. Which of the following techniques may be used to qualitatively identify immunoglobulins and light chains in a specimen?
   A. radial immunodiffusion
   B. immunoelectrophoresis
   C. immunofixation
   D. both A and B
   E. both B and C

8. Which of the following techniques may be used to quantitatively measure immunoglobulins in a specimen?
   A. radial immunodiffusion
   B. rocket immunoelectrophoresis
   C. nephelometry
   D. both A and B
   E. A, B, and C

9. Which of the following may be associated with nephelometric analysis?
   A. detector placed at a 90° angle to the light source
   B. measures light scatter
   C. direct relationship between amount of light scatter and analyte concentration
   D. both A and B
   E. A, B, and C

10. The acute-phase reactant protein that is able to inhibit enzymatic proteolysis is
    A. alpha$_1$ antitrypsin
    B. complement
    C. haptoglobin
    D. transferrin

11. The protein that is able to bind free hemoglobin in the plasma is
    A. alpha$_2$ macroglobulin
    B. complement
    C. haptoglobin
    D. prealbumin

12. Which of the following is characteristic of multiple myeloma?
    A. increased total protein
    B. presence of Bence Jones protein in urine
    C. monoclonal band in gamma-globulin region
    D. both A and C
    E. A, B, and C

13. Which of the following is characteristic of a cerebrospinal fluid specimen in multiple sclerosis?
    A. oligoclonal bands in gamma-globulin region
    B. total protein decreased
    C. albumin increased
    D. both A and B
    E. all of the above

BIBLIOGRAPHY

Amzel, L., and Poljak, R. J., "Three-dimensional structure of immunoglobulins," *Ann. Rev. Biochem.* 48:961–997, 1979.

Berzofsky, J. A., "Intrinsic and extrinsic factors in protein antigenic structure," *Science* 229:932–940, 1985.

Bienenstock, J., and Befus, A. D., "Some thoughts on the biologic role of immunoglobulin A," *Gastroenterology* 84:178–185, 1984.

Bienvenu, J. et al., "The acute-phase proteins in neonatal infection," *Lab. Manage.,* October, 1983.

Bloch, K. J., and Salvaggio, J. E., "Use and interpretation of diagnostic immunological laboratory tests," *J. Amer. Med. Assoc.* 248:2734–2758, 1982.

Calbreath, D. F., "Diagnosis of multiple sclerosis and related disorders. Part one: Analysis of CSF proteins," *Amer. Clin. Prod. Rev.,* November/December: 14–19, 1983.

Calbreath, D. F., "Diagnosis of multiple sclerosis and related disorders. Part two: Techniques and technologies," *Amer. Clin. Prod. Rev.,* February: 20–23, 1984.

Calbreath, D. F., "Diagnosis of multiple sclerosis and related disorders. Part three: Clinical aspects," *Amer. Clin. Prod. Rev.,* April: 12–16, 1984.

Clausen, J., *"Immunochemical Techniques for the Identification and Estimation of Macromolecules."* Amsterdam: North Holland Publishing Co., 1969.

Cohen, I. R., "The self, the world, and autoimmunity," *Sci. Amer.,* April: 52–60, 1988.

Deverill, I., and Reeves, W. G., "Light scattering and absorption—developments in immunology," *J. Immunol. Meth.* 38:191–204, 1980.

Finley, P. R., "Nephelometry: Principles and clinical laboratory applications," *Lab. Manage.,* September: 34–45, 1982.

Gross, W., and Maerz, W., "Immunoelectrophoretic techniques in protein analysis and quantitation," *Amer. Biotechnol. Lab.,* February, 1988.

Hamilton, R. G., "Human IgG subclass measurements in the clinical laboratory," *Clin. Chem.* 33:1707–1725, 1987.

Janik, B., "Immunofixation electrophoresis: A method for identification of monoclonal immunoglobulins," *Amer. Clin. Prod. Rev.,* January: 44–51, 1985.

Johnson, A. M., "Immunofixation following electrophoresis or isoelectric focusing for identification and phenotyping of proteins," *Ann. Clin. Lab. Sci.* 8:195–200, 1978.

Killingsworth, L. M., "Plasma protein patterns in health and disease," *CRC Crit. Rev. Clin. Lab. Sci.* 11:1–30, 1979.

Killingsworth, L. M. et al., "Protein analysis: Finding clues to disease in urine," *Diag. Med.,* May/June: 69, 1980.

Killingsworth, L. M. et al., "Protein analysis in light chain disease," *Diag. Med.,* June: 71–83, 1983.

Kyle R. A., "Monoclonal gammopathy of undetermined origin (MGUS): A review," *Clin. Haematol.* 11:123–150, 1982.

Kyle, R. A., "Monoclonal gammopathies and the kidney," *Ann. Rev. Med.* 40:53–60, 1989.

Maddison, S. E., and Reimer, C. B., "Normative values of serum immunoglobulins by single radial immunodiffusion: A review," *Clin. Chem.* 22:594–601, 1976.

Manuel, Y. et al. (ed.), *Proteins in Normal and Pathological Urine.* Baltimore: University Park Press, 1970.

Narayanan, S., "Method-comparison studies on immunoglobulins," *Clin. Chem.* 28:1528–1531, 1982.

Nichols, W. S., and Nakamura, R. M., "Antibody patterns in autoimmune disease," *Lab. Manage.,* May: 39–47, 1984.

Normansell, D. E., "Quantitation of serum immunoglobulins," *CRC Crit. Rev. Clin. Lab. Sci.* 17:103–178, 1982.

Nossal, G. J. V., "The basic components of the immune system," *New Eng. J. Med.* 316:1320–1325, 1987.

Penn, G. M., "The monoclonal gammopathies: Laboratory detection," *Lab. Manage.,* March: 30–40, 1981.

Pevzner, L. Z., and Gerson, B., "Multiple sclerosis: The state of laboratory diagnostic methods," *Lab. Manage.,* October: 37–47, 1985.

Poser, C. M. et al., "New diagnostic criteria for multiple sclerosis: Guidelines for research protocols," *Ann. Neurol.* 13:227–231, 1983.

Reckel, R., "Monoclonal antibodies: Clinical applications," *Adv. Clin. Chem.* 27:355–415, 1989.

Reimer, C. B., and Maddison, S. E., "Standardization of human immunoglobulin quantitation: A review of current status and problems," *Clin. Chem.* 22:577–582, 1976.

Rennert, O. M., "The hypogammaglobulinemias," *Ann. Clin. Lab. Sci,* 8:276–282, 1978.

Rodriguez, M., "Multiple sclerosis: Basic concepts and hypothesis," *Mayo Clin. Proc.* 64:570–576, 1989.

Royer, H. D., and Reinherz, E. L., "T lymphocytes: Ontogeny, function, and relevance to clinical disorders," *New Eng. J. Med.* 317:1136–1142, 1987.

Shoenfeld, Y., and Schwartz, R. S., "Immunologic and genetic factors in autoimmune diseases," *New Eng. J. Med.* 311:1019–1029, 1984.

Sittampalam, G., and Wilson, G. S., "Theory of light scattering measurements as applied to immunoprecipitin reactions," *Anal. Chem.* 56:2176–2180, 1984.

Stahlheber, P. A., and Peter, J. B., "Multiple sclerosis: A clearer path to a complex diagnosis," *Diag. Med.,* January: 43–48, 1984.

Chapter Eight

CLINICAL ENZYMOLOGY I: PROPERTIES OF ENZYMES

• •

UPON COMPLETION OF THIS CHAPTER, THE STUDENT WILL BE ABLE TO

1. Define the following terms:
   A. enzyme
   B. catalyst
   C. substrate
   D. product
   E. coenzyme
   F. activator
   G. inhibitor
   H. active site
   I. international unit
   J. katal unit
2. Describe the general role that enzymes play in daily body metabolism.
3. State the generic enzyme reaction equation.
4. Describe how enzyme nomenclature is determined in the EC system, using histidine decarboxylase as an example.
5. Name the six classes of enzymes.
6. Describe the type of reaction characteristic for each of the six classes of enzymes.
7. Discuss the use of NADH/NAD$^+$ and NADPH/NADP$^+$ as hydrogen transfer coenzymes.
8. Explain the role of metal ions as enzyme activators.
9. Describe the interaction that occurs between enzyme and substrate at the active site.
10. Discuss the degree of specificity that is expressed by an enzyme for its substrate.
11. Describe how measuring the rate of a reaction is used to estimate the amount of enzyme activity present in a sample.
12. Explain how the rate of an enzyme reaction is affected by each of the following parameters:
    A. pH
    B. temperature
    C. substrate concentration
    D. enzyme concentration
    E. coenzyme concentration
    F. inhibitor concentration
    G. activator concentration
    H. buffer concentration
13. Relate why substrate is present in excess when measuring the amount of enzyme present in a sample.
14. Describe how deviation from an enzyme's pH optimum may affect the integrity of either the substrate or the enzyme.
15. Describe how each of the following modes of analysis may be used to measure enzyme activity:
    A. end point
    B. multipoint
    C. kinetic
16. Illustrate the absorbance characteristics of the NADH/NAD$^+$ coenzyme system.
17. Explain the methodological significance of the NADH/NAD$^+$ coenzyme absorption patterns when quantitating enzyme activity.
18. Discuss the use of coupled enzyme reactions in clinical enzymology and of enzymes as analytical reagents.
19. Contrast how enzyme activity units have been expressed in the past with the current international unit and katal unit designations.

INTRODUCTION

As you read this page, a variety of reactions are taking place in your body. The sugar you put in your coffee is being broken down into smaller compounds and absorbed into the circulation. Fats from the margarine on your toast are being assimilated into larger molecules and stored in specific cells. Hemoglobin is being made and incorporated into red blood cells. Nerve impulses are being sent through the body. Old cells die and new cells are created to take their place. These and many hundreds of other biochemical reactions are going on within your system each moment of the day. The biochemical constituents that make these reactions occur are called enzymes.

8.1

DEFINITION OF AN ENZYME

We can describe enzymes two ways: in terms of their biochemical makeup and in terms of their function within the living system. **Enzymes** are proteins, as was described in Chapter 5. Although there are some exceptions to this rule (and we are not sure of the extent or validity of the exceptions), the basic composition of an enzyme is that of a protein, usually a globular one. The protein may have other groups (either organic or inorganic) attached to it. These prosthetic groups play important roles in helping the enzyme carry out its function.

The role of an enzyme is that of a **biochemical catalyst.** Each biochemical reaction (or group of related reactions) which occurs in the body has a specific enzyme associated with it. The enzyme makes it possible for the reaction to take place under normal physiological conditions. A bond-breaking reaction which may require high temperatures, extremes of pH, or other rather lethal reaction conditions in the test tube can be carried out at normal body temperature in dilute, aqueous solution at a moderate pH if an enzyme is present to make the reaction proceed.

8.2

ROLE OF ENZYMES IN BIOCHEMICAL PROCESSES

The Generic Enzyme Reaction

If we were to isolate a specific biochemical reaction and run it in a test tube, we would see that any reaction catalyzed by an enzyme follows the following general pattern:

$$E + S \rightleftharpoons ES \rightarrow E + P$$

In this equation, E stands for the enzyme under consideration, S is the substrate (the material acted on by the enzyme) and P is the product of the reaction. There may be more than one substrate in a given reaction and there may be more than one product produced, but the basic principles still hold.

The Role of the Enzyme/ Substrate Complex

In any reaction requiring a catalyst, one of the important steps is the attachment of the catalyst to the starting reactant (or substrate). This attachment to the catalyst must occur before the substrate can have a group added, a bond broken, or a structural rearrangement. There is no covalent bond formed when this complex is generated; the binding is usually ionic or weaker (hydrogen bonds or van der Waals interactions). This attachment of enzyme to substrate takes place very rapidly in comparison with other steps in the process. Formation of the enzyme/substrate complex is reversible; the complex can either proceed onward to product or can dissociate back to separate enzyme and substrate molecules, with no reaction taking place.

A catalyst in other types of reactions (organic or inorganic) serves mainly to orient the substrate molecule so the attacking group can more easily move into position to form the final product. The enzyme/substrate complex involves a more complicated reaction strategy. After the substrate binds (or as a result of substrate binding), the enzyme portion changes shape. This shift of three-dimensional structure puts a strain on a given bond of the substrate molecule if a bond-breaking reaction is taking place. If another group is to be added, the change in enzyme conformation may bring that attacking group into the right position to join the substrate molecule. Whatever the reaction, the enzyme's conformation changes to help the reaction go to completion.

The final process in the reaction is release of product. In some instances, the product is formed on the enzyme and then released. In other cases, as the product forms it immediately goes off the enzyme. Whatever the situation, the release of product is the rate-limiting step in the reaction. The enzyme reaction can go no faster than the product can be released.

Reversibility of Enzyme Reactions

The general enzyme reaction suggests that the process is unidirectional, that there is no back reaction involving the formation of S from P. In many instances this is the case. The reaction is not reversible, but proceeds (for all practical purposes) only in the direction written. In other situations, the reaction can go in either direction depending on the relative concentrations of substrate, product, and other factors involved. These reactions are called **reversible.** There is no good way to predict whether a given enzyme process is reversible; this property can only be determined experimentally. The ability of an enzyme to drive a certain reaction in either direction is important for metabolic control of biochemical processes and in some situations for assay of the enzyme.

Enzymes and Metabolic Pathways

In living systems we do not see isolated enzyme reactions. We cannot look at a single reaction involving only one enzyme, one substrate, and one product. Any organism has a large number of enzyme reactions taking place at once. The substrate for a reaction may have been the product of another reaction. What is really occurring is a complicated chain of enzyme reactions leading to some final product many steps down the way.

In Figure 8-1, A, B, C . . . are biochemical molecules and E_1, E_2 . . . are specific enzymes which catalyze the indicated reaction. Keep in mind that A may have been formed by a previous enzyme reaction. Substrate A is transformed to product B with the aid of enzyme E_1. However, in the reaction catalyzed by enzyme E_2, B is now considered a substrate which forms product C. No biochemical molecule can be considered a substrate or a product simply on the basis of its identity. We have to see the role it plays in a biochemical reaction before we can define it as one or the other.

You will also observe from the diagram that substrate D can form two different products: E and F. There are specific enzymes which catalyze these two reactions. At one point there may be a large amount of E formed from D and very little F. Conditions may change later so that F becomes the main product formed and little E is generated. A variety of factors contribute to determining which branch reac-

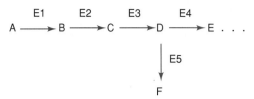

FIGURE 8-1 Metabolic pathways.

tion predominates at any given moment. The control of enzyme reactions and metabolic pathways is complicated and is one of the current areas of intensive research.

8.3

ENZYME NOMENCLATURE

Enzyme names came about in much the same way as organic nomenclature. Each enzyme was designated according to the reaction it catalyzed. Some characteristic of the reaction was identified (usually the substrate) and the suffix *-ase* was added to the name. The enzyme lipase is responsible for the metabolism of certain types of lipids (*lipid* + *-ase*). Disaccharides (a class of carbohydrates) are broken down by disaccharidases. A dehydrogenase enzyme is one which removes hydrogen from a molecule.

As long as there were only a few enzymes known, this informal system of nomenclature was adequate. However, we now know of over 1500 different enzymes, many of which catalyze very similar reactions. It became necessary for the International Union of Biochemistry (IUB, which is affiliated with the International Union of Pure and Applied Chemistry, IUPAC), to establish nomenclature rules for enzymes. The Enzyme Commission of the IUB first divided all known enzymes into six categories, based on the type of reaction they carried out. Each category (or class) was further subdivided on the basis of similarities or differences in the specific reactions. The result is a long listing (which is periodically updated) of all known enzyme reactions. Table 8-1 provides an example of how nomenclature is approached for a specific enzyme.

Table 8-1
PROCESS OF ENZYME NOMENCLATURE

Reaction catalyzed: histidine → histamine + CO_2

Enzyme: histidine decarboxylase EC 4.1.1.22
Class 4 : lyases—cleavage of $C-C$, $C-O$, $C-N$ or other bonds by means other than hydrolysis or oxidation
    4.1   carbon-carbon lyases—cleave $C-C$ bonds
    4.1.1   carboxylases—cleave $C-COO^-$ bond
    4.1.1.22 histidine carboxylase—cleavage of $C-COO^-$ bond in histidine

---

When reading the scientific literature, you will encounter a variety of ways of designating enzymes. In some places, particularly in the older literature, only the trivial, or nonsystematic, name is given. You may need to search carefully to be determine which enzyme is involved and which reaction is being catalyzed. More current articles generally name the enzyme two ways. Both the trivial and the EC nomenclature is used. Thus, you may encounter "creatine kinase" (EC 2.7.3.2, ATP:creatine *N*-phosphotransferase). This approach removes all ambiguity about the enzyme's identity.

8.4

CLASSES OF ENZYMES

As mentioned previously, the IUB Enzyme Commission categorized all enzymes into six classes based on the type of reaction catalyzed. We will look briefly at the classification and some representative examples of enzymes in each class, but we will not cover subclasses. Students interested in reading further may find more information in advanced biochemistry textbooks. Table 8–2 summarizes the various classes of enzymes.

EC Class 1: Oxidoreductases

Oxidoreductase enzymes carry out all reactions involving some type of oxidation/reduction. In biochemical systems, this usually means the addition of hydrogen to a double bond (reduction) or the re-

Table 8–2
ENZYME CLASSES

Class	Category	Type of Reaction Catalyzed
1	Oxidoreductase	Oxidation/reduction reactions
2	Transferase	Transfer of intact group of atoms from one molecule to another
3	Hydrolase	Cleavage of bonds with water
4	Lyases	Cleavage of C−C, C−O, C−N or other types of bonds; does not involve water
5	Isomerases	Convert one isomer to another
6	Ligases	Bond formation between two groups of atoms; with ATP as energy source

moval of hydrogen from a molecule to leave a double bond (oxidation). The older name for these enzymes is **dehydrogenase.** In many instances the process is simply one of hydrogen addition or loss. Some enzymes, however, couple hydrogen transfer to the loss of a carboxyl group or an amino group. The hydrogen is transferred with the use of another molecule called a **coenzyme** (to be discussed in Section 8.5). Perhaps the best known clinical example of an oxidoreductase is lactate dehydrogenase (EC 1.1.1.27, L-lactate:NAD$^+$ oxidoreductase). The reaction catalyzed by this enzyme is

$$\text{pyruvate} + \text{NADH} + \text{H}^+ \rightarrow \text{lactate} + \text{NAD}^+$$

Lactate dehydrogenase is found in a variety of tissues, and measurement of its activity can provide important information in cases of heart attacks or liver problems. Another oxidoreductase is alcohol dehydrogenase, which facilitates the conversion of ethanol to acetaldehyde in the liver.

EC Class 2: Transferases

The **transferase** enzymes move an intact group of atoms from one molecule to another. The group moved is a functional group, such as an amine or a phosphate entity. We have already mentioned one such transferase, creatine kinase. The EC name, ATP:creatine *N*-phosphotransferase, indicates more clearly what the reaction involves:

$$\text{ATP} + \text{creatine} \rightarrow \text{ADP} + \text{creatine phosphate}$$

A phosphate group from ATP (adenosine triphosphate) is transferred to the nitrogen atom of the creatine to produce ADP (adenosine diphosphate) and creatine phosphate. A more complicated transferase reaction is the transaminase process, an important means of regulating amino acid levels. In this reaction an amino group is exchanged.

amino acid I + keto acid II → keto acid I
+ amino acid II

Study of aminotransferase enzymes will give important information about liver damage.

EC Class 3: Hydrolases

Hydrolases are enzymes involved in the splitting of molecules, with water as part of the reaction process. There are a number of clinically important hydrolase enzymes. Amylase is involved with the cleavage of −C−O−C− bonds in starch. The hydrolytic enzyme lipase breaks down triglycerides to form glycerol and free fatty acids, a form of ester hydrolysis. Acid phosphatase and alkaline phosphatase remove a phosphate group from a variety of molecules.

EC Class 4: Lyases

As the name implies, **lyases** split molecules (*lysis* means "splitting"). The bonds broken may be $C-C$, $C-O$, $C-N$, or other bonds. The reactions are those other than oxidation, reduction, or hydrolysis. The enzyme aldolase (EC 4.1.2.13, D-fructose-1,6-biphosphate-D-glyceraldehyde-3-phosphate lyase) is one example of this class. Aldolase cleaves the 6-carbon molecule fructose-1,6-diphosphate to produce two 3-carbon compounds: glyceraldehyde-3-phosphate and dihydroxyacetone phosphate. This enzyme is occasionally assayed in disorders of the skeletal muscles.

EC Class 5: Isomerases

One important class of biochemical reactions involves the conversion of one isomer to another. These reactions are catalyzed by **isomerases**. Examples of transformations may include change of *cis* to *trans,* of an L-form of a compound to the corresponding D-form, or of an aldehyde to a ketone. Isomerase reactions are generally reversible. One such isomerase reaction can be seen in the glycolytic pathway, responsible for the conversion of the carbohydrate glucose to other compounds. One step in this process involves the isomerization of glyceraldehyde-3-phosphate (an aldehyde) to the corresponding ketone, dihydroxyacetone phosphate.

$$
\begin{array}{ccc}
\text{H}-\text{C}=\text{O} & & \text{CH}_2\text{OH} \\
| & \text{triose phosphate} & | \\
\text{H}-\text{C}-\text{OH} & \text{isomerase} & \text{C}=\text{O} \\
| & \xrightarrow{\hspace{1.5cm}} & | \\
\text{CH}_2-\text{OPO}_3 & & \text{CH}_2-\text{OPO}_3 \\
\text{Glyceraldehyde-3-phosphate} & & \text{Dihydroxyacetone} \\
& & \text{phosphate}
\end{array}
$$

The reaction is catalyzed by the enzyme triose phosphate isomerase (EC 5.3.1.1, D-glyceraldehyde-3-phosphate ketol-isomerase). At present, there are no isomerase enzymes assayed for diagnostic purposes in the clinical chemistry laboratory.

EC Class 6: Ligases

The **ligases** cause bond formation between two molecules to form a larger molecule. A requirement for any ligase reaction is the breakdown of ATP, which provides the biochemical energy necessary for the reaction to take place. One important set of ligases are the aminoacyl-tRNA synthetases. A representative reaction is

$$\text{ATP} + \text{L-tyrosine} + \text{tRNA} \rightarrow \text{AMP} + \text{pyrophosphate} + \text{L-tyrosyl-tRNA}$$

This class of reactions is important for activation of amino acids before they are incorporated into the growing peptide chain in the process of protein synthesis. Although ligases are important for the biosynthesis of many materials, they are not studied in the routine clinical chemistry laboratory.

8.5

COENZYMES AND OTHER ACTIVATING GROUPS

In the dehydrogenase (oxidoreductase) reaction, another material was required in addition to the substrate. A molecule called nicotinamide adenine dinucleotide (NAD) served as hydrogen acceptor in this reaction. Without NAD, the reaction would not have taken place. There are several nonprotein biochemicals or ions that are a part of the enzyme reaction but may not be explicitly included in the equation as written. If the needed substance is an organic molecule (such as NAD), it is referred to as a **coenzyme.** If it is an inorganic ionic cofactor (Mg^{2+}, Na^+), we call it an **activator.** Some organic molecules may serve as activators but may not be required for the reaction to take place. These molecules simply exert control on the rate and are valuable as metabolic regulators of a given reaction series.

Coenzymes

There are a number of coenzymes of importance in enzymatic reactions, but we will focus only on two: **NAD** and **pyridoxal phosphate.** We have already seen that NAD participates in reactions involving the transfer of hydrogen. The generalized reaction can be written as follows:

$$\text{R}-\text{H}_2 \text{ (reduced form)} + \text{NAD}^+ \rightleftharpoons \text{R (oxidized form)} + \text{NADH} + \text{H}^+$$

Note that the reaction is reversible. Depending on the circumstances, NAD^+ may accept a hydrogen atom (not a proton), or NADH may donate one. Also note that NAD as written above has a positive charge on the molecule. Quite often, for convenience the compound is designated without the plus charge.

When NAD is involved as part of an enzyme reaction, it is usually in solution, not tightly bound as part of the enzyme molecule itself. Therefore, in a reaction involving NAD or NADH, the coenzyme needs to be added to the reaction system. You cannot assume that it is present in sufficient amounts as part of the enzyme.

A second form of this coenzyme is NADP$^+$ (NADPH in the reduced form). The reaction involved is the same; the only difference is that NADP has an extra phosphate group attached to the molecule. Some biochemical systems require NADP instead of NAD, but the basic principles of the hydrogen transfer process remain the same in either case.

Another important coenzyme is **pyridoxal phosphate** (a form of vitamin B$_6$). In contrast to NAD, this molecule is fairly tightly bound to its enzyme by a covalent linkage. Pyridoxal phosphate is an important part of the aminotransferase reaction mentioned earlier. In this reaction, an amino group is transferred from one compound to another. The pyridoxal phosphate serves as the intermediary for the transfer by accepting the amino group from amino acid I and moving it to keto acid I. We illustrate the process schematically, with the following:

$$\text{amino acid I} + \text{PP} \rightarrow \text{keto acid II} + \text{PP}-\text{NH}_2$$
$$\text{keto acid II} + \text{PP}-\text{NH}_2 \rightarrow \text{amino acid II} + \text{PP}$$

The pyridoxal phosphate (PP) is recycled from the pyridoxal form through an intermediate pyridoxamine structure and then returns to the original pyridoxal at the end of the reaction cycle. In Chapter 10 we will discuss this coenzyme more as we look at the assay methods for two of the aminotransferase enzymes found in serum.

Activators

In some instances, a specific enzyme reaction may require an inorganic activator in order for that process to occur. Usually the activator is a cation, generally a +2 species. Probably the best-known system requiring a cation activator is alkaline phosphatase. This enzyme has Zn^{2+} tightly bound to the enzyme. Removal of the zinc ion results in total loss of activity. In addition, another cation (usually Mg^{2+} is employed) is needed for maximal activity. Although the Zn cation does not need to be included in the reaction mixture, we must add Mg^{2+} to the system for the reaction to take place. The only known anion activator is Cl$^-$, which is a part of the amylase enzyme reaction.

8.6

ANATOMY OF AN ENZYME

To understand how an enzyme carries out its function, we need to look closely at the structure of these proteins. By understanding how an enzyme is constructed, we can make some predictions about its behavior.

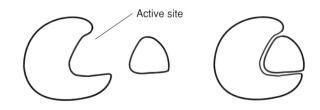

Enzyme + Substrate ⟶ Enzyme / substrate complex

FIGURE 8-2   The enzyme active site.

Concept of the Active Site

The key part of the enzyme structure is the **active site** (Fig. 8-2). This is the location on the protein where the substrate binds to the enzyme. The active site is a sort of pocket, a cleft in the enzyme into which the substrate fits. At one time it was thought that this site was a rigid structure, a concept known as the "lock and key" mechanism of enzyme action. Now we know that the active site (like the remainder of the protein) is somewhat flexible. This location can adjust somewhat to the shape of the incoming substrate molecule. Thus, we have an "induced fit" of substrate to active site.

Enzyme Specificity

The key point about the active site is that the structure allows only specific molecules to bind. A given enzyme may allow only one molecule into the active site. Another enzyme allows several molecules that are structurally very similar to attach to the site. In each case, the enzyme is exhibiting a property called **specificity.** It is important for each enzyme to show specificity; otherwise, biochemical processes would take place in a random, chaotic fashion.

Some enzymes are nonspecific, meaning they are capable of catalyzing a reaction involving a number of substrates. Other enzymes have more restrictive reactive properties (Table 8-3). We see from the table that all three enzymes are peptidases; they cleave the peptide bond formed between two amino acids. However, the specificity of each enzyme is different. Subtilisin promotes the breaking of a large number of peptide linkages, but trypsin is somewhat more selective. Thrombin is quite specific, requiring a peptide bond formed between a certain two amino acids, with the added requirement that one of the two be on the carboxy side while the other is on the amino side of the linkage. If the positions of the two amino acids were reversed, thrombin would not catalyze the reaction.

Table 8-3
SPECIFICITIES OF ENZYMES

Enzyme	Bond Cleaved
Subtilisin	Any peptide bond
Trypsin	Peptide bonds involving carboxyl group of lysine or arginine
Thrombin	Only the peptide bond between the carboxyl group of arginine and the amino group of glycine

The specificity of a given enzyme is determined by the amino acid sequence in the vicinity of the active site. There may be certain amino acid side chains which are ionized, allowing the binding of a substrate molecule with the opposite charge. The shape of the active site cleft may allow the binding of a short-chain molecule, but not a long-chain one. A particular grouping of amino acids may permit access to the active site by a cyclic molecule (phenol derivative or other), but not a straight-chain structure.

This active site portion of the enzyme molecule is not one continuous sequence of the protein (like that in Fig. 8-3 A). Because of the folding and coiling of the molecule, portions of the amino acid sequence which are far removed from one another

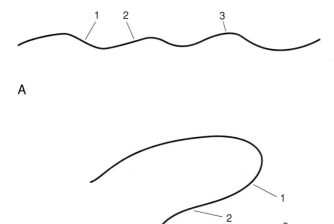

A

B

FIGURE 8-3   Enzyme conformation and active site. A. Protein in open-chain configuration. 1, 2, and 3 are amino acid residues involved in the enzyme reaction. B. Protein in folded (active) configuration.

when the protein is stretched out straight come into close proximity when the molecule folds into its proper conformation (Fig. 8-3 B).

8.7

ENZYME REACTION RATE

Measuring Enzymes

In Chapter 5 we discussed measuring the amount of protein in a body fluid. We saw that there were few (if any) good chemical methods for protein quantitation, since all proteins have the same fundamental structure. Although albumin can be specifically measured (because there was so much of this molecule present), most proteins exist in such low quantities in body fluids that they cannot be assayed easily by chemical means. However, enzymes possess an advantage over other types of proteins in that they catalyze biochemical reactions which we can observe. In addition, each enzyme catalyzes a fairly specific reaction. When we want to assess the amount of enzyme present in a tissue or body fluid, we do so indirectly by studying the ability of the enzyme to make a biochemical process take place.

Concept of Reaction Rate

When a biochemical transformation takes place, substrate is converted to product in the presence of an enzyme. The process involves a molecule of substrate binding to an enzyme molecule, the substrate being converted to product, and then the product being released from the enzyme. All of the substrate does not instantaneously change to product. There is a certain time required for each molecule to bind, be converted, and then come off the enzyme. A number of factors determine how much time is necessary for each transformation, (These factors will be discussed in later sections of this chapter). The time needed for each substrate-to-product transformation can be determined. From this information we can estimate the reaction rate of the enzyme and (indirectly) the amount of enzyme present in the biological material under study.

In measuring the rate of a reaction, we determine how much product is formed in a given time. We could look at substrate utilization, but there are some methodological problems associated with that approach. If 10 μmol of product are formed in 5 min, the rate of the reaction is 2 μmol/min. So we measure the amount of product formed and divide by the time required to form that amount of material. Details of this process will be covered later in this chapter and in Chapter 10 when we look at specific enzyme assays.

What does this have to do with determining how much enzyme is present in a sample? Look back at the basic equation for an enzyme reaction (Section 8.2). Each molecule of substrate must combine with a molecule of enzyme to form a molecule of product. In measuring enzyme activity, we set up assay conditions so that there are many more substrate molecules than enzyme. In this situation, each enzyme molecule binds substrate and converts it, then accepts another substrate molecule for further reaction. The substrate cannot be converted any faster than the number of enzyme molecules present allows. Therefore, we say that the enzyme level is **rate-limiting;** that is, the enzyme concentration alone determines how fast the reaction proceeds when substrate is present in excess.

For example, let us measure the amount of amylase activity in 0.2 mL serum (holding the sample volume constant is necessary so we can compare amounts of enzyme from different patients). We add our starch substrate and measure the product formation at a stated time (say 10 min). We find that the amount of enzyme present in sample I is sufficient to form 25 μmol of product in 10 min; the rate is 2.5 μmol/min. Sample II forms 63 μmol of sample in the same period; the rate of activity here is 6.3 μmol/min. From this information we infer that the concentration of amylase in sample I is approximately 2½ times that in sample II. By measuring the rate of reaction under standardized conditions, we can estimate the concentration of a specific enzyme without being able to quantitate that enzyme directly by means of a chemical analysis.

8.8

FACTORS AFFECTING REACTION RATE

When we study an enzyme reaction, several parameters are involved. The concentrations of substrate, enzyme, necessary cofactors—all affect how fast the reaction proceeds. Type of buffer, pH, temperature, inhibitors, and activators alter the reaction process in some way. These parameters need to be evaluated so we can learn about the properties of any given enzyme, gain some understanding of what it does in the body, and learn how to measure enzyme activity under set conditions.

Substrate Concentration

If the enzyme itself is present in high enough amounts, the rate of the reaction is determined by the concentration of substrate present. As the substrate level increases, the enzyme reaction rate also increases (Fig. 8–4). There comes a point where a

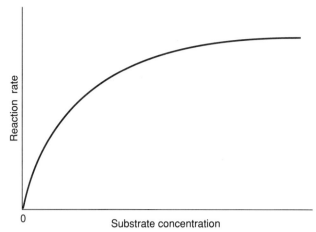

FIGURE 8–4 Effect of substrate concentration on reaction rate.

further increase in substrate concentration produces no more enhancement of the reaction rate. Keep in mind that we are running these experiments with a fixed amount of enzyme in the reaction system. At this point, the enzyme is "saturated" with substrate.

The active site concept provides a simple explanation of what is taking place (Fig. 8–5). A certain number of available active sites are present (one per enzyme molecule in our example). When we add a low concentration of substrate, each substrate molecule eventually binds to the active site of an en-

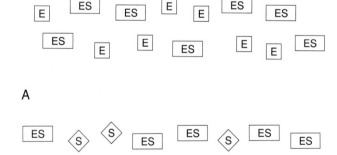

FIGURE 8–5 Enzyme/substrate interactions and reaction rate. A. No free substrate available. Enzyme in excess—substrate is rate-limiting. B. No free enzyme available. Substrate in excess—enzyme is rate-limiting. $\boxed{E}$ = enzyme, $\langle S \rangle$ = substrate, $\boxed{ES}$ = enzyme/substrate complex.

zyme molecule. Not all the enzyme molecules have substrate attached to the active site at any given time. If we increase the substrate concentration, we increase the probability of substrate colliding with enzyme to form the ES complex. When we increase these collisions, we increase the reaction rate. However, at even higher substrate concentrations, there is so much substrate that the individual molecules have to "wait in line" to attach to an enzyme active site. At this point, the enzyme is saturated with substrate. No matter how many more substrate molecules are added, there are only enough enzyme molecules to bind them at a certain rate. Further increase in substrate concentration does not result in a more rapid conversion to product.

When measuring the amount of enzyme present in a body fluid, the substrate is in excess. The rate-limiting factor (to be discussed in the next section) then becomes the amount of enzyme. The rate is determined in this case by the amount of enzyme present. The more enzyme present, the higher the rate (up to a point).

On the other hand, we can use an enzyme in the quantitation of a given material in biological fluids. By adding an excess of enzyme, the rate of reaction is proportional to the substrate concentration. With the use of calibration curves we can determine the amount of substrate originally present in the assayed material. This analytical approach will be discussed in detail later in this chapter.

Enzyme Concentration

The above discussion centered on keeping the enzyme concentration higher than the substrate level. When we measure the amount of enzyme in a body fluid, we take the opposite approach. If the substrate is present in sufficiently high amounts, the rate of reaction is a function of the enzyme concentration. As the enzyme level increases, the rate increases for a defined volume of body fluid. Again, the rate levels off when the level of enzyme gets to the point where the substrate concentration is no longer saturating. If we encounter this situation when doing enzyme determinations, we can estimate the level of the enzyme by running another assay using a smaller volume of body fluid or by diluting the sample and reassaying it (the better practice). Keep in mind the results obtained must be corrected for the change in volume or the dilution in order for the final value to be correct. In some situations, other reaction conditions may also need to be adjusted (such as activator concentration). Dilution could lead to a false increase in the amount of enzyme supposedly present in the system. The effects of dilution need to be carefully evaluated for each enzyme being measured in the clinical laboratory.

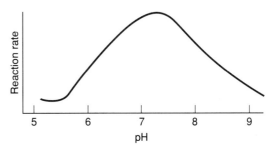

FIGURE 8-6   Effect of pH on reaction rate.

Effect of pH on Rate

A change in the hydrogen ion concentration of the reaction medium can have profound effects on the rate of an enzyme reaction. The typical bell-shaped curve is illustrated in Figure 8-6. This curve is the composite of several separate effects. Although the shape of a specific curve may be somewhat different, depending on the enzyme and buffer, some common characteristics are seen. At the pH extremes (whatever they may be), the reaction rate is rather low. The rate gradually increases to a **pH optimum,** the point at which the reaction rate is greatest for the conditions. This pH optimum is the hydrogen ion concentration at which the assay should be run for analytical purposes.

Several factors determine the pH optimum. If the substrate can be ionized at a certain pH, the degree of ionization may affect binding to active site and the resultant activity. Conversely, the active site may be able to exist in an ionized or unionized form. The presence or absence of charge on the active site affects substrate binding and reactivity. Stability of the protein molecule is also affected by pH. If the solution is too acid or too basic, the protein structure unfolds. Further extremes of pH may cause hydrolysis of portions of the enzyme, leading to loss of structural integrity and decrease in activity.

Temperature

In the general chemistry laboratory we learned that we can speed up a reaction if we heat the system. Similarly, we can increase the rate of an enzyme reaction by increasing the temperature. The data in Figure 8-7, however, suggest that we can have too much of a good thing. After a certain point, further increase in temperature leads not to further increase in rate, but to loss of enzyme activity.

As the temperature increases, the movement of molecules increases. This means that there are more collisions between enzyme and substrate molecules and more reactions taking place. Within a narrowly

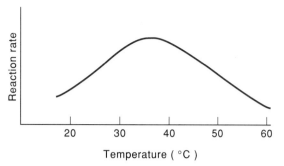

FIGURE 8-7 Effect of temperature on reaction rate.

defined range of temperatures, enzyme activity doubles (more or less) for every 10°C rise in temperature. But there comes a point where another factor begins to enter into the equation. The increase in temperature leads to an increase in the unfolding of the enzyme molecule. From its rather tight globular structure, the protein begins to spread out into a more linear straight-chain configuration. With this unfolding comes a loss of enzyme activity. Active site structure is disrupted, less substrate is able to bind and be converted to product, and the reaction speed goes down.

This process of denaturation at elevated temperature varies from one enzyme to another. Some enzymes can maintain structural integrity at higher temperatures than others. Amylase may remain reasonably stable until the temperature reaches nearly 60°C, whereas other enzymes cannot be incubated at temperatures of 40 to 45°C without reduction in catalytic function. The temperature curve for each enzyme must be determined experimentally; there is no way to predict this behavior in advance.

In the clinical laboratory, some compromises must be made when assaying enzymes. Although we can tailor a buffer-substrate-cofactor system which is reasonably specific for each enzyme, there are limits to how much we can vary temperature. A number of different assay systems may be used on the same instrument. A certain amount of time is required to go from one temperature to another in the reaction chamber. Therefore, some compromises are made when selecting the temperature for a specific enzyme assay. Most commercial systems carry out enzyme analyses at one (or perhaps two) set temperatures. For a variety of reasons, 30°C appears to be a good compromise in the majority of cases. Some systems employ 37° (either alone or with a 30° option). A few instruments may have a 25° setting, but this temperature often results in levels of activity which may be too low for adequate measurement of normal samples. In addition, if the room temperature rises above 25°C, the sample chamber may not be able to maintain the proper assay temperature without a cooling device (an expensive and impractical solution in most laboratories).

Coenzyme Concentration

Many enzymes require a coenzyme of some sort for the reaction to proceed. Probably the most commonly used coenzyme in clinical settings is NAD (or NADH). This particular molecule is not tightly bound to the enzyme, but exists free. NAD (or NADH, depending upon the direction of the specific reaction) must be present at the proper concentration for many enzyme reactions to take place. The rate of reaction for lactate dehydrogenase, for example, is regulated as much by the concentration of this nucleotide coenzyme as it is by the substrate (either lactate or pyruvate, depending on the direction of the assay).

Other enzyme systems have more complicated coenzyme requirements because the coenzyme is more tightly bound to the protein molecule. Pyridoxal phosphate is a coenzyme for aminotransferase reactions. This molecule is loosely attached to the enzyme by a covalent bond. For many years it was felt that the aminotransferase enzymes in serum contained sufficient pyridoxal phosphate to function at optimum activity even if no additional coenzyme was added to the system. In the mid-1970s a number of studies showed that it was necessary to supplement the reaction mixture with added coenzyme in order to measure accurately the serum aminotransferase levels in some disease states. We will discuss this situation further in Chapter 10.

Inhibitor Concentration

There are a number of molecules which decrease the rate of an enzyme reaction if they are present in the reaction system. These molecules are called **inhibitors.** An inhibitor may bind to the active site, blocking the access of substrate to the enzyme. Other inhibitors bind elsewhere on the enzyme molecule, causing a change in shape that interferes with substrate binding or somehow alters the rate. The use of inhibitors can give important information about the enzyme and how it carries out a specific reaction. However, inhibitors are not normally em-

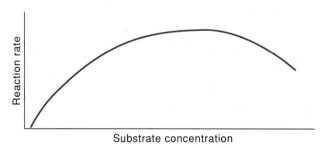

FIGURE 8-8 Substrate inhibition of enzyme.

ployed as part of an assay system for a clinically significant enzyme.

There are situations in which inhibitors can affect the enzyme rate, which may affect an assay for specific enzymes. If substrate is present in too high a concentration (substrate inhibition), a decrease in the rate may be seen (Fig. 8-8). In other instances, the product concentration could get high enough to provide product inhibition of the enzyme. In some instances, an inhibitor is added to a reaction system which specifically blocks the activity of a given isoenzyme so that the tissue source of elevated enzyme activity may be determined. Specific examples will be considered in Chapter 10 when we look at the use of enzyme measurements in clinical diagnosis.

There is increasing data to show that the body has endogenous inhibitors to some enzymes. These inhibitors are substances produced by the organism which exert regulatory control as a part of normal biochemical processes. One such inhibitor is the chemical which lowers the rate of monoamine oxidase. This enzyme has been extensively studied as a possible biochemical marker for schizophrenia. An endogenous inhibitor for monoamine oxidase has been found in human urine, plasma, and CSF. Neither the exact structure or function of this inhibitor is known at present.

Activator Concentration

Many enzymes require the presence of metal **activators** for full activity. These inorganic entities help bind the substrate to the active site by forming ionic bridges. The metal also helps orient the substrate so it can attach to the protein at the proper point and in the correct configuration. Although some metal activators are tightly bound to the enzyme (and do not need to be added to the reaction system), others are more loosely attached. In these cases, supplemental activator must be a part of the reaction mixture in order to obtain full enzyme activity. It is quite common to have Mg^{2+} or a similar divalent cation added to the assay system to provide optimum reaction conditions for a given enzyme.

Buffer

We are coming to realize that the buffer selected for our assay plays a more important role than just the regulation of pH. In many instances, the buffer itself affects the rate of the reaction (Fig. 8-9). A change in concentration of the buffer changes the rate of the enzyme reaction. A change in buffer structure may have the same effect. Probably the best known buffer contribution to enzyme reaction

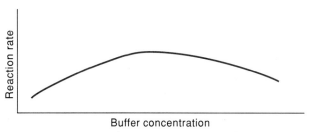

FIGURE 8-9   Effect of buffer on reaction rate.

rate is in the alkaline phosphatase system. Here the buffer serves as an acceptor of the phosphate group removed from the substrate. Without this acceptor present, the rate of the reaction is markedly lower.

8.9

MEASUREMENT OF ENZYME ACTIVITY

How do we translate this mass of information into a useful enzyme determination? What do we really want to learn when we study an enzyme in a clinical diagnostic situation? We are using the enzyme as a signal of tissue damage, a marker of cellular destruction. Chapter 5 discussed the release of proteins from a cell in various circumstances. The measurement of body fluid levels of enzymes is simply an extension of that concept. Instead of quantitating the amount of some specific protein or assessing the nonspecific release of many proteins into the circulation, we use the enzyme as a marker for the exploration of cellular turnover. It is important to remember that all the parameters discussed previously must be carefully controlled so that the only variable is the amount of enzyme present. Only by maintaining consistency in every aspect of the assay can reproducible results be obtained. Applications of the data in determining the site and extent of tissue damage will be discussed in Chapter 10.

Activity Versus Mass

In Chapter 5, we used the quantitation of specific proteins to tell us something about the dynamics of cellular metabolism in health and disease. With the use of chemical or immunochemical assays, we can determine how much of a certain protein is present in serum or other body fluids. These assays involve direct quantitation of the material under consideration. Enzyme measurement, on the other hand, is an indirect assessment of the amount of these proteins because we base our considerations on the measurement of enzyme activity, not on the direct quantitation of the specific protein. There are some techniques becoming available for direct quantitation of enzymes as proteins, but there is still controversy about the usefulness of this approach.

Substrate Disappearance or Product Formation

To determine the rate of an enzyme reaction, we measure the change in concentration of some chemical as a function of time. The material quantitated can be either substrate or product; theoretically, we could measure either component and get useful results. But what is the practical, most reliable way to approach the problem?

If we measure substrate concentration changes, we are looking at a rather small change in the concentration of total material over the time selected. Keep in mind that the enzyme assay is running with excess substrate present. Ideally, the substrate concentration is still saturating throughout the course of the reaction. This means that we might be looking at a 10% or smaller change in the substrate concentration over the time defined by the assay conditions. If we start with a substrate concentration of 100 μM (micromolar) and have a final substrate concentration of 90 μM, this 10 μM change is difficult to measure accurately in the presence of a high final substrate concentration.

Measurement of product, on the other hand, allows us to start at zero initial product level. Here we measure an increase in product concentration, from nothing up to the final product level. Looking at our previous example, we have an initial product level of 0 μmol. At the end of the assay time, 10 μmol of product have formed. This concentration change can be more accurately measured than if we were looking at change in substrate concentration. Most enzyme assays use the measurement of product formation as the principle for analysis.

Types of Enzyme Assays

Not only must our enzyme assay be reliable, it also must be relatively easy to run, require as little time as possible (so more assays can be run during that time), and be cost-effective. An assay approach which might be suitable for a research laboratory may be totally inappropriate for a routine clinical setting. Let us look at three different ways to measure the rate of an enzyme reaction.

1. end point
2. multipoint
3. kinetic

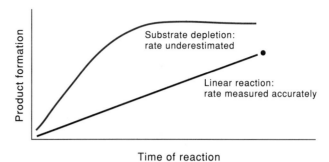

FIGURE 8–10   End-point enzyme analysis.

END-POINT ANALYSIS

End-point analysis is the simplest and most widely used technique. In this approach, the reaction is initiated by the addition of substrate and allowed to proceed for a set time. At the end of that time, the reaction is stopped and the amount of product formed during that time is measured. If the reaction time is 10 min and the amount of product formed at the end of that time is 47.2 μmol, the rate is 4.72 μmol/min for the volume of sample employed. One measurement is made for each reaction tube, allowing a large number of assays to be run at once. This system is easily automated.

A major problem arises with the end-point approach when we have a sample with high activity. If substrate depletion occurs, the rate of the reaction drops off part way through the assay. The "true" activity of the enzyme is then underestimated (Fig. 8–10). Because we measure only one point, we have no way of knowing whether the reaction rate has remained linear throughout the entire time. This problem is overcome in part by establishing the limits for linearity (using a multipoint approach). If a sample has an enzyme level above a certain point, the procedure must state that the sample is to be diluted and reassayed in order for a reliable result to be obtained. There is no direct way to assess linearity with a one-point assay system.

MULTIPOINT ASSAY

As the name implies, the **multipoint assay** measures the change in concentration at several intervals during the course of the assay. This type of assay is practical only with automated analysis systems for laboratories with a moderate to large enzyme assay workload. Some assay systems may measure only two points during the period, whereas other instruments allow the determination of a large number of data points for each sample.

Multipoint assays are most practical when a sim-

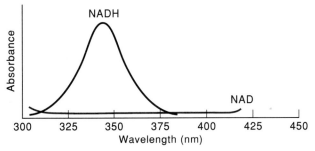

FIGURE 8-11   Absorbance spectra of NAD and NADH.

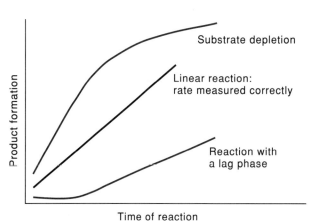

FIGURE 8-13   Multipoint enzyme assay.

ple change in absorbance can be determined and related to change in concentration. One of the most versatile systems for this kind of analysis is the oxidoreductase reaction which involves either formation or utilization of NADH. This form of the coenzyme has a strong absorbance of light at 340 nm, whereas NAD has essentially no absorbance at this wavelength (Fig. 8-11). By simply passing light of the correct wavelength through the reaction solution, the change in absorbance as a function of time can be readily determined (Fig. 8-12). This change can then be related to change in concentration and used to calculate enzyme reaction rates.

A major advantage of this approach is that it allows us to assess whether the reaction is linear throughout the entire time of the assay (Fig. 8-13). We would expect to see a straight-line relationship between concentration and time if linearity held. When substrate becomes depleted, the rate drops off and we no longer obtain a straight line, but a curved plot of data. Other enzymes, such as creatine kinase, show an initial **lag phase.** At first, the rate of reaction is very low. After a time (from seconds to minutes, depending on the enzyme), the reaction rate increases and becomes linear. Some automated systems analyze all the readings and print out a flag,

warning the operator of a nonlinear reaction when it occurs.

The multipoint assay does not lend itself well to manual procedures. An automated system is necessary to exploit this analytical approach to the fullest. Fortunately, almost all enzyme assays today can be run on some sort of automated or semiautomated analyzer.

KINETIC ASSAY

In its purest sense, the **kinetic assay** involves the continuous measurement of change in concentration as a function of time. A research laboratory using a kinetic assay places the reaction cuvet in a spectrophotometer, sets the instrument at the desired wavelength, and runs a strip chart to trace the change in absorbance as a line on the chart. Obviously, this is a very time-consuming approach to enzyme analysis. It has a value in a research setting, but is usually impractical in the clinical laboratory.

Confusion arises because many people use the term *kinetic* to describe what is actually a multipoint assay system. The situation becomes particularly unclear when the multipoint system uses time intervals only a few seconds apart. To some extent, this choice of terminology is a matter of preference, and we will not make an issue of it in our discussion.

USE OF COUPLED ENZYME REACTIONS

There are several enzymes whose substrate or product cannot be easily quantitated. To carry out assays of these enzymes is difficult, time-consuming, and expensive. In many instances, though, it is possible to assay the product of one enzyme reaction with the use of another enzyme reaction. Because

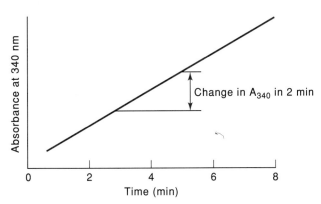

FIGURE 8-12   Change in absorbance as a function of time.

the product of the second reaction is much easier to quantitate, it allows us to measure indirectly the reaction we are interested in.

Figure 8–14 shows an assay widely used for the measurement of aminotransferase activity. None of the materials (either substrates or products) of reaction 1 can be measured easily and specifically by chemical means. It is possible to estimate the amount of oxaloacetate formed by reacting it with a second enzyme, malate dehydrogenase. This enzyme requires NADH as a coenzyme. Since NADH absorbs light strongly at 340 nm, we can follow the decrease in absorbance as NAD is formed. For every molecule of oxaloacetate generated, one molecule of NADH is used in the second enzyme reaction. Therefore, each molecule of NADH used in the malate dehydrogenase assay represents one molecule of oxaloacetate formed in the first reaction. From this data, the activity of aspartate aminotransferase can be calculated.

Coupled assays are widely used in clinical enzymology for several enzymes. Creatine kinase employs two supplementary reactions in one assay system.

8.10

UNITS OF ENZYME ACTIVITY MEASUREMENT

Discussion of units of enzyme activity calls to mind some of the great names in clinical enzymology. Unfortunately, the discussion also brings up a great deal of confusion, particularly when we look back at some of the earlier scientific literature in this area. There are many ways to indicate enzyme activity, some of them quite confusing. In recent years there has been a strong move toward an international system of enzyme units, which is bringing a great deal of order to this rather chaotic field.

Reporting Enzyme Results

Earlier in this chapter we discussed the idea of a rate of reaction. We saw that enzyme data is reported in terms of a concentration change per unit time (micromoles/minute or other units). The concentration change generated by the enzyme must also be related to the amount of sample employed in the assay system. If two assays are run and each gives a reaction rate of 15 μmol/min, we assume that each reaction mixture had the same amount of enzyme present. But let us introduce one more variable. Assume that the first reaction tube had 0.1 mL of serum in it and the second tube had only 0.05 mL

Reaction One:

$$\text{alpha–ketoglutarate + aspartate} \xrightarrow{\text{aspartate aminotransferase}} \text{glutamate + oxaloacetate}$$

Reaction Two:

$$\text{oxaloacetate + NADH} \xrightarrow{\text{malate dehydrogenase}} \text{malate + NAD}$$

FIGURE 8–14 Coupled-enzyme assay for aspartate aminotransferase.

of serum. The actual level of enzyme in the second tube would be twice that of the first. So our results must be expressed as change of concentration per unit time per sample volume, or

rate = concentration/unit time/sample volume

An example of the correct expression of a rate might be 20 μmol/min/mL sample.

Traditional Enzyme Activity Units

When we look at the "classical" methods of clinical enzymology, we see a very confusing picture. Researchers did what was convenient, defining their enzyme units according to their own preference and not in accord with some generally accepted standard. A good example of this multiplicity of units is seen in the assay for alkaline phosphatase. The reaction involves the conversion of phenyl phosphate (or p-nitrophenyl phosphate) to the corresponding phenol, with phosphate being liberated. Table 8–4 illustrates some of the various ways the reaction rate for alkaline phosphatase has been reported. Not indicated in the table is the fact that the assays did not all use the same substrate, buffer, pH optimum, or temperature for the reaction. Neither is the volume of sample indicated. Obviously, there can be little value in comparing the units of one assay with those of another.

Table 8–4
UNITS OF ALKALINE PHOSPHATASE ACTIVITY

Name	Units of Activity
Bodansky	mg phosphorus/h
Kind-King	mg phenol/15 min
Bessey-Lowry-Brock	mmol p-nitrophenol/h
Bowers-McComb	μmol p-nitrophenol/min

Standardization of Enzyme Activity Units

Because of this confusion in activity units and sample volumes, it became clear that some form of standard reporting was needed. The Commission on Enzymes of the International Union of Biochemistry and IUPAC proposed a system of units which is receiving wide international support. They defined the **international unit (IU)** of enzyme activity as that amount of enzyme which catalyzes the conversion of 1 μmol of substrate in 1 min. The volume of sample to be used in the reporting is that volume which gives "convenient" (whole-number) data. In most instances, enzyme results in the clinical laboratory are reported in international units per liter of sample (even if only 10 μL of sample are used in the assay). Often we see IU simply referred to as U (such as "4700 U/L").

More recently, the term *katal* has been recommended as a unit of enzyme activity. The **katal** is the amount of enzyme which converts 1 mol of substrate per second. This proposal has not yet received universal acceptance.

Standardization of Assay Conditions

Various international groups are involved in developing standard assay approaches for biochemical components. A major effort has been made to standardize the area of enzyme analysis. Many enzymes have a **Selected Reference Method** or an assay approach defined by an international group. These methods have been adopted after careful study and evaluation of all parameters of the reaction. By developing standard methods, it is much easier for laboratories in various parts of the world to compare their data.

8.11

QUALITY CONTROL IN ENZYME ASSAYS

Unlike other constituents in body fluids, the development of quality control materials for enzymes presents some major problems. Enzymes are inherently unstable molecules. Even in lyophilized form, they lose activity over time. After reconstitution of the quality control sample, significant activity may be lost within a few hours. Although standard methods for enzyme assay are being developed (and accepted), care must be taken when comparing results with those listed in the assay sheet or proficiency report. Standards for enzymes are not readily available. We cannot simply weigh in a stated amount of enzyme like we can glucose or uric acid. Much work remains to be done in this area.

8.12

USE OF ENZYMES IN QUANTITATIVE ANALYSIS

Enzymes as analytical reagents are rapidly supplanting many of the classical chemical methods for assay. With the use of enzymes we have been able to develop a number of sensitive and specific methods for biochemical analysis. The general approach is to employ the enzyme in amounts much greater than that of the material to be analyzed. Thus, the analyte is rate-limiting. The quantitation of the biochemical component may be done by determining the rate of the resulting enzyme reaction, with the rate being proportional to the amount of material present. This approach has been used in one widely employed method for measurement of glucose.

More often, the analyte is completely converted to product and some parameter of the reaction measured. We can assay for serum lactate by adding an excess of lactate dehydrogenase and NAD. The lactate is completely converted in the reaction, with one molecule of NADH being formed for every molecule of lactate utilized. Measurement of absorbance at 340 nm allows us to calculate the total amount of NADH formed. From these data we can then determine the quantity of lactate originally present in the sample. Glucose, cholesterol, and other materials are routinely measured using coupled enzyme reactions. Because of the specificity of the enzyme, there are usually few if any interferences from medications or other components of the biochemical system.

SUMMARY

Enzymes are biochemical catalysts, causing reactions to occur much more rapidly than they would otherwise. In the general enzyme reaction, a substrate is converted to a product through interaction with the enzyme. The intermediate enzyme/substrate complex can revert to the starting materials. All metabolic pathways consist of a series of interlocking enzyme reactions. Quite often, the product of an enzyme reaction may inhibit a much earlier step in the pathway, providing a means of regulating the rate of formation of end products.

There are six major categories of enzymes, classified by type of reaction. Oxidoreductases catalyze oxidation reduction reactions, usually involving hydrogen transfers. Transferases facilitate movement of amine groups (and others) from one molecule to another. Hydrolases carry out cleavage reactions which involve water as one of the reaction components. Lyases are concerned with nonhydrolytic splitting of molecules. Isomerase enzymes interconvert isomers. Ligases join two small molecules to form a larger molecule.

Often, a specific enzyme requires a coenzyme (organic compound) or other activator (inorganic, usually a cation) for full activity. These small molecules are loosely attached to the enzyme and can easily be separated from it.

The heart of any enzyme reaction is the active site of the molecule. This is the location for substrate (and activating group) binding. The amino acid composition of the active site determines the specificity of the enzyme. Some enzymes may catalyze a particular reaction for a broad range of structurally similar substrates, but other enzymes are much more restricted about the type of substrate with which they interact.

Enzymes are measured in body fluids by the reactions they catalyze, not by some chemical method of determining the presence of specific protein. The reaction is assessed by determining the amount of product formed (or substrate consumed) during a specified time under defined conditions by a particular amount of enzyme. By relating reaction rate to amount of body fluid assayed, we can indirectly estimate the amount of active enzyme present in the body fluid.

A variety of factors affect the reaction rate for a particular enzyme. As substrate concentration increases, the rate of the reaction increases until all the enzyme active sites are saturated. In some instances, a further increase in substrate results in a decrease in activity (substrate inhibition). Each enzyme has a pH optimum; at pH values above and below the pH optimum, the enzyme functions at a reduced rate. Enzyme activity increases as the temperature rises, until the point is reached where protein denaturation begins to take place. Concentrations of coenzymes, activators, and inhibitors all affect the reaction rate for a given enzyme. The type of buffer employed may enhance the rate by facilitating transfer of groups or could slow the reaction rate through inhibitory binding of buffer to the active site.

Reaction rates can be assessed by three major approaches. The end-point analysis is the simplest means of determining enzyme activity but provides no information about rate changes due to substrate depletion. Multipoint assays permit determination of the maximum rate for the enzyme. These assays are simple to carry out with automated equipment. Kinetic assays provide a continuous assessment of enzyme rate but are impractical for routine applications. Coupled enzyme assays have grown in popularity by providing a means of measuring activity in situations where the first product formed is difficult to assay. A second enzyme is added in excess to generate another product whose formation can be easily monitored.

Units for expression of enzyme activity have evolved slowly through the years from an unorganized collection of special units to the set defined by international convention. Enzyme activities are increasingly expressed in terms of micromoles of product formed (or substrate consumed) per minute per liter of body fluid. International organizations are also seeking to standardize assay conditions to allow true interconvertibility of results from one laboratory to another. This standardization includes quality control measures to ensure accurate measurement of pooled enzyme reference materials.

Enzymes are finding wide application in the area of quantitative analysis. The enzyme is employed as a reagent to convert substrate completely to product. By measuring a change in absorbance or some other suitable parameter, the concentration of the biochemical analyte can easily be determined. Assessing the initial rate of reaction and relating that information to substrate concentration provides more rapid throughput of samples and allows more efficient utilization of analytical instrumentation.

FOR REVIEW

Directions: For each question, choose the best response.

1. Which of the following best describes an enzyme?
   A. catalyst
   B. carbohydrate
   C. protein
   D. both A and B
   E. both A and C

2. Enzymes are divided into six categories based on
   A. product formation
   B. substrate requirements
   C. cosubstrate requirements
   D. type of reaction carried out

3. The enzyme class that is characterized by the use of the coenzyme NAD^+/NADH is
   A. hydrolase
   B. lyase
   C. oxidoreductase
   D. transferase

4. The enzyme class that is characterized by the movement of an amine or a phosphate group from one molecule to another is
   A. hydrolase
   B. isomerase
   C. lyase
   D. transferase

5. When an enzyme requires an inorganic substance such as zinc for activity, this substance is termed a(an)
   A. activator
   B. coenzyme
   C. facilitator
   D. regulator

6. When substrate is present in excess, what component acts in a rate-limiting manner?
   A. activator
   B. enzyme

    C. product

    D. substrate

7. Which of the following parameters needs to be controlled for an enzyme reaction to proceed properly?

    A. type of buffer

    B. pH

    C. temperature

    D. both B and C

    E. A, B, and C

8. Which of the following may cause denaturation of the enzyme structure?

    A. severe pH alteration (too acidic or too basic)

    B. significant increase in temperature beyond optimal

    C. presence of an inhibitor substance

    D. both A and B

    E. A, B, and C

9. Which of the following may be associated with an inhibitor of an enzyme reaction?

    A. binds in a position other than the enzyme's active site

    B. binds to the active site of the enzyme

    C. decreases the rate of the enzymatic reaction

    D. both B and C

    E. A, B, and C

10. When an enzyme assay involves the continuous measurement of change in concentration as a function of time, the mode of analysis is termed

    A. end point

    B. kinetic

    C. multipoint

    D. spectrophotometric

11. When an enzyme assay measures the change in concentration at several intervals during the course of an assay, the mode of analysis is termed

    A. end point

    B. kinetic

    C. multipoint

    D. spectrophotometric

12. What type of information is taken into consideration when defining an enzyme unit of activity?

    A. sample volume

    B. time

    C. amount of substrate acted upon

    D. both A and C

    E. A, B, and C

BIBLIOGRAPHY

American Association for Clinical Chemistry, "Guidelines for photometric instruments for measuring enzyme reaction rates," *Clin. Chem.* 23:2160–2162, 1977.

Bell, J. E., and Bell, E. T., *Proteins and Enzymes.* Englewood Cliffs, NJ: Prentice-Hall, Inc., 1988.

Bergmeyer, H. U., "Standardization of enzyme assays," *Clin. Chem.* 18:1305–1311, 1972.

Cleland, W. W., "Optimizing coupled enzyme assays," *Anal. Biochem.* 99:142–145, 1979.

Duggan, P. F., "Activities of enzymes should be measured at 37°C," *Clin. Chem.* 25:348–352, 1979.

Dybaeker, R., "Problems of quantities and units in enzymology," *Enzyme.* 20:46–64, 1975.

Engel, P. C., *Enzyme Kinetics.* New York: John Wiley and Sons, 1977.

Henderson, A. R., "Errors in measuring enzyme activity by reaction-rate methods," *Clin. Biochem.* 9:165–167, 1976.

International Union of Pure and Applied Chemistry and the International Union of Biochemistry, *Enzyme Nomenclature.* Amsterdam: Elsevier Scientific Publishing Co., 1972.

Kraut, J., "How do enzymes work?" *Science* 242:533–540, 1988.

Lott, J. A., "Practical problems in clinical enzymology," *CRC Crit. Rev. Clin. Lab. Sci.* 8:277–301, 1977.

Moss, D. W., "Dilemmas in quality control of enzyme determinations," *Clin. Chem.* 16:500–502, 1970.

Moss, D. W., "The relative merits and applicability of kinetic and fixed-incubation methods of enzyme assay in clinical enzymology," *Clin. Chem.* 18:1449–1452, 1972.

Pardue, H. L., "A comprehensive classification of kinetic methods of analysis used in clinical chemistry," *Clin. Chem.* 23:2189–2201, 1977.

Peters, Jr., T., "Making headway in enzyme standardization," *Clin. Chem.* 34:225–226, 1988.

Srere, P. A., "Complexes of sequential metabolic enzymes," *Ann. Rev. Biochem.* 56:89–124, 1987.

Chapter Nine

PRINCIPLES OF IMMUNOASSAY

• • • • • • • • • • • • • • • • • • • •

UPON COMPLETION OF THIS CHAPTER, THE STUDENT WILL BE ABLE TO

1. Define the following terms:
   A. radioimmunoassay
   B. radioactivity
   C. radioisotope
   D. background
   E. total count
   F. nonspecific binding
   G. B_0

2. State the three components required to perform the technique of radioimmunoassay.

3. Explain the general principles of the following methods:
   A. radioimmunoassay (RIA)
   B. immunoradiometric assay (IRMA)
   C. homogeneous enzyme immunoassay (EIA)

4. Discuss the importance of the separation technique as part of the radioimmunoassay procedure and the major requirements for successful separation.

5. Describe each of the following separation techniques:
   A. charcoal adsorption
   B. salt or solvent precipitation
   C. double antibody
   D. solid-phase antibody

6. Describe the following data-reduction techniques used in radioimmunoassay systems:
   A. concentration per count
   B. percentage bound per concentration
   C. logit-log

7. Discuss how each of the following factors may affect the accuracy and validity of radioimmunoassay results:
   A. equipment
   B. reagents
   C. personnel

8. State the difference in the attachment site of the radioactive label in RIA as compared with IRMA.

9. Describe the data-reduction techniques employed in IRMA and homogeneous EIA systems.

10. Discuss the inverse relationship seen in RIA with the direct relationship seen in IRMA in regard to the amount of bound isotope and the concentration of patient antigen.

11. Describe the phenomenon of the high-dose "hook" effect exhibited in IRMA procedures.

12. Discuss the difference in separation technique requirements in homogeneous EIA as compared with RIA and IRMA.

13. Explain the principles of the following enzyme-linked immunosorbent assay (ELISA) methodologies:
    A. competitive
    B. indirect
    C. double-antibody sandwich

14. State the principles of the following fluorescence immunoassay methods:
    A. direct quenching fluoroimmunoassay
    B. indirect quenching fluoroimmunoassay
    C. substrate-labeled fluorescence immunoassay
    D. fluorescence polarization immunoassay
    E. time-resolved fluorescence immunoassay
15. Explain the process of chemiluminescence.
16. Explain the firefly reaction as it relates to bioluminescence.
17. Discuss the growing success of immunochemical assays in relationship to the
    A. ability to measure a wide variety of biochemical compounds, for example, hormones, proteins, and drugs
    B. ability to quantitate low concentrations
    C. high specificity of measurements
    D. ease of measurements
    E. equipment requirements

INTRODUCTION

The need for analytical sensitivity is apparent in the clinical laboratory. With colorimetric methods we perhaps can measure as low as a few micrograms per deciliter concentration for some substances. Other materials do not have structural characteristics unique enough to allow us to employ a color reaction or fluorescence process. With the use of specific antibodies, however, precise measurement of extremely low levels of a biochemical material is now possible. Applications of immunoassays are widespread. This chapter explores some of the major approaches to the analytical use of antibodies. Specific applications for proteins were discussed in Chapter 5 and will be described for enzymes (Chapter 10), hormones (Chapters 14–16), and drug monitoring (Chapters 20 and 21).

9.1

RADIOIMMUNOASSAY

Historical Development

In the mid-1950s a research team in New York headed by Drs. Rosalyn Yalow and Solomon Berson began a study of insulin use in diabetics. They wanted to see if the diabetic patient metabolized insulin differently from the normal individual. Human insulin was labeled with radioactive iodine (^{131}I) and injected into the circulation. Yalow and Berson discovered that there appeared to be a protein in the blood of the diabetic which specifically bound insulin; this protein was lacking in the circulation of normal persons.

To study this process, an electrophoresis procedure was developed to separate the unbound radioactive insulin from that portion which had attached to this atypical protein. The researchers found that the amount of radioactive insulin which bound was proportional to the amount of the protein (later shown to be an antibody) which was present in the serum. This discovery not only led to significant contributions to our understanding of the disease diabetes mellitus, but also opened the door to a powerful analytical tool. In 1977, Rosalyn Yalow received the Nobel Prize in medicine for her contributions (Solomon Berson died in 1972 and was unable to share in this honor).

Basic Principles

The technique of radioimmunoassay (RIA) utilizes three components:

1. patient antigen—the specific compound we wish to quantitate
2. labeled antigen—the same compound as above, to which is attached a radioactive label
3. antibody—specific for the antigen we wish to quantitate

In practice, a stated amount of patient sample containing the antigen (sometimes called the **ligand**) of interest is added to a test tube which has a fixed amount of labeled antigen and an excess of antibody. After mixing, an equilibrium is established among free patient antigen (that material which does not bind to antibody), bound patient antigen (that material which attaches to antibody), free labeled antigen, bound labeled antigen, and free an-

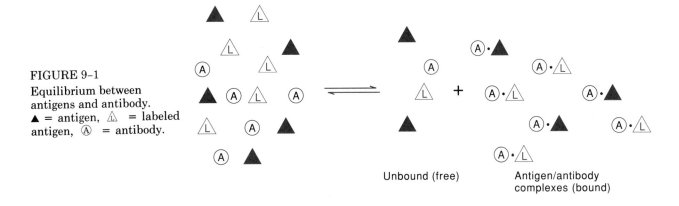

FIGURE 9–1
Equilibrium between antigens and antibody. ▲ = antigen, △ = labeled antigen, Ⓐ = antibody.

Unbound (free)

Antigen/antibody complexes (bound)

tibody. This process is illustrated in Figure 9–1. After a time (usually 1 h or less), bound and free materials are separated from one another. By determining the amount of radioactive ligand in either the bound or the free fraction, the amount of patient antigen can be estimated. In each assay, samples of known concentration are employed to establish a standard curve for use in quantitating patient materials.

An examination of Figure 9–2 helps us better understand the principle of RIA. Fixed (constant) amounts of antibody and labeled antigen are present in each assay tube. The variable in the system is the amount of patient antigen added. If a large amount of patient antigen is added relative to labeled antigen (Fig. 9–2A), the proportion of labeled antigen binding to antibody is less, since the patient antigen and labeled material compete for binding sites on the antibody molecule. Since we can detect only the radioactive label attached to the antibody (and not the patient ligand), we would observe a lower amount of label bound in this situation. If a low amount of patient material is available, more labeled antigen can bind, and the amount of radioactivity attached to antibody will be higher (Fig. 9–2B). Therefore, the concentration of patient material in the sample is inversely proportional to the amount of radioactive label attached to antibody.

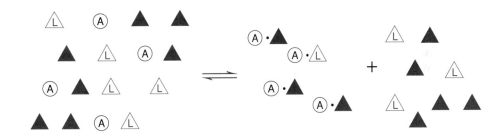

Much unbound labeled antigen

FIGURE 9–2
Patient antigen concentration and binding of labeled antigen. A. Low concentration of patient antigen. B. High concentration of patient antigen. ▲ = antigen, △ = labeled antigen, Ⓐ = antibody.

A

Little unbound labeled antigen

B

Table 9-1
SEPARATION TECHNIQUES IN RIA

Technique	Characteristics
Dextran-coated charcoal	Free ligand binds to charcoal; antigen/antibody complex in solution
Salt or solvent precipitation	Antigen/antibody complex precipitates; free ligand remains in solution
Double-antibody	Addition of second antibody precipitates antigen/antibody complex; free ligand remains in solution
Solid-phase antibody	Antibody bound to insoluble particle or wall of test tube; free ligand remains in solution

Separation Techniques

A great deal of the success of any given RIA procedure depends on the effectiveness of the separation technique. Ideally, all the bound material separates cleanly from all the free material. In practice, this goal is not usually achieved.

Any separation procedure needs to meet several criteria to contribute to accurate results. Major requirements for successful separation are as follows:

1. Separation of bound and free components must be essentially complete. Minor variations in technique must not affect results. The approach selected should not affect the equilibrium between bound and free antigen during the separation process.
2. The technique should be easily and quickly performed. The technique should not be long or complicated. All equipment and reagents need to be inexpensive and simple to use.
3. The technique must not be affected by serum or plasma. Restriction to a single type of sample limits the versatility of the method. Problems also arise in the preparation of standards, blanks, and dilutions.
4. Separation should be carried out in one tube, if possible. The procedure should be simple and lend itself to automation to minimize variability in results due to different techniques.

A wide variety of separation techniques have been developed over the years (Table 9-1), some of the more common of which we will discuss.

CHARCOAL ADSORPTION

The **charcoal adsorption** method takes its principle from a standard practice of organic chemistry. If charcoal is added to a solution, small molecules adsorb to it and can be coprecipitated from solution. The term *adsorb* means that the chemical sticks to the surface of the charcoal particle. When used in a

radioimmunoassay procedure, the "free" fraction precipitates with the charcoal, leaving the antigen/antibody "bound" fraction in solution. Precipitation occurs when the tubes are centrifuged after the charcoal is added. The supernatant containing the bound fraction is poured off (it may be saved for counting) and the precipitate (with the free labeled and unlabeled antigens) remains in the tube. This fraction may be counted to determine the amount of radioactivity present, allowing us to estimate the amount of labeled ligand in the tube.

SALT OR SOLVENT PRECIPITATION

Instead of precipitating the free fraction, adding solvent or salt results in the bound portion's becoming insoluble (due to protein precipitation) while the unbound ligand remains in solution. Use of a salt solution (usually ammonium sulfate) or a solvent (ethanol and polyethylene glycol are popular) results in a change in the solubility of the proteins in the reaction mixture. These techniques, although widely used in research applications, have not found extensive use in the routine clinical laboratory. The major advantages to **salt or solvent precipitation** is its speed and relative inexpensiveness.

DOUBLE-ANTIBODY APPROACHES

A single antigen/antibody complex by itself may still be somewhat soluble, particularly if the antigen portion is a reasonably small molecule. One of the earliest approaches to separation in RIA involves the use of a second antibody, known as the **double-antibody technique** (Fig. 9-3). Addition of this second antibody permits the formation of a complex suf-

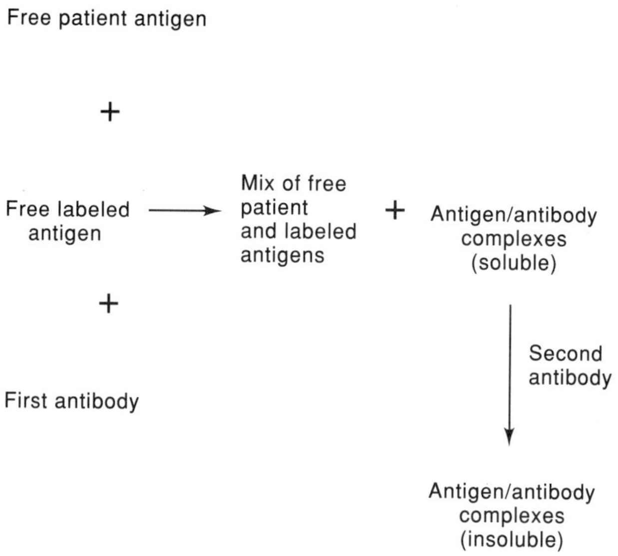

FIGURE 9-3 Principle of double-antibody separation.

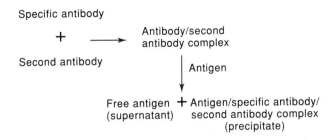

A

B

FIGURE 9-4 Double-antibody separation approaches. A. Preprecipitation. B. Postprecipitation.

ficiently large to precipitate and be removed from the reaction system by centrifugation. The second antibody can be fairly nonspecific, lowering the cost of the assay somewhat.

Depending on the characteristics of the specific assay, the second antibody may be added at the same time as or after the first antibody (Fig. 9-4). In the pre-precipitation approach (Fig. 9-4A), the first antibody (specific for the ligand of interest) and the second antibody (relatively nonspecific) are mixed together in the reaction tube. The sample containing the antigen for which we are assaying is then added to the mixture. After equilibrium has been achieved, we have a mixture of free antigen (soluble) and antigen/first antibody/second antibody complex (precipitate). Centrifugation removes the bound ligand from the solution.

The post-precipitation approach (Fig. 9-4B) reverses much of this process. Antigen and first antibody are mixed together and form an equilibrium mixture of antigen and antigen/antibody complex. The second antibody is then added to the reaction system, forming the insoluble complex described above. Separation then proceeds in the usual manner.

In both cases, the formation of precipitate may require several hours (often overnight). If polyethylene glycol is added in low concentration to the mixture, complex formation is markedly enhanced. Precipitate may form within a few minutes with this modification of the procedure. Caution must be taken that the ligand of interest does not precipitate in the presence of polyethylene glycol; some coagulation factors may manifest this type of behavior.

The major advantage of this type of separation is that it is relatively nondisruptive; there is little chance for the antigen/antibody complex to be torn apart by the separation treatment. Use of the pre-precipitation approach decreases any possibility of disturbing or reversing the binding of antigen to antibody. Other proteins have less influence on the reaction, which improves specificity.

SOLID-PHASE ANTIBODY

The techniques described up to this point have involved separation of bound and free antigen using techniques which have the potential for distorting the equilibrium between the two fractions. The physical processes employed often strip some of the bound antigen from the antibody. In other instances, the timing of the separation process is critical so that minimal amounts of the undesired fraction are precipitated. An extra pipeting step may be required if a specific precipitant is to be added. To overcome these drawbacks, the **solid-phase antibody** approach was developed.

In the traditional RIA procedure, antibody was added in solution to the reaction mixture. In the solid-phase approach, either the antibody is added in a suspension (with antibody bound to an insoluble particle) as seen in Figure 9-5, or the antibody is covalently bound to the inside wall of the reaction tube (Fig. 9-6). After equilibrium between bound and free antigen has been attained, the insoluble material may be centrifuged or the supernatant liquid may simply be poured off (if the bound material is attached to the wall of the test tube). In either case, separation is rapid and complete, with little opportunity for disturbance of the ratio between the two fractions.

A number of particles have been used in solid-phase systems. Glass beads or Sephadex particles (a large carbohydrate polymer) are sometimes employed. Whatever the solid particle, care must be taken to keep the material well stirred during the pipeting process. Otherwise, material settles out and varying amounts of antibody are added to the

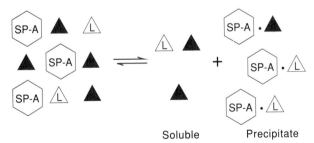

FIGURE 9-5 Solid phase separation with insoluble antibody. ▲ = antigen, △ = labeled antigen, SP-A = solid-phase antibody.

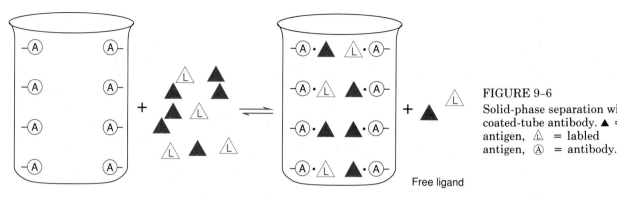

FIGURE 9-6
Solid-phase separation with coated-tube antibody. ▲ = antigen, △ = labled antigen, Ⓐ = antibody.

Free ligand

Antibody covalently attached to inside of coated tube

Ligand attached to antibody—now insoluble (bound)

different tubes. The solid-phase materials vary in their ability to stay packed after centrifugation, so decanting of supernatant must be done cautiously so as not to pour off precipitate along with the liquid fraction.

Some assays have employed a double-antibody solid-phase approach (Fig. 9-7). The antigen and first antibody are in solution and are mixed in the test tube in the usual fashion. The second antibody, however, is covalently bound to a solid phase of some sort. When this second antibody is added to the reaction mixture, it binds to the first antibody and forms a precipitate. In most instances, the post-precipitation approach is preferred for this technique. Separation of bound and free fractions is enhanced with the use of the solid phase. Precipitates pack better, with less possibility of material being stirred up during the decanting process. The problem of nonspecific interference is lessened, although it is not eliminated completely.

Another interesting variation of the solid-phase theme is one that has been widely developed commercially. The antibody is bound to a small magnetic particle. Usually a polymer containing iron

oxide is used. All the components of the reaction mixture are added and the tubes are placed on an electromagnet. During the course of the antigen/antibody reaction, the magnet is turned on and off. This change in magnetic field keeps all the material well mixed. Precipitation of bound antigen is achieved by leaving the magnetic field on. The antigen attached to the solid-phase antibody packs at the bottom of the tube, allowing supernatant containing free antigen to be poured off. The technique appears to be reliable and versatile. Some approaches to automation of RIA using this separation technique have been developed, although they are not widely used at this point.

Data Reduction

INTRODUCTION

Calculations in the immunoassay field are more complicated than the familiar spectrophotometric or fluorometric data-reduction techniques. RIA (and other immunoassay) standard curves do not pro-

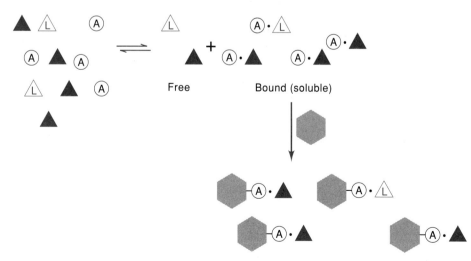

Free

Bound (soluble)

Precipitate

FIGURE 9-7
Solid-phase second antibody separation. ▲ = antigen, △ = labeled antigen, Ⓐ = antibody, ● = solid-phase second antibody.

duce straight lines; there is not a direct and linear relationship between the amount of antigen bound to antibody and the concentration of that antigen. Moreover, we must be much more specific about what we are actually measuring—bound or free antigen. Correction for nonspecific aspects of the assay must be made. Even the particular data-reduction technique employed may have some impact on the actual value obtained and its probable error.

TERMINOLOGY

A number of pieces of information are necessary if our final calculations are to be accurate (Table 9–2). **Background** is the term for the amount of radioactivity measured in the absence of any radioisotope; this value must be subtracted from the counts per minute measured in all individual tubes. The background count is obtained by counting an empty tube with no added radioisotope. **Total count** refers to the amount of radioisotope added initially. Each tube has the same amount of radioactivity added, but does not have the same amount present after separation of bound and free antigen. The total count is obtained by pipeting the specified amount of radioactivity into a tube and determining the counts per minute.

Nonspecific binding refers to the radioisotope which binds to materials in the reaction mixture other than the antibody. This parameter can be measured by setting up an assay tube containing sample, radioisotope, and all other materials except antibody. Separation is carried out in the same manner as for a regular assay tube. Nonspecific binding is usually quite low (generally less than 6–7%). Although we may not always use this value in our calculations, it is a very useful piece of quality control data and allows us to monitor the reliability of the reaction.

A related parameter, B_0, refers to the binding of radiolabeled ligand in the absence of any patient ligand. B_0 is determined by adding the appropriate amount of isotope to the reaction mixture containing antibody, but not the unlabeled antigen of interest. Since the labeled material does not have to compete with patient antigen for binding to the

antibody, the maximum amount of radioactive label attaches to antibody. The binding of isotope in this case is greater than the binding in any situation when unlabeled antigen is present. Therefore, B_0 represents a zero standard point for the standard curve. The value for B_0 in a given assay can also be followed from day to day as a quality control measure.

GENERAL PRINCIPLES

Any assay we perform requires some sort of standardization. In some instances, a standard curve may be needed only once a month. With high-volume automated chemistry systems, standardization is carried out at set intervals, usually several times a day. RIA procedures, on the other hand, require standardization each time an assay is run because of changes in the amount of radioactive label present and alterations in the ability of antibody to bind antigen.

The principle of a standard curve for an RIA method is the same as for any other procedure. Standard solutions containing known amounts of the analyte ligand are assayed in exactly the same fashion as patient samples. By determining the relationship between bound and free antigen in a standard sample containing a certain amount of ligand, we can plot a standard curve. We then measure the bound-antigen-to-free-antigen ratio in the patient material and calculate the quantity of ligand in the patient sample by reference to this standard curve. The standard solutions must approximate the patient samples as closely as possible so that factors such as nonspecific binding are essentially the same. For this reason, serum-based standards are preferred.

RIA data-reduction techniques are varied and complex, usually involving logarithmic functions and nonlinear standard curves (Table 9–3). With the incorporation of the computer into the radioisotope detection instruments, data reduction is handled automatically. In many systems, the entire proc-

Table 9–2
TERMS EMPLOYED IN RIA

Term	Definition
Background	Radioactivity measured in absence of isotope
Total count	Total amount of radioactivity added to tube
Nonspecific binding	Isotope binding to materials other than antibody
B_0	Binding of radiolabeled ligand in absence of patient ligand

Table 9–3
COMMON DATA-REDUCTION TECHNIQUES IN RIA

Type of Plot	Parameters Examined
Concentration/ count	Net cpm on y axis; concentration on x axis; nonlinear curve
Percentage bound/ concentration	Percentage bound on y axis (linear scale); concentration on x axis (log scale); nonlinear curve—several variations possible
Logit-log	$\ln\left(B/B_0/(1-B/B_0)\right)$ on y axis; concentration on x axis (log scale); most approximates linear curve

ess of data reduction, reporting of patient results, and quality control inspection is carried out by the computer.

Special care must be taken in examining patient results. Data near the extremes of an RIA standard curve usually have a larger margin of error than results in the middle portion of the curve. Although the various mathematical treatments are designed to provide linear graphs, this goal is not always achieved, and some extrapolation is necessary. As a result, high patient results may need to be rechecked at some dilution to verify the initial value. Although low results cannot be further analyzed, these data must be interpreted with due consideration of the amount of error possible in the method at these low levels of analyte.

Quality Control Aspects

Factors which affect the accuracy and validity of RIA results are equipment, reagents, and personnel. Each parameter must be assessed regularly, as is the case with all quality control programs. Regular documentation of test performance can allow the operator to identify problems quickly and take appropriate remedial action.

EQUIPMENT

Failure to obtain reliable data may be attributed to problems with centrifugation, pipeting, incubation, or the measurement of radioactivity. Poor centrifugation leads to inadequate packing of precipitate and subsequent loss of material. If the solid fraction is being counted, this loss can result in serious error in determining the amount of radioisotope present. Pipeting errors can be minimized by regular calibration of the measuring devices and a program of documented routine maintenance. Incubation temperature plays an important role in determining the ratio of bound-to-free antigen in the assay system. Shifts of only a few degrees in temperature often drastically alter this ratio. Counting equipment must be checked periodically for radioisotope contamination (leading to higher counts per minute than expected) or loss of sensitivity (indicated by lower than usual counts per minute). The use of National Bureau of Standards material (usually an isotope with an extremely long half-life) provides a ready check on this aspect of instrument function.

REAGENTS

Loss of isotope activity due to age can be monitored by following the total counts per minute added each time the assay is performed. Effectiveness of antibody is reflected in both the percentage B/B_0

and the shape of the standard curve. Even if automated data reduction is employed for calculations, these parameters should be obtained and routinely checked to assess performance of the reagent system. Contamination or deterioration of standards may be suspected if quality control material fails to give appropriate data or if the standard curve has observable anomalies in it.

PERSONNEL

Performance of the technical staff must be monitored as well as performance of equipment and reagents. A variety of approaches are available, including blind samples, repetition of the assay by another individual, monitoring of daily quality control results (both quality control material and standard curve data), and regular participation in regional proficiency testing.

Applications of RIA in Clinical Chemistry

The development of radioimmunoassay procedures led to an explosion in the capabilities of the clinical chemistry laboratory in the early 1970s. One of the first tests which became commercially available was for T4 (thyroxine). Other hormone assays employing RIA techniques quickly followed, providing increased sensitivity and specificity. In many instances RIA was the only technique available which could detect the low levels of these hormones found in blood. Therapeutic drug monitoring later yielded to RIA techniques with the development of assays for digoxin, gentamicin, tobramycin, and other pharmaceuticals. Vitamin B_{12} and folate are important nutrients that can be measured using RIA. Quantitation of ferritin and other protein components of blood allows diagnosis of a number of diseases.

However, RIA approaches to analysis often became supplanted by nonisotopic methods. The major drawback to RIA is its use of a radioisotope. Although radiation hazards are minimal in most cases, the use of radioactive material has some practical drawbacks. The major limitation is the gradual loss of activity in the isotope, which results in frequent replacement of reagents. Another significant hindrance to many laboratories is the need to acquire equipment to measure radioactivity. Although technology has advanced significantly, leading to more reliable instruments at lower cost, a piece of equipment must still be dedicated to this function.

Even with these shortcomings, the RIA field still flourishes. Some serum or plasma constituents cannot be accurately quantitated except by RIA. Other techniques may also require rather sophisticated instrumentation. Although RIA no longer occupies an overwhelmingly dominant position in the immunoassay field, it still makes an important contribution

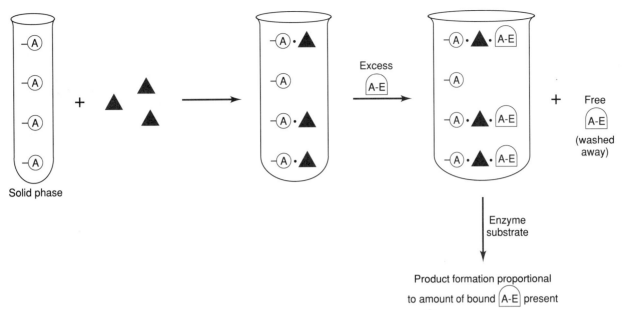

FIGURE 9–8   Principle of IRMA assay. ▲ = antigen, Ⓐ = antibody, [A-E] = enzyme-labled antibody.

in the quantitation of biochemical materials in body fluids in health and disease.

9.2

IMMUNORADIOMETRIC ASSAY

Basic Principles

In an effort to improve specificity of measurement and increase the sensitivity of RIA, an alternative approach to the traditional radioimmunoassay was developed. Instead of attaching a radioactive label to the antigen, the isotope was covalently bound to the antibody itself. This modification resulted in the establishment of the **immunoradiometric assay (IRMA).**

The basic approach to quantitation of a ligand using IRMA is illustrated in Figure 9–8. Antibody to the ligand we wish to measure is bound to a solid phase (usually the inside of the test tube itself). Sample containing ligand is added and the mixture is allowed to incubate. During this time, the ligand specifically attaches to the solid-phase antibody. Since the antibody is present in excess, essentially all the ligand becomes attached, and for all practical purposes no free antigen remains. A second antibody is then added to the reaction tube; this antibody is also specific for the ligand in question and contains a radioactive label (usually ^{125}I). In most instances, the second antibody is specific for a different part of the ligand molecule than the first antibody. This differential selectivity allows for greater specificity in the assay. The second antibody is added in excess, so that all the ligand present has this antibody attached to it. The result is a "sandwich" of solid-phase antibody, ligand, and radioactively labeled antibody (these assays have sometimes been referred to as **sandwich immunoassays**). Since the entire complex is now attached to the solid phase, simple decanting and washing are all that is necessary to remove excess radiolabeled second antibody. The amount of isotope present can then be counted and the quantity of ligand in the sample determined.

Data Reduction

Perhaps the most striking difference between RIA data and the information generated by IRMA is the standard curve. With RIA there is an inverse relationship between isotope bound to antibody and the concentration of the ligand: The higher the counts per minute detected in the tube, the lower the concentration of patient ligand. In IRMA techniques, the reverse is true: There is a direct relationship between amount of bound isotope and the concentration of the patient analyte antigen.

This relationship is easily understood when we look at the basic principle of the assay. With an excess of first (solid-phase) antibody, all of the analyte ligand binds to that antibody. The more ligand present, the more sites occupied on the first antibody. When the second antibody is added, it is also in excess. Therefore, all the bound ligand also attaches to this second antibody, which contains the radioactive label. The more bound ligand present, the more second antibody binds and the more radiolabel is detected.

As with RIA, IRMA data do not yield a simple straight line unless some mathematical manipulations are carried out. The easiest approach is to plot

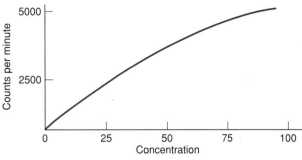

FIGURE 9–9 IRMA standard curve.

the data on semilog graph paper, with the counts per minute plotted on the linear scale against the concentration on the log scale (Fig. 9–9). A somewhat straight line results, although there is some curvature present. "Normalizing" techniques, which use calculations similar to the logit-log manipulations of RIA, produce straighter lines with steeper slopes. In most situations, these calculations are carried out automatically by the microcomputer that is a part of the isotope-counting system.

The measurement of nonspecific binding is necessary in IRMA techniques, as in other immunoassays, to correct the data for any blank interactions and as a quality control step. In IRMA, the nonspecific binding assay is carried out by adding to the assay tube a serum sample that does not contain any of the ligand of interest. The second (labeled) antibody is then added and the tube incubated, decanted, and washed in the normal manner. Whatever label remains in the tube is due to nonspecific binding of labeled antibody to some other component. In most situations, there may be a slight interaction with the glass or plastic material of the tube itself. There may be a protein component which binds slightly to the first antibody and then interacts with the second antibody, producing a low nonspecific binding of label. Generally, the amount of nonspecific binding in IRMA assays is quite small, below that seen in RIA procedures.

The High-Dose "Hook" Effect

When performing an IRMA procedure, sometimes an increase in the concentration of the patient analyte ligand results in a decrease in the amount of radiolabeled second antibody bound to the complex (Fig. 9–10). This phenomenon is called the **high-dose "hook" effect.** As the concentration of ligand increases, the amount of label bound increases (as is expected). However, a point is reached where further increase in ligand leads to the binding of less (not more) second antibody.

It is not altogether clear why this hook effect occurs. Some studies suggest that the problem is due to the use of a mixture of antibodies for the first antibody. The antibody fraction which binds more

readily to ligand takes up material until it becomes saturated. Then another antibody fraction begins to bind additional ligand, but it does not bind as tightly. With the advent of monoclonal antibodies, this shortcoming has been corrected to a great extent. Other problems may include incomplete washing of the tube after the reaction is complete.

The most likely explanation for the hook effect is that the concentration of the first antibody becomes limiting at extremely high ligand concentrations. This limitation of binding sites results in some ligand's remaining unbound to the first antibody. Therefore, when the second antibody is added, the unbound ligand competes with the bound ligand for the second antibody. If more and more ligand is unbound at higher ligand concentrations, more and more second antibody attaches to that unbound ligand and is discarded during the decanting and washing steps. Steps can be taken during the development of a specific assay to define the upper limit of reliability for the procedure and to increase the amount of first antibody present. An alternative approach is to assay all samples (or at least the suspect ones) at two different dilutions to determine if the assay capabilities of the system have been exceeded. If both samples give the same value after correction for the dilution, then the system is functioning properly. If the more dilute sample produces a higher result than the more concentrated one, then we are seeing the hook phenomenon in action, and several different dilutions of the sample must be made to determine the correct concentration of the ligand.

Comparison of IRMA and RIA Techniques

IRMA procedures offer several advantages over comparable RIA methodologies. The assay protocols are usually much simpler, involving fewer pipeting steps (and less chance for error). The nonspecific binding in an IRMA procedure is generally lower, providing less opportunity for interference. IRMA test sensitivity is frequently better, which allows for accurate measurement of very low levels of a particular analyte. It is also possible to measure a wider range of concentrations with IRMA than with RIA, a definite advantage in such assays as ferritin,

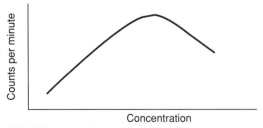

FIGURE 9–10 High-dose "hook" effect in IRMA.

where values from 5 to 1000 mg/mL may routinely be encountered. With the use of two different antibodies, each specific for a different site on the ligand molecule, very selective and specific IRMA assays can be developed.

A major drawback to the test lies in the need for those two antibodies. The cost of materials is appreciably higher for an IRMA assay because two antibodies (of high purity) are required. In addition, both antibodies must be present in the system in excess, leading to a high consumption of antibody (and increased cost). IRMA assays are, at present, limited to larger molecules such as thyroid-stimulating hormone (TSH) so that the differing specificities of the two antibodies can be utilized to maximum advantage. Modifications of the basic protocol may make it possible to measure other biochemical compounds of much smaller molecular weight.

9.3

HOMOGENEOUS ENZYME IMMUNOASSAY

Although RIA and IRMA have been widely used in clinical chemistry, there are some drawbacks associated with their use. The use of radioisotopes presents problems in monitoring and disposal. Radioactive emission often begins to degrade the antigen to which the isotope is attached, creating a mixture of materials which could adversely affect the assay system. Specialized equipment is required to measure the radioactivity, putting strain on already tight laboratory budgets. Many procedures require lengthy incubation times, causing delays in the generation of needed laboratory data. These shortcomings in RIA and IRMA methodologies led to the development of new approaches to immunoassay, involving the use of labels other than radioactive isotopes to assess the degree of binding of ligand to an antibody. One approach, first developed in 1972, used enzyme activity as the marker. This methodology has become known as **homogeneous enzyme immunoassay,** sometimes simply designated **enzyme immunoassay (EIA).** The related term **enzyme-multiplied immunoassay technique (EMIT)** is a registered trademark used to refer to materials manufactured by the Syva Corporation (Palo Alto, CA).

Basic Principle

All EIA procedures involve the competition between enzyme-bound ligand and free patient ligand for antibody specific for that ligand (Fig. 9–11). The enzyme we use for our example is glucose-6-phosphate dehydrogenase (G6PD), which catalyzes the following reaction:

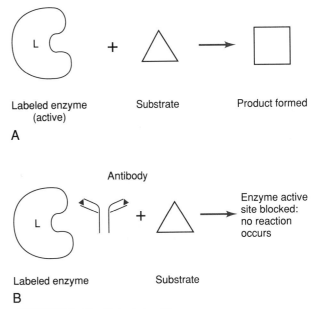

FIGURE 9–11 Principle of homogeneous enzyme immunoassay. A. Absence of antibody. B. Presence of antibody.

$$glucose\text{-}6\text{-}phosphate + NADP^+ \rightarrow$$
$$6\text{-}phosphogluconate + NADPH + H^+$$

Enzyme activity is measured by determining the increase in absorbance at 340 nm when NADPH is formed.

For our assay, we measure the concentration of theophylline (an antiasthmatic drug) in serum. Theophylline is covalently bound to G6PD, forming an enzyme/theophylline complex. If an antitheophylline antibody is added to the enzyme/theophylline complex, the antibody binds at the theophylline sites on the enzyme. Because the antibody is a large molecule, this binding blocks the active site of the enzyme and markedly decreases the amount of enzyme activity which is measured. In the absence of any free theophylline, the activity of enzyme in the reaction mixture is quite low. If we then add a patient sample which contains theophylline, some of the antibody binds to the free drug and not to the the enzyme/theophylline complex. The G6PD can then show more activity since fewer of the active sites are blocked. As we increase the concentration of theophylline in the patient samples, more antibody moves from the enzyme to bind the free drug, and the enzyme activity increases further. Therefore, we have a fairly direct relationship between the amount of enzyme activity in the assay tube and the concentration of free drug in the patient sample—the higher the enzyme activity, the higher the drug concentration.

One obvious difference between homogeneous EIA and the procedures we have discussed up to now is the absence of a separation step in the EIA assay. The assessment of the degree of antibody binding to patient ligand is made while the reaction is taking place and without the need for a preliminary separation of bound and free fractions. This approach is referred to as a **homogeneous assay,** indicating that separation does not take place during the course of the measurement. We will discuss other homogeneous assay approaches later in this chapter.

A number of enzyme systems can be used in EIA procedures. One criterion for use is that the enzyme employed in the assay should be undetectable in "normal" human serum and should not be affected (either inhibited or activated) by any component of the body fluid being studied. Malate dehydrogenase, acetylcholinesterase, glucose oxidase, peroxidase, and glucose-6-phosphate dehydrogenase have found many applications in EIA assays. Assay methodologies for these enzymes are usually simple and straightforward. In addition, both the enzymes and the assay approach have been characterized in depth so that all the relevant parameters are well defined.

Data Reduction

In contrast to the large number of publications on data-reduction techniques in RIA, little has been published on this subject for EIA procedures. To some extent, the same bound/free relationships hold that we have already discussed in RIA techniques. However, the situation is more complex than it initially appears. If we follow the formation of product of the enzyme reaction, we see that we do not get a linear change in absorbance (indicating product generation). The relationship is curved, suggesting that the rate of the reaction is changing as the ratio of bound and free ligand/antibody complexes changes. Measurement of absorbance is usually made at some stated time after the reaction is initiated (usually within 1 min). Since all measurements are made using the same interval, there is a degree of standardization, but the approach is rather empirical. The EIA technique was commercialized very quickly after its development, both in the production of reagents and in the manufacture of special microprocessor-driven spectrophotometers which performed the data-reduction steps.

Comparison of EIA and RIA Techniques

EIA approaches have several advantages over RIA, the most obvious being the elimination of the use of radioactive materials. Since no separation step is required, the EIA procedure is faster and less complex. Although both EIA and RIA can be automated, the equipment required for EIA is more general (spectrophotometer or centrifugal analyzer) than the specialized counting equipment needed for the measurement of a radioisotope. The enzyme

label in EIA is usually more stable than the isotope label. A single EIA assay requires only a few minutes time, making stat analyses feasible.

The major drawback to present EIA procedures is the limited range of analytes which can be measured by the method. Originally developed (and still widely used) for drug screening, EMIT procedures have found their major applications in the quantitation of small molecules. Procedures for monitoring drugs used for treatment and quantitating serum T4 have been very successfully adapted to EIA techniques. However, estimation of larger molecules (such as proteins) and of very low levels of analytes (many hormones) usually requires other analytical approaches. New developments in EIA chemistry may expand the capabilities of this versatile and useful system.

9.4

ENZYME-LINKED IMMUNOSORBENT ASSAY

At about the same time as the homogeneous EIA system was being developed, other methods for immunoassay which also did not depend on a radioactive label as a marker were being explored. Perhaps the most successful concept was one which drew upon the basic principles of the IRMA assay but used a chemical label instead of a radioisotope. This set of analytical approaches is called **enzyme-linked immunosorbent assay (ELISA).** Like IRMA, the principle in ELISA involves a heterogeneous methodology; bound and free materials must be separated using a series of washing steps. Like for homogeneous EIA procedures, the marker in ELISA is the presence or absence of enzymatic activity; either qualitative and quantitative determinations may be carried out.

Three major methods of ELISA are commonly used:

1. competitive
2. indirect
3. double-antibody sandwich

Each analytical approach will be considered separately. Keep in mind that all ELISA procedures are solid-phase systems, which minimizes the problems involved in separation of bound and free material. Simple washing with buffer is all that is required whenever a wash step appears in the protocol.

ELISA assays are carried out either in test tubes (similar to solid-phase RIA and IRMA procedures) or in microtiter plates. These plates are rectangular and contain a series of small wells to which sample and other reagents are added. The plates are useful when screening large numbers of materials, especially if only a positive/negative answer is needed. However, special plate-scanning devices are available which allow for spectrophotometric readout

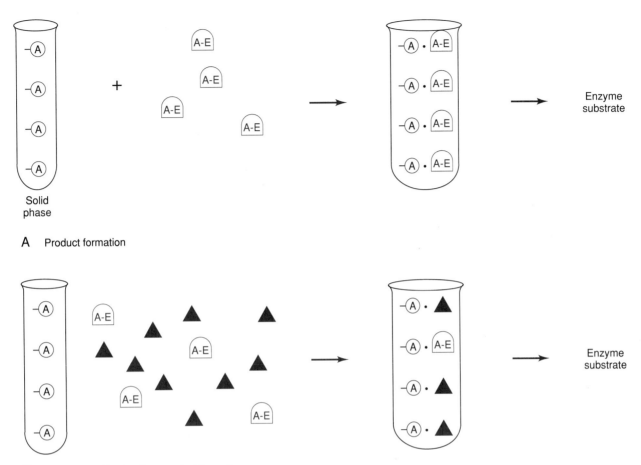

A   Product formation

B   Less bound E – Ag; less product formation

FIGURE 9–12   ELISA—competitive method. A. Absence of patient antigen. B. Presence of patient antigen. ▲ = antigen, Ⓐ = antibody, [A-E] = enzyme/antigen complex.

and quantitative results under the proper conditions. The basic procedure is the same whether tubes or plates are used; some details of manipulation may be different but do not affect our discussion of methodology.

Competitive Method

The **competitive method** can be employed for the measurement of either small molecules or large protein materials. Like RIA, the basic principle involves competition of patient antigen and labeled antigen for antibody binding sites (Fig. 9–12). An antibody to the analyte is covalently bound to the tube or well. Patient or standard samples containing the antigen to be measured are added. At the same time, antigen labeled with an enzyme marker is also added. Alternatively, the enzyme-labeled marker may be added separately. Patient antigen and enzyme-labeled antigen compete for binding sites on the solid-phase antibody. After washing to remove any unbound materials, enzyme substrate is added to the system. Following a timed period of incubation, the enzyme reaction is stopped. Measurement of product is performed (usually with a standard absorbance spectrophotometer) and enzyme activity present is calculated.

Unlike homogeneous EIA procedures, enzyme bound to antibody in all types of ELISA methodologies is active. The antibody is not binding to block the active site, but merely to attach the enzyme to the antibody protein so it does not wash off.

Data analysis for the competitive ELISA method is similar to that for an RIA procedure. The higher the concentration of patient antigen, the more competition for binding sites between patient antigen and enzyme-labeled antigen. If more patient antigen binds to antibody, less enzyme label is retained and the enzyme activity in that tube is lower. If there is a low amount of patient antigen present, a higher amount of enzyme labeled antigen attaches and a

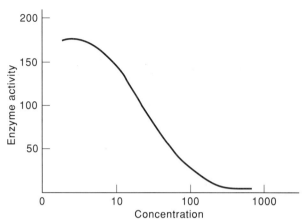

FIGURE 9–13   ELISA standard curve—competitive method.

greater amount of enzyme activity is observed. Hence, there is an inverse relationship between detected enzyme activity and the concentration of patient antigen present in a given sample (Fig. 9–13).

Indirect Method

The **indirect approach** to quantitation using ELISA involves the same principles as the IRMA technique. This methodology finds wide application in the analysis of large molecules, particularly immunoglobulins and other protein components (Fig. 9–14). Antigen to the analyte immunoglobulin is covalently bound to the inside of the tube or well. To this antigen is added the sample (or standard) containing the protein of interest. The protein forms a complex with the antigen, and other materials are removed by washing. To this mixture is added antiglobulin (antibody to the protein of interest) to which is attached an enzyme marker. After incubation, excess antiglobulin/enzyme material is removed by washing. Enzyme substrate is then added

to the system and product formation is determined under the appropriate conditions.

With the indirect method, enzyme activity is directly proportional to the concentration of the analyte. The more immunoglobulin binds, the more enzyme/antiglobulin complex is retained. Data reduction follows the same general principles as in the IRMA methodology.

Double-Antibody Sandwich Method

The **sandwich method** for analysis using ELISA (Fig. 9–15) also bears a strong resemblance to IRMA methodology. This approach is more suitable for smaller molecules and is very versatile. Antibody to the analyte antigen is covalently bound to the wall of the reaction vessel. The test solution, which contains either patient antigen or a standard, is added and antigen is allowed to bind to antibody. Since the antibody is in excess, essentially all the antigen should attach. After washing to remove extraneous materials, a second antibody is added which is labeled with an enzyme. This second antibody attaches to the antigen, forming an antibody/antigen/second antibody (labeled) complex. Another washing removes excess second antibody/enzyme. Enzyme activity is determined; activity is directly proportional to the concentration of the analyte.

ELISA and Other Immunoassay Techniques

Analytical procedures involving ELISA technology have become widely available in the last 10 years or so. One of the main advantages to the use of ELISA is its versatility. Both large and small molecules can be easily measured, unlike in homogeneous EIA, which is restricted to compounds of

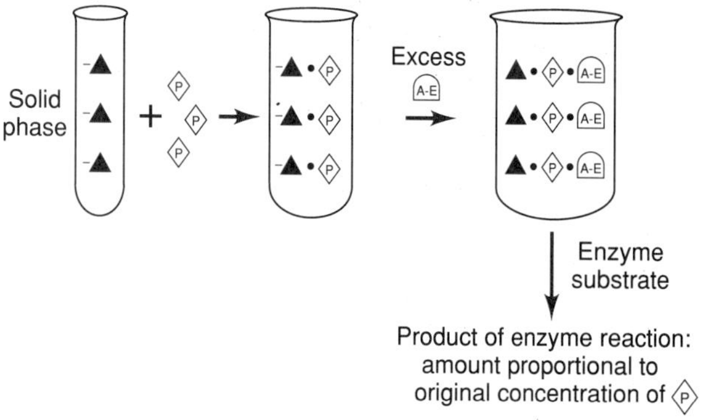

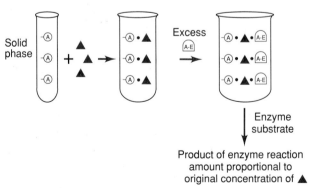

FIGURE 9–14   ELISA—indirect method. ▲ = antigen, ⬦ = protein to be measured, A-E = enzyme/antibody complex.

FIGURE 9–15   ELISA—double-antibody sandwich method. ▲ = antigen, Ⓐ = antibody, A-E = enzyme/antibody complex.

low molecular weight. The direct measurement approaches of ELISA are more sensitive and have a wider range for the standard curve. No radioisotope is required, as is the case with RIA, which eliminates problems with storage and limitations due to short half-lives of radioactive materials. Coupling of enzyme tag to the antibody involves relatively simple chemistry and produces reasonably high yields of product. The cost of enzyme and antibody may work against ELISA's being as economical as some other approaches, but the overall expense does not appear to be a major problem.

Clinical Applications of ELISA

The major clinical usage for ELISA measurements has been in microbiology, serology, and immunology. With its capability of binding to specific proteins, ELISA has been widely employed to detect the presence of viruses. This technique is heavily drawn on in assays for the presence of markers for both hepatitis A and hepatitis B. ELISA techniques are available for the measurement of carcinoembryonic antigen, a very useful tumor marker. In the environmental field, a number of ELISA methodologies have been published for the detection of very low levels of pesticides and other toxic chemicals; some of these assays may move into the clinical lab for specific applications.

9.5

FLUORESCENCE IMMUNOASSAY APPROACHES

Immunoassays utilizing some sort of a fluorescent label to tag a molecule include a mixture of techniques. The only unifying factor is the presence of a fluorescent molecule somewhere in the system which serves as a marker to differentiate bound from free material or to act as a signal in a reaction. **Fluorescence immunoassays** may be either heterogeneous or homogeneous; they may employ EIA or ELISA concepts in the assay architecture; some approaches require sophisticated instrumentation, but they all require the fluorescent label.

Direct Quenching Fluoroimmunoassay

If an antigen is labeled with a fluorescent tag, it may give a certain amount of fluorescence in the free state. When antibody binds to that antigen, the fluorescence is quenched; that is, the amount of fluorescence observed decreases (Fig. 9–16A). This process is known as **direct quenching.** Homogeneous assays are available in which a limiting amount of antibody is added to a mixture of patient ligand and

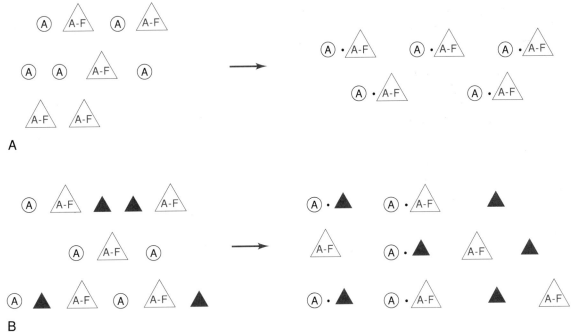

FIGURE 9–16 Direct-quenching fluoroimmunoassay. A. Absence of patient antigen. All fluorescent-labeled antigen bound to antibody; little or no fluorescence observed. B. Presence of patient antigen. Less fluorescent-labeled antigen bound to antibody; more fluorescence observed. ▲ = antigen, Ⓐ = antibody, △A-F = fluorescent-labeled antigen.

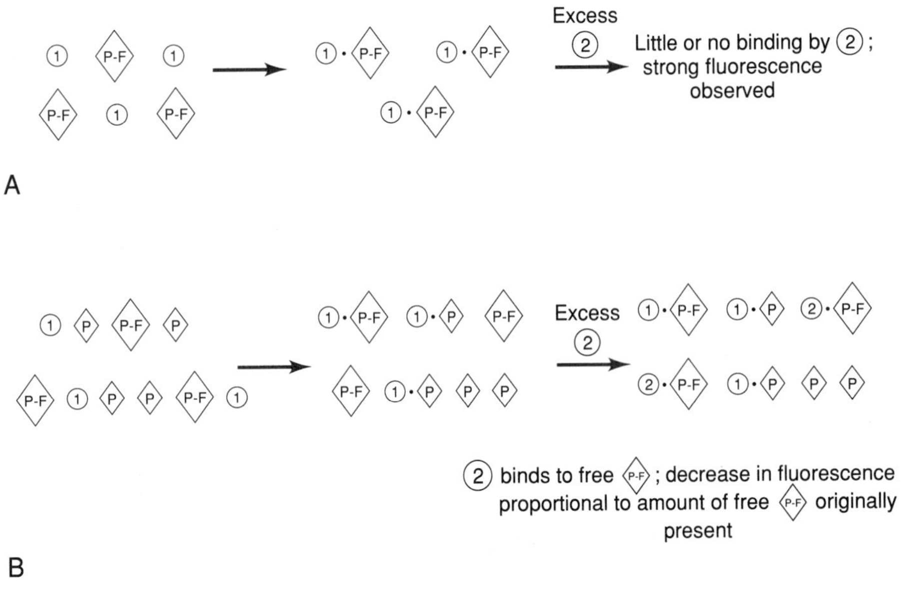

A

B

FIGURE 9–17
Indirect-quenching
fluoroimmunoassay. A. No
patient material present.
Little or no binding by
antibody specific for F
portion of the complex. B.
Assay in presence of patient
material. Antibody specific
for F portion of complex
binds to free P labeled with
fluorescent material
originally present. Ⓟ =
patient protein, ① =
antibody to Ⓟ , ② =
antibody to fluorescent
portion of complex, ⟨P-F⟩ =
Ⓟ labeled with fluorescent
material.

fluorescent-tagged ligand. As in RIA, the more patient ligand present, the less fluorescent-labeled ligand is bound to antibody (Fig. 9–16B). Therefore, the fluorescence observed in the reaction mixture is roughly proportional to the amount of patient ligand present.

Indirect Quenching Fluoroimmunoassay

The direct quenching approach works well with small molecules, since they can easily be entrapped by antibody and the molecular fluorescence diminished. However, large molecules (such as proteins) cannot easily be quantitated in this manner. An alternative procedure, called the **indirect quenching method,** has been developed for the assay of these macromolecules (Fig. 9–17). In this approach, patient antigens and fluorescent-labeled antigens are mixed with an antibody present in limiting amounts. When equilibrium is attained, a certain fraction of the patient ligand and a certain fraction of the labeled ligand have antibody attached to them. To this mixture is added an antibody specific for the fluorescent label (that tag on the protein which fluoresces, not the protein itself). If a fluorescent-labeled ligand has the first antibody attached to it, the second antibody cannot bind, and that complex fluoresces when excited (Fig. 9–17A). A fluorescent-labeled ligand which has no antibody attached binds to the second antibody and the fluorescence of that molecule is blocked (Fig. 9–17B). Therefore, fluorescence of the mixture is again proportional to the amount of patient ligand present.

Substrate-Labeled Fluorescence Immunoassay

Substrate-labeled fluorescence immunoassay (SLFIA) is widely used and has some similarities in concept to the EIA system, since it both is a homogeneous assay and employs enzyme activity as the marker to assess concentration of the analyte. Using the SLFIA technique, both small molecules (such as drugs) and large molecules (immunoglobulins) can be quantitated. The analyte ligand is covalently attached to the substrate of an enzyme (unlike homogeneous EIA, where the ligand is attached directly to the enzyme). The enzyme reaction employed is one in which the product of the reaction fluoresces, and the substrate does not. If antibody to the ligand is added to the system, antibody binds to the ligand and blocks the ability of the substrate to attach to the enzyme for reaction (Fig. 9–18A). In a mixture of patient ligand and substrate-labeled ligand (Fig. 9–18B), the addition of limiting amounts of antibody produces a mixture of patient ligand/antibody and substrate-labeled/ligand antibody. The more patient ligand present, the less antibody binds to substrate-labeled ligand. If we then add enzyme to the system, all free substrate-labeled ligand reacts with enzyme to produce the fluorescent product. Therefore, the higher the enzyme activity, the higher the concentration of patient ligand.

The SLFIA approach to fluorescence immunoassay has been commercially developed with special instrumentation and the use of dipstick reagent strips for a solid-phase assay. The method is simple and versatile, with results available within 10 min. The apparatus and methodology are well suited for

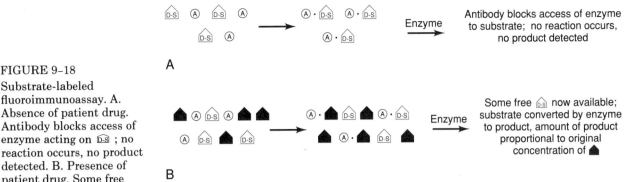

FIGURE 9–18
Substrate-labeled
fluoroimmunoassay. A.
Absence of patient drug.
Antibody blocks access of
enzyme acting on D-S ; no
reaction occurs, no product
detected. B. Presence of
patient drug. Some free
substrate/drug complex now available. Substrate converted by enzyme to product; amount of product is proportional to
the original concentration of the drug in the patient's body fluid. A = antibody to free ▲, ▲ = drug in patient's
body fluid, D-S = substrate/drug complex.

field use, such as clinics or physician's office practices.

Fluorescence Polarization Immunoassay

To understand **fluorescence polarization** methodology, we first need to discuss some theory about polarized light, fluorescence, and molecular motions. We laid out the principles of fluorescence previously and need not repeat them here, but we do want to examine polarization more closely.

Recall that light can be described as a wave. Normally, these waves occur in all planes (Fig. 9–19). With the use of special materials, we can obtain light of only one plane, or **polarized light.** The light is passed through the polarizing material, which blocks all the rays except those in a specific plane. There is nothing different about the wavelength of the light; only the direction of the plane has been changed from light going in all planes to light going in only one plane.

The principle of fluorescence polarization is relatively simple. When polarized light of a specific wavelength is shined into a solution containing molecules which fluoresce at that wavelength, a certain number of the molecules interact with the light and themselves fluoresce. Since the materials in solu-

tion are not static, they rotate at varying speeds; small molecules rotate fairly rapidly, and large ones turn more slowly. Only molecules with the proper orientation in solution at any given moment interact with the polarized light and fluoresce.

Using these principles, let us look at a typical fluorescence polarization immunoassay. The components are ones with which we are already familiar: patient ligand (the material we want to quantitate), antibody to the ligand, and ligand labeled with a molecule which will fluoresce when light of the proper wavelength strikes it. If we mix these three components with antibody in limiting amounts, equilibrium is achieved between bound and free ligands. Keep in mind that the higher the concentration of patient ligand, the lower the amount of fluorescent-labeled ligand that binds to the antibody.

When the polarized beam of light is shined into the system, it interacts with the fluorescing molecule attached to ligand. If the fluorescent ligand is free (not bound to antibody), it rotates rapidly. At any given moment, only a very few molecules are properly oriented to interact with the polarized light and fluoresce. If this labeled ligand is bound to antibody, it becomes part of a large molecular complex and rotates much more slowly. More fluorescent molecules are properly oriented to fluoresce when stimulated by polarized light. Therefore, the higher the fluorescence polarization, the more labeled ligand is bound to antibody. In the presence of patient ligand, part of the antibody is tied up through antigen/antibody complexes with the molecules from the patient sample. Less antibody is available to bind with labeled ligand and less polarization results. The higher the concentration of patient ligand, the lower the observed fluorescence polarization.

Instrumentation and materials for fluorescence polarization are commercially available. Primary applications are in the quantitation of small molecules. A large number of kits have been developed for therapeutic drug monitoring as well as for mea-

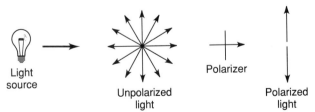

FIGURE 9–19 "Regular" and plane-polarized light waves.

surement of T4. Because of the principle of the test, fluorescence polarization is expected to find wide application in the quantitation of high-molecular-weight molecules such as protein fractions.

Time-Resolved Fluorescence Immunoassay

Fluorescent probes have shortcomings which hinder widespread use in the assay of biochemical constituents. Many of the probes may not provide the desired level of sensitivity. Frequently, there is high natural background fluorescence from the body fluid being assayed. Other components present in the assay system may quench the fluorescence. These limitations suggest the need for a different methodology for fluorescence immunoassays.

If a fluorescent material is excited by a single pulse of light, the light emitted rises rapidly to a peak and then drops back to zero (we see continuous fluorescence when we use a constant beam of light and not a single sharp burst). The time required in this **time-resolved fluorescence** technique for peak emission and drop-off is very short, usually about 10 ns less. Materials can be added to the system which have much longer periods of fluorescence. If this added chemical component is present, the decay time (relating to the rate of drop-off of fluorescence) is quite long, often more than 100 ns (Fig. 9-20). We can use this property in a fluorescence immunoassay system to eliminate problems due to background interference and to increase the sensitivity and specificity of the assay.

Complexes of the rare-earth elements europium and terbium demonstrate long fluorescence decay times. In addition, they fluoresce strongly. These materials can be attached to antibodies and used in heterogeneous immunoassay systems. In a typical assay, antibody to the analyte ligand is bound to a solid phase (test tube, well, or other). Patient sample is incubated with this system and all patient ligand binds to the antibody. After washing to remove extraneous materials, a second antibody is added which is tagged with the rare-earth complex. A sec-

ond washing eliminates extra material which did not bind to the initial antibody/ligand complex. The result is a sandwich of the same type we saw in IRMA and ELISA assays. In this case the signal is the long-term fluorescence produced by the rare-earth material. When placed in a special fluorescence instrument and stimulated by a short flash of light, the rare-earth atoms are excited and fluoresce with a long decay time. This curve is measured and the intensity of light is determined at a given time. The amount of fluorescence is directly proportional to the concentration of ligand present in the patient sample.

Other Fluorescence Immunoassay Techniques

With the exception of time-resolved fluorescence immunoassay, all the approaches we have discussed have been homogeneous methods—no separation of bound and free materials have been required. A number of heterogeneous approaches are also available using fluorescent molecules as labels. Their methodologies are very similar in concept to RIA, IRMA, and ELISA procedures, using a fluorescent tag in place of a radioisotope or an enzyme. These methods are effective, but subject to the same limitations as the analogous techniques involving other labels.

Advantages and Limitations of Fluorescence Immunoassays

Fluorescence immunoassays have the advantage over RIA and ELISA techniques of being simple—measurement of fluorescence does not require the use of radioisotopes or complex enzyme assays. Present instrumentation for some of the more exotic techniques (such as fluorescence polarization or time-resolved fluorimetry) is limited, which hinders widespread use of these methodologies. Some fluorescent probes are less sensitive than their isotopic or enzymatic counterparts. Background fluorescence can provide serious interference problems in some homogeneous assays, but contributes little or no interference to heterogeneous systems. Quenching of fluorescence may interfere in some systems.

Fluorescence immunoassays continue to contribute greatly to the application of immunoassay techniques in clinical analysis. Several systems, both reagents and instrumentation, are commercially available. The simplicity and versatility of fluorescence immunoassay methods make this analytical technique one of great and ongoing value to clinical chemistry.

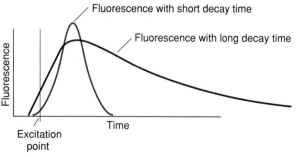

FIGURE 9-20   Time course of fluorescence decay.

9.6

APPLICATIONS OF CHEMILUMINESCENCE AND BIOLUMINESCENCE

Research continues in the immunoassay field into the ability to detect low levels of materials. Many of the markers used—whether radioisotopes, enzymes, or fluorescent tags—are not sensitive enough to be used in the quantitation of components which may be present at levels of 10^{-10} mol/L or less. Two phenomena which appear to overcome this problem may have widespread applications in clinical chemistry in the future: chemiluminescence and bioluminescence.

Chemiluminescence

The process by which light is generated as a result of a chemical reaction is **chemiluminescence.** In the presence of hydrogen peroxide or other "active" oxygen derivatives, certain organic compounds (such as luminol) give off a flash or a glow of light (see formula below).

The intensity of light is affected by many factors, among which are the concentration of the compound giving the reaction, the amount of oxygen derivative present, and the composition of the reaction mixture. These reactions take place at room temperature and require no energy input. The light produced is generated by the reaction between the luminol and the active oxygen compound.

Bioluminescence

Some enzymes produce light as one of the products of the reaction series. The firefly glows on summer nights because of a reaction in its abdomen between enzyme and oxygen, generating "cold" light—no heat is produced. Some bacteria also have the capability of generating light through a bioluminescent reaction.

The firefly reaction is:

1. luciferin + luciferase + ATP → activated luciferin
2. activated luciferin + luciferase + O_2 → oxyluciferin + light

Luciferase is the enzyme involved in this reaction sequence. The first step involves the activation of luciferin (a biochemical compound with a structure similar to luminol) with ATP. The activated luciferin then reacts with molecular oxygen to form another derivative plus a burst of light. Any reaction which produces oxygen can be followed with this assay. Some systems have been developed using this method to detect ATP; space capsules have employed this technique to look for life in other areas of our solar system.

A related reaction is carried out by several marine microorganisms:

1. FMN + NAD(P)H → $FMNH_2$ + NAD(P)
2. $FMNH_2$ + aldehyde + O_2 → acid + FMN + light

The coenzyme flavin mononucleotide (FMN) is reduced to $FMNH_2$ enzymatically. Oxygen is involved in the second reaction (which is catalyzed by the enzyme luciferase) to reoxidize the reduced coenzyme back to FMN as well as oxidize the aldehyde to an acid, producing light in the process. Again we see that oxygen plays an important role in the reaction sequence.

Use of Luminescence Reactions in Quantitative Analysis

If we look at the above light-generating systems (both chemiluminescent and bioluminescent), we see that all the reactions involve oxygen or an oxygen derivative in some way. If we can carry out a chemical or an enzymatic reaction which generates O_2 or H_2O_2, we can couple this product to a luminescent reaction to produce light. In each case, the quantity of light produced is proportional to the

luminol monoanion + H_2O_2 + OH^- → 3-aminophthalate + H_2O + light

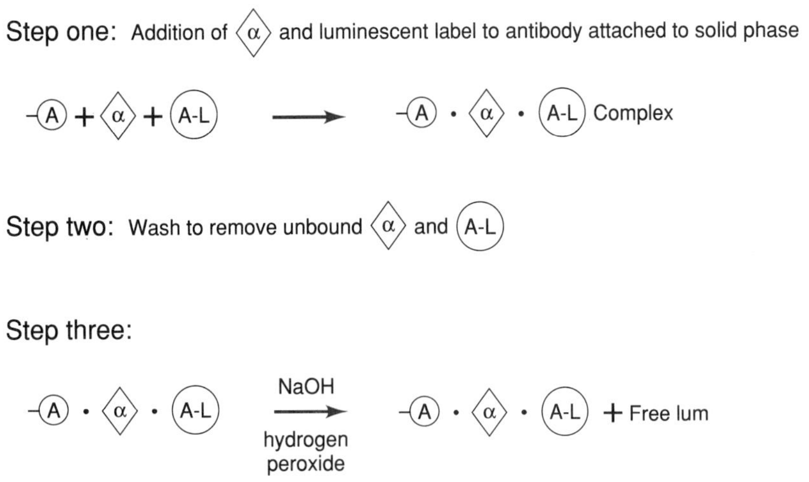

Step one: Addition of $\langle\alpha\rangle$ and luminescent label to antibody attached to solid phase

$$-\text{(A)} + \langle\alpha\rangle + \text{(A-L)} \longrightarrow -\text{(A)} \cdot \langle\alpha\rangle \cdot \text{(A-L)} \text{ Complex}$$

Step two: Wash to remove unbound $\langle\alpha\rangle$ and (A-L)

Step three:

$$-\text{(A)} \cdot \langle\alpha\rangle \cdot \text{(A-L)} \xrightarrow[\substack{\text{hydrogen}\\\text{peroxide}}]{\text{NaOH}} -\text{(A)} \cdot \langle\alpha\rangle \cdot \text{(A-L)} + \text{Free lum}$$

FIGURE 9-21
Chemiluminescence immunoassay of alpha fetoprotein. The free luminescent-labeled antibody will not emit light as detected by the luminometer. The amount of light produced is proportional to the amount of labeled antibody bound to alpha-fetoprotein in the original reaction system. (A) = antibody to alpha fetoprotein, $\langle\diamond\rangle$ = alpha fetoprotein, (A-L) = antibody with attached luminescent material.

amount of initial enzyme present. One role for luminescence assays today is in the measurement of very low levels of enzyme activity in tissues and body fluids. Reactions involving production of NAD(P)H can also be followed if an appropriate bioluminescent system is used.

Luminescence Measurements and Immunoassays

Since both chemiluminescent and bioluminescent reactions are very sensitive, they should provide good tags for use with an immunoassay system. Much work has been done adapting various immunoassay approaches to methods which can be used with this type of label. Almost every technique we have discussed so far (EMIT, ELISA, and homogeneous and heterogeneous methodologies) has been utilized in some fashion to produce a luminescent immunoassay system. We will illustrate with only one example, the quantitation of alpha fetoprotein.

The basic reaction sequence is outlined in Figure 9-21. Antibody to the protein is coupled to a solid phase. To this system is added the sample containing alpha fetoprotein and antibody labeled with a chemiluminescent compound. While coupled to antibody, the compound does not display chemiluminescence. After incubation and washing, extraneous materials are discarded and the antibody/protein-labeled antibody sandwich remains bound to the tube. A basic solution of hydrogen peroxide is added, the luminescent label is cleaved from the antibody, and the light generated by the free label is detected and measured with a luminometer (an instrument which measures light production from a sample). The amount of light emitted is proportional to the amount of labeled antibody bound and thus to the quantity of alpha fetoprotein present in the initial sample.

Considerations in Using Luminescence Immunoassays

When we look at the amount of light yielded in a luminescence reaction, we quickly see that there are striking differences in the yields from chemiluminescent and bioluminescent processes. The bacterial and firefly systems produce almost 100% of the theoretical yield; they are very efficient light-generating systems. On the other hand, organic compounds usually yield 1–2% of theoretical predictions; chemiluminescent systems are very inefficient. The overall light yield can be increased by coupling several molecules of the luminescent compound to the antibody used in the assay.

Instrumentation has been another problem. Very few companies manufacture equipment which measures light-production rates (since we are looking at an ongoing process, not an end point). As these methods become more popular, manufacturers are developing more ways to measure the light production.

9.7

CONCLUSIONS

The immunoassay field is growing rapidly. These procedures are widely used in screening for substance abuse, for therapeutic drug monitoring, in the assay of many hormones, and for the measurement of enzyme levels. A large number of markers in the microbiology area (particularly in the monitoring of hepatitis and AIDS) are measured with the use of ELISA and other related procedures. The very popular (and controversial) "home testing" market employs immunoassay technology extensively. The local pharmacy now carries pregnancy test kits and materials for monitoring ovulation (to aid in achieving pregnancy), which are available from a number of companies and are based on immunoassay techniques discussed in this chapter (usually ELISA methods). RIA may be diminishing as a method of choice (although it is still widely used), but other immunoassay approaches are finding more and more use in the clinical laboratory setting. From the simple qualitative protein precipitation to the more complex methods, the antigen/antibody reaction has proved to be of enduring value.

SUMMARY

One initial application of immunochemical techniques to quantitative analysis was radioimmunoassay (RIA). Patient ligand and radiolabeled ligand compete for binding sites on antibody. The higher the concentration of patient ligand, the lower the amount of binding of radioisotopically tagged ligand to the antibody. A variety of separation techniques are available, the most selective being double-antibody and solid-phase systems. Data reduction is complex and usually handled by computer. Graphical analysis indicates an inverse relationship between patient ligand concentration and binding of labeled ligand to antibody. RIA has been widely employed for assay of hormones, proteins, and drugs.

Immunoradiometric assay (IRMA) involves binding of labeled antibody to patient antigen. Antibody is bound to a solid phase (either the test tube wall or an insoluble particle). Patient ligand is added, which binds to the antibody. A sandwich is formed by adding a second antibody to the analyte ligand; this second antibody is labeled with a radioisotope. The amount of radiolabel bound is directly proportional to the amount of patient ligand present. A hook effect may be seen at high concentrations of ligand, requiring sample dilution and reassay.

A variety of nonradioactive labels have replaced radioisotopes in many assay systems. The homogeneous enzyme immunoassay uses an enzyme attached to the analyte ligand. When antibody to the ligand is added, this protein blocks access of substrate to enzyme, allowing little enzyme activity to be detected. If free patient ligand is then added, some of the antibody binds to patient ligand, freeing the ligand/enzyme complex. The more patient ligand present, the less ligand/enzyme complex bound to antibody. Patient ligand concentration is then directly proportional to the amount of enzyme activity detected in the system.

A variety of enzyme-linked immunosorbent assays (ELISA) are available. These analytical approaches resemble IRMA in concept, but the radiolabel is replaced by an enzyme tag. Antibody covalently bound to enzyme binds to ligand attached to solid-phase first antibody. The amount of enzyme detected is related to the amount of ligand present. Competitive, indirect, and double-antibody sandwich assays are available for ELISA. This analytical approach has found wide application in the areas of microbiology, serology, and immunology. Detection of viral components (such as those seen in hepatitis A and B) is easily performed with ELISA methodology.

The common theme of fluorescence immunoassays is the use of a fluorescent label somewhere in the system. Fluorescence immunoassays exploit a wide range of assay techniques, overlapping the competitive-binding approaches similar to RIA and the enzyme assay technologies of EIA and ELISA. Fluorescence polarization has proven to be a sensitive and specific analytical tool. Time-resolved fluorescence immunoassay uses a rare-earth tag with a long fluorescence decay time. Measurement of the residual fluorescence in the system allows detection of the amount of label present.

Both chemiluminescence and bioluminescence are growing as markers for immunoassays. Chemiluminescence is produced when certain organic compounds react with oxygen, producing a burst of light. Bioluminescence is the product of specific enzyme reactions. The reactions that produce luminescence can be coupled with immunoassay techniques to produce measurements with a high degree of specificity.

Hormones, proteins, drugs—in fact, a wide variety of biochemical compounds—can be measured at low concentrations and with high specificity with various immunoassay techniques. The field of immunochemical measurements is a growing one with increasing numbers of applications in clinical chemistry.

FOR REVIEW

Directions: For each question, choose the best response.

1. All of the following are required in order to perform a radioimmunoassay *except*
   A. spectrophotometer
   B. antibody
   C. labeled antigen
   D. patient antigen

2. In radioimmunoassay the tag used to label the antigen is a(an)
   A. chemiluminescent material
   B. fluorescent material
   C. isotope
   D. enzyme

3. Which component is actually measured by the instrument in radioimmunoassay?
   A. free patient antigen
   B. free antibody
   C. antibody bound to patient antigen
   D. free or bound labeled antigen

4. Techniques used in radioimmunoassay to isolate the bound label from the free label include all of the following *except*
   A. charcoal adsorption
   B. double-antibody technique
   C. urea precipitation
   D. salt or solvent precipitation

5. The RIA separation technique where antibody is covalently bound to the inside wall of the reaction tube best describes the technique known as
   A. adsorption
   B. dialysis
   C. double antibody
   D. solid phase

6. In RIA the term *background* refers to the amount of radioactivity
   A. left in the tube at the end of the separation step
   B. measured in the absence of any isotope
   C. added initially to each tube
   D. that binds to materials in the reaction mixture other than the antibody

7. Immunoradiometric assay differs from radioimmunoassay in that in IRMA the radioactive label is originally attached to the
   A. antibody
   B. antigen
   C. patient antigen
   D. antigen/antibody complex

8. In enzyme immunoassay the tag used to label the antigen is a(an)
   A. chemiluminescent material
   B. fluorescent material
   C. enzyme
   D. isotope

9. Which of the following best describes the separation process used in homogeneous enzyme immunoassay?
   A. solid-phase technique
   B. double-antibody technique
   C. dialysis technique
   D. separation step not necessary

10. All of the following are associated with the enzyme-linked immunosorbent assay *except*
    A. heterogeneous assay
    B. enzyme label
    C. qualitative determinations only
    D. solid-phase system

11. Which of the following may be associated with the fluorescence immunoassay system?
    A. heterogeneous
    B. homogeneous
    C. employs enzymes
    D. fluorescent label
    E. all of the above

12. Which of the following may be associated with fluorescence polarization immunoassay?
    A. the more free fluorescent ligand, the lesser the fluorescence polarization
    B. the more labeled ligand bound to antibody, the greater the fluorescence polarization
    C. the greater the patient ligand present, the greater the fluorescence polarization
    D. both A and B
    E. A, B, and C

13. When an enzyme catalyzes a reaction where light is one of the products formed, this type of assay is referred to as
    A. bioluminescence
    B. chemiluminescence
    C. fluorescence
    D. photometric

BIBLIOGRAPHY

Alfthan, H., and Stenman, U-H., "Time-resolved fluoroimmunoassays," *Amer. Biotech. Lab.* August: 8–13, 1988.
Bakerman, S., "Enzyme immunoassays," *Lab. Manage.* August: 21–29, 1980.

Bakerman, S., "Fluorescence polarization immunoassay," *Lab. Manage.* June: 16–18, 1983.
Bakerman, S., "Substrate-labeled fluorescence immunoassay," *Lab. Manage.* October: 13–16, 1983.
Bellet, N., and Wagman, B. S., "Running homogeneous enzyme immunoassays on clinical analyzers," *Amer. Clin. Prod. Rev.* June: 10–15, 1986.

Borlaug, G., "RIA: Basic quality control of the gamma scintillation system," *Lab. Med.* 10:346–348, 1979.

Carey, R. N. et al., "Performance characteristics of some statistical quality control rules for radioimmunoassay," *J. Clin. Immunoassay.* 8:245–252, 1985.

Carter, J. H., "Enzyme immunoassays: Practical aspects of their methodology," *J. Clin. Immunoassay.* 7:64–72, 1984.

Cernosek, Jr., S. F., "Data reduction in radioimmunoassay: Computerized data reduction," *J. Clin. Immunoassay.* 8:203–212, 1985.

Chait, E. M., and Ebersole, R. C., "Clinical analysis: A perspective on chromatographic and immunoassay technology," *Anal. Chem.* 53:682A–692A, 1981.

Craine, J. E., "Latex agglutination immunoassays," *Amer. Biotech. Lab.* May/June: 34–41, 1987.

De Grella, R. F., "Fluorescence polarization: A review of laboratory applications," *Amer. Biotech. Lab.* August: 29–33, 1988.

Diamandis, E. P., "Immunoassays with time-resolved fluorescence spectroscopy: Principles and applications," *Clin. Biochem.* 21:139–150, 1988.

Diamond, B., and Scharff, M. D., "Monoclonal antibodies," *J. Amer. Med. Assoc.* 248:3165–3169, 1982.

Dito, W. R., "Nonisotopic immunoassay," *Diag. Med.* October: 10–18, 1981.

Dudley, R. A. et al., "Guidelines for immunoassay data processing," *Clin. Chem.* 31:1264–1271, 1985.

Ehrhardt, V. et al., "Mechanization of heterogeneous immunoassays," *J. Clin. Immunoassay.* 11:74–80, 1988.

Fackrell, H. B. et al., "ELISA data reduction: A review," *J. Clin. Immunoassay.* 8:213–219, 1985.

Geokas, M. C. et al., "Peptide radioimmunoassays in clinical medicine," *Ann. Intern. Med.* 97:389–407, 1982.

Gerson, B., "Fluorescence immunoassay," *J. Clin. Immunoassay.* 7:73–81, 1984.

Gochman, N., and Bowie, L. J., "Automated systems for radioimmunoassay," *Anal. Chem.* 49:1183A–1190A, 1977.

Goldsmith, S. J., "Radioimmunoassay: Review of basic principles," *Semin. Nuclear Med.* 5:125–152, 1975.

Gordon, D. S., "Fast track for monoclonal antibodies," *Med. Lab. Observer.* July: 53–60, 1983.

Gorus, F., and Schram, E., "Applications of bio- and chemiluminescence in the clinical laboratory," *Clin. Chem.* 25:512–519, 1979.

Grayeski, M. L., "Chemiluminescence analysis," *Anal. Chem.* 1243A–1256A, 1987.

Hamilton, R. G., "The clinical immunology laboratory: Reflections of an immunoassayist," *J. Clin. Immunoassay.* 7:297–301, 1984.

Hamilton, R. G., "Application of personal computer integrated software to the immunoassay laboratory," *J. Clin. Immunoassay.* 8:220–229, 1985.

Heineman, W. R., and Halsall, H. B., "Strategies for electrochemical immunoassay," *Anal. Chem.* 57:1321A–1331A, 1985.

Hemmila, I., "Fluoroimmunoassays and immunofluorometric assays," *Clin. Chem.* 31:359–370, 1985.

Henderson, D. R., "CEDIA®, a new homogeneous immunoassay system," *Clin. Chem.* 32:1637–1641, 1986.

Herner, A. E., and Clayton-Hopkins, J. A., "Troubleshooting radioligand assays," *Clin. Chem.* 24:1275–1280, 1978.

Hoffman, K. L., "Optimization of the sandwich immunometric assay," *J. Clin. Immunoassay.* 8:237–244, 1985.

Homsher, R., "Effect of data reduction on accuracy assessment in an enzyme immunoassay," *J. Clin. Immunoassay.* 8:230–233, 1985.

Howanitz, J. H., "Immunoassay: Innovations in label technology," *Arch. Pathol. Lab. Med.* 112:775–779, 1988.

Howanitz, P. J., "Immunoassay: Developments and directions in antibody technology," *Arch. Pathol. Lab. Med.* 112:771–774, 1988.

Ishikawa, E., "Development and clinical application of sensitive enzyme immunoassay for macromolecular antigens—a review," *Clin. Biochem.* 20:375–385, 1987.

Joustra, M., "Current trends in solid phase RIA," *Protide Biol. Fluids.* 24:711–714, 1976.

Kemp, H. A. et al., "Labeled antibody immunoassays," *Ligand Quart.* 5:27–34, 1982.

Lentrichia, B. B. et al., "Immunoassay by centrifugal analysis," *Amer. Biotech. Lab.* May/June: 17–29, 1987.

Leuvering, J. H. W., "SPIA: A new immunological technique," *Amer. Clin. Prod. Rev.* September: 36–39, 1986.

Linthicum, D. S. et al., "Idiotypes and anti-idiotypes: Significance in immunoassays," *J. Clin. Immunoassay.* 11:31–36, 1988.

Lunte, C. H. et al., "Electrochemical immunoassay: An ultrasensitive assay technique," *Lab. Manage.* August: 51–58, 1986.

Maciel, R. J., "Standard curve fitting in immunodiagnostics: A primer," *J. Clin. Immunoassay.* 8:98–106, 1985.

Marsden, D. S., "Radioimmunoassay separation techniques," *Lab. Manage.* March: 31–34, 1977.

Marshall, D. L., and Bush, G. A., "Latex particle enzyme immunoassay," *Amer. Biotech. Lab.* May/June: 48–53, 1987.

Monroe, D., "ELISA: A versatile chemical tool," *Amer. Clin. Prod. Rev.* May/June: 22–27, 1983.

Monroe, D., "Enzyme immunoassay," *Anal. Chem.* 56:920A–931A, 1984.

Monroe, D., "Liposome immunoassay: A new ultrasensitive analytical method," *Amer. Clin. Prod. Rev.* December: 34–41, 1986.

Monroe, D., "Potentiometric immunoassay," *Amer. Clin. Prod. Rev.* March: 31–39, 1987.

Nichols, A. L., "Radioimmunoassay," *Lab. Manage.* November: 44–56, 1975.

O'Donnell, C. M., and Suffin, S. C., "Fluorescence immunoassays," *Anal. Chem.* 51:33A–40A, 1979.

O'Sullivan, M. J. et al., "Enzyme immunoassay: A review," *Ann. Clin. Biochem.* 16:221–240, 1979.

Polsky-Cynkin, R., "Automated immunoassay: Why has success been so elusive?" *J. Clin. Immunoassay.* 11:69–73, 1988.

Pourfarzaneh, M. et al., "Production and use of magnetizable particles in immunoassay," *Ligand Quart.* 5:41–47, 1982.

Rodbard, D., and Feldman, Y., "Kinetics of two-site immunoradiometric ('sandwich') assays—I. Mathematical models for simulation, optimization, and curve fitting," *Immunochem.* 15:71–76, 1978.

Rodbard, D. et al., "Kinetics of two-site immunoradiometric ('sandwich') assays—II. Studies on the nature of the high-dose 'hook' effect," *Immunochem.* 15:77–82, 1978.

Sanville, C. "Luminescence as an analytical tool," *Amer. Clin. Prod. Rev.* June: 28–32, 1985.

Scall, Jr., R. F., and Tenoso, H. J., "Alternatives to radioimmunoassay: Labels and methods," *Clin. Chem.* 27:1157–1164, 1981.

Scharpe, S. L. et al., "Quantitative enzyme immunoassay: Current status," *Clin. Chem.* 22:733–738, 1976.

Schuurs, A. H. W. M., and van Weemen, B. K., "Enzyme-immunoassay," *Clin. Chim. Acta* 81:1–40, 1977.

Sevier, E. D., "Monoclonal antibodies: Expanded potential for labeled antibody ligand assays," *Amer. J. Med. Tech.* 48:651–653, 1982.

Shamel, L. B., "The immunoradiometric assay," *Amer. Clin. Prod. Rev.* November/December: 44–52, 1983.

Smith, D. S. et al., "A review of fluoroimmunoassay and immunofluorometric assay," *Ann. Clin. Biochem.* 18:253–274, 1981.

Smith, S. W., "Application of the Scatchard plot to radioimmunoassay I: Theoretical considerations," *J. Clin. Immunoassay.* 8:52–56, 1985.

Soini, E., and Hemmila, I., "Fluoroimmunoassay: Present status and key problems," *Clin. Chem.* 25:353–361, 1979.

Spierto, F. W., and Shaw, W., "Problems affecting radioimmunoassay procedures," *CRC Crit. Rev. Clin. Lab. Sci.* 7:365–372, 1977.

Stanley, C. J., "Enzyme amplification: A new technique for enhancing the speed and sensitivity of enzyme immunoassays," *Amer. Biotechnol. Lab.* May/June: 48–53, 1985.

Thore, A., "Luminescence in clinical analysis," *Ann. Clin. Biochem.* 16:359–369, 1979.

Van Lente, F., and Galen, R. S., "Luminescence assays," *Diag. Med.* January/February: 71–75, 1980.

Voller, A. et al., "Enzyme immunoassays with special reference to ELISA techniques," *J. Clin. Pathol.* 31:507–520, 1978.

Walker, W. H. C., "An approach to immunoassay," *Clin. Chem.* 23:384–402, 1977.

Weeks, I., and Woodhead, J. S., "Chemiluminescence immunoassay," *J. Clin. Immunoassay.* 7:82–89, 1984.

Weilde, C. E., and Ottewell, D., "A practical guide to gammacounting in radioimmunoassay," *Ann. Clin. Biochem.* 17:1–9, 1980.

Weiss, A. J., and Blankstein, L. A., "Membranes as a solid phase for clinical diagnostic assays," *Amer. Clin Prod. Rev.* June: 8–19, 1987.

Whitehead, T. P. et al., "Analytical luminescence: Its potential in the clinical laboratory," *Clin. Chem.* 25:1531–1546, 1979.

Wiedbrauk, D. L. et al., "Rapid enzyme immunoassays for the clinical laboratory," *Amer. Clin. Lab.* May: 16–19, 1989.

Wisdom, G. B., "Enzyme-immunoassay," *Clin. Chem.* 22:1243–1255, 1976.

Yadav, R. N., "Applications of radioreceptor analysis," *Amer. Lab.* February: 65–76, 1980.

Yalow, R. S., "Radioimmunoassay: A probe for the fine structure of biologic systems," *Science.* 200:1236–1245, 1978.

Yalow, R. S., "Radioimmunoassay—a historical perspective," *J. Clin. Immunoassay.* 10:13–19, 1987.

Chapter Ten

CLINICAL ENZYMOLOGY II: USE OF ENZYMES IN DIAGNOSIS

• •

UPON COMPLETION OF THIS CHAPTER, THE STUDENT WILL BE ABLE TO

1. Explain why under normal circumstances some enzymes are found in higher concentrations in the circulation, and other enzymes are present at very low levels.
2. Explain how proteases and renal filtration can affect enzyme levels in the blood.
3. Discuss how pathological conditions can change enzyme levels in the blood.
4. List several enzymes and their corresponding tissue site(s) to illustrate high, moderate, and low tissue specificity.
5. Discuss the diagnostic usefulness of measuring enzyme levels and determining isoenzyme patterns.
6. Define the term *isoenzyme*.
7. Discuss the following information for each enzyme of clinical interest*:
   A. type of reaction catalyzed
   B. tissue distribution
   C. specimen collection requirements
   D. principles of measurement methods
   E. assay interferences
   F. reference ranges and units
   G. clinical significance
   > *The enzymes of clinical interest are:
   > acid phosphatase, total and prostatic
   > alkaline phosphatase
   > alanine aminotransferase
   > aspartate aminotransferase
   > amylase
   > cholinesterase, true and pseudo
   > creatine kinase
   > gamma-glutamyltransferase
   > lactate dehydrogenase
   > lipase
8. Discuss the following information for each of the isoenzymes of clinical interest listed below*:
   A. molecular structure
   B. tissue distribution
   C. clinical significance
   D. principles of measurement methods
   E. electrophoretic pattern (if applicable)
   > *The isoenzymes of clinical interest are:
   > alkaline phosphatase
   > creatine kinase
   > lactate dehydrogenase
9. Describe the clinical significance associated with detecting acid phosphatase in rape cases.
10. Describe the clinical significance associated with measuring urine amylase and determining the amylase/creatinine clearance ratio.
11. Describe the molecular structure of macroamylase and its clinical significance.

179

12. Describe the clinical application of cholinesterase measurements in screening for pesticide exposure.
13. Describe the electrophoretic pattern and the clinical significance of detecting mitochondrial CK and macro-CK.
14. List the enzyme and/or isoenzyme tests that would be useful in diagnosing the following conditions:
    A. myocardial infarction   D. bone disease
    B. muscle disease   E. acute pancreatitis
    C. liver disease
15. Describe the enzyme and isoenzyme patterns that would characterize a myocardial infarction.

INTRODUCTION

Our knowledge of enzyme reactions and the sources of enzymes in tissues allows us to detect changes in body function and assess the presence or absence of specific tissue damage. Knowing where certain enzymes are concentrated in the body and how their levels in body fluids change during the course of disease are powerful diagnostic tools which reveal important information about our health.

10.1

SOURCES OF CIRCULATING ENZYMES

Intracellular Formation of Enzymes

Most enzymes detected in body fluids are not formed there, but rather are released from cells. The amount of a given enzyme within a cell varies from one moment to the next as the metabolic demands on the cell change. Some enzymes are formed specifically for release into the circulation, since this is where they carry out their metabolic functions. The various coagulation factors are enzymes; these materials are found in high concentrations in the bloodstream since the clotting process takes place within the circulation. Enzymes responsible for the metabolism of proteins within the circulation are found in higher levels in blood than in most tissues. A high concentration of an enzyme in a body fluid of healthy individuals suggests a specific function for that enzyme in the fluid.

Release into Body Fluids

The levels of activity of most enzymes in blood or other body fluids are quite low. The body fluid concentration of a given enzyme reflects a variety of processes. One major factor is the amount of cellular **turnover**—how rapidly the old cells are dying and breaking down. As a cell reaches the end of its life span, it disintegrates and releases its contents into the surrounding tissues. The enzymes within the cell escape, and a certain amount of this material enters the bloodstream, urine, or CSF. Since most cells turn over farily slowly (for example, the average life span of a red blood cell is 120 days), only low levels of the enzyme are observed in the body fluid under normal circumstances.

Metabolism and Excretion

Once the enzyme enters the body fluid, further changes in concentration take place. In blood (and, to a lesser extent, other fluids) there are a number of **proteases**—enzymes whose role is to break down proteins into smaller components. These proteases attack the various enzymes and metabolize them at different rates. The rate of decrease in activity depends on the specific enzyme and the processes involved in its clearance from the circulation.

The circulating levels of some enzymes are affected through excretion by the kidney. Proteins with a molecular weight of less than approximately 60,000 are filtered by the kidney and excreted in the urine. One such enzyme is amylase, responsible for the digestion of starch. This enzyme has a molecular weight of 40,000 or so and is readily excreted by way of the kidneys. Quite often, the level of amylase in urine is higher than its level in blood.

10.2

ENZYME CHANGES IN PATHOLOGICAL CONDITIONS

Changes in Levels in Body Fluids

The routine turnover of cells and the accompanying steady level of an enzyme in a body fluid is altered in a striking fashion in many disease states. Instead of a fairly constant concentration of the enzyme, the level may rise markedly over a short time and then subside fairly rapidly to the normal

Table 10-1
TISSUE SPECIFICITY OF ENZYMES

	Enzyme	Principal Tissue(s)
High specificity	Acid phosphatase	Erythrocyte, prostate
	Alanine aminotransferase	Liver
	Amylase	Pancreas, salivary glands
	Lipase	Pancreas
Moderate specificity	Aspartate aminotransferase	Liver, heart, skeletal muscles
	Creatine kinase	Heart, skeletal muscles, brain
Low specificity	Alkaline phosphatase	Liver, bone, kidney
	Lactate dehydrogenase	All tissues

amount for that enzyme. In other situations, where the "normal" concentration is high, we may see an abrupt drop in response to a physiological change in the organism. These drastic alterations of the normal enzyme level provide valuable diagnostic clues as to the underlying pathological state in the body.

In many disease states, one of the consequences is the rapid destruction of tissue. Let us use the case of a **myocardial infarction** (heart attack) as an example. As a result of changes in heart beat, loss of oxygen, decreased blood flow, or other factors, heart tissue begins to break down very quickly. Instead of the usual slow turnover of cells, the rate of cell breakdown rises, and greatly increased quantities of certain enzymes are released into the circulation. The blood levels of these enzymes increase over a very short time. As the process of cell death slows, less enzyme is released from the dying cells. At some later point, cell turnover returns to normal, metabolism (and excretion) of the excess enzyme has occurred, and the usual low level of the enzyme in the circulation is restored.

Enzyme Level and Tissue Specificity

Each enzyme has a defined tissue specificity. There are tissues in the body which contain higher concentrations of one enzyme than another (Table 10-1). By knowing the tissue specificity of any enzyme, we can better ascertain where the damage occurred in the body. Enzymes of high specificity (amylase, acid phosphatase) are found predominantly in one type of tissue. Those enzymes which have moderate specificity are more widely distributed in the body. Some enzymes (alkaline phosphatase and lactate dehydrogenase) are ubiquitous—they are found everywhere. Even with this very wide distribution, these enzymes can provide valuable diagnostic information if we study the proper parameters.

Use in Clinical Diagnosis

By combining data from tissue specificity studies and from research on enzyme changes in various disease states, we can assemble a useful package of information which allows us to interpret enzyme levels in body fluids in a reliable manner. If we know that a certain enzyme is located primarily in the liver, changes in the blood level of that enzyme should alert us to the possibility of liver damage. The amount of increase and the duration of the increase also provide data about the extent of the damage. A high enzyme level over a long time suggests chronic (ongoing) damage to the tissue, instead of the acute single event which a rapid rise and fall of enzyme concentration suggest.

Some precautions must be used concerning the application of enzyme data to the diagnostic process (Table 10-2). There is no truly "tissue-specific" enzyme; all enzymes are found in more than one tissue. A single enzyme measurement is usually of little value in assessing tissue damage. Serial measurements over the course of several days permit the development of any pattern of increase or decrease, as well as giving a better look at the extent of change over a given period. "Negative" results (normal values), when properly utilized, are useful in ruling out the presence of tissue damage. Enzyme data cannot be interpeted by itself; we must look at other laboratory results and other pertinent clinical information before a diagnosis can be made.

Table 10-2
PRINCIPLES OF ENZYME DATA INTERPRETATION

1. There is no truly "organ-specific" enzyme.
2. Serial measurements provide most useful data; a single measurement can be misleading.
3. "Negative" (normal) results are useful.
4. Enzyme data must be integrated with other information.

10.3

ISOENZYMES

Up to this point, we have assumed that an **enzyme** is a protein composed of one polypeptide chain and consisting of a definite, specific structure. Many enzymes, however, consist of two or more polypeptide chains. In addition, not all molecules of these multichain enzymes have an identical structure. These enzyme molecules, which are referred to as **isoenzymes,** all catalyze the same reaction but have slightly different structures. Another important feature of isoenzymes is that they are more specifically located in tissues than enzymes are; that is, although a particular enzyme might be found in a number of tissues, the site of a specific one of its isoenzymes may be restricted to only one tissue.

We can best illustrate the concept of isoenzymes with a specific example. **Creatine kinase** (CK) is an enzyme made up of two polypeptide chains. Each specific chain is referred to as a subunit of the complete enzyme. The two chains must interact with one another for enzyme activity to occur. The two chains in CK found in skeletal muscle are identical and have a specific structure; this chain is referred to as the M chain (or subunit). The CK found in brain tissue is also composed of two identical chains, but these peptides (B subunits) have a structure different from that of the M chain. Creatine kinase from heart muscle is composed of one M subunit and one B subunit. These three isoenzymes carry out the same reaction but have different structures. These structural variations allow us to separate and study the amount of each isoenzyme present in body fluids during various clinical situations, greatly expanding the use of enzyme measurements in clinical diagnosis.

10.4

ACID PHOSPHATASE

Reaction Catalyzed

Acid phosphatase (orthophosphoric monoester phosphohydrolase, acid optimum, EC 3.1.3.2) promotes the hydrolysis of a number of orthophosphate esters. The generic reaction is:

$$R-OPO_3 + acceptor \rightarrow R-OH + acceptor-PO_3$$

The R may be one of a number of substrates; at present there is no clearly defined specific physiological substrate for the enzyme. A variety of compounds have been employed for assay, and they have been selected mainly because the product of the reaction can be easily detected by either colorimetric or fluorometric techniques.

Tissue Distribution

Almost every body tissue contains some amount of acid phosphatase, usually located in lysosomes. Both erythrocytes and platelets contain significant amounts of this enzyme. Much of the enzyme in serum or plasma in healthy individuals is believed to derive from platelet turnover. In adult males, approximately one-third to one-half of the acid phos-

phatase measured in plasma comes from the prostate gland, a rich source of the enzyme. Seminal fluid also contains significant levels of acid phosphatase, a finding of importance in rape cases. Some acid phosphatase may be released from specific bone cells and leukocytes. Cerebrospinal fluid does not appear to have any measurable amounts of this enzyme.

Sample Collection

Acid phosphatase is a rather labile enzyme, losing activity rapidly if not stored at appropriate pH and temperature. To minimize contamination by platelets, plasma should be the sample of choice. Citrate is the preferred anticoagulant, buffered to a pH of 6.2 to 6.6. The enzyme is reasonably stable for a few hours at room temperature if left on the clot but loses activity rapidly after separation. If refrigerated, little activity is lost in a week. When stored frozen ($-20°C$), the enzyme is stable for a couple of months. To avoid problems, assay should be carried out as soon as possible after sample collection. Hemolysis is to be avoided, since the acid phosphatase content of erythrocytes is high.

Measurement of Total Activity

GENERAL PRINCIPLES

Quantitation of acid phosphatase in plasma or other materials is complicated by the fact that we do not know the physiological role for this enzyme in the body. Therefore, we cannot select a substrate which is specific for the enzyme and that better reflects its biochemical function. As a result, the assays which have been developed over the years have focused primarily on ease of measurement, reproducibility, and ability to detect product. Several of the substrates employed are listed in Table 10–3. Some major efforts have been devoted to the specific quantitation of the prostatic acid phosphatase isoenzyme because of the importance of this fraction in the diagnosis of prostate cancer.

Another result of this lack of specific substrate is the confusing number of different units of measurement found in the literature. Frequently, values for acid phosphatase (and for many other enzymes) are expressed in units named after the individual(s) who developed the particular assay. These units differ in the substrate employed, the product measured, the time of the reaction, the temperature, the buffer, and the pH of the reaction system. As a result, a wide variety of confusing units exist which cannot be directly compared with any degree of confidence. When reading an article or book dealing with enzyme measurement data, be sure that activities are comparable and note the limitations in the numbers presented.

Table 10-3
SUBSTRATES FOR ACID PHOSPHATASE

Reaction Name	Substrate Used	Comments
Bodansky	Beta-glycerophosphate	Lengthy assay Nonspecific
Gutman, King- Armstrong	Phenylphosphate	Nonspecific
Hudson	p-Nitrophenylphosphate	Rapid, nonspecific
Babson and Reed	Alpha-naphthylphosphate	Complicated, less sensitive
Roy	Thymolphthalein monophosphate	More specific for prostatic form
Rietz, Guilbault	4-Methylumbelliferonephosphate	Fluorescent, some improved sensitivity

p-NITROPHENYLPHOSPHATE AS SUBSTRATE

Perhaps the most commonly employed substrate for acid phosphatase assays is **p-nitrophenylphosphate** (Fig. 10-1A). Because the product (p-nitrophenol) yields a bright yellow color in alkaline solution, measurement of the amount of hydrolysis is straightforward. At the end of the incubation period, the addition of concentrated base (Fig. 10-1B) stops the reaction and shifts the pH to the alkaline range where the p-nitrophenolate anion has a strong absorbance at 410 nm. The major drawback to the assay is the sample size (0.1 mL or more) and an incubation period of at least 30 min. Because the reaction takes place in acid medium (acetate buffer, pH 4.5-5.5), a kinetic determination is not possible, since the pH must be above 10 for the yellow color of the product to appear at maximum intensity. It is preferable to run a substrate blank with each set of assays, since p-nitrophenylphosphate hydrolyzes slowly at acid pH and all enzyme measurements can be falsely elevated if this blank correction is not made.

THYMOLPHTHALEIN MONOPHOSPHATE AS SUBSTRATE

Since the major reason for assaying acid phosphatase is to detect prostatic carcinoma, researchers have been searching for a "specific" substrate—one that reacts only with the particular isoenzyme found in patients with the disease and not with other acid phosphatase isoenzymes. **Thymolphthalein monophosphate** was first developed for use in measuring alkaline phosphatase but soon found application in the assay of acid phosphatase. This substrate shows a fairly high specificity for the prostatic acid phosphatase, with little hydrolysis being seen with other isoenzymes. Although the structure of thymolphthalein monophosphate is much more complicated than that of p-nitrophenylphosphate, the color of the product results from the same type of phenolate anion formation in basic solution.

The reaction is carried out in citrate buffer at pH 6.0. After incubation of thymolphthalein monophosphate with 0.2 mL of serum for 30 min, sodium hydroxide is added to stop the reaction and enhance color development. In basic solution, thymolphtha-

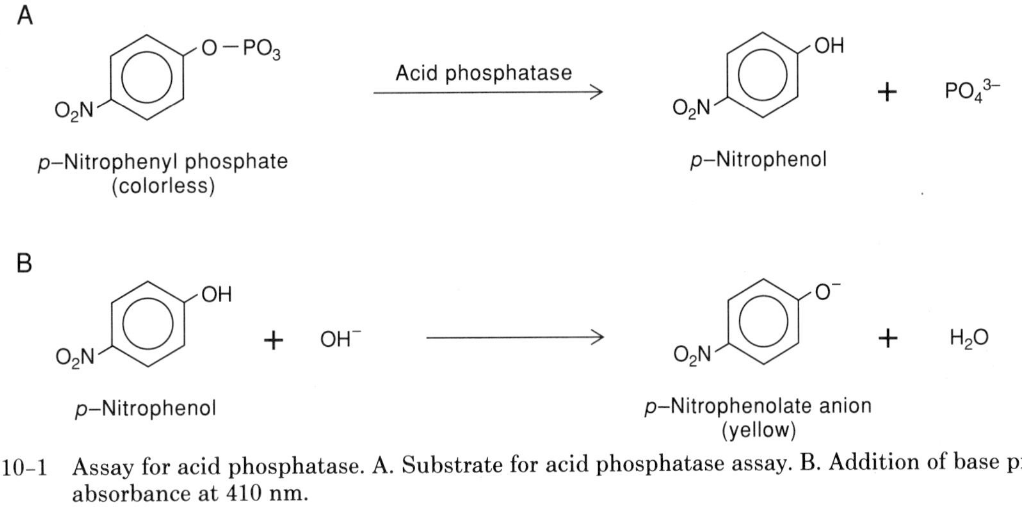

FIGURE 10-1 Assay for acid phosphatase. A. Substrate for acid phosphatase assay. B. Addition of base produces absorbance at 410 nm.

lein (the product) has an intense absorbance at 590 nm. The intensity, on a mole-for-mole basis, is much stronger for thymolphthalein than for other substrates. In addition, by measuring the absorbance at 590 nm, we eliminate interference due to absorbance by hemoglobin or bilirubin, both of which absorb strongly in the 410–450 nm range. Because of the linearity of the reaction, higher levels of acid phosphatase can be measured with this substrate than with other materials.

In studies of tissue specificity, strong activity was seen for prostatic acid phosphatase using thymolphthalein monophosphate as substrate. Little or no acid phosphatase activity was detected in hemolysates or platelet extracts. However, if other substrates were used, both platelet and erythrocyte preparations showed high acid phosphatase activity. Therefore, thymolphthalein monophosphate appears to be quite specific for the prostatic acid phosphatase.

More recently, the ammonium salt has been used instead of the sodium salt (employed in the original studies). Reaction kinetics are the same with both salts, but the ammonium salt remains stable at room temperature in the dry form, whereas the sodium derivative is stable only when refrigerated. Both salts can be obtained highly purified and contain less than 0.5% free thymolphthalein, which results in very low reagent blanks.

INTERFERENCES WITH THE ASSAY OF TOTAL ACID PHOSPHATASE

A variety of factors produce false low levels of acid phosphatase. Fluoride inhibits the enzyme. Selection of a proper anticoagulant is important, since both oxalate and heparin have been shown to produce decreased activity. As mentioned previously, storage conditions are critical since changes in pH and prolonged storage at room temperature both result in loss of enzyme activity.

The major factor producing false elevations is hemolysis. Since the erythrocytes contain significant amounts of acid phosphatase, loss of enzyme from these cells strongly influences the value obtained from a serum or plasma sample. Failure to use an anticoagulant results in release of enzyme from platelets, contributing to an increase in the amount of enzyme measured. In methodologies that measure product formation at 410 nm (or near this wavelength), hemoglobin and bilirubin in high concentrations contribute to the absorbance and yield falsely elevated enzyme values.

Clinical Significance: Total Activity

REFERENCE RANGES

Acid phosphatase levels vary according to age. Newborns have values two to three times as high as those for adults. During childhood, values remain noticeably higher than adult levels. Beginning in early adolescence, serum or plasma activity begins to decline and reaches adult levels in the late teen years. Acid phosphatase values then remain reasonably stationary throughout adulthood. The actual reference range varies greatly, depending on the method employed. If thymolphthalein monophosphate is the substrate, the adult reference range is 0.5–1.9 U/L (units per liter). The Bessey-Lowry-Brock assay with p-nitrophenylphosphate shows a reference range of approximately 1–12 U/L.

ENZYME ELEVATIONS IN VARIOUS DISEASE STATES

The major clinical situation for which measurement of acid phosphatase is valuable is prostatic carcinoma, in which serum acid phosphatase levels are elevated and increase further as the disease progresses (we will discuss this medical problem more fully later in this section). A significant number of women with malignant neoplasms of the breast also show an increase in serum acid phosphatase, as do patients with a variety of other types of cancer. Increases of the enzyme have been reported in patients with bone disease, including osteoporosis, multiple myeloma, Paget's disease, and other related clinical conditions. In Gaucher's disease, a metabolic disorder characterized by an inability to utilize certain sphingolipid compounds, we frequently observe increases in acid phosphatase levels. In this disorder, the specific isoenzyme can be identified by electrophoresis or column chromatography, allowing better characterization of the disease. Some kidney disorders are accompanied by an increase in acid phosphatase. Elevations of the enzyme in urine are frequently observed, although this has not been exploited clinically to any significant degree. The enzyme is also increased in some cases of liver disease or biliary obstruction.

Isoenzymes of Acid Phosphatase

DESCRIPTION

The native form of acid phosphatase is a dimer of approximately 100,000–125,000 mol wt; the two subunits making up the intact enzyme appear to be identical in structure. Attached to the enzyme are a number of molecules of **sialic acid** (a complex carbohydrate). The number of sialic acid molecules bound to protein seem to determine, in part, how many isoenzyme fractions are detected.

There does not appear to be good agreement at present as to exactly how many acid phosphatase isoenzymes really exist. We can say with some certainty that there are specific isoenzymes found in prostatic tissue, erythrocytes, platelets, and "tissues

Table 10-4
ACID PHOSPHATASE ISOENZYME
TECHNIQUES

Technique	Comments
Chemical inhibition	Cumbersome, not specific
Electrophoresis	Not easily reproduced, too complicated for routine clinical use
Immunoassay	Best approach for prostatic isoenzyme

other than prostate," a generic phrase indicating that the isoenzyme under discussion is not found in prostate tissue. Most recent clinical studies on the use of acid phosphatase isoenzymes in diagnosis have focused on prostatic acid phosphatase and prostatic cancer.

ASSAYS

Numerous studies have demonstrated the presence of several isoenzymes of acid phosphatase. To some extent, these various forms are the result of different protein structures. In other cases, the differences reside in the number of sialic acid residues attached to the protein. In this latter situation, what may appear to be several isoenzymes is actually one protein form when the carbohydrate groups are removed from the molecules. Because of the importance of being able to differentiate elevations of acid phosphatase in specific disease states, several techniques for quantitation of the various isoenzymes have been developed (Table 10-4).

Chemical Methods

The prostatic acid phosphatase is strongly inhibited by tartrate (the ionized form of tartaric acid), whereas most other fractions show full activity in the presence of this compound. However, this specificity is not completely reliable. Isoenzyme fractions in thrombocytes, leukocytes, and platelets are also inhibited to a great extent by tartrate. Obviously, chemical inhibition does not provide a reliable means of differentiating between elevations of acid phosphatase due to the prostatic fraction and increases caused by a rise in acid phosphatase from some other tissue source.

Electrophoretic Separation

Electrophoresis studies using starch gel, polyacrylamide gel, and other separation media show the existence of up to eight isoenzyme forms. These forms differ greatly in electrophoretic mobility: the isoenzyme from erythrocytes does not move from the origin, but fractions from the prostate, kidney, bone,

and a form from leukemic cells demonstrate great mobility. Based on electrophoresis studies, at least two prostatic isoenzyme forms exist. The separations do not yield clear-cut fractions that allow good quantitation of the different isoenzymes, so this technique has not been developed into a routine clinical assay for acid phosphatase isoenzymes.

Immunoassay Approaches for Prostatic Acid Phosphatase

The major clinical reason for the assay of acid phosphatase is to detect prostatic cancer. If an elevation of total activity is seen, it is very important to determine the source of the increase in order to detect this disease as early as possible. Because of the need for reliable assessment of the prostatic isoenzyme, a number of immunoassay procedures have been developed for this fraction. Both RIA and enzyme immunoassay approaches are commercially available which follow standard formats for immunoassay procedures and need not be described in detail.

CLINICAL UTILITY OF ACID PHOSPHATASE ISOENZYMES

Prostatic carcinoma is probably one of the most common types of cancer to develop in elderly males. The disease begins with small islands of cancer within the prostate (stage I). In stage II, nodules of the cancer begin to develop within the prostate; the cancer has not yet begun to spread to other tissues in most instances. Involvement of nearby areas in the pelvic region in the spread of tumor characterizes stage III. The final stage (IV) is indicated by metastasis to other tissues, mainly lymph nodes and bone.

Early diagnosis is critical to successful treatment without having to resort to surgery. If the disease is detected at stage I, the survival rate is equivalent to that of someone in the normal population. Even after development of stage II prostatic carcinoma, the 5-year survival rate is 70-90%. Survival drops markedly in stage III (40-70%) and even further in stage IV (16-25% after 5 years). Obviously, the earlier the cancer is detected, the sooner treatment can begin and the better the chance of long-term survival.

The classic means of detecting prostatic cancer has been during the physical examination where the physician palpates the prostate. Unfortunately, this exam rarely detects a stage I carcinoma and very often misses a stage II situation. Only in stages III and IV will the physical examination prove to be positive in most instances of prostatic carcinoma.

Very few patients with either stage I or stage II disease display marked increases in serum acid phosphatase. Only 12% of patients with the first stage of the disease and 15% of those with stage II show noticeable elevations in serum enzyme activity. Even at stage III, approximately 70% of patients studied demonstrate normal acid phosphatase levels by enzyme activity measurement. Only when stage IV is reached do a significant majority of patients (roughly 60%) manifest clear increases in acid phosphatase.

With the development of immunoassay techniques for measurement of acid phosphatase mass (*not* activity), some improvement in the reliability of the test has been seen. Between 70 and 80% of patients with stage II or III disease are characterized by increased levels of acid phosphatase by immunoassay (the exact figures depend on the specific assay being discussed). Somewhat over 90% of patients with stage IV disease have elevated amounts of acid phosphatase. The difficulty lies in the results for stage I carcinoma: only 30–35% of patients in this category are detected by immunoassay techniques. Complicating the situation further, between 5 and 10% of patients with nonprostatic carcinomas (both male and female) demonstrate increased amounts of acid phosphatase in the circulation as assayed by immunochemical techniques.

Although the measurement of acid phosphatase in serum or plasma has aided greatly in the diagnosis of prostatic carcinoma, this is obviously not the final answer to this thorny diagnostic problem. Other tests are currently under investigation in hopes that one day a truly specific assay for the detection of prostatic carcinoma in the early, treatable stage can be developed.

VAGINAL ACID PHOSPHATASE AND RAPE CASES

Because of the high concentration of acid phosphatase in seminal fluid, detection of this enzyme in the vagina has long been used as part of the physical evidence in establishing the existence of recent sexual intercourse. In cases of alleged rape, a vaginal swab is obtained as quickly as possible after the incident and examined both for the presence of sperm cells and for acid phosphatase activity. Vaginal levels of acid phosphatase in the absence of recent coitus are essentially undetectable, whereas the presence of seminal fluid produces striking elevations of acid phosphatase activity for several hours after intercourse. Although an increase in enzyme activity is presumptive evidence of coitus, the absence of increased levels cannot rule out sexual activity since it has been shown that a significant number of rapes do not result in ejaculation by the rapist.

As is the case with most forensic matters, the laboratory personnel are probably not involved in the collection of material and are almost never concerned with the analysis since the testing is carried out by a special crime laboratory. However, some awareness of the procedures available through the hospital or other sources may prove helpful if you are called upon for assistance.

10.5

ALKALINE PHOSPHATASE

Reaction Catalyzed

Alkaline phosphatase (orthophosphoric monoester phosphohydrolase, alkaline optimum, EC 3.1.3.1) catalyzes the hydrolysis of a number of phosphate esters, transferring the phosphate group to an acceptor molecule. The general reaction is:

$$R-O-PO_3 + acceptor \rightarrow$$
$$R-OH + acceptor-PO_3$$

As the name implies, the pH optimum for the reaction is in the alkaline range, around pH 10. Each isoenzyme of alkaline phosphatase has a slightly different pH optimum as well as different substrate preferences and concentrations for maximum activity.

It has proven to be very difficult to identify unambiguously exactly what function is played by alkaline phosphatase in the organism. Several theories have been proposed, each of which deals with the role of a specific isoenzyme. The liver fraction is believed to be involved with transport processes, although we do not know at present exactly what materials are being carried by this enzyme and where they are being taken. Bone formation is thought to be enhanced by the bone isoenzyme of alkaline phosphatase through the enzymatic hydrolysis of orthophosphate, an inhibitor of calcium deposition. The intestinal fraction is felt to be involved in some manner with fatty acid transport and calcium absorption. Phosphate absorption is believed to be facilitated by both the intestinal and kidney isoenzymes. All the fractions show some ability to regulate the synthesis of DNA. No specific function has yet been suggested for the placental fraction, although some studies suggest it is a part of the process of nutrient transport to the fetus.

Tissue Distribution

Alkaline phosphatase is located in a wide variety of tissues. Significant amounts of the enzyme are found in liver, bone, intestine, kidney, and placenta. There is essentially no alkaline phosphatase located in erythrocytes, so hemolysis does not produce a false elevation of enzyme activity. Platelets contain very little of this enzyme.

Sample Collection

Samples should be collected as serum or heparinized plasma. Use of other anticoagulants lowers values, because they form complexes with the metal in the enzyme. Physical activity, for the most part, does not affect the level of the enzyme. Lengthy bed

Table 10-5
SUBSTRATES FOR ALKALINE PHOSPHATASE

Reaction Name	Substrate Used	Comments
Sinowara-Jones-Reinhart	Beta-glycerophosphate	Long incubation time; high blank values
King-Armstrong	Phenylphosphate	End point; requires protein removal
Bessey-Lowry-Brock	p-Nitrophenylphosphate	End point or kinetic; rapid
Bowers-McComb	p-Nitrophenylphosphate	Uses phosphate-accepting buffer; reference method

rest also appears to have little or no effect on activity. If a sample is left standing on the clot for a long time, alkaline phosphatase activity may actually increase up to 20 to 30% over the initial level. This change in activity is believed to be due to the gradual development of a more basic pH in the system as CO_2 is lost.

Measurement of Total Activity

GENERAL PRINCIPLES

Methods of the measurement of alkaline phosphatase closely parallel those for assay of acid phosphatase. Because of the similarity of the reactions (with pH optimum being the major variable), a substrate which produced good results for one enzyme was often explored as a possible means of measuring the activity of the other enzyme (Table 10-5).

The selection of the buffer is critical to accurate measurement of this enzyme. The reaction proceeds much more rapidly in the presence of a phosphate acceptor. All current assays have incorporated a buffer with one or more $-OH$ groups in the buffer molecule. These hydroxyl substituents form reasonably stable phosphate derivatives after the substrate is cleaved by the enzyme. The presence of an acceptor results in higher enzyme activities and permits the use of smaller samples and shorter reaction times.

p-NITROPHENYLPHOSPHATE AS SUBSTRATE

Bessey, Lowry, and Brock developed an assay, using p-nitrophenylphosphate, which could be adapted to a kinetic approach. The product of this hydrolysis, p-nitrophenol, gives a strong yellow color in alkaline solution. Therefore, the formation of this product can be monitored directly by measuring the change in absorbance at 400 nm as a function of time. Some interference occurs if bilirubin or hemoglobin are present in high concentrations, since these two molecules also absorb strongly near 400

nm. If the assay is run as a kinetic method and not an end-point procedure, these interferences are minimized. The **Bessey-Lowry-Brock method** (with modifications) has proven to be a very popular approach to the measurement of alkaline phosphatase in both manual and automated methodologies.

The original Bessey-Lowry-Brock method was run in glycine buffer, which does not serve as a phosphate acceptor. Bowers and McComb modified this approach by using a transphosphorylating buffer. They selected 2-amino-2-methyl-1-propanol (sometimes referred to as AMP buffer) at pH 10.5 at 30°C (keep in mind that buffer pH changes as the temperature changes). With the change in buffer, significantly higher reaction rates were seen, allowing lower sample sizes and shorter reaction times. This method has been widely adopted for clinical use. In fact, many people who say they are using the Bessey-Lowry-Brock method are actually employing the **Bowers-McComb** modification. Several international groups have proposed this approach, using diethanolamine as buffer and with different assay temperatures, as a standard reference method for alkaline phosphatase.

It should be noted that selection of buffer may be critical to the reliable measurement of the activity of alkaline phosphatase. Although all buffers function as phosphate acceptors, there appears to be some buffer *specificity* in the same manner as the substrate specificity manifested by enzymes. Some buffers activate a specific isoenzyme of alkaline phosphatase more than other fractions. In clinical situations, this phenomenon could produce misleading information about the presence of elevated serum enzyme activity or the extent of that increase.

Clinical Significance: Total Activity

REFERENCE RANGES

Alkaline phosphatase activity in serum is high during the first year of life as rapid skeletal growth takes place. During childhood the values drop somewhat but are still increased over adult levels. A sharp spurt in activity is seen during adolescence, again due to rapid bone growth. This increase is seen sooner for girls than for boys, reflecting the earlier physical maturation of females during puberty.

In adults, males generally have higher serum levels of alkaline phosphatase than females. With *p*-nitrophenylphosphate as substrate, males have values of 90 to 190 U/L, whereas females show values in the range 85-165 U/L. However, this trend reverses after about age 50, with women showing increases in enzyme activity and male levels remaining unchanged. The late rise in female values is thought to be a reflection of changes occurring during menopause, where diminished synthesis of estrogen results in subclinical deterioration of bone structure in many women. The rise in serum alkaline phosphatase values at this time reflects the increased bone turnover.

Diet has an interesting effect on alkaline phosphatase levels in normal individuals. A high-fat meal results in noticeable increases in the enzyme activity in serum within a few hours after eating. This rise is believed to be due mainly to increases in the intestinal enzyme fraction, a phenomenon seen to a greater extent with individuals with type O blood. The elevation may not be seen with all substrate used to assay for enzyme activity. Intake of carbohydrates also produces some increase, but only if the person has been on a high-carbohydrate diet for a long time.

ELEVATIONS IN VARIOUS DISEASE STATES

Serum total alkaline phosphatase is somewhat elevated in a variety of liver diseases, but is not a sensitive indicator of damage to this organ. Patients with viral hepatitis frequently show increases only twice the upper limit of normal. Little change may be seen in alcoholics, even in the face of marked increases in the activities of other "liver-specific" serum enzymes. Obstruction of bile ducts may be accompanied by a mild increase in total alkaline phosphatase activity.

Elevations of alkaline phosphatase in bone disease are quite common. Increases are seen in Paget's disease, rickets, hyperparathyroidism (since calcium deposition and bone formation are profoundly affected by the amount of parathyroid hormone present), and when there are extensive fractures. Cancers of the bone and tumors which have spread to the bone from other locations result in elevations of total serum activity of alkaline phosphatase.

Other causes for increased serum activity are hyperthyroidism (with the increase seen in the bone fraction) and diabetes mellitus (possibly due to changes in enzyme production by the liver). Most elevations seen in cancer patients are due to metastasis to bone, with resulting bone deterioration. Some preparations of albumin produce false elevations of serum alkaline phosphatase if the albumin has been obtained from placental material.

Isoenzymes of Alkaline Phosphatase

GENERAL PRINCIPLES

Currently, it is believed that there are three naturally occurring isoenzymes of alkaline phosphatase in human serum: liver, bone, and intestinal fractions. An additional placental fraction is observed in women during the third trimester of pregnancy. Other alkaline phosphatase isoenzymes have been detected in certain cancer tissues, but would not be considered normal physiological forms of the enzyme.

ASSAYS

A variety of approaches have been developed over the years for fractionation of the various alkaline phosphatase isoenzymes (Table 10-6). Heat inactivation, urea denaturation, the use of specific inhibitors, and electrophoretic separation have all been carefully characterized regarding their ability to fractionate the isoenzymes. Column chromatography and high-performance liquid chromatography have been explored, but do not lend themselves well to situations where large numbers of samples must be processed fairly quickly, as is the case in most clinical laboratories.

Heat Stability

The various alkaline phosphatase isoenzymes have differing stabilities at elevated temperatures. When a sample is incubated at 56°C, the bone fraction of the enzyme rapidly loses activity, whereas the liver and intestinal isoenzymes are somewhat more stable. This somewhat selective loss of the bone isoenzyme becomes very apparent if the sample is subjected to electrophoretic fractionation after

Table 10-6
ALKALINE PHOSPHATASE ISOENZYME TECHNIQUES

Technique	Comments
Heat inactivation	Temperature control difficult
Inhibitors	Isoenzyme inhibitions overlap; somewhat useful when employed with electrophoresis
Electrophoresis	Can distinguish major fractions; not quantitative
Immunological	Data not reproducible

the heat treatment. The bone fraction disappears, leaving the liver (and any other fraction) still visible on the strip. The placental form appears to be quite stable at 56°C, losing little activity after a 15-min incubation at this temperature. Quantitation of the placental fraction has been performed by heating a serum sample for 10 min at 60°C, which almost completely inactivates all other forms of the enzyme. Whatever activity is detected following this treatment is believed to be due to the placental isoenzyme.

There are severe shortcomings to the use of heat inactivation by itself. Time and temperature must be carefully controlled; loss of activity is quite variable even when these parameters are held to close standards. The number of sialic acid residues on the isoenzyme affect heat stability. Some fractions may actually be more heat-stable if the sialic acid content is reduced. Although some information may be gained through heat-inactivation studies, it is not a preferred way to identify the isoenzyme(s) responsible for increased alkaline phosphatase activity in a sample.

Inhibitors

A number of compounds are somewhat selective in their inactivation or inhibition of specific alkaline phosphatase isoenzymes. Urea at a concentration of 2 M inactivates the bone isoenzyme more than the liver fraction. L-phenylalanine inhibits the intestinal and placental isoenzymes, lowering activity by some 70% or more. If enzyme activity loss in a serum sample incubated with L-phenylalanine is 10% or less, no conclusions can be drawn about the isoenzyme's composition. Levamisol (a broad-spectrum antiparasite medication used in animals) inhibits mainly the liver and bone fractions, with less effect on the placental and intestinal isoenzymes. Variant alkaline phosphatase isoenzymes from cancer cells behave much like the placental fraction. L-homoarginine also inhibits the activities of liver and bone fractions more than other isoenzymes which may be present in the sample.

Many of the studies on selective inhibition have been carried out on purified fractions of alkaline phosphatase isolated from specific tissues. When these inhibition techniques are used on patient samples, the data are often difficult to interpret. The best application of specific inhibitors would be in conjunction with electrophoretic separation to clarify ambiguous patterns on the film.

Electrophoresis

A variety of electrophoresis techniques have been explored for the characterization of alkaline phosphatase isoenzymes. The two most commonly used methodologies are cellulose acetate and agarose. Polyacrylamide gel gives somewhat improved sepa-

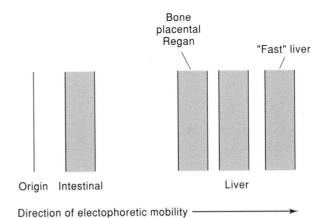

FIGURE 10–2 Alkaline phosphatase isoenzyme fractionation.

ration, but is not practical in most routine clinical situations. Isoelectric focusing yields 12 or more bands of enzyme activity, but the patterns are often very difficult to characterize. Each isoenzyme may yield multiple bands (owing to differences in sialic acid content on the protein), and some overlap of fractions does occur. With further development, however, this technique may prove to be a useful addition to the clinical laboratory.

If we look at the pattern obtained on conventional electrophoresis (Fig. 10–2), we see that the liver fraction moves the farthest from the origin. In some instances, a second "fast liver" fraction is observed ahead of the normal liver isoenzyme. Directly behind the liver fraction is the bone isoenzyme; frequently it is impossible to tell where one fraction ends and the next one begins. Separation of a heat-fractionated sample (with the bone portion losing activity and not appearing on the gel) allows differentiation of the two forms. The placental isoenzyme (and those characteristic of some cancers) comigrate with the bone fraction. Heat fractionation and incubation with one or more of the inhibitors mentioned above permit more specific characterization. Located close to the origin is the slow-migrating intestinal fraction, quite clearly separated from the other isoenzymes.

A number of colorimetric and fluorometric methods have been employed for development of bands after electrophoresis. The substrate *p*-nitrophenylphosphate gives a yellow band after separation, but does not produce an intense color. Perhaps the most reliable substrate for this purpose is naphthyl-AS-MXC-phosphate. This material is hydrolyzed by the enzyme to yield a fluorescent product. Unlike other isoenzyme fractionations, there is little or no useful information to be gained from densitometric quantitation of the bands. The fractions obtained are diffuse and often overlap one another. A simple visual inspection is usually sufficient to identify the source of elevated serum total alkaline phosphatase activity.

Immunological Characterization

Some efforts have been made to develop specific antibodies to the various isoenzymes of alkaline phosphatase, but with little success. Because of the complexities of the structure of the enzyme and the varying amounts of sialic acid attached to the protein molecule, no immunological approaches have proven to be of value in the study of this enzyme system.

CLINICAL UTILITY OF ALKALINE PHOSPHATASE ISOENZYMES

Liver Disease

Increases in the amount of the liver isoenzyme are seen in most cases where the elevation of serum total alkaline phosphatase is due to liver damage from hepatitis, cirrhosis, or biliary obstruction. A fast-liver fraction has also been observed in some cases of hepatitis and alcoholic cirrhosis. Some studies suggest it may be an indicator for the presence of metastasis of cancer to the liver.

Bone Disease

Any damage to bone which produces an increase in serum total activity of alkaline phosphatase results in elevation of the bone isoenzyme. In addition, this fraction increases in children (with striking elevations during early life and throughout adolescence) and in adults over 50 years old.

Changes During Pregnancy

The placental isoenzyme is detected during the third trimester of pregnancy and remains elevated for as long as a month or so after birth. If placental infarction (tissue death resulting from impaired blood supply) occurs earlier in the pregnancy, the placental isoenzyme appears in the serum and may be diagnostically useful.

Isoenzyme Studies in Cancer Patients

There are several alkaline phosphatase isoenzymes which appear to be characteristic of cancer. These fractions are usually named after the individual in whom they were first discovered. Electrophoretic migration is to the same location as the placental isoenzyme. The first of these atypical fractions to be identified was the **Regan isoenzyme,** found in a patient with a particular type of lung cancer. This protein has also been seen in some women with cancer of the ovary and in patients with breast cancer. The **Nagao isoenzyme** is very similar in its biochemical properties and identical in mobility to the placental isoenzyme. Patients with adenocarcinoma sometimes demonstrate the presence of this isoenzyme. Another potential tumor marker is the **Ka-sahara isoenzyme** (formerly referred to as the Regan variant). First observed in a patient with hepatoma, it has also been seen in serum of individuals with tumors of the gastrointestinal tract.

Unfortunately, none of these marker proteins appear in high enough frequency to be used specifically for the detection of cancer. They are present on occasion and need to be characterized further if detected (through heat inactivation and specific inhibitor studies), but cannot presently be employed in any sort of screening test for a particular type of tumor.

Alkaline Phosphatase Complexes with Proteins

Occasionally, an atypical electrophoretic pattern is seen when alkaline phosphatase isoenzymes are being studied. These unusual patterns may be produced by a complex between the enzyme and other protein materials in the sample. In most instances, a complex has formed between alkaline phosphatase and an immunoglobulin, resulting in the detection of a very slow-moving band which does not migrate far from the origin. In other instances, interaction with lipoprotein-X has been demonstrated. Frequently, the serum total activity for alkaline phosphatase is somewhat elevated, since the enzyme cannot clear from the bloodstream as rapidly when it is in the complexed form. Although these unusual forms do not seem to have any clinical significance that we know of, they do appear on rare occasions and must be included as possibilities when interpreting an abnormal alkaline phosphatase isoenzyme pattern.

10.6

AMINOTRANSFERASES

Reaction Catalyzed

There are two closely related enzymes involved in the reversible transfer of an amino group to a keto acid, which are of some clinical significance. Because of the similarities in reactions, assay methods, and clinical utility, we will consider these two aminotransferases together.

Alanine aminotransferase (L-alanine:2-oxoglutarate aminotransferase, EC 2.6.1.2) has traditionally been referred to as **serum glutamic pyruvic transaminase.** In the older literature, this enzyme is often abbreviated as SGPT or GPT; currently the preferred abbreviation is **ALT.** The other enzyme of the duo is **aspartate aminotransferase** (L-aspartate:2-oxoglutarate aminotransferase, EC 2.6.1.1). Formerly, this enzyme was called **serum glutamic oxaloacetic transaminase** (SGOT or GOT) and is now abbreviated as **AST.**

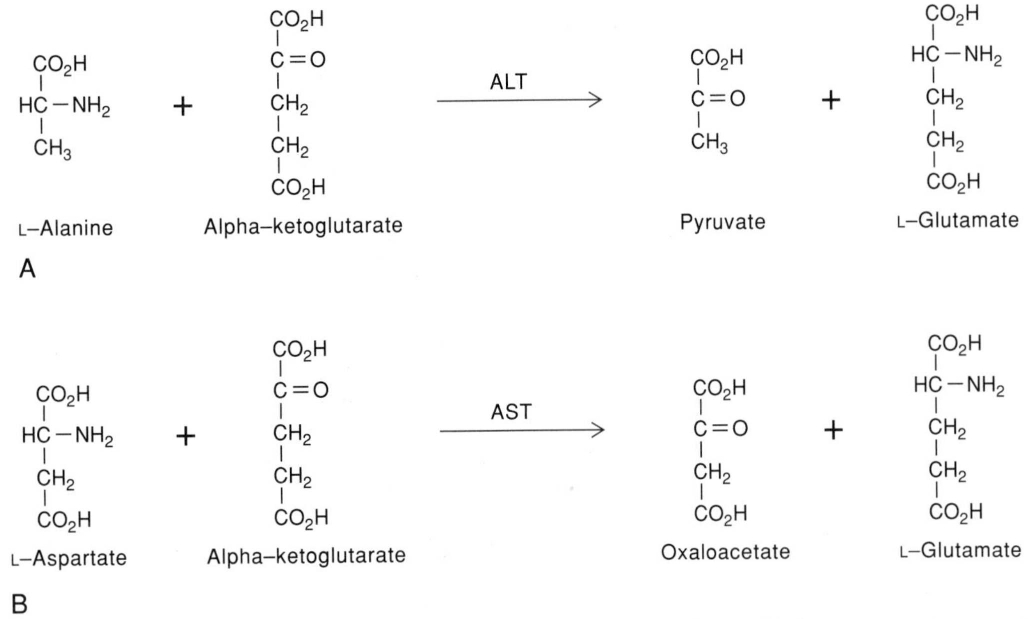

FIGURE 10-3   Aminotransferase reactions. A. Alanine aminotransferase. B. Aspartate aminotransferase.

In each case, the reaction catalyzed by the enzyme is the reversible transfer of an $-NH_2$ from an amino acid to a keto acid (Fig. 10-3). In the body, the processes serve to provide sources of nitrogen for the urea cycle. In the living system, the reaction goes in the direction written in Figure 10-3. The pyruvate formed is metabolized further through the citric acid cycle to provide biochemical energy or is involved in the synthesis of fatty acids. Oxaloacetate is also utilized in the citric acid cycle. In each reaction, glutamate is formed which then is deaminated to produce ammonia (for use in the urea cycle) and to regenerate alpha-ketoglutarate.

Tissue Distribution

Tissue levels of AST are highest in heart and liver (one of the early enzyme markers for a myocardial infarct was an elevated serum level of AST). Significant amounts are found in skeletal muscle and kidney, with lower levels in pancreas, spleen, and lung. Low levels are seen in erythrocytes and serum. However, the red cell concentration of AST is approximately 10 to 15 times that of serum, so hemolysis should be avoided when samples are collected.

The highest amount of ALT is found in liver, with much lower levels in heart and skeletal muscle. Significant concentrations can be detected in kidney tissue. The erythrocyte contains five to eight times as much ALT activity as does the serum, so hemolysis must be avoided.

Sample Collection

Serum or plasma may be used for either enzyme assay. None of the commonly employed anticoagu-

lants inhibit the enzymes. Hemolysis should be avoided since falsely elevated results may be obtained. Turbid or icteric (pertaining to jaundice) samples create some difficulties with blanking.

Measurement of Total Activity

GENERAL PRINCIPLES

Assays for both AST and ALT utilize the same types of reactions for detection of product. Some basic principles are applicable to both measurement systems. The three major approaches are (1) reaction with dinitrophenylhydrazone, (2) coupling with diazonium salts, and (3) coupled enzyme assay. All reactions involve measurement of the formation of the keto acid produced in the reaction.

REACTION WITH 2,4-DINITROPHENYLHYDRAZONE

The colorimetric assay **(Reitman-Frankel)** utilizing dinitrophenylhydrazone involves a reaction between the color reagent and the keto acid formed (Fig. 10-4). A dinitrophenylhydrazone derivative is

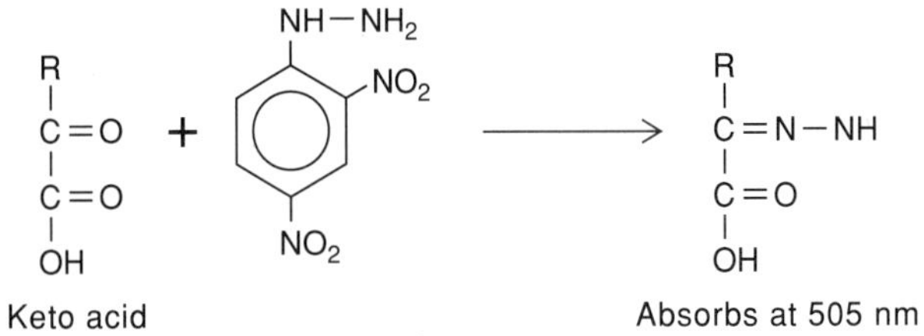

FIGURE 10-4   Spectrophotometric assay for aminotransferases.

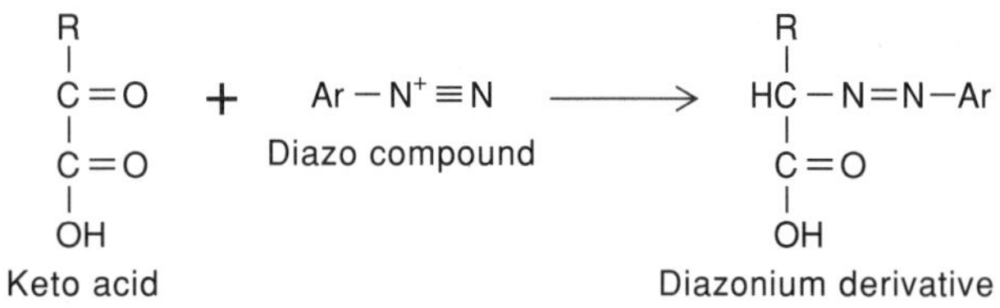

FIGURE 10-5   Reaction of keto acid with diazonium salt.

formed which gives a strong blue color, measured at 505 nm. One major problem with this assay is its lack of specificity. Any keto compound present contributes to the color reaction, giving rise to falsely elevated values. In addition, the substrate keto acid reacts with the color reagent to some extent.

REACTION WITH DIAZONIUM SALTS

The coupling of keto acid product with a diazonium salt (Fig. 10-5) is subject to some of the same problems. Although the procedure is relatively simple and straightforward, interferences do exist. Some determination of blank values is necessary so that the final activity measured may be corrected for these other materials which react with the color reagent.

COUPLED ENZYME ASSAY

The preferred method for assaying both ALT and AST is the enzyme-coupled system, in which the keto acid formed by the aminotransferase reacts in a system using NADH (Fig. 10-6). The coenzyme is oxidized to NAD and the decrease in absorbance at 340 nm is measured. For ALT, the pyruvate formed in the reaction is converted to lactate by lactate dehydrogenase. Malate dehydrogenase is used to reduce oxaloacetate to malate in the AST system. In both cases, NADH is oxidized to NAD with a resultant decrease in 340 nm absorbance. The major drawback to this system is the high initial absorbance readings due to NADH.

ROLE OF PYRIDOXAL PHOSPHATE

A necessary coenzyme for the transaminase reaction is **pyridoxal phosphate.** This molecule functions to carry the amino group from one acid to the other. In the generic aminotransferase reaction, illustrated below, we can see better the role played by this coenzyme:

Overall reaction:

amino acid I + keto acid II →
        amino acid II + keto acid I

The process is stepwise; the first amino acid transfers the $-NH_2$ group to pyridoxal phosphate, making it pyridoxamine phosphate. The coenzyme then gives that amine group to the keto acid, converting it into the corresponding amino acid.

Steps in transamination reaction:

amino acid I + pyridoxal →
        keto acid I + pyridoxamine
keto acid II + pyridoxamine →
        amino acid II + pyridoxal

For many years, the transaminase assays were carried out in the absence of added pyridoxal phosphate. The assumption was that the enzyme in the serum already had sufficient coenzyme bound to it and did not require additional pyridoxal. In the 1970s, however, a number of studies showed that a large percentage of samples were deficient in coenzyme. Addition of pyridoxal to the reaction mixture before the substrate was added resulted in a significant increase in enzyme activity for many samples.

This modification of assay protocol has made the measurement of both AST and ALT more reliable and sensitive. Samples which may have given normal results using the previous protocol now frequently show measurable increases in activity once the coenzyme is added to the reaction mixture. One practical implication of these findings is that much of the premodification data is suspect. Reassessment of changes in liver disease and other medical problems associated with changes in serum levels of the two enzymes had to take place. Although the overall pattern has not changed, we now have a better idea of the extent of serum enzyme changes associated with disease states.

Clinical Significance: Total Activity

Normal adult values for ALT are less than 55 U/L, whereas the expected values for AST are approximately 5–35 U/L. Both sets of values are based on the assay mixture being supplemented with pyridoxal phosphate.

Both enzymes increase in many disorders related to liver damage. Patients with viral hepatitis demonstrate marked increases in activities of the two enzymes, frequently before clinical symptoms of the

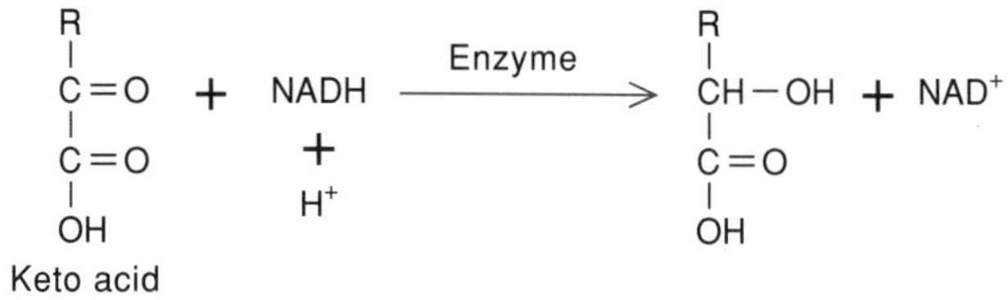

FIGURE 10-6   Coupled enzyme assay for aminotransferases.

disease become apparent. ALT is more elevated than AST in various inflammatory conditions of the liver, reflecting its greater specificity as a liver disease marker. As mentioned earlier, AST levels rise shortly after a myocardial infarction, with a peak at about 24 h and a decline to normal within 4 to 5 days. Because liver damage is often secondary to myocardial infarction, monitoring the activity of this enzyme is not recommended if a reliable assessment of cardiac damage is desired. ALT serum values are usually unchanged during a heart attack. AST values in serum frequently increase in cases of pulmonary embolism and muscular dystrophy.

Recent interest has been shown in the use of ALT measurements as a screen in blood banking. Since this enzyme is considered to be more specific for liver damage than AST, there are suggestions that an elevated ALT level in donated blood may be an indication of the presence of non-A,non-B hepatitis. Although sensitive tests for hepatitis A and hepatitis B have been available for years, many complications of blood transfusions arise from inadvertent transmission of a third form of hepatitis which has not yet been characterized. One known symptom is liver damage, which would be reflected in an increase in liver-specific enzymes such as ALT. This process has been employed in many areas and has proven a useful means of detecting possible cases of heptatitis. However, the cost-effectiveness of this approach is still being debated as well as its effect on donors (both present and prospective).

Isoenzymes of the Aminotransferases

Isoenzymes of ALT and AST have been demonstrated in serum and from various tissues. There are apparently two forms, one soluble, found in the cytoplasm, and one located in the mitochondria. Separation of the isoenzymes has been carried out using either electrophoresis or column chromatography. Some studies have been done to look at the clinical utility of isoenzyme determination, but little of clinical significance has been found to date. At present, there is no value in the determination of isoenzymes for either aminotransferase.

10.7

AMYLASE

Reaction Catalyzed

Amylase (alpha-1,4,-glucan-4-glucanhydrolase, EC 3.2.1.1) is primarily responsible for the digestion of starch. The enzyme cleaves this polysaccharide into a variety of smaller carbohydrates—maltose being a major end product. Hydrolysis ceases when a branch point in the starch is reached. **Limit dex-**

tran is the starch backbone with very short carbohydrate branches, usually one monosaccharide in length, left over after hydrolysis occurs.

Tissue Distribution

Several tissues have been shown to contain amylase, but two major sources are salivary glands and the pancreas (where amylase is produced for release into the gastrointestinal tract for digestive purposes). Enzyme activity has also been observed in extracts from lung, some muscle tissues, testes, ovary, semen, and adipose tissue, as well as being found in tears and breast milk. For our purposes, only the enzyme located in the pancreas and salivary glands will be considered.

Sample Collection

Either serum or plasma may be used for assay. Neither citrate, oxalate, nor ethylenediaminetetraacetic acid (EDTA) can be used as an anticoagulant because these materials remove calcium, which appears to be essential for activity.

Measurement of Total Activity

GENERAL PRINCIPLES

A variety of approaches to the detection of amylase activity in body fluids have been developed over the years, but most of the classic methods are cumbersome, time-consuming, and rather insensitive. The natural substrate (starch) is a large molecule with a wide range of sizes. The products are either difficult to detect easily (shorter chain polysaccharides, maltose) or are naturally present in body fluids in fairly high amounts (glucose from maltose hydrolysis). However, several currently employed methods provide rapid and reproducible estimations of amylase activity.

AMYLOCLASTIC TECHNIQUES

The classic method for determination of amylase activity has been to monitor the disappearance of the starch substrate. If starch reacts with elemental iodine (I_2), a blue color forms, the intensity of which is proportional to the amount of starch still present in the reaction mixture. Early approaches involved a visual estimation of the color, with the end point of the reaction being the time at which the color finally disappeared. Since color disappearance is much more difficult to estimate than color formation, this method demonstrated wide variability in results. Current modifications involve a spectrophotometric determination of the amount of blue color present after a given time. The advantage of the **amyloclastic** (or iodometric) method is its need for

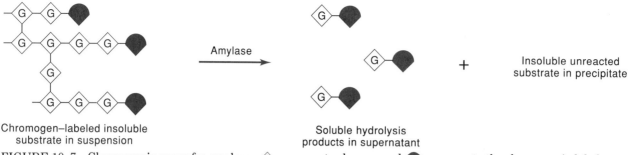

Chromogen–labeled insoluble substrate in suspension

Soluble hydrolysis products in supernatant

Insoluble unreacted substrate in precipitate

FIGURE 10–7 Chromogenic assay for amylase. ⬙ represents glucose, and ◆ represents the chromogenic label, which absorbs light in the visible region.

only a simple colorimeter or spectrophotometer and inexpensive reagents. The major drawback to the method is the inability to standardize the substrate from one laboratory to the next.

SACCHAROGENIC TECHNIQUES

Since measurement of substrate disappearance as an index of enzyme activity has some very real drawbacks, much effort has gone into developing methodologies which assess the amount of some product of starch hydrolysis. A mixture of carbohydrates is obtained when amylase attacks starch, including the disaccharide maltose and the monosaccharide glucose. Saccharogenic techniques measure either the glucose formed directly or the amount present after treatment of the reaction material with maltase, which cleaves maltose to produce two molecules of glucose. The glucose formed can be measured by a variety of enzymatic approaches (discussed in Chapter 12). This method can be easily automated, allowing for the rapid analysis of a large number of samples.

The major problem with the use of saccharogenic methods lies in the fact that glucose is normally present in body fluids in significant concentration. Any analytical approach to the determination of amylase which employs glucose quantitation must have built into it some means of correcting for the endogenous glucose already in the sample. One modification involves the use of a blank, with the glucose present in the sample being measured directly. This amount is then subtracted from the results obtained when substrate is incubated with the enzyme to produce glucose as product. High blank values are usually obtained, which makes the spectrophotometric determination of the total amount of glucose somewhat difficult.

Alternatively, a sample pretreatment can be carried out where any glucose originally present is converted by an enzyme system to another material which does not interfere with the reaction. The many different enzyme reactions taking place make this approach somewhat complicated in theory and more expensive in practice.

Several commercially available amylase assays use saccharogenic methodologies, but with artificial substrates. Instead of starch, small carbohydrates are employed which often consist of four or five glucose molecules. Amylase cleaves these small chains sequentially, forming the requisite number of glucose units as final products. Since more molecules of product are produced per substrate molecule, the reaction can be run for a shorter time and/or with a smaller sample size while still obtaining reliable results.

CHROMOGENIC TECHNIQUES

In an attempt to simplify the assay of amylase in body fluids, a number of companies have produced substrates which have dye molecules attached to them. As the starch is cleaved by amylase, small fragments are released in the usual manner. However, these fragments have a brightly colored dye as a part of their structure. By measuring the intensity of color in the supernatant, an estimation of amylase activity can easily be obtained (Fig. 10–7). The serum is incubated with a substrate suspension and allowed to react for a set time. The reaction is terminated by the addition of a suitable material (usually an acid of some sort). After centrifugation to remove the insoluble substrate, the absorbance of the supernatant is determined with a spectrophotometer and is proportional to enzyme activity. A reagent blank must be run with each batch to assess the degree of nonenzymatic breakdown of substrate. These methods are quite popular and can be run fairly quickly (making them good for stat testing), but cannot be easily automated due to the insoluble nature of the substrate.

Clinical Significance: Total Activity

REFERENCE RANGES

There is little standardization of amylase assay technology at present. All reference range values are extremely method-dependent. Amylase values are somewhat low at birth but gradually rise to adult levels within a year. Women have serum amylase values about 15% higher than those for men. During pregnancy, serum amylase levels increase somewhat, peaking at about 25 weeks. The values for women during the second and third trimester frequently are above the levels observed in men and nonpregnant women.

VALUES IN VARIOUS DISEASE STATES

The major clinical reason for increased serum amylase is acute pancreatitis, an inflammatory process. Serum levels may rise markedly within a few hours, providing a clear marker for the disorder. Viral disorders (such as hepatitis) frequently produce pancreatic inflammation as a secondary phenomenon, resulting in elevation of serum amylase. In chronic pancreatitis, serum levels increase somewhat and remain elevated (with some fluctuation), in contrast to the sharp rise and fall seen in the acute stage.

Several cancers result in elevations of serum amylase. The source of the enzyme appears to be from the tumor itself: ectopic production of amylase by the tumor followed by release into the circulation. This phenomenon has been demonstrated with a variety of cancers, including those of the lung, ovary, and colon. In some instances, the ectopic enzyme has been shown to have a molecular weight of approximately 54,000; this result is lower than the corresponding values of 61,000–64,000 mol wt for the salivary amylase or 60,000 mol wt for the pancreatic isoenzyme.

Serum amylase is frequently increased in patients with diabetes mellitus, particularly in cases of diabetic ketoacidosis. Both the clinical symptoms and the sharp rise and fall of serum amylase values within a few days indicate that these patients are experiencing an attack of acute pancreatitis. However, isoenzyme studies suggest that the pancreas is not the source of elevated enzyme activity in a number of instances. This situation is still under investigation.

Serum amylase has a low molecular weight and is readily filtered by the glomerulus to be excreted in the urine. Situations where there is renal impairment (such as kidney failure) result in a decrease in the rate of urinary excretion of this enzyme, with a consequent increase in serum levels. Elevation of serum amylase in this situation does not indicate increased production by the pancreas, but rather a diminished ability of the body to remove the enzyme from the circulation at a normal rate.

A variety of other clinical problems may produce increases in serum amylase values. Any clinical situation involving the salivary glands (mumps, stones, surgery) results in release of the salivary amylase in excessive amounts. Patients with various liver diseases often have elevated amylase levels. Biliary tract disease and intestinal obstruction can also lead to increased amounts of amylase in the serum.

A number of drugs increase serum amylase. These elevations are not artefacts of the assay, but true increases due to the effect of the drug on the pancreas. Codeine, morphine, and related medications produce spasms of the sphincter of Oddi, leading to release of amylase from the pancreas. Glucocorticoids, dexamethasone, and oral contraceptives frequently induce increase in serum amylase, either by causing transient pancreatitis or by some impairment of liver function. Some x-ray contrast media may cause elevation of the enzyme.

Isoenzymes of Amylase

Although we know that that there are several amylase isoenzymes, there is little agreement as to exactly how many exist. Depending on the tissues used as enzyme source, some seven or eight amylase fractions can be identified. Only two of these components, the pancreatic and the salivary isoenzymes, have been routinely observed in human serum. These two fractions are not single isoenzymes, but each consists of a mixture of at least two amylase components.

Although electrophoresis has been the major means of characterizing these fractions, clinical use of this tool has not been explored to any extent. The patterns obtained on electrophoretic fractionation are often difficult to interpret, and quantitation is not very reliable. A variety of inhibitor studies have been employed to assess the major isoenzyme(s) present in various clinical situations. There are a number of natural inhibitors of amylase available which specifically inhibit either the salivary isoenzyme or the pancreatic fraction. More recently, monoclonal antibodies have been developed which have specific inhibitory properties. A better assessment of the source of an elevated amylase total activity can be obtained with these inhibitors, but more work needs to be done before they are employed routinely.

The situation is complicated further by the fact that amylase is not the only enzyme which metabolizes the substrate present in the assay tube. Other starch-cleaving enzymes have been detected in serum in some clinical conditions. For example, phosphorylases *a* and *b* have been observed in patients with myocardial infarction or viral hepatitis. These other enzymes react with the starch substrate and manifest enzyme activity which cannot be distinguished from amylase activity under normal circumstances. The use of specific inhibitors allows better differentiation of amylase activity from that of similar enzymes.

Amylase isoenzyme studies have been useful in a

number of situations. Atypical isoenzymes of amylase have been identified in several types of cancer; in many instances, it has been shown that the specific isoenzyme was produced by the cancer tissue. In liver disease, evidence suggests the production of a specific isoenzyme different from both the salivary and pancreatic forms. As more work is carried out to develop better means of fractionation, amylase isoenzyme studies will become a useful tool in the clinical laboratory.

Urine Amylase and the Amylase/Creatinine Ratio

Amylase has a molecular weight of 40,000–50,000. Because of this small size, amylase can be easily filtered at the glomerulus and excreted in the urine. In many cases, the pancreas releases a large amount of amylase as a result of damage to the organ, but the serum level does not rise markedly because of the rapid excretion of amylase in the urine. Two approaches have been developed to minimize this complication: measurement of urine amylase in a timed specimen and the determination of the amylase/creatinine clearance ratio.

A number of studies have shown that a timed urine collection frequently gives a better assessment of amylase release by the pancreas than does a serum sample. The sample can be collected over a period of a few hours, eliminating the need for a 24-h collection. Normal values are very method dependent and are affected to some extent by the state of kidney function and the amount of urine collected. A major drawback to the use of a timed urine sample is the delay required to collect the specimen. Often, a 2-h or longer wait is necessary, a situation which is unsatisfactory if stat results are needed.

An alternative approach has been widely used which eliminates the need for a timed urine specimen. By relating the amylase clearance (the amount of enzyme excreted in the urine in a specified time as a function of the serum concentration of the enzyme) to the creatinine clearance, correction can be made to take into account the degree of hydration of the patient and the ability of the kidney to function. Clearance studies require a timed urine specimen and a serum sample collected at the start of the urine collection. The ratio of urine concentration to serum concentration is calculated and related to time and the total urine volume.

Both the amylase clearance and the creatinine clearance have a unit of time in the measurement; results are routinely expressed as volume per minute per unit of body surface. If each clearance were being determined separately, a timed specimen would be needed. However (without going through all the math), if we look at the ratio of amylase clearance to creatinine clearance, the time units cancel out. Therefore, it is possible to use a random urine

specimen for measurement of the urine values and eliminate the need for a timed collection. We simply measure the amylase and creatinine concentrations in urine and serum and calculate the clearance ratio according to the following formula:

$$\text{Ratio (\%)} = \frac{[\text{Urine amylase}] \times [\text{Serum creatinine}]}{[\text{Serum amylase}] \times [\text{Urine creatinine}]} \times 100$$

The resulting value provides a value for amylase excretion which can be related to kidney function.

Many studies have been carried out using the amylase/creatinine ratio in the diagnosis of acute pancreatitis. Since the ratio is independent of the method used to analyze amylase activity, normal values from different laboratories are much easier to relate to one another. Usually, the reference range is approximately 1–5%, with elevated ratios indicating increased production and excretion of amylase.

The amylase/creatinine ratio has been questioned in terms of reliability and specificity in the diagnosis of acute pancreatitis. The original studies were carried out on patients who were known to have the disease. In patients where the diagnosis was not clear, the ratio often did not provide useful information. Other disorders also produce an elevated ratio. Patients with diabetic ketoacidosis, renal insufficiency, severe burns, or those recovering from thoracic surgery frequently show an elevated amylase/creatinine ratio. As with any single diagnostic test, the determination of the ratio and the use of the data must be integrated with other facts about the patient before a reliable diagnosis can be made.

Macroamylase

As is true for several other enzymes, amylase occasionally combines with other proteins in the circulation to form a complex called **macroamylase.** Binding occurs either with an IgG or an IgA molecule; no attachment to other antibody classes has been detected to date. The result is a complex with a molecular weight of anywhere from 150,000 to over a million. This macromolecule remains in the circulation for a long time, in contrast to the rapid clearance of unbound amylase. As a result, serum amylase levels appear elevated and the possibility of pancreatitis is raised.

About 1% of randomly selected patients show the macroamylase complex and it has been observed in almost 3% of patients with elevated serum amylase levels. There is no known cause for this situation and it has not been related to any particular disease state. Measurement of amylase with the amylase/creatinine clearance ratio can usually detect the complex. Patients with macroamylasemia have very low clearance ratios (well below 1%). Confirmation of the presence of the complex can be made by column chromatography using Sephadex (a synthetic carbohydrate used to separate proteins on the basis of size) to assess the molecular weight of the material.

10.8

CHOLINESTERASE

Reaction Catalyzed

GENERAL REACTION

Cholinesterase catalyzes the hydrolysis of a number of esters which contain the molecule choline. The general reaction is:

$$
\begin{array}{c}
\text{O} \\
\parallel \\
R-C-O-CH_2-CH_2-N^+(CH_3)_3 \rightarrow \\
\text{choline ester} \\
\text{O} \\
\parallel \\
R-C-OH + HO-CH_2-CH_2-N^+(CH_3)_3 \\
\text{acid} \qquad\qquad\qquad \text{choline}
\end{array}
$$

In the body, this reaction is an important part of the process for the transmission of nerve impulses. Acetylcholine is released at a presynaptic membrane when a nerve impulse is to be passed. This compound goes across the nerve synapse to another membrane where it binds to a specific portion of the postsynaptic membrane. When it attaches, the new section of nerve is stimulated and an impulse is passed further down the nerve. The stimulation continues until the acetylcholine is hydrolyzed by the cholinesterase enzyme.

"TRUE" VERSUS "PSEUDO" CHOLINESTERASES

There are several cholinesterase enzyme forms in the body. They can be divided into two categories: the **"true" cholinesterase,** which uses only acetylcholine as a substrate, and the **"pseudo" cholinesterases,** which hydrolyze a variety of choline esters. There are also isoenzyme forms for each type of cholinesterase.

The true cholinesterase (acetylcholine acetyl-hydrolase, EC 3.1.1.7) has high substrate specificity. The primary compound broken down by this enzyme is acetylcholine, the chemical responsible for the transmission of the nerve impulse. Often we will refer to this form as "acetylcholinesterase." The primary location for this enzyme is the synapse of the nerve. It can also be detected in erythrocytes.

The pseudocholinesterase (acylcholine acyl-hydrolase, EC 3.1.1.8) readily hydrolyzes a variety of substrates. High concentrations of this cholinester-

ase are found in serum and in the white matter of the central and peripheral nervous system. The role of this enzyme is still somewhat unclear, but there is increasing evidence that it serves a protective function in the body. A number of choline esters other than acetylcholine are believed to be formed in the body and are capable of inhibiting acetylcholinesterase. The pseudocholinesterase (which we will refer to simply as cholinesterase) hydrolyzes these esters and releases the acetylcholinesterase from their blocking effect.

Tissue Distribution

Acetylcholinesterase is found predominantly in erythrocytes, lung, brain, spleen, and nerve endings. This cholinesterase does not appear in the serum. The pseudocholinesterase is located in heart, liver, pancreas, and brain. Enzyme synthesized by the liver is released in high concentrations into the serum.

Sample Collection

Assays may be run on either serum or heparinized plasma. Mild hemolysis does not create any significant problems if assays are being carried out for the more general cholinesterase, since acetylcholinesterase does not react with these substrates.

Measurement of Total Activity

GENERAL PRINCIPLES

The analysis of cholinesterase in body fluids has historically presented some difficulties. The products of the reaction do not have any distinguishing characteristics in the way of spectral absorbance or fluorescence, so a coupling reaction must be used for quantitation of enzyme activity (Table 10–7). Other approaches have involved measuring the change in pH as the reaction proceeds.

MEASUREMENT OF pH CHANGE

When the choline ester is hydrolyzed, a proton is released, promoting a drop in pH. This change can be monitored with a pH meter and an automatic titration device. Alternatively, pH indicator dyes can be employed and the reaction monitored spectrophotometrically. These techniques are cumbersome and do not find much utility in the routine clinical chemistry laboratory, although they are useful in some research studies.

Table 10-7
ASSAY TECHNIQUES FOR CHOLINESTERASE

Reaction Name	Technique	Comments
Michel	pH change	Temperature-sensitive; much variability among different labs
—	Change in UV absorption; hydrolysis of benzoylcholine	Need special equipment
Ellman	Colorimetric; thiocholine derivatives	Sensitive, rapid, recommended method

SPECTROPHOTOMETRIC ASSAYS

Spectrophotometric methods for the assay of cholinesterase have been developed and are much more applicable to clinical laboratory usage. The most widely used method for clinical laboratories is the **Ellman technique,** which has been modified by several workers. The substrate is a thiol ester, which produces a thiol as product instead of an alcohol. A common reaction is

$$CH_3CH_2 - \overset{\overset{\displaystyle O}{\displaystyle \|}}{C} - S - CH_2CH_2 - N^+(CH_3)_3 \longrightarrow$$
propionylthiocholine

$$CH_3CH_2COOH + (CH_3)_3N^+ - CH_2 - SH$$
propionic   thiocholine
acid

The thiocholine then reacts with a disulfide called 5,5'-dithiobis-(2-nitrobenzoic acid) to form a colored compound with an absorption maximum at 410 nm. By measuring the increase in absorbance at this wavelength, enzyme activity can easily be determined.

Several precautions must be taken when performing this assay. All solutions should be stored in glass bottles; some plastic bottles contain inhibitors of the enzyme, which could be leached out. A weak buffer is required since activity drops somewhat as buffer strength increases. Phosphate buffer appears to be preferable since nonenzymatic hydrolysis of substrate is lower in this buffer and pH changes due to temperature are less than in other buffer systems. If acetylcholinesterase is to be measured, the substrate concentration may need to be adjusted. This enzyme has a much lower substrate optimum than the pseudocholinesterase and could be inhibited by the substrate level usually employed.

Clinical Significance: Total Activity

REFERENCE RANGES

Cholinesterase is an atypical enzyme as far as the clinical laboratory is concerned. When we assay for most enzymes in serum, a normal result is usually a rather low value; sometimes little or no enzyme activity is detected. These results indicate that little tissue damage is occurring, normal cell turnover is taking place, and there is no pathological problem. With cholinesterase, the data are reversed. Normal serum levels of this enzyme are quite high, reflecting its continual synthesis and release by the liver, among other sources. Pathological levels are decreased from the normal values, sometimes by as much as 80–90%. Cholinesterase levels vary in a number of clinical situations and could provide more useful information than is currently the case. Two specific areas where information on cholinesterase levels play an important role in diagnosis and management are pesticide poisoning and inhibition by anesthetic.

Reference values for cholinesterase vary markedly, depending on the exact method employed. Although normal adult levels of cholinesterase are quite high (some methods have reference ranges of 6,000 to 12,000 U/L), the values in newborns and infants are much lower. Women in the third trimester of pregnancy have decreased levels of the serum enzyme. There are rare instances where a patient has an inherited variant enzyme with a very low activity. Treatment with a number of drugs results in low levels of the enzyme. Medications such as oral contraceptives, monoamine oxidase inhibitors (used either as antihypertensives or antidepressants), some drugs used in cancer treatment, and some x-ray contrast media have been observed to lower the serum cholinesterase values.

VALUES IN VARIOUS DISEASE STATES

Serum cholinesterase values decrease noticeably in liver disease. This enzyme may provide an important assessment of liver function, particularly in monitoring the progress of patients treated surgically for cirrhosis or those who have undergone liver transplant. Decreases have been observed in myocardial infarction, possibly due to the secondary liver damage also seen in this disorder. Other medical problems where lowered levels of cholinesterase have been reported are muscular dystrophy, malnutrition, myxedema, acute infections, and a number of chronic disease states.

Increased levels are somewhat rare but have been observed. Good documentation of elevated levels requires some knowledge of the normal values for the individual. There are rare electrophoretic variants of serum cholinesterase which produce elevated values when total activity is determined. High values have also been reported in cases of extreme obesity, thyrotoxicosis, and hypertension. Because of the role of cholinesterase enzymes in transmission of nerve impulses, it is interesting to note that elevated values of the serum enzyme have been reported in patients with anxiety, alcoholism, or schizophrenia. The significance of these data are not known at present.

ACTIVITY IN PATIENTS WITH PESTICIDE EXPOSURE

A major clinical application of cholinesterase measurements is in screening for possible pesticide exposure. A number of these agricultural chemicals produce inhibition of the serum and red cell enzymes. If acetylcholinesterase activity is blocked by a pesticide, impairment of nerve transmission occurs. Although some inhibition is reversible (Carbaryl and other carbamate derivatives), the organophosphate compounds covalently bind to the enzyme, causing irreversible inhibition. These materials do not clear from the bloodstream, but rather are removed only when the erythrocyte is destroyed by normal metabolic processes.

When collecting samples to assess cholinesterase activity in a suspected pesticide exposure case, several precautions need to be taken. The blood sample should be chilled immediately upon collection so as to block the destruction of the inhibitor through enzymatic activity. Dilution of sample should be avoided as much as possible since the degree of inhibition is altered somewhat by the dilution. Clear separation of serum and red cells is important to avoid cross-contamination by the two different cholinesterase enzymes.

If an inhibitor is present, a marked decrease in the level of serum cholinesterase is observed. Activity as low as 50% or less of normal is quite common in these situations. Employees in jobs where exposure to pesticides is great are often screened for cholinesterase activity on a regular basis. If the serum enzyme level shows a noticeable decline from normal, the employee is often moved to another task in order to minimize or eliminate pesticide exposure. Follow-up measurements should be done on a regular basis until the serum cholinesterase activity is restored to normal.

INHIBITION BY SUCCINYLCHOLINE

Succinylcholine (sometimes referred to as suxamethonium) is a short-acting muscle relaxant used in clinical anesthesia. Because it is structurally similar to cholinesterase substrates, it can be hydrolyzed by the serum enzyme and has a short duration of action (several minutes) before it is inactivated.

However, there are individuals with a variant form of cholinesterase which cannot hydrolyze succinylcholine efficiently. As a result, the relaxant remains in the circulation for a long time, requiring maintenance of respiration by artificial means until the drug is cleared from the system (often several hours are required). Special screening techniques are available to help detect some of these genetic abnormalities.

SCREENING FOR NEURAL TUBE DEFECTS

The congenital disorder spina bifida is characterized by open segments of the spinal cord. These defects develop before birth and can cause severe problems in the newborn, including death. Although we have improved our ability to diagnose and treat this disease, many physicians and patients choose abortion when the disorder is detected.

If spina bifida is suspected in a fetus (from ultrasound studies or other fetal monitoring techniques), an amniocentesis is done. Fluid is collected from the amniotic sac and is analyzed for alpha fetoprotein, elevated in cases of spina bifida. There are often difficulties with this analysis, requiring confirmation of the clinical situation by measurement of acetylcholinesterase. Unfortunately, a high false positive rate (almost 20% in one study) suggests this approach needs further refinement before it is more widely used as a confirmatory test.

10.9

CREATINE KINASE

Reaction Catalyzed

Creatine kinase (ATP:creatine N-phosphotransferase, EC 2.7.3.2) catalyzes a reaction responsible for the formation of ATP in tissues:

$$ATP + creatine \rightleftharpoons ADP + creatine\ phosphate$$

When muscle contraction occurs, ATP is hydrolyzed to ADP to produce chemical energy for the contraction process. An important step in the regeneration of ATP is the reaction with creatine phosphate. There is some reason to believe that the same enzyme may be involved with some of the phosphorylation processes which occur in mitochondria.

Tissue Distribution

Creatine kinase (CK) is widely distributed in skeletal muscle, brain, and cardiac muscle. Some information suggests the enzyme may be detected in nerve tissue, testical tissue, amniotic fluid, and in certain malignant tissues. The clinical significance of these data is not yet fully known.

Sample Collection

Serum samples are to be used for CK assays; anticoagulants inhibit enzyme activity. The enzyme is light-sensitive and loses significant amounts of activity when exposed to light for long periods.

Measurement of Total Activity

GENERAL PRINCIPLES

Most approaches to the assay of CK require some type of coupled reaction system, since neither substrates nor products of the main reaction have any unique structural characteristics which would allow easy detection by spectrophotometric or fluorometric means.

One of the interesting characteristics of the enzyme is that it contains essential thiol groups. These $-SH$ components of the enzyme must be present in the reduced form for the enzyme to manifest full activity. Over time, the CK in a sample has the $-SH$ groups on the enzyme oxidize to form $-S-S-$ dimers. When this process occurs, the enzyme loses activity. Most assays deal with this problem by including some type of cysteine derivative in the reaction mixture. By means of an exchange reaction the enzyme can be reactivated:

$$\text{cysteine} - SH + \text{enz} - S - S - \text{enz} \rightarrow$$
$$\text{cys} - S - S - \text{cys} + \text{enz} - SH$$

The $-SH$ groups help maintain the three-dimensional structure of the enzyme so that full activity can be expressed. Unlike the situation with many other enzymes, these groups in CK are not involved at the active site with the actual enzyme-catalyzed reaction.

TANZER-GILVARG ASSAY

Two major approaches have been developed using coupled enzyme assays for the quantitation of creatine kinase. Both systems involve measurement of some aspect of the ADP/ATP reaction system. The **Tanzer-Gilvarg method** assesses the rate of the forward reaction in which creatine is converted to creatine phosphate (Fig. 10–8).

$$\text{Creatine} + \text{ATP} \longrightarrow \text{Creatine phosphate} + \text{ADP}$$

$$\text{ADP} + \text{Phosphoenolpyruvate} \xrightarrow{\text{Pyruvate kinase}} \text{ATP} + \text{Pyruvate}$$

$$\text{Pyruvate} + \text{NADH} + \text{H}^+ \xrightarrow{\text{Lactate dehydrogenase}} \text{Lactate} + \text{NAD}^+$$

FIGURE 10–8   Tanzer-Gilvarg assay for creatine kinase.

The initial reaction in the sequence is the phosphorylation of creatine by creatine kinase, with the formation of ADP. This molecule then participates in a reaction with phosphoenolpyruvate, catalyzed by the enzyme pyruvate kinase, regenerating ATP (which can be recycled back to the first reaction) and producing pyruvate. The indicator reaction in this sequence is the reaction of pyruvate with NADH to form lactate and NAD, catalyzed by the enzyme lactate dehydrogenase. Since NADH absorbs strongly at 340 nm (and NAD does not absorb light at this wavelength), a decrease in absorbance is the indicator of enzyme activity. The more rapid the drop in 340-nm absorbance with time, the higher the amount of enzyme in the sample.

One operational drawback to the Tanzer-Gilvarg assay is that the initial 340-nm absorbance in the system is rather high. Lower levels of activity are not measured as accurately since small changes in absorbance are more difficult to detect.

OLIVER-ROSALKI ASSAY

The reverse reaction, in which creatine is produced from creatine phosphate, is the basis of the **Oliver-Rosalki technique** (Fig. 10–9). In this assay system, the initial process is the formation of ATP and creatine, catalyzed by creatine kinase. The ATP

$$\text{Creatine phosphate} + \text{ADP} \rightleftharpoons \text{Creatine} + \text{ATP}$$

$$\text{ATP} + \text{Glucose} \xrightarrow{\text{Hexokinase}} \text{ADP} + \text{Glucose-6-phosphate}$$

$$\text{Glucose-6-phosphate} + \text{NADP}^+ \xrightarrow{\text{Glucose-6-phosphate dehydrogenase}} \text{6-Phosphogluconate} + \text{NADPH} + \text{H}^+$$

FIGURE 10–9
Oliver-Rosalki assay for creatine kinase.

is then used in a reaction with glucose and the enzyme hexokinase to regenerate ADP and glucose-6-phosphate. Once again, the indicator reaction involves an NAD derivative; in this case, the material is NADP. Glucose-6-phosphate is oxidized by the enzyme glucose-6-phosphate dehydrogenase to form 6-phosphogluconate, and the NADP accepts the hydrogen atoms and is reduced to NADPH. Since NADPH absorbs strongly at 340 nm and NADP has no measurable absorbance at that wavelength, an increase in 340-nm absorbance as a function of time is used to calculate the creatine kinase activity in this system. The advantages of measuring an increase in 340-nm absorbance (instead of a decrease) were discussed above.

ACTIVATOR REQUIREMENTS

In both assay systems, a sulfhydryl activator is required. At present there is no general agreement about the best compound to use. Materials such as cysteine, N-acetyl cysteine, glutathione, dithiothreitol ("Cleland's reagent"), and monothiolglycerol have been employed. Some workers recommend that serum samples also have activator added to them to minimize irreversible loss of activity upon storage. This practice has some merit, but has not yet become standard in the clinical laboratory.

ADENYLATE KINASE AND CREATINE KINASE ASSAYS

The measurement of creatine kinase by these coupled assays is complicated by the presence of adenylate kinase in serum samples. This enzyme catalyzes the direct conversion of ADP to ATP and AMP:

$$2\ ADP \rightleftharpoons ATP + AMP$$

Obviously, this process distorts the measurement of CK because of the changes in the concentrations of ATP and ADP generated by adenylate kinase. To minimize this problem, various inhibitors of adenylate kinase are added to the assay system. AMP, fluoride, and diadenosine-5-pentaphosphate all inhibit the reaction considerably. Perhaps the best combination of inhibitors is AMP plus the diadenosine derivative, which is used in the reference method for creatine kinase.

Clinical Significance: Total Activity

REFERENCE RANGES

Reference ranges for total CK activity in serum of adults are up to 160 U/L for males and up to 130 U/L for females when a standard coupled enzyme assay is used following the Oliver-Rosalki method. This sex-related difference apparently is related to muscle mass. Individuals with large muscle mass have higher CK values than persons with less muscle mass, when matched for age and sex.

During the neonatal period, serum total activity may be twice the adult level, but it declines fairly rapidly. Serum CK values gradually rise in pregnant women and are somewhat elevated, especially during the last several weeks of pregnancy. These increases gradually disappear after delivery, returning to normal within 5 days or so.

There are apparently some racial differences in CK normal values. African American men have values considerably higher than those of men from other genetic heritages; the same pattern is seen in women. There is no readily apparent explanation for this difference in values. Interpretation of results must be conditioned on complete knowledge of the individual patient.

A problem sometimes arises in the interpretation of CK values in the elderly. Frequently, abnormal isoenzyme patterns are observed in older patients with apparently normal levels of total CK activity. It is becoming increasingly apparent that the normal adult ranges cited earlier cannot be applied to values obtained from elderly individuals. Due to decreases in activity and a diminished muscle mass in these persons, the normal range for CK is lower than expected. It is helpful to obtain a reference range for senior citizens as part of the laboratory's local data base.

ELEVATIONS IN VARIOUS DISEASE STATES

Creatine kinase levels increase in a variety of disorders in which cardiac or skeletal muscle tissue is affected. Clinical situations such as myocardial infarction, muscular dystrophy, polymyositis, malignant hyperthermia, and related disorders produce striking elevations in serum CK levels. Individuals who undergo strenuous physical exertion or who receive intramuscular drug injections manifest elevated values, which are usually transient. There are a number of infections (such as trichinosis) involving skeletal muscle tissue, that result in elevations of serum CK activity.

Although the focus is often on muscle tissue and the ability to assess trauma to these tissues, CK measurement can also reflect damage to brain tissue. Patients with malignant neoplasms of the brain frequently exhibit increased serum CK levels. Values also rise in cases of cerebral thrombosis, brain infarction, and cerebral embolism. One of the more puzzling findings has been the increases in total creatine kinase levels in manic-depressive patients prior to lithium treatment and in patients who exhibit various psychoses (including schizophrenia). At present, there is no good explanation for these findings.

Isoenzymes of Creatine Kinase

DESCRIPTION

Creatine kinase exists as a dimer, composed of two polypeptide chains. There are two possible structures for the protein subunit, an M chain and a B chain. The two subunits can combine to form three different CK isoenzymes. The MM isoenzyme is found in skeletal muscle and cardiac muscle; this isoenzyme is the one detected in the serum of patients with no clinical problems. Significant amounts of the MM fraction are also found in lung, thyroid, liver, spleen, and placenta. The MB isoenzyme is found in cardiac muscle, constituting about 30% or so of the CK activity in that tissue. Approximately 5% of the skeletal muscle CK is of the MB variety. The BB isoenzyme has been detected in brain tissue, where it is the only CK isoenzyme present. Most tissues other than skeletal and cardiac muscle contain high amounts of the BB isoenzyme.

ASSAYS

Electrophoresis

Perhaps the most widely used technique for the separation of creatine kinase isoenzymes is electrophoresis. Separation on cellulose acetate or agarose is easily achieved using the same methodology as is employed for serum proteins. Development of the individual isoenzyme fractions on the film is accomplished by rolling a thin layer of enzyme assay solution over the support medium and allowing the system to incubate for a set time.

To obtain a visual end product, some modification of the Oliver-Rosalki assay method is used. The NADPH formed can be detected when a fluorescent light is shown on the system. With an appropriate densitometer, the degree of fluorescence can be quantitated and the relative amounts of each isoenzyme fraction determined. The fractions can be visualized with a colorimetric assay. The addition of nitroblue tetrazolium or a similar compound and phenazine methosulfate result in a reaction with the NADPH present. A band of color is produced, the intensity of which is proportional to the amount of enzyme activity present in that fraction. Unfortunately, the colorimetric method does not appear to be as sensitive as the fluorometric one and may underestimate the amount of CK-MB and CK-BB present in a sample.

The separation of the various CK fractions is illustrated in Figure 10-10. In keeping with international convention, the isoenzymes are numbered according to mobility, with CK-1 (CK-BB) moving the farthest in the time allotted. CK-2 (CK-MB) is intermediate mobile, and CK-3 (CK-MM) does not travel far from the origin. Other bands which may appear are macro-CK, with a mobility between that of CK-2 and CK-3, and the mitochondrial form of

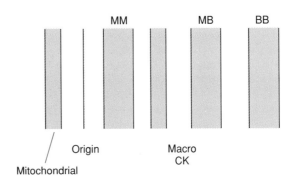

FIGURE 10-10 Creatine kinase isoenzyme fractionation.

the enzyme, which migrates slightly in the opposite direction. On occasion, a spurious band of activity may be observed in the mitochondrial CK region. If adenylate kinase is not sufficiently inhibited by the assay mixture used to detect the isoenzymes, it shows a small band in this region and can be misinterpreted as creatine kinase activity.

One major drawback to electrophoresis is related to the heat sensitivity of the isoenzymes, particularly CK-MB and CK-BB. Some of the procedures for developing the bands after substrate is added call for temperatures (60°C) which may partially inactivate these two fractions. Thus, there is some loss of activity and the two fractions are underestimated. This problem is of special concern when quantitation of the MB fraction is required in the diagnosis of a myocardial infarction.

Occasionally, a fluorescent artefact may appear on the electrophoresis strip. It is possible that this false positive band could be confused with the presence of a clinically significant fraction of CK; frequently, this atypical band migrates in the region of CK-BB. If known samples are run and mobilities compared, these atypical bands can often be identified. The artefact may be naturally fluorescent and could be detected by performing an electrophoretic separation of the sample, followed by direct examination under fluorescent light without adding the substrate mixture to measure enzyme activity.

Immunoinhibition

Since the electrophoretic method has some drawbacks, other techniques have been explored to assess their usefulness in measuring the various isoenzymes easily and reliably. One method which has been developed and is commercially available is the immunoinhibition assay (Fig. 10-11). Although a number of variations have been reported, the basic approach involves the use of antibodies to the M subunit of creatine kinase. If the antibody is added to the assay system, it binds to any M subunit and inhibits the enzyme. In normal individuals with

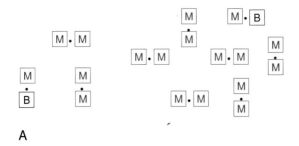

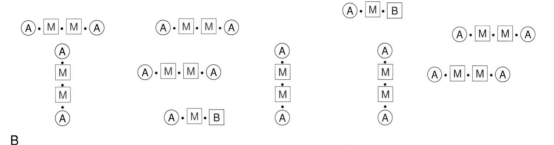

FIGURE 10–11   Creatine kinase (CK) isoenzyme assay by immunoinhibition. A. CK isoenzyme in serum. B. Reaction with Ⓐ to Ⓜ. Note that Ⓑ is still accessible to the substrate because it is not inhibited by Ⓐ. Ⓜ = M subunit, Ⓐ = Antibody, Ⓑ = B subunit.

only the CK-MM fraction present, addition of such an antibody should result in essentially complete loss of activity. In reality, a small residual amount of enzyme activity is still detected in most systems, suggesting less than complete inhibition by the antibody.

In patients with a myocardial infarction, CK-MB is present in addition to the normal CK-MM fraction. If the M subunit is inhibited, the MM isoenzyme should not show any measurable activity. Antibody only partially inhibits the MB isoenzyme, since the B subunit still remains active. The argument is then made that activity seen after antibody is added is due to B subunit and therefore indicates the amount of CK-MB fraction present. The method is rapid and sensitive to the presence of B subunit, more so than the electrophoretic approach. An assay can be performed in a few minutes using a centrifugal analyzer or one of the newer random-access systems currently available. This approach has been applied to a nephelometric system to determine the mass of enzyme present instead of its activity. This data raises some interesting questions about the relationship of active and inactive enzyme present in the circulation.

If the immunoinhibition method is employed, special precautions must be taken to see that the adenylate kinase reaction is completely inhibited. Any trace of this side reaction seriously affects the results obtained, since we are studying low levels of activity to begin with.

The flaw in the immunoinhibition approach is in the assumption that B subunit activity equals CK-MB activity. There are data available which suggest

that the activity being measured in some patients may include (or be made up entirely of) the CK-BB isoenzyme. In the absence of any way to actually determine the constituents of the enzyme activity being seen, we cannot assume that any B subunit activity we detect is truly due to CK-MB. If BB isoenzyme is present, it will show activity. If the antibody does not inhibit completely, a small amount of the MM fraction will register activity. If there is cross-reactivity of the antibody with the B-subunit, a lower amount of activity may be detected than is actually present. When the immunoinhibition approach is adopted in a laboratory, the staff must be aware of the problems and questions which can arise.

Column Chromatography

Because of shortcomings in the electrophoretic method of CK isoenzyme separation, other attempts were made to develop a system which would allow direct quantitation of each CK isoenzyme fraction. A column separation procedure was proposed which became quite popular for a time and is still used in a number of laboratories. Several companies manufacture the materials and reagents, so some degree of standardization has been achieved.

The procedure involves the use of a DEAE (diethyl amino ethyl)-Sephadex A-50, or similar anion exchange resin. At the pH of the system (pH 8.0), all the proteins have a net negative charge and bind to the column. These attachments are weak ones; the proteins can be removed if a stronger anion is added to the system. In addition, the various isoenzymes have a somewhat different charge (as we saw in electrophoresis), so the degree of attachment to the resin in the column is different for each CK isoenzyme. By changing the concentration of the NaCl solution used to remove the proteins, we can selectively elute each isoenzyme and then determine the activity of that fraction by itself.

One possible advantage of the column system over electrophoretic methods has to do with the adenylate kinase interferant present. This interfering enzyme appears to elute completely with the CK-MM fraction and does not interfere with the measurement of CK-MB. We still need to add inhibitor to the assay system so that the CK-MM activity is not overestimated.

This system has been modified to allow simultaneous measurement of CK isoenzymes and some specific isoenzyme fractions of lactate dehydrogenase (LD). This latter enzyme (to be discussed later in this chapter) is also of great importance in following the course of a myocardial infarction. The lactate dehydrogenase isoenzymes of interest elute in the same fraction as CK-MB, so one eluate can be used for two different enzyme measurements.

A major drawback of the column elution method is that the fractions are often incompletely separated. In some instances, carryover of the MM fraction into the MB area may occur, especially if the total activity is high. When following the usual procedure, an excess amount of CK-MM begins to elute in the MB fraction and falsely elevates the measured amount of the cardiac isoenzyme. Frequently, there may be poor separation of CK-MB and CK-BB (if the brain isoenzyme is present). Again, the CK-BB partially elutes off the column in the same fraction as CK-MB, increasing the amount of the MB isoenzyme detected. These false positive values seriously compromise the utility of CK isoenzyme determinations and provide misleading information.

In addition to the separation problems, the procedure is somewhat time-consuming. Although a single column assay can be run in less time than an electrophoretic separation, the column method quickly becomes cumbersome when a large number of samples need to be analyzed. More technician time must be spent in monitoring the columns and whatever speed is obtained initially is quickly lost when several samples are assayed simultaneously by the column method.

Immunoassay

An extension of the immunoinhibition approach uses the same types of antibodies to measure the mass of CK, rather than the enzyme activity. In most approaches using RIA, an antibody is prepared against the B subunit of the enzyme. The competitive ligand in the assay is frequently CK-BB from an animal source and is labeled with an isotope (generally ^{125}I). Patient serum containing creatine kinase and the labeled CK-BB are mixed with antibody to the B subunit and separation of bound and free materials is carried out as described previously. Quantitation of B subunit allows an estimation of the amount of CK-MB present.

An inherent difficulty in this approach comes if CK-BB isoenzyme is also present for some reason, either in addition to (or instead of) the CK-MB fraction. The antibody cannot distinguish between the two isoenzymes and so detects both if they are in the sample. Overestimation of CK-MB levels can occur in this situation.

At least one manufacturer has attempted to get around this difficulty with a sandwich assay employing both B-subunit and M-subunit antibodies (Fig. 10–12). The antibody to CK-B is attached to a solid phase. When sample is added, all fractions containing a B subunit bind to the antibody; the MM fraction can then be washed away. The antibody to the M subunit is added and attaches only to the CK-MB present. This antibody is labeled with a radioisotope or some other tag so that its degree of binding can be assessed. The amount of M-subunit antibody which attaches is an estimate of the quantity of CK-MB present. CK-BB does not interfere, and there is no overestimation of the MB isoenzyme in this case.

Estimation of CK isoenzymes by radioimmunoassay calls for a shift in our understanding of the time frame for changes in enzyme levels. When creatine kinase is assayed by an enzyme activity approach, we are measuring only functional enzyme. There may be other inactive CK protein present, but we do not detect it. However, when we measure CK mass (as is the case with radioimmunoassay), we are seeing all the protein which reacts with the antibody, both active enzyme and that protein still present in the circulation but no longer displaying enzyme activity. The interpretation of data, both in terms of the changes in the levels of various enzyme fractions and in the time frame for those changes, is altered markedly. In a patient with a myocardial infarct, total CK activity (measured as active enzyme) rises and drops back down again within 2 to 3 days. However, if we look at enzyme mass only (both active and inactive enzyme), the decline is much slower since enzyme becomes inactivated long before the protein is metabolized and cleared from the circulation (Fig. 10–13). Interpretation of data requires that we be very clear about how our information was obtained before we can use it in the proper manner.

CLINICAL UTILITY OF CREATINE KINASE ISOENZYMES

Creatine Kinase-MB and Cardiac Disease

The major value of CK-MB measurements is in the diagnosis and monitoring of myocardial infarc-

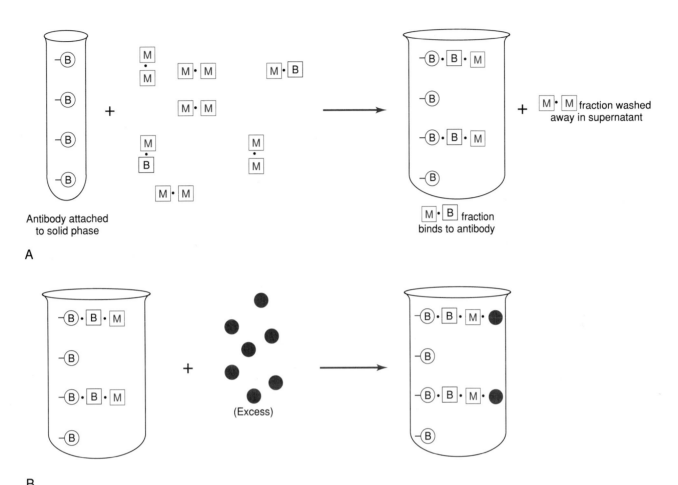

FIGURE 10-12   Immunoradiometric assay for CK-MB. A. Attachment of $\boxed{M}$ · $\boxed{B}$ fraction to ⑧. B. Attachment of ● to $\boxed{M}$. $\boxed{M}$ = M subunit, $\boxed{B}$ = B subunit, ⑧ = antibody to the B subunit, ● = radiolabeled antibody to the M subunit.

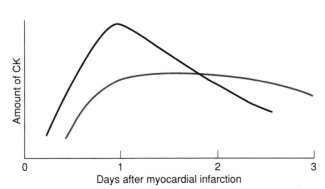

FIGURE 10-13   Comparison of activity versus mass for CK estimation.

tion (MI). There are some drawbacks to relying on CK isoenzyme measurements, but the data obtained have provided invaluable information over the years in the early detection of cardiac damage.

When cardiac muscle is damaged, the cells release enzymes and other materials into the circulation. Because CK-MB is present in high amounts in cardiac tissue, the level of this isoenzyme fraction increases after an infarct. The total CK activity rises rapidly, peaking about 18 to 24 h after the event and then returning to normal values within another day or so (Fig. 10-14). The amount of CK-MB in serum shows a parallel rise and fall. In many instances, peak MB isoenzyme amounts are as much as 30–40% or more of the total creatine kinase activity in serum. If there is an extension (or reinfarction) a few days after the initial attack, the CK total activity and MB isoenzyme levels both increase again, providing a sensitive marker for the new event.

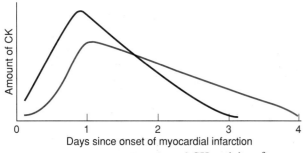

FIGURE 10-14   CK-MB and total CK activity after myocardial infarction.

How reliable are CK isoenzyme studies in detecting an MI? Depending on the data surveyed, a large percentage of patients with an MI (up to 40% or so in some studies) have nondiagnostic electrocardiograms. This test assesses the electric conductivity of the heart, giving a direct indication of some aspects of cardiac function. In contrast, almost every patient with significant amounts of CK-MB (5% or more of total activity) present in the serum had an MI. The measurement is particularly valuable when serial assays are done. If a sample is collected on admission and subsequent samples are collected at 8-h intervals, a clear pattern of activity can be determined and the rise and fall in both total activity and isoenzyme levels is readily observed.

In recent years a number of studies have found that a subpopulation of patients with MI show significant amounts of CK-MB in the serum, yet their total CK activities do not rise above the "normal" range. These patients tend to be elderly individuals, with the decreased muscle mass expected in this group. The presence of CK-MB has been documented both by electrophoresis and by radioimmunoassay techniques, with the occurrence of an infarct confirmed by electrocardiogram. We find baseline values to be on the very low end of the normal range for serum total CK activity. In this group of patients, the positive indicator is the presence of elevated (greater than 5%) MB, with the total activity increasing over base line in the expected fashion. The reference range for creatine kinase activity in serum undoubtedly needs to be reassessed in order to interpret data from geriatric patients properly.

A more sensitive early indicator of an MI may be found in the subfractionation of the CK isoenzymes identified by routine electrophoresis. Creatine kinase-MM can be further divided into three different enzyme fractions if isoelectric focusing is employed instead of agarose. The subfraction CK-MM3 is released from tissue and is subsequently converted to CK-MM1 by a carboxypeptidase in the circulation. The amount of release and conversion of this fraction is increased during an MI (Fig. 10-15). In the first several hours afterward, the rise in CK-MM1

appears to be a more reliable indicator of an MI than does the presence of CK-MB. In addition, the MM1 fraction appears somewhat earlier in the course of cardiac damage. However, beyond about 9 h after the initial event, the usefulness of CK-MM3 measurement drops considerably and the CK-MB determination becomes the parameter to follow.

Creatine Kinase-MB in Other Clinical Situations

Increased amounts of CK-MB have frequently been observed in conditions other than those related to cardiac damage. Small elevations of this isoenzyme are often seen in patients with muscle diseases of various sorts (muscular dystrophy, polymyositis, and other related disorders). In these individuals, the increase in CK-MB is frequently less than 5% of the total activity, and the pattern is one of sustained increase, not the rise and fall associated with an MI. Patients who experience skeletal muscle trauma also show some CK-MB in the serum. Although their total activity indicates an increase in CK followed by a decrease, the CK-MB is usually 5% or less. With proper monitoring of data and close investigation of the clinical situation, these cases can fairly easily be distinguished from those with a myocardial infarct.

More problematic are those individuals in whom CK-MB is quite elevated (sometimes 15% or more of the total activity) and yet do not show symptoms of a heart attack. In many instances, the increase in CK-MB has been shown to be due to cancer. Apparently, there is ectopic production of this isoenzyme of creatine kinase by the cancer tissue itself, with release into the circulation. One clue which allows differentiation between these cases and MIs is the sustained increase over a several-day period. Where the patient with a heart attack shows a rise and subsequent drop in CK-MB activity over the course of 1 to 2 days, the levels of an individual with ectopic production remain high, although some fluctuation may be seen from day to day.

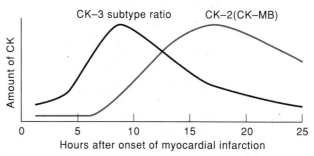

FIGURE 10-15   CK-3 subtypes and myocardial infarction.

Creatine Kinase-BB

The CK-BB fraction has proven to be both an enigma and a challenge to clinical chemists for years. Present in high concentrations in the brain, its passage through the blood-brain barrier into the serum is not well understood. Of the three major CK isoenzymes, the BB fraction has the shortest half-life in the circulation (about 2 to 3 h compared with the 12 h for CK-MB and 15 h for CK-MM). First considered an artefact in many assays, the CK-BB isoenzyme has achieved recognition in its own right as a clinical marker, but we are not sure exactly what it is a marker for.

Perhaps the most common (and easiest) approach to measurement of CK-BB is the electrophoretic technique. If the BB isoenzyme is present, it is detected as a discrete band with a greater migration than the other two isoenzymes. Column methods are too cumbersome for quantitative measurement of this fraction, since additional elutions must be made and care must be taken to avoid overlap with the CK-MB fraction during the eluting process. Immunoassay and immunoinhibition procedures are unable to measure only the BB fraction (although a sandwich assay could perhaps be developed).

The BB isoenzyme has been detected in a variety of clinical situations. In some gastrointestinal disorders with extensive intestinal necrosis, the CK-BB fraction is elevated (the intestine apparently contains a significant amount of CK-BB). This isoenzyme is detected in serum after coronary bypass surgery, since there seems to be some CK-BB present in the saphenous vein. Increases in CK-BB activity have been correlated with the presence of various cancers. Both the serum and the CSF BB fraction may be increased in several neurological diseases. Since the CK-BB possibly derives from the large amount of this enzyme in brain tissue, this marker should be explored more thoroughly as a tool for the difficult diagnoses which often need to be made by the neurologist. There have been reports that CK-BB is elevated in some patients with schizophrenia, but the data seem inconclusive right now.

Mitochondrial Creatine Kinase

The CK isoenzymes we have discussed to this point have all been located within the cytoplasmic portion of the cell. When cell damage occurs, these fractions are readily released into the circulation. There is another CK isoenzyme, which is contained within the mitochondria of the cell and which does not appear in the circulation unless more severe cellular damage takes place. This mitochondrial creatine kinase is a dimer, but does not have any structural characteristics that would relate it to the other isoenzymes. There is no reactivity with antibodies to CK-B or CK-M subunits. Mixtures of subunits of the mitochondrial isoenzyme do not form hybrids with B or M subunits, suggesting a very different structure for the protein. On electrophoresis, the mitochondrial isoenzyme can migrate cathodally or remain very close to the origin. The mitochondrial enzyme not only exists as a dimer, but can also form large-molecular-weight aggregates which then appear in the serum as a type of macro-CK. The clinical utility of these data have not yet been explored sufficiently to allow any firm conclusions regarding the diagnostic usefulness of the mitochondrial isoenzyme.

Macro-Creatine Kinase

Creatine kinase has the ability to form large complexes referred to as **macroenzymes.** One type of CK complex is with an immunoglobulin, usually IgG. Most of the time, the complex is between two molecules of CK-BB and a molecule of IgG. Occasionally, CK-MM and IgA may be involved in a complex. This type of macro-CK usually migrates between the MM and the MB fractions on electrophoresis. Approximately 3–6% of hospitalized patients display macro-CK in the serum. There is no clear pattern of disease related to the presence of this fraction, so it cannot be used as a diagnostic marker at this time.

As mentioned above, a second type of macro-CK has been observed in which the mitochondrial isoenzyme forms aggregates. In some instances, it is believed that fragments of the outer wall of the mitochondrion are also involved as part of this complex. This form of macro-CK undergoes several conversions while in the circulation. Sometimes the differentiation between the dimer form and the larger complex of mitochondrial CK is difficult to detect. Although this fraction is often seen in patients with severe disease (particularly those with malignancies), its diagnostic usefulness has not been firmly established.

10.10

GAMMA-GLUTAMYLTRANSFERASE

Reaction Catalyzed

The enzyme **gamma-glutamyltransferase** (EC 2.3.2.2) **(GGT)** plays a unique role in the metabolism of amino acids. The basic reaction catalyzed by this enzyme (often referred to as gamma-glutamyl-transpeptidase) is

$$\text{gamma-glutamyl-cysteinyl-glycine} \xrightleftharpoons{\text{GGT}}$$
$$+ \text{ amino acid (peptide)}$$

$$\text{gamma-glutamyl-amino acid (peptide)}$$
$$+ \text{ cysteinyl-glycine}$$

This basic transpeptidation reaction is significant in some aspects of the control of protein synthesis. The synthesis of glutathione and certain components of the movement of amino acids into cells are, in part, regulated by the activity of gamma-glutamyltranspeptidase.

Tissue Distribution

Gamma-glutamyltransferase is widely distributed in a number of tissues. The kidneys contain the highest amount of this enzyme, with significant amounts also detected in pancreas and liver. Although most enzymes are located within the cell, gamma-glutamyltransferase is somewhat unique in that it is found on the outside surface of the cell membrane.

Sample Collection

Serum is the preferred specimen, preferably without hemolysis (which produces interference with the measurement of product absorbance). Citrate, oxalate, and fluoride inhibit enzyme activity. Heparin causes turbidity. Samples can be frozen and remain stable for several weeks.

Measurement of Total Activity

GENERAL PRINCIPLES

The initial assays exploited the capability of the enzyme to catalyze a coupling reaction between two molecules of substrate (one of the minor physiological roles for GGT). Two molecules of gamma-L-glutamyl-alpha-naphthylamide were converted by the enzyme into a molecule of the coupled product gamma-L-glutamyl-gamma-L-glutamyl-alpha-naphthylamide and a molecule of free naphthylamine. The free amine derivative could be assayed using a diazo reaction to form a colored compound. The assay was not very sensitive and required precipitation of protein before the color step could be run. More recent approaches involve the use of an acceptor molecule for transfer of the gamma-glutamyl group.

THE SZASZ ASSAY

When the acceptor glycylglycine was introduced into the assay, the sensitivity and ease improved significantly. As in the case of acid and alkaline phosphatases, the enzyme now had a molecule to transfer a group to (similar to the normal physiological reaction). The substrate of choice then became gamma-L-glutamyl-p-nitroanilide. The gamma-glutamyl portion of the molecule would be transferred to the glycylglycine, and the free p-nitroaniline could be measured by spectrophotometer. This product has a strong absorption maximum in the range 405–420 nm, and is usually detected in a kinetic assay. A more recent substrate is gamma-L-glutamyl-2-carboxy-4-nitroanilide, a compound similar in structure to the previously mentioned substrate. One product of the reaction, 5-amino-2-nitrobenzoate, provides better absorbance at 410 nm and improved sensitivity. This approach, known as the **Szasz assay,** is (with some small modifications) the reference assay for gamma-glutamyltransferase.

Some precautions must be observed in running the assay. The p-nitroanilide substrate is somewhat unstable and tends to hydrolyze under acid conditions. For kinetic assays, this property does not cause any difficulty, but a blank measurement needs to be taken if a fixed-point method is used.

Clinical Significance: Total Activity

REFERENCE RANGES

The actual reference ranges for gamma-glutamyltransferase depends to some extent on the substrate used and other parameters of the assay. Males have somewhat higher levels (up to 40 U/L) than females (up to 25 U/L). Newborns have marked elevations of GGT compared with adults, but these levels decline to normal by about 6 months after birth. From that time on, serum levels for this enzyme are reasonably constant in the healthy individual and do not display any fluctuations during adolescence. Gross obesity results in some increase in serum values for GGT. There is no change in serum enzyme activity seen in strenuous exercise. Patients on some medications, particularly anticonvulsants, manifest marked increases in serum GGT levels.

GAMMA-GLUTAMYLTRANSFERASE AND LIVER DISEASE

Gamma-glutamyltransferase has been shown to be a useful marker for liver damage. The enzyme is elevated in obstructive liver disease, inflammation of the liver, or obstruction of the biliary tract. Since this enzyme does not show the marked shift in values during adolescence that we see with alkaline phosphatase, it is especially useful in studying liver disease during the teen years. Measurement of GGT can give a good clue as to the source of an elevated alkaline phosphatase value. If the GGT level is increased also, some type of liver disorder is strongly suspected. If the GGT level is normal while the alkaline phosphatase value is high, some other tissue (such as bone) must be involved in the disease state.

In patients on long-term medications (anticonvulsants in particular), we frequently see an increase in GGT levels over time. In this case, the medication has affected protein synthesis by the liver through stimulating the excess production of proteins, including the induction of GGT. These excess protein levels are not a manifestation of hepatic damage, but a response to the medication. Careful examination of the patient history must be made so that this factor might be included in the interpretation of data when appropriate.

GAMMA-GLUTAMYLTRANSFERASE LEVELS IN OTHER DISORDERS

Elevated serum levels of GGT have been reported in a number of different types of cancer. However, this enzyme does not appear to be useful as a diagnostic marker for cancer. The increases are more likely due to hepatic damage seen as a result of metastasis of the primary tumor. Marked increases in GGT serum levels are almost invariably seen in patients with acute pancreatitis, although the reason for this elevation is unclear. Some increases in chronic pancreatitis may be due to biliary tract obstruction or inflammation. In diabetics there is a correlation between the increase in triglyceride levels and an elevation of serum GGT. As treatment is initiated, both levels drop noticeably. The increase in GGT seen here may be related to induction of hepatic enzymes as a result of stimuli to produce greater amounts of triglycerides.

GAMMA-GLUTAMYLTRANSFERASE AND ALCOHOLISM

Perhaps the most useful application for GGT measurements is in the detection of alcoholism and the monitoring of alcohol consumption by these patients during treatment. Heavy long-term consumption of ethanol stimulates the synthesis and release of higher amounts of GGT from the liver in response to the enzyme-inducing properties of the alcohol. Individuals who abuse alcohol have obvious elevations of this enzyme. This information is valuable in confronting the person who denies they have a drinking problem and in following the course of therapy after such a problem has been identified.

Although the initial values for serum GGT in the alcoholic are high, the levels of the enzyme begin to decline after alcohol intake is discontinued. Periodic monitoring of serum GGT gives a ready indicator of the faithfulness of the patient in following the treatment program. Any heavy intake of alcohol results in another marked increase in the serum level and is a sign of relapse. Of course, if the alcohol consumption problem is serious enough, there may already be hepatic damage and GGT levels may remain increased. Proper monitoring by a physician is necessary to distinguish these two cases.

In contrast, the social drinker shows no elevation of serum GGT. Occasional consumption of moderate amounts of alcoholic beverages does not measurably affect the level of GGT in a given individual. What is required to produce the elevation is chronic and excessive intake of alcohol.

Isoenzymes of Gamma-Glutamyltransferase

Two and sometimes three bands of GGT activity have been seen after electrophoresis. To date, there has been little in the way of application of GGT isoenzyme studies to specific clinical situations. Some immunoassay studies indicate that the protein from serum and liver and the protein from kidney and urine were identical in each case.

Urine Gamma-Glutamyltransferase

Significant amounts of GGT have been observed in urine. This is not surprising since the kidney contains large concentrations of this enzyme due to its importance in reabsorption of amino acids. Some studies have demonstrated the possible clinical utility of measuring urine GGT, since the enzyme level increases markedly in renal disease. Patients treated with various aminoglycoside antibiotics have been shown to have increased urine GGT levels, which correlates fairly well with the degree of kidney damage produced by the antibiotic. Complications to the widespread use of urine GGT assays include the observation that there are inhibitors to the enzyme present in urine. These inhibitors must be completely removed before assay. The separation procedure itself sometimes results in a measurable loss of enzyme activity.

10.11

LACTATE DEHYDROGENASE

Reaction Catalyzed

Lactate dehydrogenase (LD; L-lactate:NAD oxidoreductase, EC 1.1.1.27) catalyzes the readily reversible reaction involving the oxidation of lactate to pyruvate with NAD serving as coenzyme:

$$CH_3-\overset{\overset{O}{\|}}{C}-\overset{\overset{O}{\|}}{C}-OH + NADH + H^+ \rightleftharpoons$$
$$\underset{\text{pyruvic acid}}{}$$

$$CH_3-\overset{\overset{OH}{|}}{CH}-\overset{\overset{O}{\|}}{C}-OH$$
$$\underset{\text{lactic acid}}{}$$

In the body, pyruvate is formed as an end product of the **Embden-Meyerhof glycolytic pathway,** which converts glucose to energy. Pyruvate is a key branch point in this process and can be further metabolized to a number of different products. When the oxygen level in the system is low, part of the pyruvate is converted to lactate.

Tissue Distribution

Lactate dehydrogenase is widely distributed in the tissues of the body. High concentrations are found in heart and liver. Significant amounts are present in erythrocytes; hemolysis produces an artefactual increase in the serum level of the enzyme. Skeletal muscle and kidney also contain considerable concentrations of LD. Overall, tissue concentrations are some 500-fold greater than the serum levels under normal circumstances. Even small amounts of tissue damage are reflected in measurable elevations of total lactate dehydrogenase activity.

Sample Collection

Serum is the sample of choice, since some anticoagulants (such as oxalate) inhibit the enzyme. Hemolysis must be avoided since the erythrocyte contains high concentrations of the enzyme. Freezing may distort the isoenzyme distribution markedly.

Measurement of Total Activity

GENERAL PRINCIPLES

The major assay approaches for the determination of lactate dehydrogenase in serum or other body fluids are based on the detection of NADH in the reaction. This material has a strong absorbance at 340 nm, which permits direct spectrophotometric measurement. The reduced form of nicotinamide adenine dinucleotide can also be detected by a coupled reaction with a tetrazolium salt. One other approach utilizes p-nitrophenylhydrazine to form a colored complex with pyruvate, but this method is imprecise and is not employed in many laboratories.

THE WACKER METHOD

The **Wacker method** for quantitation of LD activity utilizes the lactate → pyruvate reaction with formation of NADH from NAD. The 340-nm absorbance can be read directly, allowing kinetic assays to be performed. If a suitable spectrophotometer is not available, the reaction can be monitored with a colorimetric measurement. A mixture of phenazine methosulfate and nitroblue tetrazolium (or similar material) reacts with the NADH to produce a blue-purple color. The absorbance of this product is proportional to the amount of NADH present. This reaction is widely used for the detection of LD isoenzymes fractionated by electrophoresis.

THE WROBLEWSKI-LADUE METHOD

The **Wroblewski-LaDue method** employs the reverse reaction: pyruvate → lactate. In this situation NADH serves as a cosubstrate and is consumed during the course of the reaction. If kinetic measurement of activity is carried out, a decrease in 340-nm absorbance with time is observed. Although the pyruvate → lactate reaction is somewhat faster than the lactate → pyruvate process, the initial high absorbance of the NADH creates some difficulties in accurate measurement. Some problems have been observed with the presence of inhibitors in the NADH preparations commercially available.

SOME GENERAL ASSAY CONSIDERATIONS

Both methods must be run over somewhat short periods due to the presence of product inhibition. As lactate or pyruvate product accumulates, the reaction rate decreases. This is not a significant factor if kinetic assays are performed, since the rate curve can be monitored. The effect of temperature on the rate seems to be more pronounced for lactate dehydrogenase than for some other systems. Close control over temperature is very important in achieving accurate assays.

The selection of buffer type and pH requires some compromises. Since there are five isoenzymes of LD (to be discussed later), there is some variation in activity for a given serum sample, which depends on the isoenzyme concentrations. Several different buffers are employed in the many commercial kits and reagent packages for manual and automated systems. If the same buffer is not used for both automated and manual back-up methods, there are discrepancies in the results, and data gathered on a specific patient over time cannot be correlated well.

Clinical Significance: Total Activity

REFERENCE RANGES

Normal lactate dehydrogenase values depend on the assay technique employed. For the pyruvate → lactate method, the reference range is approximately 95–200 U/L at 30°C. If the assay is run in the opposite direction (lactate → pyruvate), the reference range is lower: 35–90 U/L. There is apparently no difference in the normal values for this enzyme between men and women and no significant age-related differences.

ELEVATIONS IN VARIOUS DISEASE STATES

Lactate dehydrogenase total activity is elevated in a wide variety of disease states. In most cases, much more useful information is obtained through the use of isoenzyme fractionation studies. Detection of increased LD activity is quite useful in the diagnosis and monitoring of myocardial infarction. Total activity is elevated in a number of situations associated with liver damage. A high percentage of patients with various cancers demonstrate increases in serum LD activity. Since the enzyme is present in high amounts in red blood cells, patients who demonstrate increased erythrocyte breakdown (due to megaloblastic anemia or deficiencies of vitamin B_{12} or folate) may have high serum values for total LD. Skeletal muscle involvement (either due to trauma or to diseases such as muscular dystrophy) produce increased amounts of LD in the serum. A significant number of patients with kidney damage also show elevations in LD activity.

Measurement of LD activity in CSF can sometimes give useful information about the clinical situation. Patients with bacterial meningitis frequently have elevated levels of LD activity. Increases in cerebrospinal fluid LD values are seen in patients with cerebral hemorrhage or thrombosis. On occasion, an elevated LD value in the CSF may be an indicator of a malignant neoplasm in the brain.

Although the data are rarely (if ever) used for diagnosis, increases in total LD activity are often seen in urine samples. Malignant neoplasms of the prostate, bladder, or kidney frequently produce elevations of urine LD activity. High values are also seen

Table 10–8
COMPOSITION OF LD ISOENZYMES

Polypeptides	Isoenzyme	Relative % in Serum
H · H · H · H	LD-1	18–33
H · H · H · M	LD-2	28–40
H · H · M · M	LD-3	18–30
H · M · M · M	LD-4	6–16
M · M · M · M	LD-5	2–13

in some cases of glomerulonephritis, nephrotic syndrome, and renal failure.

Isoenzymes of Lactate Dehydrogenase

DESCRIPTION

Normal human serum contains five different isoenzymes of lactate dehydrogenase. Each isoenzyme is composed of four protein subunits. The subunits are of two different structures: the H (heart) polypeptide and the M (muscle) polypeptide. These two protein chains combine in various ways to create the five isoenzymes normally detected (Table 10–8). In accordance with international nomenclature, the LD-1 migrates the farthest from the origin on electrophoresis, whereas the LD-5 has the slowest migration.

There is some tissue specificity for the various isoenzymes. LD-1 (and some LD-2) is found predominantly in heart muscle. Skeletal muscle and liver contain high amounts of LD-5. Although erythrocytes contain high concentrations of lactate dehydrogenase, there is some disagreement about the exact isoenzyme distribution; it is not clear whether LD-1 exists in greater concentration than LD-2. Kidney tissue seems to have a significant amount of LD-4 and LD-5, whereas the major LD component of lung is LD-3.

ASSAY

Electrophoresis

The primary approach to the analysis of LD isoenzymes is electrophoretic separation on agarose or cellulose acetate (Fig. 10–16). Buffer, pH, and other separation conditions are the same as those used for serum protein fractionation; a number of commercially available systems can be purchased which provide reliable results. After separation, the film is treated with substrate and other reagents to develop the specific bands of activity. The Wacker (lactate → pyruvate) reaction is used for this purpose. A solution containing buffer, lactate, and NAD is gently

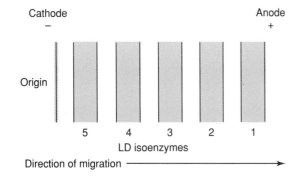

Cathode −

Anode +

Origin

5 4 3 2 1

LD isoenzymes

Direction of migration ⟶

A

5 4 3 2 1

B

FIGURE 10–16   Lactate dehydrogenase (LD) isoenzyme fractionation by electrophoresis. A. Separation of LD isoenzymes. B. Relative amounts of LD isoenzyme fractions.

rolled onto the surface of the film, which is then incubated at either 30°C or 37°C for a period (30 min or more). The NADH which forms as a result of the enzyme reaction is detected by treating the film with a solution of nitroblue tetrazolium and phenazine methosulfate. Whatever NADH is present reduces the nitroblue compound to form a colored band in each area of the film where lactate dehydrogenase activity is found. The film is then fixed and dried for scanning with a densitometer. The relative percentages of the LD isoenzyme fractions are determined and reported along with the total activity of the sample. Frequently, the actual electrophoresis strip itself is mounted to the report for later visual inspection.

The use of electrophoresis allows identification and quantitation of each of the five fractions. However, the procedure is rather time-consuming (a single run may require 2 h or more) and is somewhat complex. Depending on the apparatus and materials employed, eight or more samples may be separated on a single electrophoresis film. Some work has been done in automating the process, but this approach is not feasible unless the laboratory has to perform a very large daily volume of fractionations. In spite of its complexity, electrophoresis provides more useful information for the money than any of the other methods for separation and quantitation of LD isoenzymes.

Column Chromatography

Separation systems have been developed which use a column to isolate LD-1 and LD-2 along with the CK-MB fraction. The rationale for this approach is that the CK-MB fraction and (predominantly) the LD-1 fraction represent those isoenzymes which are elevated in the serum of patients with a myocardial infarct. If these isoenzymes can be isolated and measured, useful clinical information can be obtained without the need for electrophoretic fractionation of all the isoenzymes. A sample is applied to an ion-exchange column and unwanted fractions (CK-MM and LD-3, -4, and -5) are eluted with Tris buffer containing a low concentration of NaCl. The desired CK and LD fractions are then removed from the column by increasing the amount of sodium chloride in the buffer. After removal, the total activity for creatine kinase and lactate dehydrogenase in this second fraction are determined.

This minicolumn approach has some significant shortcomings. The sample volume required (often as much as 0.5 mL) is quite large compared with the 1 μL or so needed for electrophoresis. Our previous discussion of CK isoenzyme fractionation already outlined some of the difficulties in cleanly separating the various creatine kinase fractions. In using only the first two LD isoenzyme fractions, much useful information is lost regarding damage to other tissues, particularly lung and liver. Although the minicolumn method has been commercially developed, it is not widely used. The limitations described here, as well as the difficulty of efficiently processing large numbers of samples, severely limit the applications of this method.

Immunoinhibition

An interesting immunochemical approach to the assay of LD-1 by itself has been developed and is commercially available (Fig. 10–17). Antibody to the M subunit of lactate dehydrogenase is added to a serum sample. Any isoenzyme in the sample containing an M-subunit binds antibody to it, rendering that protein inactive. Since only LD-1 does not contain M subunits, this fraction should be the only one showing enzyme activity when the treated sample is assayed for total lactate dehydrogenase activity. By comparing the total activity with and without antibody present in the sample, the relative percentage of LD-1 can easily be determined. This approach lends itself well to automation, since the enzyme activity determination can be carried out using any standard methodology after the antibody has been added to the original sample.

The immunoinhibition method has been criticized because only LD-1 activity is assessed. There are situations in which the increase in total LD activity is due to elevation of some other LD isoen-

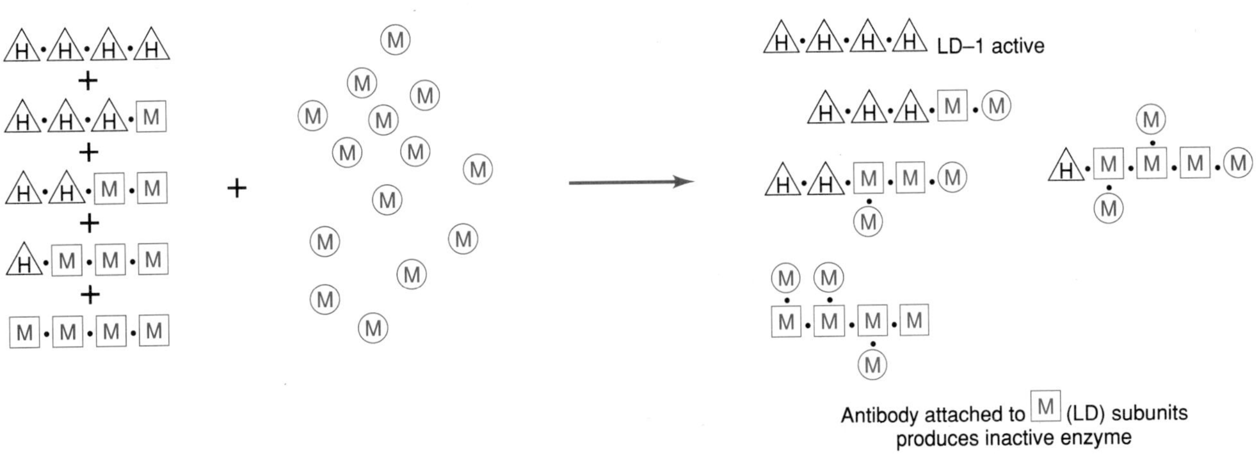

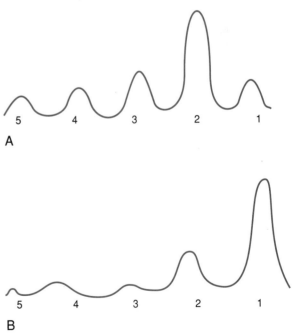

FIGURE 10-17   Immunoinhibition assay of LD-1 activity. ⚠ = heart subunit with no isoenzyme, Ⓜ̲ = muscle subunit, Ⓜ = antibody to the M subunit. Note that M attaches to Ⓜ̲ (LD subunits) to produce an inactive enzyme.

zyme; this method does not shed light on the cause of lactate dehydrogenase elevations in serum in those cases. However, the procedure is reasonably quick and the cost is in the same range as the electrophoretic procedure. Some data suggest that LD isoenzyme changes associated with a myocardial infarction can be detected several hours earlier with the immunoinhibition method than with measurement of the relative amounts of LD-1 and LD-2 by electrophoresis.

to indicate a reinfarction shortly after the initial event.

There has long been a sense that the inverted (LD-1 > LD-2) pattern was also seen in hemolysis and could not be considered a definite laboratory diagnostic sign of an infarct. More recent data suggest that the pattern in hemolyzed samples is never as pronounced as the shift in isoenzyme distribution seen in the patient with an MI. Further work may need to be done to clarify this issue.

CLINICAL UTILITY OF LACTATE DEHYDROGENASE ISOENZYMES

Use in Myocardial Infarction

The primary clinical role for determination of LD isoenzymes is in the case of a myocardial infarction. A characteristic pattern develops within 24 h after the onset of cardiac damage, which correlates well with other data to indicate a heart attack. In a normal LD isoenzyme pattern, LD-1 is present in a lower amount than LD-2 (the isoenzyme observed in highest concentration). Because of release of isoenzymes from cardiac tissue, a characteristic pattern develops in a heart attack in which LD-1 is the predominant isoenzyme detected. When the LD-1 level exceeds that of LD-2, the isoenzyme pattern is said to be **flipped** and the ratio of LD-1:LD-2 exceeds unity (Fig. 10–18). When this flipped pattern is seen in conjunction with a significant amount of CK-MB, there is better than 95% probability that the patient has experienced a myocardial infarct. The abnormal pattern persists for another day or two and then gradually reverts to normal. The changes in LD isoenzymes are not quick enough or sensitive enough

FIGURE 10–18   LD isoenzyme pattern in myocardial infarction. A. Normal serum LD isoenzyme pattern. B. LD isoenzyme pattern in myocardial infarction.

Use in Cancer Diagnosis

Total lactate dehydrogenase activity is markedly elevated in many patients with different types of cancer. Beyond this finding, there is no clear indication that cancer patients have a distinctive LD isoenzyme profile. Elevations of LD-1, LD-3, or LD-5 appear to be most common. An "isomorphic" pattern has often been described in which the total activity is increased, but the relative percentages of the five isoenzymes does not change noticeably. It is interesting that a number of researchers did not bother to characterize the isoenzyme patterns and only reported an elevation in total LD activity. More work needs to be carried out in this area to clarify a very confused situation.

Use in Other Disease States

Determination of LD isoenzymes can be of value in assessing the presence of either lung or liver damage. An increase in LD-3 is often seen in patients with pulmonary infarction (which is sometimes confused with a myocardial infarction). Lactate dehydrogenase-5 elevations are indicative of hepatic damage, although this approach does not give much indication about the specifics of the problem. Very frequently, an increase in LD-5 is seen as part of the developing pattern characteristic of a myocardial infarction. In this case the liver damage is secondary to that experienced by the heart and is usually a result of the changes in circulatory dynamics after the heart attack.

"Atypical" Lactate Dehydrogenase Isoenzymes

On occasion, six or more bands of LD activity have been reported in the serum of some patients. One patient had 13 different LD isoenzyme fractions present. These extra isoenzymes are frequently the result of changes in the primary structure of a subunit of the enzyme molecule. At present, no reproducible clinical abnormality can be linked to the appearance of extra bands in an isoenzyme pattern.

Complexes with Immunoglobulins

Apart from the presence of extra LD fractions arising from the differences in protein structure described above, an occasional patient demonstrates an atypical LD fraction as a result of a complex formed with proteins. Usually, the complex consists of an LD molecule loosely attached to an immunoglobulin. There appears to be little specificity; either IgA or IgG is found as part of the complex. The major effect of these complexes is on the circulating level of the enzyme. Since the immunoglobulin/enzyme complex is not broken down as rapidly in the bloodstream, high levels of total activity are observed over time. These high amounts of LD activity can provide misleading information, suggesting some kind of pathological process when there is no tissue damage. Electrophoresis allows detection of these enzyme/immunoglobulin complexes and helps clarify the situation. Although the complexes have been observed in a number of patients (and some normal individuals), there is no clear relationship at present between the appearance of the atypical enzyme form and a specific clinical problem.

10.12

LIPASE

Reaction Catalyzed

Pancreatic lipase (triacylglycerol acylhydrolase, EC 3.1.1.3) is an enzyme of fairly broad specificity which is responsible for triglyceride metabolism. The enzyme functions as an esterase, removing fatty acids in a sequential manner from the triglyceride molecule. The initial step in the reaction is:

$$\begin{array}{l} \overset{\displaystyle O}{\overset{\displaystyle \|}{}} \\ CH_2-O-C-R \\ | \\ CH-\text{fatty acid} \xrightarrow{\text{lipase}} \\ | \\ CH_2-\text{fatty acid} \end{array}$$

$$\begin{array}{l} CH_2-OH O \\ | \| \\ CH-\text{fatty acid} + R-C-OH \\ | \\ CH_2-\text{fatty acid} \end{array}$$

The fatty acid is represented schematically on the top carbon and not drawn out in detail on the second and third carbons. Removal of each fatty acid by lipase results in the formation of an alcohol on the glycerol skeleton and a free fatty acid.

The major source for lipase is the pancreas. When food enters the small intestine, it is mixed with a large number of materials released by the pancreas and gall bladder. Bile salts (synthesized by the liver and stored in the gall bladder) mix with the food and neutralize the acid present (HCl from the stomach). Then lipase breaks down the triglycerides into smaller components which can be absorbed across the small intestine. After these metabolic products cross the intestinal wall, they are then resynthesized into triglycerides and transported to cells where the fats are stored until needed.

Tissue Distribution

The highest concentrations of lipase are found in pancreatic tissue and secretions. There is a measurable amount of enzyme activity in gastrointestinal mucosa, erythrocytes, and leukocytes. Essentially no lipase activity can be found in muscle, liver, kidney, or lung.

Sample Collection

Serum is the preferred specimen for lipase assays. If the specimen is contaminated with bacteria, false elevations of lipase activity are seen.

Measurement of Total Activity

A variety of assay approaches for the measurement of serum lipase have been developed, but none have proven really satisfactory. The most common method involves incubating the serum sample with an olive oil substrate. Since the oil (the source of triglycerides) is not soluble in an aqueous medium, a turbid mixture is formed when the oil and buffer are mixed. Hydrolysis of the triglycerides present is initiated by the addition of the serum sample. As triglyceride breakdown proceeds, the turbidity decreases since the products of the reaction are much more water soluble than the substrate. Enzyme activity is determined by measuring the absorbance of the reaction mixture initially and at some fixed time after the reaction was started. The more rapid the decrease in the absorption of the mixture, the more enzyme activity present.

Other methods have been developed which lend themselves more to research situations than to routine clinical use. Some early approaches involved titrating with NaOH the amount of acids formed after a set period of hydrolysis. Several methods used copper salts to determine the amount of acid released. The copper ion forms a complex with the fatty acid generated by hydrolysis. This fatty acid/copper complex dissolves in water, and the unhydrolyzed substrate does not. The products are removed from the substrate by water extraction of the contents of the assay tube. Colorimetric measurement of copper indicates the amount of fatty acids present and thus shows the level of enzyme activity.

Clinical Significance: Total Activity

The variability in lipase methodology results in a lack of standardization for the enzyme assay. Reference ranges are extremely method dependent. The primary diagnostic role for lipase is in the assessment of acute pancreatitis. After the initial attack, the lipase levels in serum gradually rise over a period of 2 to 4 days. This increase in serum enzyme activity frequently parallels the changes seen with amylase. More recently, we have learned that lipase levels in many patients with pancreatitis change almost as rapidly as amylase concentrations. However, lipase remains elevated in the circulation for a much longer time. Elevated values may be seen as long as 2 weeks after the attack of pancreatitis.

If chronic pancreatitis exists, lipase determinations are of little value. The loss of pancreatic tissue results in decrease in the ability to produce and release enzyme into the bloodstream. In this situation, lipase levels in the normal range are expected.

Serum lipase may be increased in some disease states. Approximately 10% of patients with mumps demonstrate somewhat elevated levels of lipase. There may be slight increases of serum lipase in the early stages of pancreatic cancer, although the levels quickly decline to normal (or below normal). Drugs which induce spasm of the sphincter of Oddi (particularly opiates or codeine) cause a transient increase in serum lipase. The enzyme is frequently elevated in acute renal failure. Some 40% or more of patients with bone fractures may have increases in serum lipase values.

Role of Colipase

The lipase molecule possesses some complexities of structure which are not readily apparent. In the early 1960s, chromatography studies showed there were actually two proteins involved with this enzyme: the lipase protein itself and another protein now referred to as **colipase.** In the intact system, there is a 1:1 ratio between these two molecules. If the lipase and colipase are separated from one another, no enzyme activity is seen. Once the two proteins are remixed, full enzyme activity is recovered.

The role of colipase is still somewhat unclear, but the molecule apparently provides some protection for lipase against inactivation. The lipase enzyme portion is water-soluble and stable in aqueous solution. However, the triglycerides are not soluble in water and require interaction with bile salts to be metabolized. When the lipase molecule comes in contact with this nonaqueous environment, it begins to denature and lose activity. If colipase is present, lipase is much less likely to denature and is therefore much more stable. Apparently, the colipase binds to lipase in such a way as to change the conformation and somehow stabilize the enzyme against breakdown.

At present, no clinical condition is known which

is related to a deficiency of colipase or to an abnormal structure for this protein. Understanding this interaction may increase our knowledge of the function of lipase and perhaps allow for improved methods of analysis.

10.13

ENZYME PROFILES

General Concepts

Each enzyme we have discussed so far has a certain tissue specificity. The enzyme is concentrated more in some tissues than others. Therefore, damage to those tissues results in the release of enzyme into the circulation. When we see this increase in serum level of an enzyme, we interpret the change as the result of specific tissue necrosis.

This concept, however, is most reliable when the enzyme is specific for only one tissue. The enzymes employed as part of the clinical laboratory diagnosis are not tissue-specific. Each enzyme can become increased in serum or other body fluid as a result of tissue damage in more than one organ. To resolve this seeming dilemma, we resort to more data than simply the level of a given enzyme in the bloodstream. One approach has been to study the isoenzyme distribution; this technique is particularly effective when exploring the sources of creatine kinase and lactate dehydrogenase increases. Another type of data analysis involves the concept of the enzyme organ profile.

By studying the amount of several enzymes in a body fluid, we often gain valuable information about which specific organ has experienced damage even when a single enzyme could be released by several tissues. The enzyme profile is especially useful in situations where analysis of isoenzymes of a specific

enzyme is not technically feasible. By utilizing our knowledge of the varying tissue specificities of a number of enzymes, we can gain much useful information about the patient diagnosis and more easily monitor treatment progress.

Cardiac Profile

Perhaps the most useful combination of enzyme data is the cardiac profile. The combination of total CK and LD activities monitored over a period of several days, along with determination of isoenzymes for the two proteins, provides a rapid, relatively inexpensive, and extremely reliable means of detecting and following a myocardial infarct.

The initial data are usually furnished by the CK analyses, with more long-term information supplied by way of LD measurements. Samples for total activity and isoenzyme studies for both enzymes are collected upon admission to the hospital and at timed intervals after admission. Usually the CK samples are collected at intervals of approximately 8 h, and samples for LD measurements are obtained at 24-h intervals. The pattern of total activity and isoenzyme measurements allows for a reliable diagnosis of a myocardial infarction and its extension, if such occurs.

In a large majority of cases, the activity of both enzymes on admission is within normal limits. Stat measurement of total activity and isoenzyme fractionation is not warranted since there is such a low yield of useful information at this point. The CK level rises rapidly after an MI, with peak activities appearing within 24 h after the event (Fig. 10–19). Creatine kinase is seen within a few hours after the attack and peaks somewhere within the same time frame as the peak of total CK activity. Changes in LD levels are somewhat slower, with a peak in total activity not appearing for 1 to 2 days after the initial attack. However, the LD-1:LD-2 flip may appear within 24 h, confirming the presence of cardiac damage. If the patient has delayed seeking medical attention, the CK levels may return to normal by the time sampling is initiated. In this case, the LD total activity and isoenzyme patterns provide useful information about the diagnosis. Although the LD values themselves are not sensitive indicators of an extension (or reinfarction), the CK levels again increase and the CK-MB fraction reappear if a second MI occurs within a few days after the first one.

Hepatic Profile

A variety of enzymes have been measured in serum over the years as we attempt to find a useful

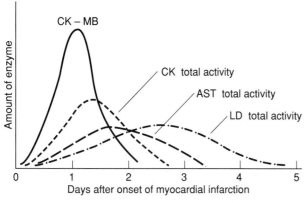

FIGURE 10–19 Cardiac enzyme pattern after myocardial infarction.

set of markers to detect and differentiate the manifold problems of liver disease. Although many enzymes have been investigated, few have been chosen as reliable indicators of hepatic damage. Some enzymes have low reliability because of lack of specificity. Lactate dehydrogenase is elevated in liver disease, but also in a variety of other disorders. Isoenzyme studies allow some differentiation, but prove to be more expensive than other approaches. Aspartate aminotransferase is usually increased in liver disease, but also in cases of cardiac damage and some skeletal muscle problems. Alkaline phosphatase levels are high in patients with bone disease as well as in those with hepatic disorders.

In spite of the difficulties associated with non-specificity of any given enzyme, the liver profile is still a very useful approach to assessing hepatic disorders. Increases in AST, ALT, and alkaline phosphatase (particularly if these results are coupled with elevations in GGT) usually point to some type of liver disorder. From this point, however, further differentiation proves difficult. Obstructive processes can often be distinguished from other types of hepatic disorders, but the data are not yet clear enough to make many useful generalizations.

Pancreatic Profile

The two enzyme studies which appear to be a great value in assessing pancreatic damage in the case of acute pancreatitis are those of amylase and lipase. Serum amylase levels rise rapidly after an attack, showing significant elevations within 23 h or less. The decline is also fairly rapid, which allows detection of any recurrent problems. Although serum lipase levels are less frequently requested (in part due to the length of time needed to perform the assay), this data is at least as reliable as amylase measurements in assessing the presence of damage to the pancreas. It has long been thought that lipase values rose much more slowly than amylase levels, but data indicate that lipase in serum may also rise within 12 to 24 h after the onset of acute pancreatitis.

The use of the serum and urine amylase values allows for the calculation of the amylase/creatinine clearance ratio. In many instances, this information permits a more refined assessment of the situation. Since amylase is an enzyme with a low molecular weight, it can clear from the bloodstream fairly quickly through excretion into the urine. Frequently, the serum level of this enzyme is lower than expected, but the clearance ratio is elevated, giving a better picture of the actual clinical situation. Determination of amylase isoenzymes, either by column chromatography or by isoelectric focusing, provides the most reliable data but must be considered a research approach at this time. Future improvements in isoenzyme methodology should bring a wider application for amylase fractionation in the detection of acute pancreatitis.

SUMMARY

Enzymes are formed within cells and released into the circulation during cell destruction. In the bloodstream, the enzymes undergo metabolism by proteases. Some low-molecular-weight enzymes may be excreted by the kidney. When cell destruction is the result of disease, higher amounts of certain enzymes enter the bloodstream. The change in circulating enzyme level is often a reflection of damage to a particular tissue. These changes can provide valuable clues to specific clinical situations. Evaluation of the isoenzyme distribution also furnishes useful information about the tissue source.

Acid phosphatase catalyzes the hydrolysis of a variety of organic phosphate derivatives at pH 5-6. The enzyme is found in high concentrations in erythrocytes, platelets, and prostate tissue. The usual method of assay involves the use of p-nitrophenylphosphate as substrate, measuring the formation of p-nitrophenol. The major clinical utility for acid phosphatase assays is in the detection of prostatic cancer. Specific immunochemical assay for the prostatic isoenzyme often provides additional valuable clinical information.

Alkaline phosphatase is responsible for the hydrolysis of organic esters under basic conditions. This enzyme is found in a variety of tissues. The substrate p-nitrophenylphosphate is used for the kinetic assay, measuring the rate of formation of p-nitrophenol at pH 10. Total alkaline phosphatase activity is elevated in diseases related to liver or bone damage, and in some cancer states. Isoenzyme fractionation by electrophoresis furnishes additional information about the tissue source for elevated total alkaline phosphatase activity.

The aminotransferases, ALT and AST, catalyze the transfer of an $-NH_2$ group from an amino acid to a keto acid. Both enzymes are found in liver and skeletal muscle. Total activity is determined with either colorimetric assay of the keto acid or a coupled reaction with a dehydrogenase. The aminotransferases can provide useful information in the case of hepatic damage. Isoenzyme studies do not furnish any additional data of value.

Starch is hydrolyzed by the enzyme amylase, found predominantly in the pancreas and salivary glands. Assay procedures measure the glucose formed during starch breakdown or detect starch fragments labeled with a chromogenic material. Elevated levels are useful in the diagnosis of acute pancreatitis. Measurement of urine amylase and creatinine and determination of the amylase/creatinine ratio often provide more sensitive detection of this disease state.

Cholinesterase is responsible for the hydrolysis of acetylcholine and related esters. True cholinesterase is specific for acetylcholine; pseudocholinesterase attacks a wide variety of substrates. The "pseudo" enzyme is synthesized by the liver and found in high concentration in the circulation, whereas acetylcholinesterase is found in erythrocytes. Measurement of the decrease in serum cholinesterase levels is indicative of possible exposure to various pesticides.

Creatine kinase promotes the interconversion of creatine and ATP to creatine phosphate and ADP, furnishing biochemical energy for muscle contraction. The amount of ADP or ATP formed is measured by the two major assay methods by coupling the CK reaction with subsequent enzyme steps. Total CK activity is elevated after damage to skeletal or cardiac muscle. Isoenzyme fractionation (by electrophoresis, column chromatography, or immunochemical approaches) is of great value in detecting the increased CK-MB fraction after a myocardial infarction.

Gamma-glutamyltransferase catalyzes the transfer of a gamma-glutamyl group, important in some aspects of protein metabolism. This enzyme is predominantly found in liver. Assay involves use of synthetic substrate and acceptor molecules, with a nitroaniline derivative being formed by the hydrolytic reaction. Elevations of GGT can be useful in identifying liver damage. Measurement of this enzyme is particularly helpful in assessing the presence of alcohol abuse.

Lactate dehydrogenase promotes the interconversion of pyruvic and lactic acids. The enzyme is found in a wide variety of tissues. Assay usually measures NADH formation or disappearance. Although total LD activity is elevated in a number of different disease states, isoenzyme fractionation by electrophoresis can provide additional data about the specific site of tissue damage. Since LD-1 is usually increased after a myocardial infarct, measurement of this fraction by column chromatography or immunochemical methods indicates the amount of the isoenzyme present.

Lipase catalyzes the hydrolysis of triglycerides and is produced predominantly by the pancreas. Assays for this enzyme usually measure the disappearance of turbidity as triglycerides are cleaved to glycerol and fatty acids. Lipase values in blood are increased during the course of acute pancreatitis.

A number of combinations of enzyme determinations (enzyme profiles) often prove helpful in ascertaining specific tissue damage. Both CK and LD total activity increase after a myocardial infarction, with CK rising sharply a few hours after the event. Liver damage is reflected by increases in AST, ALT, and alkaline phosphatase total activity. Increases in amylase and lipase strongly suggest acute pancreatitis.

FOR REVIEW

Directions: For each question, choose the best response.

1. The majority of enzymes detected in the plasma function primarily
   A. in the plasma
   B. in cerebrospinal fluid
   C. in whole blood
   D. within cells

2. Which of the following may be associated with kidney excretion of enzymes?
   A. enzymes with a molecular weight less than 60,000 filtered and excreted by the kidney
   B. amylase is filtered and excreted by the kidney
   C. all enzymes are filtered and excreted by the kidney
   D. both A and B
   E. both B and C

3. Which of the following are considered primary tissue sources of alkaline phosphatase?
   A. liver, bone, kidney
   B. liver, heart, kidney
   C. kidney, pancreas, heart
   D. prostate, liver, bone

4. Excess amounts of acid phosphatase activity in the serum is associated with
   A. cirrhosis
   B. hepatitis
   C. myocardial infarction
   D. prostatic cancer

5. Which of the following is considered a primary tissue source of amylase?
   A. bone
   B. heart
   C. liver
   D. pancreas

6. Which of the following enzymes exhibits the least tissue specificity?
   A. acid phosphatase
   B. aspartate aminotransferase
   C. creatine kinase
   D. lactate dehydrogenase

7. Which of the following parameters must be carefully monitored to preserve the integrity of an acid phosphatase specimen?
   A. maintain pH at 6.2–6.6
   B. avoid hemolysis of the specimen
   C. draw blood in citrate
   D. both A and B
   E. A, B, and C

8. Due to its high concentration in seminal fluid, the detection of which enzyme is useful in establishing recent sexual intercourse in cases of alleged rape?

   A. acid phosphatase
   B. alkaline phosphatase
   C. amylase
   D. lipase

9. The enzyme that acts on several phosphate ester substrates and exhibits low tissue specificity is
   A. acid phosphatase
   B. alkaline phosphatase
   C. creatine kinase
   D. lactate dehydrogenase

10. In which tissue is alanine aminotransferase found in the greatest concentration?
   A. heart
   B. liver
   C. lung
   D. skeletal muscle

11. In several colorimetric assays, the quantitation of enzyme activity is based on the reaction of dinitrophenylhydrazine with a(an)
   A. amine
   B. ester
   C. keto acid
   D. phosphate group

12. Which of the following functions as a coenzyme in transaminase reactions?
   A. $NAD^+/NADH$
   B. $NADP^+/NADPH$
   C. pyridoxal phosphate
   D. both A and B
   E. both A and C

13. Following a myocardial infarction, the enzyme whose serum activity generally remains within a normal level is
   A. alanine aminotransferase
   B. aspartate aminotransferase
   C. creatine kinase
   D. lactate dehydrogenase

14. The amylase method that measures the disappearance of starch substrate is
   A. amyloclastic
   B. chromogenic
   C. saccharogenic
   D. both A and C
   E. both B and C

15. Which enzymes are generally quantitated to help diagnose acute pancreatitis?
   A. alanine aminotransferase and creatine kinase
   B. alkaline phosphatase and lactate dehydrogenase
   C. amylase and lipase
   D. alkaline phosphatase and cholinesterase

16. In pesticide poisoning, a decrease in enzyme activity is associated with
   A. alkaline phosphatase
   B. cholinesterase
   C. creatine kinase
   D. lactate dehydrogenase

17. The appearance of the LD-1/LD-2 flipped pattern and the presence of the CK-MB isoenzyme on electrophoresis is highly suggestive of
    A. acute pancreatitis
    B. cirrhosis
    C. muscular dystrophy
    D. myocardial infarction

18. In an adult, a normal serum gamma-glutamyltransferase level with an elevated serum alkaline phosphatase level may be suggestive of
    A. bone disease
    B. heart disease
    C. liver disease
    D. skeletal muscle disease

19. Which lactate dehydrogenase isoenzyme migrates the farthest from the origin on electrophoresis?
    A. LD-1
    B. LD-2
    C. LD-3
    D. LD-4
    E. LD-5

20. Which lactate dehydrogenase isoenzyme is found primarily in skeletal muscle and liver?
    A. LD-1
    B. LD-2
    C. LD-3
    D. LD-4
    E. LD-5

21. As part of a hepatic profile, which of the following enzymes would be useful to quantitate?
    A. alanine aminotransferase
    B. alkaline phosphatase
    C. gamma-glutamyltransferase
    D. both A and B
    E. A, B, and C

22. Which of the following may be associated with creatine kinase?
    A. catalyzes the conversion of creatine and ATP to creatine phosphate and ADP
    B. total CK activity elevated in skeletal muscle disease
    C. the subunits of CK can combine to form six different CK isoenzymes
    D. both A and B
    E. A, B, and C

23. Which of the following may be associated with the creatine kinase isoenzymes?
    A. CK-1 referred to as CK-MM
    B. CK-3 remains close to the origin during electrophoresis
    C. CK-MM is found in skeletal muscle
    D. both B and C
    E. A, B, and C

24. Which of the following may be associated with macro-CK?
    A. complex between CK-BB and IgG
    B. complex between CK-MM and IgA
    C. migrates between the MM and MB fractions on electrophoresis
    D. both A and C
    E. A, B, and C

BIBLIOGRAPHY

Abbott, L. B., and Van Lente, F., "Methods for CK-MB analysis used to monitor cardiac patients," *J. Clin. Immunoassay.* 8:147–151, 1985.

Arvanitakis, C., and Cooke, A. R., "Diagnostic tests of exocrine pancreatic function and disease," *Gastroenterol.* 74:932–948, 1978.

Bais, R., and Edwards, J. B., "Creatine kinase," *CRC Crit. Rev. Clin. Lab. Sci.* 16:291–335, 1982.

Ban, Y., and Murphy, G. P., "Antigen marker assays in prostate cancer," *Lab. Manage.* April:19–24, 1983.

Bates, H. M., "Neural-tube defects can be identified by measuring amniotic-fluid acetylcholinesterase," *Lab. Manage.* August: 12–13, 1979.

Bates, H. M., "GGTP and alcoholism: A sober look," *Lab. Manage.* March: 17–19, 1981.

Bates, H. M., "Macroamylasemia," *Lab. Manage.* June: 33–37, 1981.

Batsakis, J. G., "Serum alkaline phosphatase. Refining an old test for the future," *Diag. Med.* May/June: 25–33, 1982.

Bauman, D. J., "Creatine phosphokinase isoenzymes and the diagnosis of myocardial infarction," *Postgrad. Med.* 67:103–116, 1980.

Bergmeyer, H. U., et al., "Provisional recommendations on IFCC methods for the measurement of catalytic concentrations of enzymes. Part 2. IFCC method for aspartate aminotransferase," *Clin. Chem.* 23:887–899, 1977.

Bowers, Jr., G. N., and McComb, R. B., "A unifying reference system for clinical enzymology: Aspartate aminotransferase and the international clinical enzyme scale," *Clin. Chem.* 30:1128–1136, 1984.

Briere, R. O., "Alkaline phosphatase isoenzymes," *CRC Crit. Rev. Clin. Lab. Sci.* 10:1–30, 1979.

Brown, S. S., et al., "The plasma cholinesterases: A new perspective," *Adv. Clin. Chem.* 22:2–123, 1981.

Burke, M. Desmond, "Hepatic function testing," *Postgrad. Med.* 64:177–185, 1978.

Campbell, J., and Sanderson, J. A., *Aminotransferases and PLP Activation.* Indianapolis: Dow Chemical Company, 1979.

Choe, B.-K., "New dimensions for prostatic acid phosphatase," *Diag. Med.* May/June: 81–91, 1982.

Committee on Enzymes, Scandinavian Society for Clinical Chemistry and Clinical Physiology, "Recommended method for the determination of gammna-glutamyltransferase in blood," *Scand. J. Clin. Lab. Invest.* 36:120–125, 1976.

Crofton, P. M., "Biochemistry of alkaline phosphatase isoenzymes," *CRC Crit. Rev. Clin. Lab. Sci.* 16:161–194, 1982.

Das, P. K., "On genetically determined human serum cholinesterases," *Enzyme* 21:253–274, 1976.

Demers, L. M., "Prostatic acid phosphatase: Assay comparisons," *Lab. Manage.* August: 45–51, 1981.

Dietz, A. A., et al., "Colorimetric determination of serum cholinesterase and its genetic variants by the propionyl-thiocholine-dithiobis(nitrobenzoic acid) procedure," *Clin. Chem.* 19:1309–1313, 1973.

Dietz, A. A. et al., "Inherited low ChE plays role in apnea," *Clin. Chem. News.* October, 1980.

Fishman, W. H., "Perspectives on alkaline phosphatase isoenzymes," *Amer. J. Med.* 56:617–650, 1974.

Foreback, C. C., and Chu, J.-W., "Creatine kinase isoenzymes: Electrophoretic and quantitative measurements," *CRC Crit. Rev. Clin. Lab. Sci.* 15:187–230, 1981.

Galen, R. S., "How far can immunoassay of enzymes take us?" *Diag. Med.* September/October: 93–95, 1980.

Gawoski, J., "Creatine kinase and lactate dehydrogenase isoenzymes," *Diag. Clin. Test.* August:40–42, 1988.

Gittes, R. F., "Serum acid phosphatase and screening for carcinoma of the prostate," *New Eng. J. Med.* 309:852–853, 1983.

Goetz, W., "Recent advances in gamma-GT determination," *Amer. Clin. Prod. Rev.* March/April:32–34, 1983.

Goldberg, D. M., "Structural, functional, and clinical aspects of gamma-glutamyltransferase," *CRC Crit. Rev. Clin. Lab. Sci.* 12:1–58, 1980.

Gowenlock, A. H., "Tests of exocrine pancreatic function," *Ann. Clin. Biochem.* 14:61–89, 1977.

Helzberg, J. H., et al., "LFTs test more than the liver," *J. Amer. Med. Assoc.* 256:3006–3007, 1986.

Henderson, A. R., and Nealon, D. A., "Enzyme measurements by mass: An interim review of the clinical efficacy of some mass measurements of prostatic acid phosphatase and the isoenzymes of creatine kinase," *Clin. Chim. Acta.* 115:9–32, 1981.

Kane, K. K., "A review of enzyme catabolism," *Ann. Clin. Lab. Sci.* 5:318–324, 1977.

Kiyasu, J. Y., "Current status of detecting CK-MB for patient management," *Amer. Clin. Prod. Rev.* October: 29–31, 1985.

Kutty, K. M., "Biological function of cholinesterase," *Clin. Biochem.* 13:239–243, 1980.

Kwong, T. C., and Arvan, D. A., "How many creatine kinase "isoenzymes" are there and what is their clinical significance?" *Clin. Chim. Acta.* 115:3–8, 1981.

LaMotta, R. V., and Woronick, C. L., "Molecular heterogeneity of human serum cholinesterases," *Clin. Chem.* 17:135–144, 1971.

Lang, H., and Wuerzburg, U., "Creatine kinase, an enzyme of many forms," *Clin Chem.* 28:1439–1447, 1982.

Lott, J. A., "Practical problems in clinical enzymology," *CRC Crit. Rev. Clin. Lab. Sci.* 8:277–301, 1977.

Lum, G., "Serum alkaline phosphatase: Sources of increased activity," *Lab. Manage.* May:55–66, 1985.

MacQueen, J., and Plaut, D., "A review of clinical applications and methods for cholinesterase," *Amer. J. Med. Technol.* 39:279–288, 1973.

Mahan, D. E., "Immunologic methods for the quantitation of prostatic acid phosphatase," *J. Clin. Immunoassay.* 6:221–227, 1983.

Markel, S. F., "Clinical enzymology in cancer," *CRC Crit. Rev. Clin. Lab. Sci.* 9:85–104, 1978.

Meyer-Sabellek, W., "Alkaline phosphatase. Laboratory and clinical implications," *J. Chromatog.* 429:419–444, 1988.

Moossa, A. R., "Diagnostic tests and procedures in acute pancreatitis," *New Eng. J. Med.* 311:639–643, 1984.

Moss, D. W., "Alkaline phosphatase isoenzymes," *Clin. Chem.* 28:2007–2016, 1982.

Moss, D. W., "Diagnostic aspects of alkaline phosphatase and its isoenzymes," *Clin. Biochem.* 20:225–230, 1987.

Nemesanszky, E., and Lott, J. A., "Gamma-glutamyltransferase and its isoenzymes: Progress and problems," *Clin. Chem.* 31:797–803, 1985.

Patton, J. S., and Carey, M. C., "Watching fat digestion," *Science.* 204:145–148, 1979.

Pesce, M. A., "The CK isoenzymes: Findings and their meaning," *Lab. Manage.* October: 25–37, 1982.

Pesce, M. A., "The CK and LD macroenzymes," *Lab. Manage.* November, 1984.

Posen, S., "Turnover of circulating enzymes," *Clin. Chem.* 16:71–84, 1970.

Posen, S., and Doherty, E., "The measurement of serum alkaline phosphatase in clinical medicine," *Adv. Clin. Chem.* 22:165–245, 1981.

Ransom, J. H. C., "Etiological and prognostic factors in human acute pancreatitis: A review," *Amer. J. Gastroenterol.* 77:633–638, 1982.

Riose, N. R., et al., "Laboratory diagnosis of prostatic cancer," *Lab. Manage.* November, 1978.

Roberts, R., "Where oh where has the MB gone?" *New Eng. J. Med.* 313:1081–1083, 1085.

Roe, C. R., "Diagnosis of myocardial infarction by serum isoenzyme analysis," *Ann. Clin. Lab. Sci.* 7:201–209, 1977.

Romas, N. A., et al., "Acid phosphatase: New developments," *Human Pathol.* 10:501–512, 1979.

Rosalki, S. B., "Gamma-glutamyltranspeptidase," *Adv. Clin. Chem.* 17:53–107, 1975.

Rosoff, Jr., L., and Rosoff, Sr., L., "Biochemical tests for hepatobiliary disease," *Surg. Clin. N. Amer.* 57:257–273, 1977.

Salt II, W. B., and Schenker, S., "Amylase—Its clinical significane: A review of the literature," *Medicine.* 55:269–289, 1976.

Schwartz, M. K., "Enzymes in cancer," *Clin. Chem.* 19:10–22, 1973.

Schwartz, M. K., "Enzymes in Cancer," *J. Clin. Immunoassay.* 6:247–252, 1983.

Shah, V., "Laboratory tests for isoforms and isoenzymes of creatine kinase," *Amer. Clin. Lab.* January/February: 28–32, 1990.

Shaw, L. M., "The GGT assay in chronic alcohol consumption," *Lab. Manage.* May: 56–63, 1982.

Shaw, L. M., "Keeping pace with a popular enzyme: GGT," *Diag. Med.* May/June: 59–78, 1982.

Silk, E. et al., "Assay of cholinesterase in clinical chemistry," *Ann. Clin. Biochem.* 16:57–75, 1979.

Solomon, A. R., "The value of the amylase/creatinine clearance ratio in the diagnosis of acute pancreatitis," *CRC Crit. Rev. Clin. Lab. Sci.* 9:367–380, 1978.

Thompson, L. S., "Acute pancreatitis," *Clin. Chem. News.* February, 1983.

Tomashefsky, P., et al., "Prostatic acid phosphatase: Methods and their clinical utility," *Lab Manage.* May: 33–38, 1980.

Toskes, P. P., and Greenberger, N. J., "Acute and chronic pancreatitis," *Disease-a-Month.* 29:5–81, 1983.

Van Lente, F., "Diagnosing acute pancreatitis the enzymatic way," *Diag. Med.* July/August: 50–56, 1982.

Van Lente, F., and Shamberger, R. J., "Defining a role for enzymes in cancer diagnosis," *Diag. Med.* June: 47–56, 1983.

Warshaw, A. L., "The kidney and changes in amylase clearance," *Gastroenterol.* 71:702–704, 1976.

Watson, R. A., and Tang, B., "The predictive value of prostatic acid phosphatase as a screening test for prostatic cancer," *New Eng. J. Med.* 303:497–499, 1980.

Whittaker, M., "Plasma cholinesterase variants and the anaesthetist," *Anaesthesia.* 35:174–197, 1980.

Williamson, T. "The estimation of pancreatic lipase—A brief review," *Med. Lab. Sci.* 33:265–279, 1976.

Wills, J. H., "Blood cholinesterase: Assay methods and considerations," *Lab. Manage.* April: 53–63, 1982.

Witkin, G. B. et al., "The clinical laboratory evaluation of hepatic function," *Bull. Lab. Med.* issue 94, 1986.

Wu, A. H. B., "CK-MB assay methods: A comparison," *Lab. Manage.* January: 44–51, 1985.

Yam, L. T., "Clinical significance of the human acid phosphatases: A review," *Amer. J. Med.* 56:604–616, 1974.

Chapter Eleven

LIVER AND KIDNEY FUNCTION

UPON COMPLETION OF THIS CHAPTER, THE STUDENT WILL BE ABLE TO

1. List examples of the synthetic, metabolic, detoxification, and excretory functions of the liver.
2. Describe how bilirubin is formed from the breakdown of hemoglobin.
3. Explain why different forms of bilirubin may be soluble or insoluble in water.
4. Name the three bilirubin fractions which together make up the total bilirubin concentration.
5. State the specimens of choice and the proper handling of specimens for bilirubin analysis and ammonia analysis.
6. Explain the chemical/physiological relationship that the following forms of bilirubin have to each other:
   A. direct
   B. indirect
   C. conjugated
   D. unconjugated
   E. delta
7. State the principles of the Evelyn-Malloy and Jendrassik-Grof bilirubin methods.
8. Describe how hemoglobin interferes with some bilirubin methods.
9. List three general causes of an elevated serum bilirubin level.
10. Name two sources that contribute to the presence of ammonia in the blood.
11. Discuss the clinical significance associated with detecting abnormal levels of the following in the blood:
    A. ammonia
    B. bilirubin
    C. creatinine
    D. urea
    E. uric acid
12. State the principles of the cation exchange resin and enzymatic methods for quantitating plasma ammonia levels.
13. List the five serum enzymes that are frequently elevated in liver disease.
14. List four types of hepatitis.
15. Describe the laboratory findings that are seen in the following conditions:
    A. hepatitis
    B. alcoholic liver disease
    C. Reye's syndrome
    D. primary biliary cirrhosis
    E. pregnancy
16. Describe the filtration and reabsorption processes of the kidney.
17. List examples of the synthetic, metabolic, and excretory functions of the kidney.
18. Calculate a creatinine clearance (corrected for body surface) when given the required information.
19. State the purpose for measuring creatinine clearance and its reference range.
20. Explain the biochemical formation of creatinine, urea, and uric acid.
21. State the principles of the Jaffe (end-point and kinetic) and the enzymatic methods for measuring creatinine.

22. State the principles of the colorimetric and enzymatic methods for quantitating urea and uric acid.
23. State the reference ranges for the following:
    A. serum creatinine
    B. serum urea
    C. serum uric acid
    D. plasma ammonia
24. Describe the laboratory findings seen in the following conditions:
    A. nephrotic syndrome
    B. glomerulonephritis
    C. acute renal failure

INTRODUCTION

The complex processes of biosynthesis, metabolism, intake, and excretion all work together to assist in sustaining life. The body is a highly complex organism capable of carrying out thousands of processes simultaneously. These reactions do not occur in isolation, but interact with one another and affect one another in a highly complicated fashion. Two of the major organs involved in this biochemistry are the liver and the kidney. Both tissues can perform a wide variety of biochemical reactions and also play major roles in the uptake and excretion of materials. Many clinical situations alter the functions of one or both of these organs. A basic knowledge of liver and kidney biochemistry and physiology are vital to the study of clinical chemistry.

11.1

OVERVIEW OF BIOSYNTHETIC AND METABOLIC PROCESSES

The liver and kidney have similar, but different, capabilities for biosynthesis, metabolism, uptake, and excretion. The liver is a major synthetic and metabolic organ of the body, concerned with the conversion of carbohydrates, lipids, amino acids,

and proteins (Table 11-1). Although the kidney's scope of synthetic function is not as broad as the liver's, synthesis of some hormones and a few key biochemical interconversions do take place in the kidney. Excretion is an important, but relatively minor, role for the liver. The kidney, on the other hand, serves as perhaps the major tissue for excretion and reabsorption of a wide variety of materials. Maintenance of proper water and electrolyte balance is a major function of the kidney.

11.2

ANATOMY OF THE LIVER

The liver is located in the abdomen under the diaphragm. A hepatic duct connects this organ to the gallbladder. Blood is supplied from the portal vein (bringing absorbed material from the gastrointestinal tract) and the hepatic artery, providing oxygen needed for many of the metabolic processes carried out by the liver. Other blood vessels carry metabolites and waste materials away from the liver.

11.3

ROLE OF LIVER IN BIOCHEMICAL PROCESSES

Synthetic Functions

The liver is the chief source of several important biochemicals, including cholesterol and bile salts. Using materials formed from carbohydrate and fatty acid metabolism, this tissue synthesizes most of the cholesterol found in the body. The cholesterol can then serve as the starting material for all the steroid hormones (produced in various tissues). Bile salts are also a major synthetic product of the liver. These cholesterol derivatives dissolve in both aqueous and lipid materials. After their formation in the liver, the bile salts are transported to the gallbladder and then to the small intestine where they play an im-

Table 11-1
SUMMARY OF LIVER FUNCTIONS

Function	Examples
Synthesis	Proteins—albumin, cholinesterase, coagulation proteins, cholesterol, bile salts, glycogen
Metabolism	Glucose to acetyl-CoA, gluconeogenesis, amino acid conversions, fatty acids
Detoxification	Bilirubin, drugs, ammonia
Excretion	Bile acids

portant role in the digestion of triglycerides and other lipids.

Glycogen, the storage form of carbohydrate, is synthesized by the liver. This process is stimulated by the hormone glucagon and accomplished by a complex series of enzyme reactions. The glycogen which is formed is stored in the liver, available when the body requires a quickly accessible source of biochemical energy.

A major synthetic role of the liver is the production of a wide variety of plasma proteins. The coagulation factors required for blood clotting are produced in the liver. The predominant plasma protein, albumin, is synthesized by this organ. The liver is responsible for the formation of the various lipoproteins needed to transport lipids in the bloodstream. A number of hormones (thyroid hormones and most steroid hormones) also require transport proteins which are made by the liver.

Metabolic Functions

A wide variety of biochemical conversions take place in the liver. The complex process of carbohydrate formation from amino acids **(gluconeogenesis)** is mediated by an enzyme system in liver cells. In addition to its role in protein formation, the liver also regulates protein levels by metabolizing proteins, converting them back to their constituent amino acids. The end product of amino acid metabolism (urea) is formed in the liver and then excreted by the kidneys.

Detoxification Functions

Many of the enzyme reactions carried out by the liver have a protective effect. Ammonia formed from amino acid metabolism or microbial breakdown of urea in the gastrointestinal tract can be toxic if high enough concentrations form in the circulation. Conversion to nontoxic substances is accomplished by the liver. Bilirubin produced by catabolism of hemoglobin is converted to a water-soluble form by the liver so it can be excreted. A wide variety of medications are transformed to less toxic and more readily excreted derivatives by enzyme systems in the liver.

Excretory Functions

The major excretory function of the liver is the formation and excretion of bile acids. Although other metabolic processes may result in a material being excreted (usually via the kidney), bile acids synthesized from cholesterol in the liver are directly excreted to the gallbladder. There the salts are stored until needed for lipid digestion in the small intestine.

11.4

BILIRUBIN AND LIVER FUNCTION

Biosynthesis and Metabolism of Bilirubin

The bile pigment **bilirubin** is formed from the breakdown of the hemoglobin molecule (Fig. 11–1). Hemoglobin consists of four protein chains, each of which contains a porphyrin ring and an iron atom. When the hemoglobin molecule is metabolized, the iron is removed and recycled to make more hemoglobin. The protein chains are hydrolyzed back to the constituent amino acids. A portion of the porphyrin ring breaks and the cyclic structure becomes the open-chain tetrapyrrole derivative biliverdin. Further reduction by the enzyme biliverdin reductase (with NADPH as coenzyme) leads to the formation of **unconjugated bilirubin.**

When initially formed, bilirubin is insoluble in water. This molecule must be carried through the circulation bound to albumin. Most of the bilirubin molecules are loosely attached to the albumin carrier in a noncovalent fashion. However, a small fraction of the bilirubin (called **delta bilirubin**) forms a covalent bond to the protein and circulates in that fashion.

When the protein/bilirubin complex reaches the liver, bilirubin passes into the liver cells through what is thought to be an active transport process.

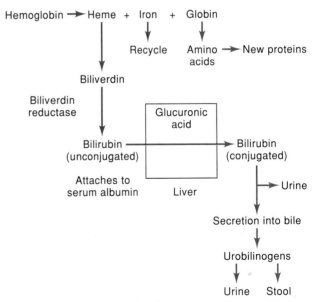

FIGURE 11–1 Bilirubin formation and metabolism.

After the molecules enter the hepatocytes, they attach to ligandin and Z protein, two soluble transport proteins. One or two glucuronic acid molecules attach to each bilirubin molecule in a reaction mediated by UDP-glucuronyltransferase (UDP stands for uridine diphosphate). Formation of the glucuronyl conjugate now makes the bilirubin complex soluble in water, in contrast to the insoluble unconjugated forms attached to protein. The **conjugated bilirubin** can then be excreted into bile and into the small intestine.

Once in the small intestine, there is little reabsorption of the bilirubin molecules. The glucuronic acid portion is hydrolyzed in a variety of processes, leaving the newly unconjugated bilirubin to be metabolized by bacteria in the small intestine. Bilirubin reduction occurs, producing a group of molecules known as **urobilinogens.** These materials may reenter the hepatic circulation and later be excreted in the bile. A small fraction of the urobilinogens is filtered by the kidney and excreted in the urine. Oxidation products of urobilinogens are also formed in the small intestine. These end products are excreted in the feces, providing its characteristic color.

Bilirubin Analysis

Because bilirubin is photosensitive and breaks down on exposure to light, samples for bilirubin

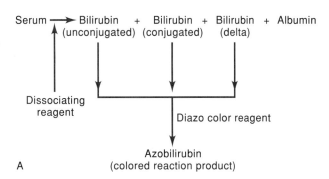

A

B

FIGURE 11-2   Approaches to bilirubin assay. A. Total bilirubin. B. Direct bilirubin. Note that the serum contains all three bilirubin fractions.

analysis should be kept in dim light or in the dark as much as possible. Serum samples are preferred in order to minimize possible turbidity produced by plasma proteins. Hemolysis must be avoided since hemoglobin produces a rather complex set of interferences with the assay.

Two components of the total amount of bilirubin can be analyzed from serum, which contains all three bilirubin fractions (Fig. 11-2). The conjugated form (containing one or two attached glucuronic acid molecules) reacts directly with the color reagent. This fraction is also referred to as **direct bilirubin.** Unconjugated bilirubin (noncovalently attached to albumin) does not react with the color reagent until the bilirubin is first dissociated from the protein. The term **indirect bilirubin** is sometimes used to describe this fraction. Although the delta bilirubin fraction is covalently attached to protein, it reacts directly with the color reagent and contributes to the direct, or conjugated, bilirubin value.

The color reaction for quantitation of bilirubin was discovered in the early part of this century and has seen little fundamental change since its inception (Fig. 11-3). In the presence of diazotized sulfanilic acid (formed by reacting sulfanilic acid with sodium nitrite and HCl), bilirubin forms a colored complex. One molecule of bilirubin splits to produce two molecules of azobilirubin. The diazo product has a red or reddish-purple color in acid pH with an absorption maximum in the region of 560 nm **(Evelyn-Malloy assay).** In alkaline pH, the color shifts to blue and becomes much more intense, with maximum absorbance around 600 nm **(Jendrassik-Grof method).** The same color reaction is used to measure the direct form without prior dissociation from protein or to assay the total bilirubin content after the unconjugated fraction is removed from albumin.

In addition to the pH difference employed in the two bilirubin assays, the Evelyn-Malloy and the Jendrassik-Grof procedures differ in their approach to dissociating the unconjugated bilirubin from protein. In the Evelyn-Malloy method, methanol serves as a dissociating agent. This reagent is fairly effective, but often produces some turbidity caused by precipitation of proteins in the organic solvent. A variety of dissociating agents have been employed in the Jendrassik-Grof procedure. The commonly accepted combination is a mixture of caffeine and sodium benzoate. Separation of unconjugated bilirubin from albumin occurs readily with little interference from precipitating proteins.

Hemoglobin is a major interferent with the Evelyn-Malloy method, but creates less of a problem with the Jendrassik-Grof approach. The absorbance maximum in the Evelyn-Malloy method (560 nm) is much closer to an absorbance band for hemoglobin than the 600-nm peak seen in the Jendrassik-Grof method.

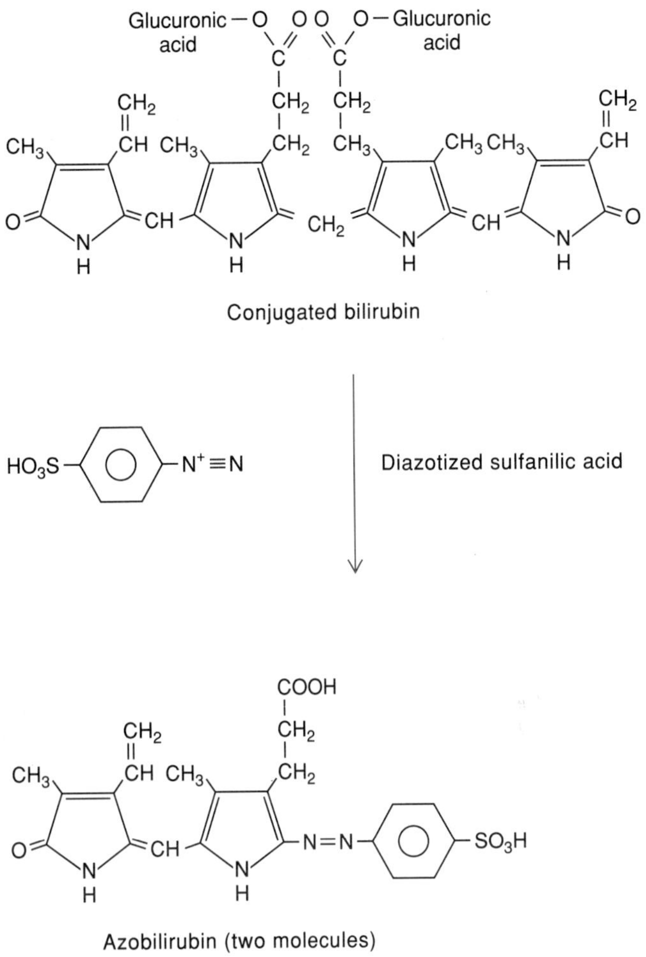

FIGURE 11-3

Diazo reaction for bilirubin assay. Unconjugated bilirubin reacts the same way as conjugated bilirubin.

Other methods for quantitation of bilirubin include fractionation by high-pressure liquid chromatography (HPLC) (good for research, but very impractical for routine analyses), enzymatic methods, and direct spectrophotometry. The enzyme method involves oxidation of bilirubin to a colorless product and measurement of the change in absorbance. Interference by other species absorbing at the wavelength maximum can be eliminated by making readings before and after the oxidation. Direct-reading spectrophotometry is not practical for adults, but has value in the monitoring of bilirubin in newborns (see Chapter 17). Eastman Kodak has a slide system available in which the bilirubin binds to a cationic polymer, producing a shift in the wavelength of maximum absorption. Measurement of absorbance at this new wavelength allows quantitation of direct bilirubin. Modifications permit quantitation of total bilirubin and the delta fraction.

Urine bilirubin can be estimated using commercially available dipstick methodology. Different diazo compounds are employed, but the basic chemistry is the same as that outlined above. These procedures measure only conjugated bilirubin, since only this fraction can be excreted in the urine. Samples must be shielded from the light as much as possible to avoid breakdown of bilirubin.

Clinical Significance of Altered Bilirubin Levels

Bilirubin in serum can be elevated for a variety of reasons (Table 11-2):

1. increased formation of bilirubin from red cell turnover
2. decreased conversion of unconjugated to conjugated forms of bilirubin
3. impairment of bilirubin excretion

An increase in serum bilirubin is an indicator of some medical problems and can shed some light on the probable cause, but does not serve as a primary diagnostic indicator.

Increased erythrocyte turnover, resulting in elevations of unconjugated bilirubin, can occur for a variety of reasons. In any one of several hereditary hemolytic anemias, unstable red cells hemolyze at a

Table 11–2
CAUSE OF INCREASED SERUM BILIRUBIN

Elevated Fraction	Clinical Situations
Unconjugated only	Hemolytic anemias
	Newborns
	Hereditary alteration of rate of conversion
	Medications
Conjugated only	Bile duct obstruction
	Some cases of hepatitis
	Medications
Both fractions	Hepatitis

much higher rate than normal. Newborns turn over their red cells at a higher rate than adults, leading to some increase in bilirubin formation. In these situations, the conjugated bilirubin level usually does not rise appreciably, since this fraction can be quickly excreted once it is formed.

Several conditions can result in diminished conversion of unconjugated to conjugated bilirubin. In the newborn, the glucuronyltransferase enzyme is not active for several days after birth, causing low conversion of bilirubin to the conjugated form. A number of hereditary disorders alter the uptake of bilirubin by the liver or impair its conversion after it enters the hepatic cells. Medications may cause inhibition of the enzymes and proteins responsible for either transport into the cell or metabolism.

Conjugated bilirubin rises in serum because of some obstructive process which hinders the excretion of this water-soluble form of the pigment. Hepatitis produces liver damage, slowing the secretion of conjugated bilirubin. Blockage of the bile duct results in accumulation of bilirubin and impaired ex-

cretion. This blockage is frequently caused by stones or by the presence of a carcinoma. Medications may also interfere with the excretory process by producing spasms or constriction of the duct.

It is not uncommon to see both the conjugated and the unconjugated fractions of bilirubin simultaneously increase in serum. When both fractions are elevated, a combination of increased bilirubin formation and decreased excretion are occurring. The bilirubin results may point the way, but cannot be used as conclusive evidence for a specific diagnosis.

11.5

AMMONIA AND THE LIVER

Source of Plasma Ammonia

Most ammonia in the circulation derives from two sources (Fig. 11–4):

1. breakdown of amino acids obtained from dietary protein in the intestine
2. hydrolysis of urea, an end product of metabolism

The primary site of ammonia production is the small intestine. Dietary proteins are hydrolyzed to their constituent amino acids by a variety of proteases. Although a large fraction of these amino acids are absorbed into the bloodstream and utilized for further protein synthesis, a significant portion are metabolized by bacterial action in the intestine. Many amino acids can undergo conversion to glutamate, a dicarboxylic amino acid. Glutamate is then converted to alpha-ketoglutarate and ammonia

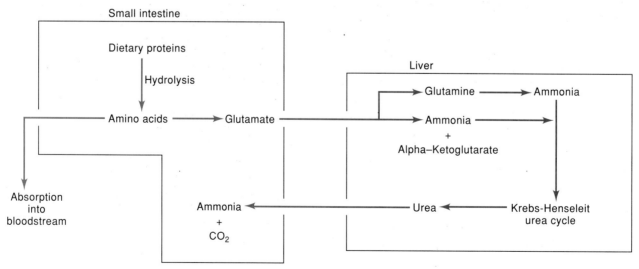

FIGURE 11–4 Ammonia formation and metabolism.

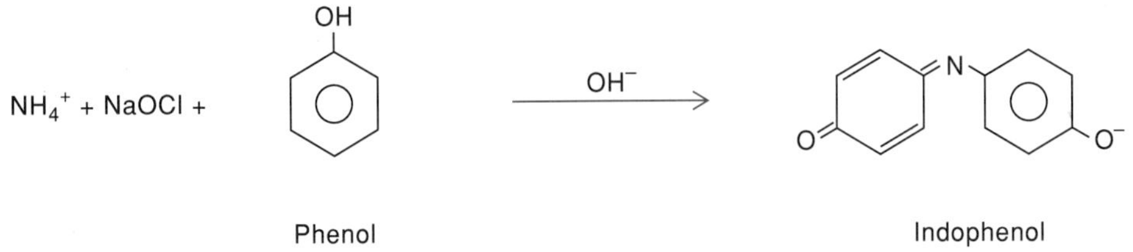

FIGURE 11-5
Colorimetric assay for ammonia.

$$NH_4^+ + NaOCl + \text{Phenol} \xrightarrow{OH^-} \text{Indophenol}$$

Phenol   Indophenol

by the enzyme glutamate dehydrogenase, and the ammonia enters the circulation.

The other route to ammonia formation involves bacterial hydrolysis of urea, also taking place in the small intestine. Urea is an end product of a series of reactions which utilize ammonia and carbon dioxide as initial materials. Although most of the urea synthesized is excreted by the kidney, a portion reenters the small intestine to undergo microbial breakdown. Some of the ammonia produced diffuses into the bloodstream.

Ammonia exists in the circulation in an equilibrium between the ionized and unionized forms of the molecule. The reversible combination of ammonia with a proton leads to the formation of the ammonium ion. The exact amounts of the two forms depends on the blood pH. A lower pH results in more H^+ being present. The equilibrium then shifts in the direction of greater formation of the ammonium ion:

$$NH_3 + H^+ \rightleftharpoons NH_4^+$$

Ammonia   Ammonium ion

The kidney excretes the ionized form of the molecule.

Since high concentrations of ammonia can be toxic, the liver removes this material using a metabolic sequence known as the **Krebs-Henseleit urea cycle.** The initial reaction in this cycle involves a reaction between ammonia and HCO_3^- (bicarbonate) to form carbamyl phosphate. This molecule then undergoes further transformation resulting in the formation of urea and ornithine (which is then recycled through the pathway). Several hereditary disorders exist in which a specific enzyme in the pathway is present at very low levels. Since the cycle cannot function efficiently, ammonia accumulates in the system to toxic proportions.

Analysis of Plasma Ammonia

Plasma samples are preferred for *blood ammonia* determinations since rapid separation from cells introduces less error. Analysis must be performed as soon as possible to eliminate problems with endogenous formation of ammonia while the sample is standing. If the sample is on ice or refrigerated, this process slows somewhat. Smoking causes marked increases in plasma ammonia values.

One popular method for analysis of plasma ammonia involves a two-step process (Fig. 11-5). The ammonia in the sample binds to a cation exchange resin and is isolated from the remainder of the material. Following elution from the resin, a reaction with phenol-hypochlorite takes place to produce indophenol, a blue compound. Quantitative estimation can be achieved by measurement of absorbance at 630 nm.

More recently, an enzymatic approach has been developed (Fig. 11-6). Ammonia in the sample reacts with alpha-ketoglutarate and NADPH in the presence of the enzyme glutamate dehydrogenase (the reverse of the metabolic process leading to ammonia formation in the body). Glutamate and $NADP^+$ form as products. Since NADPH has a strong absorbance at 340 nm, and its oxidized product does not absorb at this wavelength, the decrease in absorbance is proportional to the concentration of ammonia in the sample. Variations of this method have been adapted to several automated chemistry systems.

Clinical Situations

Reference ranges for plasma ammonia are approximately 20–80 μg/dL, depending on the method. Enzymatic methods tend to yield somewhat lower ranges than colorimetric approaches. A major reason

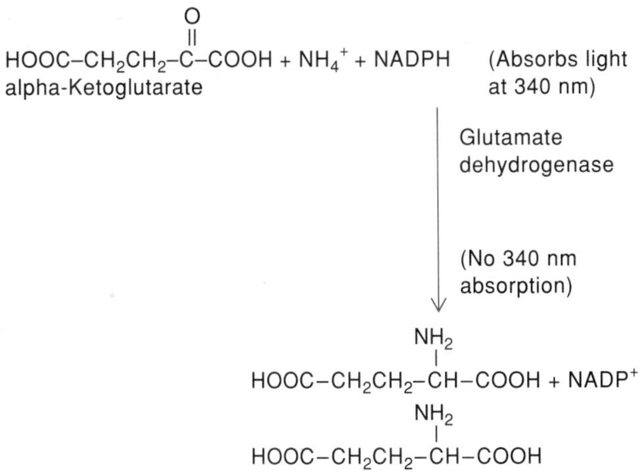

$$\underset{\text{alpha-Ketoglutarate}}{HOOC-CH_2CH_2-\overset{O}{\overset{\|}{C}}-COOH} + NH_4^+ + NADPH \quad \text{(Absorbs light at 340 nm)}$$

Glutamate dehydrogenase

(No 340 nm absorption)

$$HOOC-CH_2CH_2-\underset{NH_2}{\overset{|}{CH}}-COOH + NADP^+$$

FIGURE 11-6   Enzymatic assay for ammonia.

Table 11-3
CAUSES OF ELEVATED PLASMA
AMMONIA

Hepatic encephalopathy	Reye's syndrome
Pulmonary problems	Hereditary disorders in urea cycle enzymes

for increased ammonia levels in adults is hepatic encephalopathy, a complicated disorder in which venous blood does not pass through the liver in large enough quantity to detoxify ammonia and other materials. Personality changes develop, neurological alterations appear, and coma is often the end result of the liver damage. The underlying liver disease may involve liver necrosis or cirrhosis.

Individuals with renal failure may also manifest elevated ammonia levels due to the inability of the kidney to excrete sufficient urea and ammonia. Patients with pulmonary problems often have elevated ammonia levels, possibly secondary to altered blood pH, which distorts the equilibrium between ammonia and the ammonium ion. One characteristic of Reye's syndrome (to be discussed later in this chapter) is an increase in plasma ammonia concentration. Any abnormality in the Krebs-Henseleit urea cycle that results in impaired metabolism increases ammonia in the system. Clinical situations producing increased plasma ammonia levels are summarized in Table 11-3.

11.6

HEPATIC ENZYMES IN HEALTH AND DISEASE

Several serum enzymes are usually elevated in cases of hepatic damage (Table 11-4). These en-

Table 11-4
HEPATIC ENZYMES

Enzyme	Clinical Utility in Liver Disorders
Alkaline phosphatase	Elevated primarily in obstructive processes
Aminotransferases	Elevated in variety of liver diseases; ALT more sensitive indicator
Gamma-glutamyltransferase	Some increase in liver diseases; sensitive indicator of ethanol intake
Cholinesterase	Normally quite high; values decrease in liver disorders
Lactate dehydrogenase	Elevated in wide variety of situations; clinical utility low in absence of isoenzyme studies

zymes include alkaline phosphatase, alanine aminotransferase (ALT), aspartate aminotransferase (AST), gamma-glutamyltransferase (GGT), and lactate dehydrogenase (LD). Serum cholinesterase values are quite high in healthy individuals relative to activities of the other enzymes listed. In situations involving liver damage, the cholinesterase values drop. This pattern represents one of the very few cases in clinical enzymology (outside of the hereditary enzymopathies) in which a clinically significant change in enzyme level is seen as a decrease, not an increase, in activity.

The usual "hepatic profile" consists of measurements of alkaline phosphatase and the two aminotransferases. Increased values for alkaline phosphatase usually reflect an obstructive process more than release of enzyme from damaged cells in the liver. In many nonobstructive clinical situations, alkaline phosphatase values may be only slightly increased.

A better indicator of cellular destruction in the liver can be obtained by assay for both ALT and AST. These two aminotransferases rise earlier than other enzymes and generally increase more dramatically for a given amount of hepatic damage. In most cases, the level of ALT is a more sensitive indicator of damage to this organ. Assessment of activity of both enzymes in viral hepatitis provides a reliable monitor of the progress of the disease. Levels are quite elevated during the infective stage of the virus, followed by a gradual decline in serum enzyme activity as the patient recovers. Continued high values for these enzymes indicate chronic disease, whereas a decline and subsequent increase strongly suggest a recurrence of infection. Isoenzyme studies provide little (if any) further useful information.

Measurement of serum gamma-glutamyltransferase activity to assess liver damage often provides little more useful information than can be obtained from aminotransferase studies alone. On occasion, the GGT value may rise more than other enzymes in cases of biliary tract disorders, but not often enough to make measurement of this enzyme a routine part of a hepatic profile. GGT assays do play an important role, however, in monitoring alcohol consumption (a frequent cause of liver damage). This enzyme increases in serum shortly after even moderate alcohol intake, thus serving as a fairly sensitive check on patients with drinking problems.

Lactate dehydrogenase measurements provide little useful information when assessing liver damage. This ubiquitous enzyme is elevated in a wide variety of clinical situations, many of which are not associated with hepatic disorders. Isoenzyme fractionation (preferably by electrophoresis) is a sensitive indicator of the source of any elevated LD in serum. If the LD-5 isoenzyme is increased, there is a strong possibility that the liver is involved in some manner.

11.7

LABORATORY DIAGNOSIS OF LIVER DISEASE: GENERAL CONCEPTS

Liver function tests (Table 11–5) are a group of laboratory assays which may or may not actually relate to what the liver can do. Several enzyme assays are commonly carried out to assess hepatic damage in some fashion. Quantitation of albumin and determination of coagulation parameters (particularly the prothrombin time) reflect the synthetic capability of the liver. Measurement of bilirubin fractions further assesses the ability for hepatic synthesis, as well as providing insights into the excretory function of this organ. Other assays (such as the measurement of alpha$_1$ antitrypsin, various antibodies, or viral markers) are useful as biochemical markers for specific liver disorders, but would not be considered liver function tests in the broader sense.

Increases in the levels of several serum enzymes are usually considered indicative of hepatic damage. The aminotransferases (both ALT and AST) rise in cases of hepatic damage, with ALT considered a much more specific sign of liver disease. AST values can be elevated in a variety of other disorders, including myocardial infarction or skeletal muscle injury. Even ALT is not as specific as many consider it to be, since elevated values for this enzyme also appear in patients with muscle damage, disseminated cancer, acute pancreatitis, and some cases of renal disease. Alkaline phosphatase can be elevated in cases of liver damage, altered bone biochemistry, during pregnancy, and in some types of cancer. Isoenzyme fractionation is often necessary for full utility of this enzyme measurement. The wide distribution of lactate dehydrogenase in tissues renders the diagnostic value of an increased level almost useless in the absence of isoenzyme determination.

Although an increase in bilirubin may reflect some type of hepatic abnormality (either involved with conjugation or in excretion), a wide variety of other medical problems are also associated with elevated bilirubin. Increase in bilirubin formation due to hemolysis is fairly common. Bilirubin excretion may be impaired for some reason. Hereditary defects in the intrahepatic transport or metabolism of bilirubin are not associated with overall liver damage, but with deficiencies of specific proteins.

Lowered serum albumin and prothrombin times may be due to liver damage or to some other factor. Albumin concentrations are affected only in part by the ability of the liver to synthesize this protein. Dietary intake of protein plays a significant role in providing the amino acids essential for albumin production. Impaired absorption in the gastrointestinal tract also decreases the supply of building blocks for albumin synthesis. Albumin loss by the kidney contributes to decreased serum albumin concentrations. Nutrition and drugs play important roles in the regulation of production of clotting factors. In addition, if disseminated intravascular coagulation is occurring, the levels of blood-clotting components decreases.

11.8

HEPATITIS

The term **hepatitis** can encompass a wide variety of liver disorders. The common thread for all these related problems is an inflammatory process which leads to the death of liver cells and affects all areas of the liver. Major causes of hepatitis are the hepatitis A and B viruses. Drugs and alcohol can produce liver damage which is associated with hepatitis. On rare occasions, other viral origins of hepatitis can be seen, including (but not restricted to) cytomegalovirus, infectious mononucleosis, and yellow fever.

Clinical descriptions of hepatitis can be found as far back as the fifth century B.C. in the Babylonian Talmud. The famous Greek physician Hippocrates described some of the diagnostic signs of hepatitis in his writings. Although not as widely known as the epidemics of plague and cholera which swept Europe during the Middle Ages, hepatitis contributed greatly to the death toll with massive outbreaks during that time.

Because of its association with poor sanitary practices, hepatitis has had a significant impact on military operations throughout history. The British army in the early 1700s had an epidemic, as did Napoleon's forces in Egypt. Soldiers in the American Civil War were widely affected by the disease. During World Wars I and II, huge epidemics were re-

---

Table 11–5
CHANGES IN SERUM PARAMETERS IN LIVER DISEASE

Change in Liver Disease	Other Possible Causes
Bilirubin (increase)	Hemolysis
	Hereditary erythrocyte enzymopathies
Alkaline phosphatase (increase)	Bone cancer, pregnancy
ALT (increase)	Muscle damage, cancer, pancreatitis, renal disease
AST (increase)	Myocardial infarction, skeletal muscle damage
Cholinesterase (decrease)	Inhibition by pesticides
Lactate dehydrogenase (increase)	Myocardial infarction, skeletal muscle damage, renal disease, cancer

ported. In World War II, over 5 million German military and civilian personnel were affected. The disorder had a similar impact on Allied forces and served to influence the overall military strategy in some instances. Hepatitis also had a profound effect on military operations in Korea during the 1950s, during the Israeli war for independence (1948), in the various Arab-Israeli conflicts (1956 and 1967), and during the Vietnam war in the 1960s and 1970s.

Types of Hepatitis

There are several different types of hepatitis known, including hepatitis A, hepatitis B, at least two types of non-A/non-B hepatitis, and a new variant of hepatitis B referred to as delta hepatitis. In each case, a virus appears to be responsible for the disease. Although we understand the etiologies of types A, B, and delta hepatitis fairly well, the causative agent(s) for the non-A/non-B disorders are not known with any certainty.

Clinical diagnosis is difficult because the symptoms are quite variable and mimic a variety of other common flulike diseases. Nausea, vomiting, fever, and loss of appetite are common early signs. Since hepatitis is an extremely infectious disease, prompt diagnosis must be made, both for the benefit of the patient and for the safety of those in contact with the infected individual.

HEPATITIS A

Hepatitis A is caused by a small RNA-containing virus of the family Picornaviridae. The incubation period for this virus is relatively short (approximately 28 days on the average), but somewhat variable from one person to another. Transmission is almost always by the fecal-oral route. Children in day-care centers, particularly those with poor toilet and sanitary habits, are at high risk for the disease and can bring it home to other family members. Hepatitis A has been very prevalent among homosexual populations. A striking rise in the number of cases of hepatitis A has been seen in substance abusers who administer drugs intravenously. Frequently, news items appear about hepatitis A outbreaks among patrons of a restaurant. Invariably, the source of infection is traced back to an employee who had the disease and transmitted it to customers through the food they ate. Transmission of the hepatitis A virus by blood transfusion is extremely rare, probably due to the short survival time of the virus.

In many cases, exposure results in mild illness and there is no suspicion of hepatitis A. Biochemical markers which indicate past exposure to the virus are present in approximately 20% of the U.S. population by the age of 20. Roughly half the U.S. population exhibit these same markers by age 50. In some undeveloped nations, a high prevalence of markers is evident at an early age, presumably indicating problems with sanitation. After the disease has run its course, immunity usually develops. It is extremely rare for a person to have a second episode of hepatitis A or to exhibit a chronic form of the disease.

HEPATITIS B

Characterization of the biochemical parameters of hepatitis B has been a long and complicated quest. Beginning with the discovery of an antigen in serum of patients who had received multiple blood transfusions (initially called the **Australia antigen** since it was first detected in the serum of an Australian aborigine), the search resulted in international honors for two scientists. Baruch S. Blumberg and D. Carleton Gajdusek were awarded the Nobel Prize in Physiology in 1976 for their work in characterizing the infectious agent for hepatitis B.

The virus responsible for this form of hepatitis consists of a double shell of DNA surrounded by protein, known as the **Dane particle.** The virus must be intact for infection to occur; the various components of the Dane particle are not infectious by themselves. A widely variable incubation period is seen after infection, ranging from 1 to 26 weeks in length. Approximately 50% of those infected with hepatitis B virus manifest the disease in some form. Some 5–10% of those with symptoms go on to be chronic carriers of the disease. Death occurs in nearly 1% of the acute cases.

The primary modes of transmission of hepatitis B are needle-sticks and sexual activity. Intravenous substance abusers have a high incidence of this disease, as do homosexuals and prostitutes. The disease is frequently transmitted to unsuspecting sexual partners. At one time, a significant number of hepatitis B cases were traced back to contaminated blood supplies. With the advent of widespread screening for biochemical markers of the disease, the incidence of transfusion-related hepatitis B cases has dropped significantly, to be replaced by

Table 11–6
COMPARISON OF HEPATITIS TYPES A AND B

Parameter	Hepatitis A	Hepatitis B
Incubation period	Relatively short (2–6 weeks)	Variable (1–26 weeks)
Mode of transmission	Usually oral-fecal; infrequently by needle-stick	Usually by needle-stick or sexual transmission
Carrier state?	Rare	Some 10% of those infected

the presence of non-A/non-B hepatitis. Health-care personnel are at high risk for hepatitis B as they come in contact with needles and other sharp objects contaminated with the virus from patients with the disorder.

Some differences between the clinical aspects of hepatitis types A and B are summarized in Table 11–6.

NON-A/NON-B HEPATITIS

Although screening blood products for biochemical markers for hepatitis B has been mandatory since 1972, the incidence of posttransfusion hepatitis has not been greatly affected by these safety precautions. Only about 7–10% of the cases of posttransfusion disease have been shown to be associated with hepatitis B virus. The remainder are referred to as **non-A/non-B hepatitis,** lately called hepatitis C, to indicate the mystery surrounding the cause of this medical problem. Somewhere between 5 and 18% of those who receive blood transfusions develop this form of the disease. Patients on hemodialysis and those who have received kidney transplants are at high risk for the disease. In addition, approximately 25% of the cases of sporadic hepatitis appear to have the non-A/non-B variety.

The clinical picture for non-A/non-B hepatitis is confusing. The incubation period is variable and lies somewhere between the ranges for hepatitis A and B viruses. A high percentage of these patients develop some form of chronic hepatitis, although remission frequently occurs. Biochemical changes (serum enzymes and bilirubin) are not as pronounced as they are in cases of hepatitis B. Some studies suggest that at least two forms of non-A/non-B hepatitis presently exist.

To date, no specific biochemical marker has been identified as being associated with the non-A/non-B form of hepatitis. Although a wide variety of immunochemical approaches have been explored, there is no consistent information linking this disease with the presence of any particular antigen, as is the case with other forms of hepatitis. Screening to detect elevated levels of the liver enzyme ALT is being explored, but no clear cutoff point has been established which reliably identifies individuals with non-A/non-B hepatitis.

DELTA HEPATITIS

Since 1977 there has been intensive study of a new form of hepatitis which appears to coexist with hepatitis B. The discovery by Italian researchers of the delta antigen has led to an increased understanding of a complex disease. In type D (delta) hep-

atitis, the responsible virus cannot replicate by itself. To reproduce, this virus must interact with the surface antigen protein from hepatitis B virus. Hepatitis D will be seen either as a coinfection (appearing simultaneously with hepatitis B) or as a superinfection (in which the D virus occurs in a chronic hepatitis B case).

Seen in high incidence in southern Italy, some Middle Eastern nations, parts of Africa, and in some areas of South America, the virus appears to be transmitted by sexual contact. In the United States, Europe, and Southeast Asia, the major route of transmission appears to be through needle-stick or by blood transfusions. A high incidence of hepatitis D has not yet been seen in high-risk groups such as homosexuals.

This form of hepatitis can produce either acute or chronic states. The severity of delta hepatitis is greater and its mortality rate is much higher than that seen with hepatitis B infection alone. The mortality rate for hepatitis D can range from 2 to 20% (by comparison, hepatitis B has a mortality rate of approximately 1%). Hepatic complications can develop in almost 80% of patients with chronic delta hepatitis as compared with the 15–30% of patients with hepatitis B who develop ongoing liver problems. Although coinfection with hepatitis D virus rarely leads to long-term complications, superinfection can produce the chronic disease state.

ALCOHOLIC AND DRUG-INDUCED HEPATITIS

Chronic alcohol abuse eventually produces symptoms of liver disease which may be indistinguishable from other forms of hepatitis. Inflammation, fatty liver, fever, jaundice, abdominal pain—all these symptoms are typical of alcoholic hepatitis as well as a host of other liver diseases. A detailed laboratory exploration of hepatitis yields nothing in the way of specific biochemical abnormalities other than the expected increases in bilirubin and liver enzymes. Identification of a pattern of ongoing alcohol abuse gives more insight into the disease than do laboratory studies in most cases.

A number of medications produce hepatic damage in some patients, leading to hepatitis symptoms. These chemical agents include halothane (an anesthetic), acetaminophen, phenytoin (an anticonvulsant used in treatment of epilepsy), and alphamethyldopa (an antihypertensive). Again, the laboratory picture is inconclusive. Bilirubin and hepatic enzyme values are increased, the physical symptoms are typical of hepatitis, and the screening for viral markers is negative. In this instance, as with alcoholic hepatitis, a careful medication history is important for elucidating the cause of the liver problem.

Hepatitis and Liver Damage

Whatever the cause of hepatitis, a common constellation of laboratory results is seen. Bilirubin levels are increased, usually up to about 10 mg/dL; values up to 20 mg/dL have been reported, but are rare. Both conjugated and unconjugated bilirubin fractions may be elevated, depending on the extent of hepatic damage. Serum albumin decreases, as does serum cholesterol. Inflammation elevates iron-binding capacity and increases the gamma-globulin fraction of serum proteins.

The hepatic enzymes show variable elevations, depending on the severity of the disease. Alkaline phosphatase may rise moderately, or remain within normal limits. Both alanine and aspartate aminotransferases manifest elevations, often up to 10 times or more the upper limit of the reference range. Gamma-glutamyltransferase is elevated regardless of the cause of the hepatitis. During the acute stage, this lack of specificity may make it difficult to ascertain the contribution alcohol abuse has made to the GGT increase.

Laboratory Monitoring of Hepatitis

After the causative agent of hepatitis B was discovered, an explosion of laboratory information occurred in the 1970s. Two avenues of research contributed to our present understanding of the laboratory diagnosis of hepatitis: fundamental studies of the components of the viruses and rapid developments in the immunochemistry field leading to sophisticated diagnostic techniques for the identification of the viral markers in the various disease states. Although serum enzymes and bilirubin changes indicated that some type of liver damage was taking place, the investigation of specific biochemical markers allowed identification of the causative agent and permitted some degree of prognosis for a given patient.

From the early (and relatively insensitive) hemagglutination procedures for the detection of markers for hepatitis B (first generation), the laboratory then moved to counterimmunoelectrophoresis (second generation) to detect the presence of this protein. Current technology involves the use of radioimmunoassay (RIA) (third generation) and enzyme-linked immunoassay approaches to identify a number of specific viral components and markers for both hepatitis A and hepatitis B. The wide variety of immunochemical techniques developed for other types of assays have been successfully employed in the hepatitis screening field. Since we already discussed these approaches in Chapter 9, we will focus on the laboratory results and the diagnostic information provided by screening for the different antigens and antibodies produced during the course of viral hepatitis.

SCREENING FOR HEPATITIS A

Laboratory screening for the presence of hepatitis A is reasonably straightforward, in contrast to the complexity of markers for hepatitis B (Fig. 11–7). At present, no specific component of the A virus can be detected through routine laboratory testing. Instead, detection of the antibody to hepatitis A virus (anti-HAV) indicates the existence of hepatitis A, either presently or at some time in the past. A positive anti-HAV result by itself does not allow us to distinguish between these two possibilities.

Subclassification of anti-HAV provides further useful diagnostic information. Two antibodies exist, one of the IgM class and the other of the IgG class. The IgM antibody appears during the acute infection and persists for several months afterward, gradually declining to undetectable levels. As the amount of IgM antibody decreases, the IgG antibody level begins to rise. Long after the infection has run its course, IgG antibody is still detectable. This component usually remains in the bloodstream for the remainder of the patient's life. In some instances, the concentration of IgG antibody may decline over several years. If the IgM antibody is present, an active infection exists. Detection of IgG antibody simply indicates that the person was exposed to hepatitis A virus at some point in the past, but does not suggest the presence of a current infection with hepatitis A virus.

HEPATITIS B DETECTION AND MONITORING

Structure of the Hepatitis B Virus

The hepatitis B virus is a rather complex structure consisting of two layers of protein surrounding the nucleic acid interior (Fig. 11–8). The outer protein layer is referred to as the hepatitis B surface antigen (HBsAg) and was the first biochemical

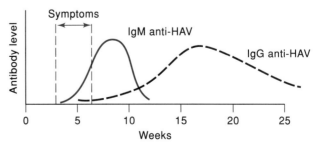

FIGURE 11-7   Time course of markers for hepatitis A.

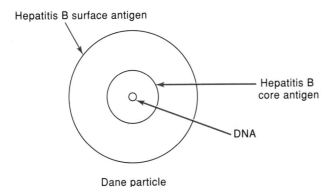

Dane particle

FIGURE 11-8 Anatomy of the hepatitis B virus.

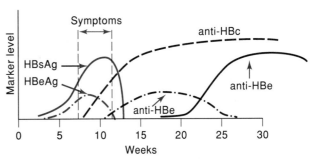

FIGURE 11-9 Time course of markers for hepatitis B.

marker to be linked with this disorder (Table 11-7). The inner protein layer contains the hepatitis B core antigen (HBcAg) and possibly the e antigen (HBeAg). Inside this double protein core is circular double-stranded DNA and the DNA polymerase enzyme which makes viral replication possible. During the course of infection, viral fragmentation evidently occurs at a high rate. The various components of the virus are detected at concentrations as much as a million times higher than that of the intact virus itself. Assays for the presence of these fragments provide much valuable information for the diagnosis and management of hepatitis B.

Laboratory Markers

Partway through the incubation period, the concentrations of the surface antigen and the e antigen begin to increase in the blood (Fig. 11-9). The peak levels for both antigens is seen during the time that clinical symptoms and the rise in bilirubin and liver enzymes occur. Both HBsAg and HBeAg then decline even though symptoms are still present.

Development of antibodies to the various antigens begins as the antigens themselves decline. Levels of antibody to hepatitis B core antigen (anti-HBc) slowly increase and persist for a long time. If laboratory studies are not initiated in time, anti-HBc may be the only positive sign pointing to the diagnosis of hepatitis B. Subsequent to the development of anti-HBc, we see a transient increase and then decline in the amount of antibody to the e antigen (anti-HBe). Immunity to the disease is indicated by

development of the presence of antibody to the surface antigen (anti-HBs), which then persists in the bloodstream.

In some 10% or so of the patients who develop acute hepatitis B, the disease may shift to the chronic state. In this situation, monitoring for biochemical markers over time reveals the persistent presence of HbsAg and anti-HBc. The continuing presence of the surface antigen indicates the patient is a carrier for hepatitis B. The e antigen may also persist, indicating a higher degree of infectiousness than a patient who is negative for this antigen. Patients with chronic hepatitis who are positive for HBeAg usually have more severe liver damage and more serious complications.

DETECTION OF DELTA ANTIGEN

Research in the measurement of components of the delta virus has led to assay for the antigen itself (HDVAg) as well as measurement of the antibodies to the delta antigen. These procedures are becoming commercially available as newer and more sensitive methodologies are developed.

Two patterns of delta antigen and antibody presence are seen in delta hepatitis (Fig. 11-10). In coinfection, where the delta virus and the hepatitis B virus infect at the same time, the delta antigen appears during the time symptoms are being seen. Serum aminotransferase levels also begin to rise at the onset of antigen presence. Approximately 1-2 weeks later, antibodies to the virus (anti-HDV, or

---

Table 11-7
GLOSSARY OF MARKERS FOR HEPATITIS B

Abbreviation	Name
HBsAg	Hepatitis B surface antigen
HBeAG	Hepatitis B e antigen
anti-HBc	Antibody to hepatitis B core antigen
anti-HBe	Antibody to hepatitis B e antigen
anti-HBs	Antibody to hepatitis B surface antigen

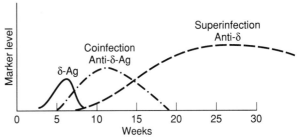

FIGURE 11-10 Time course of markers for delta hepatitis.

anti-δ-Ag) develop. In a pattern reminiscent of hepatitis A immunochemistry, the IgM fraction is present and then declines in concentration as the IgG antibody begins to be seen. Persistence of only the IgG antibody signifies the occurrence of acute delta hepatitis at some time in the past.

Superinfection by the delta virus on top of an existing hepatitis B infection usually results in development of chronic delta hepatitis. Presence and disappearance of HDV antigen follows the pattern described above. However, persistent increases in serum aminotransferases are seen instead of the rise and fall observed in the acute state. In addition, the IgM antibody level may decline somewhat (but never disappears), even after IgG antibody is fully developed. The prognosis for patients with chronic HDV infection is very poor, with higher incidences of liver necrosis, inflammation, and subsequent complications.

11.9

ALCOHOL AND THE LIVER

Ethanol and Liver Damage

Over 10 million people in the United States are alcoholics. The medical problems associated with alcohol abuse cost this country billions of dollars each year. In addition to the highway deaths associated with drinking, alcohol-associated liver disease is the fourth leading cause of death among adult males.

The impact of ethanol on liver function is not altogether clear, although some strong associations between chronic consumption of ethanol and hepatic damage do exist. In addition, factors related to poor nutrition and impaired absorption of vitamins and other needed materials must be considered.

Ethanol exerts its deleterious effects in at least two ways. Metabolism of ethanol leads to the formation of toxic materials which can impair a number of biochemical processes and induce tissue damage. The second process may involve the direct effect of ethanol on cell structure. Since this alcohol is lipid-soluble, it may concentrate in the cell membrane, producing alterations in the integrity of the membrane and setting the stage for further deterioration.

The primary site of metabolism for ethanol is the liver. This organ contains the enzyme alcohol dehydrogenase, which converts ethanol to acetaldehyde:

$$CH_3CH_2OH + NAD^+ \xrightarrow{\text{alcohol dehydrogenase}}$$

ethanol

$$CH_3CHO + NADH + H^+$$

acetaldehyde

Table 11-8
STAGES OF ALCOHOLIC LIVER DISEASE

Stage	Characteristics
Steatosis	Triglyceride deposits in liver
Steatonecrosis	Further fat accumulation
	Inflammation, fibrosis, necrosis
Cirrhosis	Extensive fibrosis
	Further inflammation
	Hepatocellular carcinoma

Acetaldehyde has been shown to produce a variety of toxic reactions in humans, including increased peroxidation of lipids, alteration of liver structure, development of fibrosis, and production of liver inflammation. These symptoms do not occur in all patients, suggesting a complicated process which is yet poorly understood.

Liver damage due to chronic alcohol intake occurs in three stages (Table 11-8). In the first stage (**alcoholic steatosis**), fatty deposits develop in the liver. These deposits arise due to the effect of ethanol on lipid metabolism, decreasing fatty acid oxidation to smaller molecules and increasing the formation of triglycerides (a storage form of fat in the body). Although steatosis does not invariably lead to further liver damage, it must contribute somehow to the development of more serious liver injury.

Alcoholic **steatonecrosis** (alcoholic hepatitis) is the second stage of liver destruction. Further fat accumulation takes place, along with inflammation and development of fibrosis. In **alcoholic cirrhosis** (the third stage), extensive development of fibrous tissue takes place. There is strong evidence to suggest that acetaldehyde stimulates the formation of collagen (a major structural protein component of fibrosis) by hepatic cells. Further cellular destruction takes place, often with the development of inflammation. Hepatocellular carcinoma (liver cancer) is quite common in the later stages of the disease. Development of cirrhosis appears to occur almost twice as often in women as it does in men.

Laboratory Findings in Alcoholic Liver Disease

The aminotransferase enzymes are often elevated in this disorder (Table 11-9), although the extent of increase is not striking (perhaps eight times the upper limit of normal). Usually, the AST value is more elevated than the result for ALT. In most cases, the degree of increase in serum enzyme values does not reflect accurately the intensity and extent of actual liver damage.

Alterations of other biochemical parameters may reflect a direct impact of ethanol on the body or may be secondary to poor dietary habits. Glucose levels

Table 11-9
SERUM CHEMISTRY ALTERATIONS IN
ALCOHOLIC LIVER DISEASE

Increased	Decreased
Aminotransferases	Glucose
Lipoproteins	Albumin
Bilirubin	Transferrin
Ketone bodies	
Triglycerides	

may be low, since ethanol can inhibit the process of glucose synthesis from amino acids. Albumin often decreases, either due to impaired hepatic synthesis of this protein caused by ethanol or because of poor diet. Decreased formation of the protein transferrin (which aids in the transport of iron) is common in alcoholics. Bilirubin values may be increased if hepatic damage has occurred. In contrast to the impaired production of several other proteins, the synthesis of lipoproteins is increased, helping produce elevated triglyceride levels in serum. Ketone body formation is enhanced, which produces ketoacidosis and an altered acid/base balance.

11.10

REYE'S SYNDROME

In 1963, a group of Australian researchers first defined the syndrome now named after the leader of the team. **Reye's syndrome** is a rare disease, primarily affecting small children. The cause is unknown, but appears to be related to some prior viral illness. In the first few days, a mild illness occurs, usually associated with a viral disorder of the respiratory or upper gastrointestinal tract. The initial symptoms subside, only to be followed in a few days by repeated vomiting and delirium. Very frequently, the child becomes comatose. If death occurs, it is generally the result of increased fluid in the brain, resulting in elevated intracranial pressure.

The cause of the disease is unknown, although the target tissue appears to be the liver. A wide variety of hepatic abnormalities are seen, reflected in both the clinical chemistry changes and on direct observation of liver tissue sections. Serum aminotransferase values increase, and coagulation abnormalities develop which are consistent with hepatic damage. A hallmark of the disorder is an increased plasma ammonia level. The higher the ammonia value above the reference range, the poorer the prognosis for survival.

An interesting (but unexplained) link among viral infection, consumption of aspirin, and Reye's syndrome has been seen recently. Children who have influenza or chickenpox are at great risk for later development of Reye's syndrome if they take aspirin.

The biochemical basis for this relationship is not known at present.

11.11

PRIMARY BILIARY CIRRHOSIS

Clinical Description

The major clinical problem in primary biliary cirrhosis is a gradual development of inflammation and bile duct destruction. This deterioration generally leads to cirrhosis and liver failure. Decrease in the number of bile ducts results in impaired production and release of bile salts into the intestine. Digestion of lipids is impaired, enhancing excretion of fats in the stool. Absorption of fat-soluble vitamins (such as vitamins A and D) can lead to vitamin deficiency and bone deterioration. Some 95% of the cases are seen in women. The cause(s) of primary biliary cirrhosis are presently unknown.

Laboratory Findings

Laboratory studies for this disease are characterized by two typical phenomena (Table 11-10). Serum enzymes indicative of an obstructive process are increased. Both the alkaline phosphatase and the gamma-glutamyltransferase values are quite elevated, whereas the aminotransferase results are within normal limits or slightly increased. In later stages of the disorder, serum albumin is lowered and the bilirubin value is increased (both direct and indirect fractions). As the situation deteriorates, the concentrations of coagulation factors decrease due to the progressive hepatic damage.

Primary biliary cirrhosis appears to be an autoimmune disorder. A wide variety of immunological abnormalities are observed, although there does not yet appear to be a distinct and specific marker for the disease. Serum immunoglobulin levels are increased, particularly the IgM fraction. The presence of antimitochondrial antibodies has been seen in over 95% of the reported cases. In addition, antinuclear antibodies, antithyroid antibodies, and others have been detected in a large number of patients.

Table 11-10
SERUM CHEMISTRY ALTERATIONS IN
BILIARY CIRRHOSIS

Increased	Decreased
Alkaline phosphatase	Albumin
Gamma-glutamyltransferase	Coagulation factors
Bilirubin (both fractions)	
Immunoglobulins (particularly IgM)	
Antimitochondrial antibodies	

11.12

LIVER DISEASE IN PREGNANCY

During pregnancy, biochemical changes take place which reflect the changing status of the unborn child as he or she develops (Table 11-11). Changes in nutrition and body reservoirs of protein, alterations of enzyme levels caused by growth of the fetal/placental unit, and shifts in antibody concentrations are some of the variations which occur during the 9-month course of a normal pregnancy. These biochemical markers are often also altered during the course of liver disease. An awareness of similarities and differences in the laboratory status of healthy and ill patients during this time is essential for appropriate diagnosis and management.

As the pregnancy progresses, some hepatic biochemical parameters of interest remain unaltered. The values for serum total bilirubin, the aminotransferase enzymes, gamma-glutamyltransferase, and prothrombin time remain normal, although an increase in conjugated bilirubin may be observed. Total protein and albumin values tend to decrease, either due to dilutional factors or because of increased need for protein by the developing baby. Serum alkaline phosphatase steadily rises during the course of pregnancy, often reaching values at term which are double those of the nonpregnant population. This enzyme increase is due to release of alkaline phosphatase by the placenta as it grows. The serum levels of alpha and beta globulins rise somewhat, whereas the gamma-globulin fraction declines. Changing immunological interactions between mother and child may be responsible for at least some of these shifts in concentration.

11.13

ANATOMY AND PHYSIOLOGY OF THE KIDNEY

The kidney carries out a wide variety of functions necessary to our health. This organ is responsible for the maintenance of proper fluid and electrolyte balance in the body. Most waste products of metab-

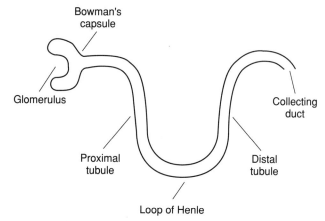

FIGURE 11-11   Anatomy of the nephron.

olism and toxic materials are excreted by the kidney. An important kidney function is its role in the synthesis of such materials as the hormones involved in the formation of red blood cells and the regulation of electrolyte excretion.

Anatomy

The human body has two kidneys, located in the back abdominal area. Each kidney connects to the **bladder** (where urine is stored) by a tube known as a **ureter.** The role of the ureter is simply to provide drainage from kidney to bladder during the process of urine formation. The bladder is composed of smooth muscle, which contracts to expel the fluid from the body.

The kidney itself is composed of nephrons (over a million in each kidney). The **nephron** (Fig. 11-11) is the basic structure involved in the filtration, secretion, and reabsorption of materials in the body. Each nephron consists of a **glomerulus** (blood supply at the initial site of filtration), **Bowman's capsule** (the initial filtration site), proximal and distal **tubules** (where secretion and reabsorption occur) and **collecting duct** (which drains the urine into the ureter and then to the bladder).

Table 11-11
LIVER MARKER CHANGES DURING PREGNANCY

Increase	Decrease	Unchanged
Bilirubin (conjugated)	Albumin	Bilirubin (total)
Alkaline phosphatase	Total protein	Aminotransferases
Alpha globulins	Gamma globulins	Gamma-glutamyltransferase
Beta globulins		Prothrombin time

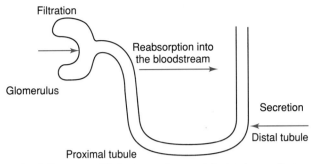

FIGURE 11–12 Movement of materials in the nephron.

Filtration and Reabsorption of Materials

The glomerulus is composed of a number of small arterioles, bringing large quantities of blood to each nephron (Fig. 11–12). Low-molecular-weight materials in the blood (anything other than most proteins) pass across the capillary wall through a thin membrane (known as the **basement membrane**) and then through the lining of Bowman's capsule. This filtration process is a passive operation, driven by the higher pressure on the blood vessel side of the system which forces water, electrolytes, and other small molecules through the membrane. The system acts in the same way as a piece of filter paper or a dialysis tube. No biochemical energy is required for this glomerular filtration portion of the urine-forming process.

As the fluid passes through the proximal tubule, reabsorption is the primary process occurring. Although the kidneys may filter some 170–190 L of fluid each day, only 1–2 L are excreted as urine. The remainder of the filtered fluid is reabsorbed through the proximal tubule back into the capillary bloodstream which surrounds each tubule. In addition to water, a large fraction of the sodium, potassium, bicarbonate, glucose, and amino acids which had originally been filtered out at the glomerulus are reabsorbed. This process requires biochemical energy and is regulated by specific hormones.

The **loop of Henle** connects the proximal and distal tubules. This portion of the tubule system also provides some reabsorption of electrolytes (but not of water). In addition, the process of secretion begins in the loop. Chloride ions (and some Na^+) are actively transported from the bloodstream into the tubule.

In the distal tubule, a small amount of reabsorption continues, but secretion of materials becomes much more important as the fluid and dissolved materials move to the collecting ducts. Hydrogen ions and K^+ are secreted into the tubule while a small amount of Na^+ is reabsorbed. Secretion of ammonia also takes place in the distal tubule as part of the process which regulates acid/base balance in the body (to be discussed in Chapter 18). Reabsorption

and secretion in the distal tubule are under hormonal control.

11.14

ROLE OF THE KIDNEY IN BIOCHEMICAL PROCESSES

Synthetic Functions

Although not as rich a source of biochemical compounds as the liver, the kidney nevertheless provides several much-needed materials for the body (Table 11–12). The major synthetic role for the kidney is in the manufacture of several hormones with wide-ranging metabolic effects. **Erythropoietin** formed by the kidney is involved in the regulation of red blood cell synthesis. **Renin,** a polypeptide, plays an important role in the renin/angiotensin system, regulating water and sodium balance by the kidney. A variety of **prostaglandins** are synthesized by the kidney and exert effects upon blood pressure, fluid and electrolyte balance, and other less clearly defined biochemical processes.

Metabolic Functions

Conversion of several biochemical materials occurs (in part) in the kidney. Breakdown of several hormones (aldosterone, glucagon, and insulin) to produce inactive fragments is accomplished by the kidney. Activation of vitamin D metabolites from inactive precursors is an important function for the kidney. Formation of creatine, required for some reactions in the muscle contraction process, takes place in the kidney as well as in some other tissues.

Excretory Functions

The major function for the kidney is excretory. All the toxic end products of metabolism (such as urea and uric acid) must be eliminated from the body,

Table 11–12
BIOCHEMICAL FUNCTIONS
OF THE KIDNEY

Function	Example
Synthesis	Erythropoietin, renin, prostaglandins
Metabolism	Inactivation of aldosterone, glucagon, insulin
	Activation of vitamin D
	Formation of creatine
Excretion	Ammonia, urea, uric acid
	Several minerals
	Toxic substances

since accumulation of these materials can be deleterious. Many minerals are consumed in excess in the diet; the amounts above what is needed for proper biochemical function are eliminated in the urine. The kidney also serves as a major route for excretion of ingested toxic substances, either consumed accidentally or in regulating drug levels during pharmacological treatment. (Chapter 20 covers the regulatory function in more detail).

11.15

ASSESSMENT OF RENAL CLEARANCE

Concept of Renal Clearance

The primary functions of the kidney are to excrete unwanted materials from the body and to retain those chemicals necessary for proper function. When the kidney malfunctions, one or both processes are altered. Measurements of the ability of this organ to carry out its major processes provide useful information regarding normal and abnormal renal function.

The excretion/retention roles of the kidney are carried out in three ways:

1. passive excretion (glomerular filtration)
2. reabsorption from the tubule back into the circulation
3. secretion from the circulation into the tubule

The most useful parameter to explore in any initial evaluation of kidney function is the excretion capability. Excretion can be assessed by determining the renal clearance of some substance, either occurring naturally in the bloodstream or administered to the patient. For reasons of safety and simplicity, measuring some material already present in the circulation is preferable to testing for an extraneous compound given to the patient.

This excretion capability is referred to as **clearance** and is defined as the (theoretical) amount of blood which can be cleared of the substance being measured in a specific unit of time. In essence, we determine a filtration rate by asking the question: how much of a specific material can the kidney remove from a given volume of blood in a stated period? To assess this parameter, a blood sample is collected along with a timed urine specimen (usually a 24-h collection). The clearance is calculated by measuring both serum and urine concentrations of the specific chemical and using the following formula:

$$\text{Clearance} = \frac{\text{Urine concentration} \times \text{Urine volume/Time}}{\text{Plasma concentration}}$$

The concentrations of material in plasma and urine must be expressed in the same units. The clearance then becomes a value expressed in milliliters per minute, indicating the volume of blood theoretically cleansed of the specific chemical each minute.

To be useful as a biochemical marker for clearance studies, the material being measured should ideally be excreted only through passive filtration at the glomerulus, and essentially no reabsorption or further secretion must take place in the tubules. The substance must also exist in the same biochemical form in blood and urine, requiring that no metabolic transformation take place before urinary excretion occurs. To date, the best naturally occurring biochemical marker is creatinine. Measurement of creatinine clearance provides useful information about the excretory capability of the kidney.

Creatinine

SOURCE

Creatinine is the anhydride of creatine, which is formed through a series of enzymatic reactions in the liver (Fig. 11-13). Creatine and adenosine tri-

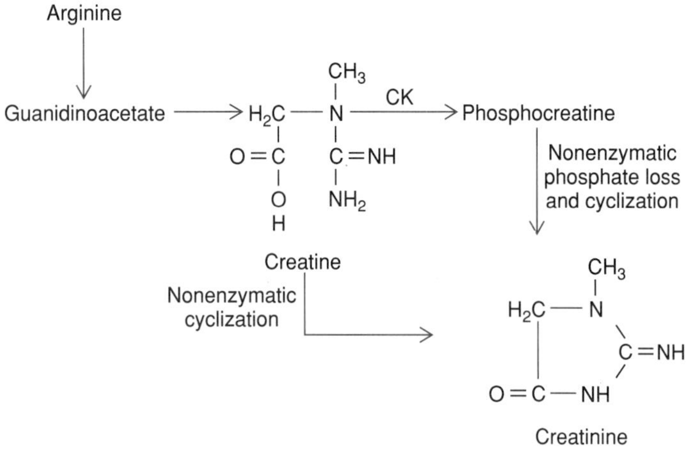

FIGURE 11-13
Formation of creatinine.

phosphate (ATP) are involved in the contractile process in skeletal muscle, mediated by the enzyme creatine kinase. As creatine is formed from phosphocreatine, a portion of this product spontaneously cyclizes to produce the anhydride, creatinine. This end product is readily filtered at the glomerulus and does not undergo any significant tubular reabsorption or secretion (although some 6–8% is eliminated by tubular secretion). The output of creatinine can be better related to muscle mass in an individual than total body mass. An obese individual can be expected to have a lower creatinine production rate than a nonobese individual with the same body mass.

ASSAY

The classic assay for serum and urine creatinine has been the reaction with picric acid in alkaline solution to form a red-orange complex with an absorbance at approximately 490–505 nm (known as the **Jaffe reaction**). In spite of extensive research, the exact structure of this complex is not known. The assay is somewhat temperature-sensitive. As the temperature increases above 30°C, other components begin to react with the picrate, forming chromogens which produce falsely high creatinine values. Time is also a significant factor in the assay. Extended incubation, even at room temperature, creates more nonspecific colored products. Protein must be removed before the reaction is carried out, either through manual precipitation or by dialysis in automated procedures.

A major drawback with the Jaffe assay is its lack of specificity. The presence of glucose, acetoacetate, protein, or ascorbic acid gives rise to falsely elevated values, sometimes as much as 0.3–0.4 mg/dL. Modifications to the Jaffe reaction have been explored, including the trapping of other materials with aluminum silicate or ion-exchange resins to remove reactive impurities before the alkaline picrate solution is added.

One alternative to the end-point Jaffe reaction is a kinetic assay. Although some interfering substances rapidly form complexes with the picrate, others require a fairly long time before they contribute significant absorbance to the system. By measuring the rate of color formation in the "window" between these two sets of reactions, the color contributed by creatinine can be more accurately assessed. Many studies recommend taking absorbance data between 25 and 60 s after the reaction has been initiated. Even then, some interferences are still present.

Other approaches to minimize interference involve applying enzymes to the assay. Creatinine iminohydrolase breaks down creatinine to produce ammonium ion. This material can be determined either colorimetrically or with an ion-selective electrode.

Table 11–13
APPROACHES TO ASSAY OF CREATININE

Method	Comments
Colorimetric: end point	Simple, nonspecific
Colorimetric: kinetic	Rapid, increased specificity
Enzymatic	Measure ammonia colorimetrically or with ion-selective electrode

Enzyme-based analysis can be employed to determine indirectly the amount of interferences present. Estimation of the nonspecific color-producing materials in the sample can be assessed by first destroying all the creatinine present (using enzymes). Any color which develops in the subsequent Jaffe reaction can be assumed to be due to interfering materials. A second assay can be performed with no enzyme pretreatment, with the value obtained being corrected for these interferences.

Analytical approaches to the measurement of creatinine are summarized in Table 11–13.

CLINICAL UTILITY

The reference range for serum creatinine is somewhat different for men and women, 0.9–1.5 mg/dL for men and 0.7–1.3 mg/dL for women. Values for children are lower, but begin to rise toward adult levels during adolescence. Serum creatinine levels do not begin to increase significantly until significant impairment of kidney function has already occurred. A more sensitive assessment of the renal system must be obtained through measurement of creatinine clearance.

Calculation of creatinine clearance is done after measurement of serum creatinine, and the creatinine in a timed urine specimen (usually 24-h output). Some studies have suggested a timed 2-h specimen provides adequate information since the creatinine production and excretion is reasonably constant throughout the day and night in a given individual. More reliable information can be obtained by correcting the clearance for body surface. This correction can be made using the following formula:

$$\text{Corrected clearance} = \frac{\text{Obtained clearance} \times 1.73}{\text{Body surface in square meters}}$$

Surface area can be calculated from standard tables by using the height and weight of the patient. The corrected clearance is expressed in milliliters per minute per 1.73 m². Values for men are roughly 85–145 mL/min/1.73 m², and women show values of approximately 75–130 mL/min/1.73 m². Clearance levels for both men and women decline somewhat with age.

Reduction of the creatinine clearance rate suggests renal impairment, but does not provide any information about the origin of the problem. Significant impairment is suggested by values which are 20–40% of normal. Values 40–60% of normal may indicate a moderate loss of kidney function, whereas results that are 60–80% of normal suggest minimal renal involvement.

Urea Nitrogen

SOURCE

Urea is a product of metabolism in the **Krebs-Henseleit urea cycle** (Fig. 11–14). This biochemical series of reactions begins with carbon dioxide and ammonia, which are coupled in an enzymatic reaction with ATP to form carbamoyl phosphate. Combination of this product with ornithine initiates a series of reactions which ultimately regenerate ornithine, producing urea as a by-product. Urea is predominantly excreted by the kidney through glomerular filtration, with essentially no reabsorption or secretion taking place. Measurement of **serum urea,** often referred to as **urea nitrogen** or **blood urea nitrogen (BUN),** can give information related to the capabilities of the kidney.

ASSAY

There are two major approaches to the assay of urea nitrogen: colorimetric and enzymatic (Table 11–14). The colorimetric method employs a reaction with diacetyl monoxime to form a colored complex which absorbs at 540 nm (Fig. 11–15). In solution,

Table 11–14
APPROACHES TO ASSAY OF UREA NITROGEN

Method	Comments
Colorimetric: diacetyl	Inexpensive
	Lacks specificity
Enzymatic: NH_3 formation	Greater specificity
	More expensive

diacetyl monoxime and acid form the unstable compound diacetyl, which is the actual species reacting with urea. Protein does not interfere with the reaction, so does not need to be removed prior to assay.

Enzymatic methods for urea measurement are quite popular and have been adapted to a variety of automated equipment. These assays involve an initial step which hydrolyzes urea to produce carbon dioxide and ammonia. In water, the products are carbonate anion and the ammonium cation. The ammonia can then be measured with the indophenol reaction previously described. An alternative method is to couple the ammonia production with the enzymatic assay of ammonia, using glutamate dehydrogenase. As alpha-ketoglutarate is converted to glutamate by the addition of the ammonium ion, NADPH is oxidized to NADP. The decrease in absorbance at 340 nm when NADPH reacts is proportional to the ammonia concentration.

CLINICAL UTILITY

The reference range for urea nitrogen is approximately 7–18 mg/dL in adults; children show somewhat lower levels. Values tend to increase somewhat with age. Urea nitrogen increases in a wide variety of disorders which produce renal impairment. Un-

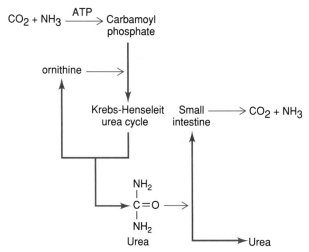

FIGURE 11–14   Source and excretion of urea.

FIGURE 11–15   Colorimetric assay for urea.

fortunately, specific levels at elevated urea nitrogen are not associated with any particular renal disease. On occasion, the ratio of serum urea nitrogen to serum creatinine may prove helpful, but it is still not a very specific indicator of particular types of kidney disease. Keep in mind that the urea nitrogen level may be somewhat low if liver damage is present or in cases of decreased protein intake, further complicating the interpretation of the urea nitrogen/creatinine ratio.

Uric Acid

SOURCE

Uric acid is the end product of purine nucleoside metabolism (Fig. 11–16). When compounds such as ATP lose phosphate groups to form the monophosphate derivatives, these intermediates are often further broken down to the original purines. Hypoxanthine is one common intermediate, which can then be converted to xanthine and finally to uric acid. The uric acid is either excreted as a waste product in the urine or transported to the intestine for further conversion.

The major route of uric acid elimination from the body is the kidney. Essentially all the uric acid is filtered at the glomerulus and appears in the proximal tubule. There a two-fold process occurs in which almost complete reabsorption takes place, accompanied by tubular secretion of approximately 50% of the total amount of uric acid in the circulation. In the distal tubule, some 40% of the uric acid is reabsorbed through active transport processes, leaving roughly 6–12% to be excreted in the urine. Approximately 400–500 mg uric acid may be excreted each day in the urine; another 200–500 mg is eliminated in the stool.

Clinical problems associated with elevated serum uric acid levels arise from the limited solubility of this compound. Once the concentration of uric acid in serum rises much above 6.5 mg/dL, it begins to precipitate out of solution. Gout is a condition related to the deposition of uric acid crystals in the joints. Renal abnormalities may result, including formation of stones in the kidney.

ASSAY

Colorimetric assay for serum uric acid is accomplished by a reaction with phosphotungstic acid. The uric acid present reduces the phosphotungstic acid to **tungsten blue,** a blue-colored derivative with an absorption maximum at 660 nm. Proteins must be removed by precipitation or dialysis prior to addition of the color reagent. Turbidity can be a problem, particularly if the pH is not strongly alkaline. Aspirin and its metabolites, acetaminophen, caffeine, and theophylline also produce color in this reaction, giving falsely elevated results.

Since the colorimetric methods lack specificity, enzymatic approaches to analysis have become very popular (Table 11–15). These assays are based on the **uricase reaction,** in which uric acid is converted to allantoin. Hydrogen peroxide is a second product in the reaction. Since uric acid absorbs light strongly in the region around 285 nm, the disappearance of this compound is indicated by a decrease in light absorbance. The decrease is proportional to the concentration of uric acid originally present. The major drawback to this assay is the need for a spectrophotometer which reads in the ultraviolet range and special quartz cuvets (glass does not transmit light much below about 340 nm).

A variation on the enzymatic assay which allows work in the visible region of the spectrum involves the measurement of hydrogen peroxide formed in the reaction. By coupling the H_2O_2 with an indicator compound, a colored product can be produced with an absorbance maximum in the visible region. However, a number of reducing substances also react with hydrogen peroxide, producing interference in this assay for uric acid.

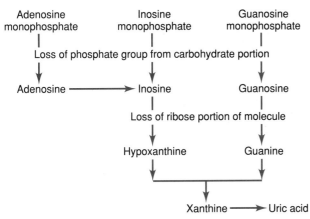
FIGURE 11–16 Formation of uric acid.

Table 11–15
APPROACHES TO ASSAY OF URIC ACID

Method	Comments
Colorimetric	Problems with turbidity / Several common drugs interfere
Enzymatic: UV	Need special instrumentation and optical cells
Enzymatic: H_2O_2 production	Interference by reducing substances

CLINICAL UTILITY

Reference ranges for serum uric acid differ between men and women and are somewhat method-dependent. If colorimetric analysis with phosphotungstic acid is employed, men show serum values of 4.5–8.2 mg/dL, and women have values of 3.0–6.5 mg/dL; the situation is more complicated for children. Values using uricase methodologies tend to be approximately 1 mg/dL lower for both men and women than those obtained using a colorimetric method.

Hyperuricemia has several causes. In diseases such as leukemia, increased cellular breakdown provides high levels of purines and increases purine turnover. Production of uric acid as end product of this metabolism increases. Several diuretics cause uric acid to increase in serum, either due to dehydration as more body water is lost or because of blockage of the tubular secretion of uric acid. Patients with diabetic ketoacidosis, renal failure, or rapid weight loss have impaired tubular secretion of uric acid, leading to elevated serum levels. In some situations, particularly those disorders related to bone disease, the cause of uric acid increase is not known.

11.16

LABORATORY DIAGNOSIS OF RENAL DISEASE

A number of common kidney diseases may be detected with fairly simple laboratory techniques (Table 11–16). Urinalysis often provides the first clue to a renal dysfunction if protein is increased

Table 11–16
CLINICAL DISORDERS OF IMPAIRED KIDNEY FUNCTION

Syndrome	Significant Laboratory Findings
Nephrotic syndrome	Striking proteinuria
	Decrease in serum albumin and total protein
	Relative increases in serum alpha$_2$- and beta-globulin fractions
	Increases in serum creatinine, BUN, and uric acid
	Lipids in urine
Glomerulonephritis	Elevated urine protein
	Increases in serum creatinine and BUN
	Red cells and casts in urine
Renal failure	Decrease in urine output
	Increases in serum BUN and creatinine
	Electrolyte and acid/base alterations
Tubular defects	Specific alterations in excretion of amino acids, electrolytes, or other specific biochemicals

and red cells are observed. Measurement of creatinine, BUN, total protein, albumin, and electrolytes frequently provides valuable clues to the presence of kidney problems. Unfortunately, these tests cannot point to a specific diagnosis. Rather, they indicate some sort of kidney problem which must then be explored with other techniques of medical diagnosis.

Nephrotic Syndrome

Loss of glomerular integrity produces an increased urinary excretion of proteins, a hallmark of the **nephrotic syndrome.** This cluster of kidney disorders is associated with increased permeability of the glomerulus, initially manifest by proteinuria. The nephrotic syndrome can be caused by a wide variety of disorders, including both renal diseases and other medical problems such as diabetes mellitus, malignancies, infectious diseases, and circulatory problems.

Significant laboratory findings include striking increases in urinary protein excretion (several grams per day instead of the usual 100 mg or less per day) and lowered serum protein concentrations. The decrease in serum albumin is the most obvious blood change. Albumin concentrations may drop to below 2.0 g/dL (reference range is 3.5–5.0 g/dL) with concurrent lowering of total protein to less than 4.0 g/dL (reference range: 6.0–8.0 g/dL). Serum electrophoresis shows lowered albumin with a relative increase in the percentage of alpha$_2$- and beta-globulin fractions. The increase in these two high-molecular-weight components is caused (in part) by a greater retention of these proteins in the serum relative to the lower molecular-weight albumin and other fractions. The beta fraction also rises in conjunction with the elevated cholesterol and lipoprotein concentrations seen in this disease.

Other laboratory findings include an increase in serum BUN and creatinine in approximately half the patients initially seen with this disorder. Uric acid values in serum are also increased in greater than 50% of patients. Lipids in the urine are observed upon staining. Urinary values for several enzymes (including alkaline phosphatase and lactate dehydrogenase) may be elevated, but cannot be considered as diagnostic. Routine assessment of urinary enzyme concentrations is not recommended. Little or no useful clinical information is gained from such analyses.

Glomerulonephritis

The clinical and laboratory findings for nephrotic syndrome and glomerulonephritis overlap considerably. In addition to the proteinuria observed in the nephrotic syndrome, patients with **glomerulonephritis** also demonstrate hematuria and red cell casts in the urine. The presence of these casts is a strong indicator of glomerular inflammation, since they rarely appear in other types of kidney diseases.

Glomerulonephritis seems to arise from dysfunction of the immune system. Whether the cause is primary (production of antibodies to the host tissue) or secondary to a wide variety of infectious or immune-system disorders (such as systemic lupus erythematosus), the precipitating step is the formation of an antibody which is foreign to the tissue. Two situations can then arise in which the formation of an antigen/antibody complex produces kidney damage. A preformed complex could deposit in the glomeruli of the kidney. Alternatively, an immune complex could form in the glomerulus through a reaction with a specific tissue antigen. In either case, the net result is damage to the glomeruli and alteration of the filtration, reabsorption, and secretion processes of the kidney.

The laboratory picture for glomerulonephritis is very similar to that for the nephrotic syndrome. Increases in urine proteins with decreases in serum albumin and total protein are seen. Both BUN and creatinine rise in the serum as a result of impaired renal function. The key diagnostic laboratory finding is the presence of hemoglobin and red cell casts in the urine.

Acute Renal Failure

There are three major causes of **renal failure.** Prerenal failure involves a decrease in urine output due to low blood pressure or fluid loss. Patients with cardiac problems (impaired circulation), heatstroke (dehydration), burns (fluid loss), or serious blood loss demonstrate prerenal failure. Renal failure is due to tubular injury which impairs the reabsorption/secretion processes, circulatory obstruction, or a blockage of the tubule by a stone of some type. Postrenal failure occurs as a result of bladder (or related) tumors, enlarged prostate, or stones. Many of the laboratory findings are similar, regardless of the cause of the disorder.

Urine output may be normal at first, but gradually declines. One noticeable finding is the daily rise in serum creatinine, indicating the progressive development of the disease. BUN levels also increase early in the development of renal failure. Urine protein may increase during the initial stages of the disease.

Dialysis is often employed for treatment of acute renal failure. Monitoring serum levels of sodium, potassium, and bicarbonate, along with pH measurement, should enable adequate assessment of treatment as well as providing early warning of any developing medical problems.

Renal Tubular Disorders

A variety of kidney disease exist which are associated with specific alterations in the ability of the tubule to carry out the reabsorption/secretion processes. Many of these disorders involve impaired transport of specific amino acids and are called **aminoacidurias. Cystinuria** (excessive loss of cysteine, a sulfur-containing amino acid) is one common tubular defect. These disorders are rarely seen in routine clinical practice. Specialized testing is required to identify the cause of the tubular problem. Other disorders arise because of changes in the regulatory biochemistry affecting electrolyte excretion. Hormonal alterations which produce effects on the excretion of sodium, potassium, and calcium will be dealt with in Chapters 18 and 19.

SUMMARY

The liver plays an important role in a wide variety of processes in the body. It is the primary site of synthesis for albumin, coagulation and transport proteins, and several hormones. A number of important metabolic processes occur in hepatic cells related to glucose synthesis and amino acid metabolism. Conversion of toxic compounds to less toxic metabolites and excretion of materials are the other functions of the liver.

Bilirubin is formed as a breakdown product of hemoglobin. This water-insoluble porphyrin derivative is transported by albumin to the liver where it is conjugated with glucuronic acid prior to excretion. Colorimetric measurement of bilirubin (both total concentration and water-soluble fraction) can be accomplished with a diazo reaction employing sulfanilic acid. Serum bilirubin levels are elevated in several diseases of the liver.

Ammonia is produced during the course of amino acid metabolism by the liver. The majority of the ammonia formed in the liver is either eliminated in the urine or detoxified by hepatic conversion to urea, which is then excreted. Measurement of ammonia with either colorimetric assay or enzymatic methods permits reliable assessment of blood ammonia concentrations. In hepatic encephalopathy, increased levels of serum ammonia can provoke neurological problems, including coma.

Laboratory diagnosis of liver disease is a nonspecific index of liver function, not usually pinpointing the cause of the disorder. Increases in serum bilirubin and aminotransferases generally point to some sort of hepatic damage. Elevated serum alkaline phosphatase may suggest an obstructive process. Lactate dehydrogenase measurements provide little or no useful information, although an increase in LD-5 may suggest hepatic involvement.

A major cause of liver disease is viral hepatitis. At least four types exist: type A, type B, and two types of non-A/non-B hepatitis. These diseases can be distinguished on the basis of serum measurement of viral markers and clinical history. Type A and type B hepatitis have distinct patterns of viral markers, but the non-A/non-B varieties are diagnosed by exclusion of other possibilities.

Alcoholic liver disease is a major health problem. Fatty liver, hepatic impairment, and cirrhosis are hallmarks of the disease. Although no specific biochemical marker is available, measurement of aminotransferases can provide some indication of the extent of tissue damage. Serial assays for gamma-glutamyltransferase are of some value in monitoring alcohol intake.

Reye's syndrome and biliary cirrhosis are somewhat rare disorders which affect the liver. The characteristic finding in Reye's syndrome is a striking increase in blood ammonia, often accompanied by coma. In children, Reye's syndrome is frequently preceded by a viral infection and can also be precipitated by aspirin ingestion. Biliary cirrhosis is an autoimmune disorder in which elevated levels of antimitochondrial antibodies are almost invariably detected.

The kidney plays a major role in fluid balance and in excretion of waste products. Materials are filtered at the glomerulus, with a complex process of reabsorption and tubular secretion taking place as the filtrate travels through the tubule to the collecting ducts. In addition, the kidney is involved with the synthesis of three hormones. Metabolic inactivation of insulin and conversion of inactive vitamin D to biochemically active metabolites are also carried out by the kidney.

Renal function is frequently assessed by measurement of serum creatinine, urea, or uric acid. Creatinine is an end product of muscle metabolism. This molecule is formed and excreted at a very regular rate by a healthy individual. Urea is synthesized as an end product of protein metabolism; the level of this compound in the blood depends to some extent on dietary factors as well as proper liver function. Uric acid is produced from nucleic acids released during cellular turnover and undergoes a complex filtration/reabsorption/secretion process in the kidney. All three parameters can be assayed by either colorimetric or enzymatic methods.

Specific kidney diseases cannot be diagnosed using particular testing procedures. Proteinuria, decreased serum levels of albumin and total protein, and increases of serum creatinine and BUN are seen in any one of the major syndromes affecting the kidney. Although these tests indicate some problem in kidney function, other diagnostic approaches are necessary to obtain a specific diagnosis.

FOR REVIEW

Directions: For each question, choose the best response.

1. Which of the following may be classified as the major tissue for excretion and reabsorption of materials?
   A. kidney
   B. liver
   C. lung
   D. spleen

2. Which of the following may be classified as being a function of the liver?
   A. detoxification of drugs
   B. excretion of bile acids
   C. metabolism of glucose
   D. synthesis of proteins
   E. all of the above

3. Which of the following may be associated with bilirubin metabolism?
   A. formed from breakdown of hemoglobin
   B. transported via albumin to the liver
   C. conjugated with glucuronic acid
   D. both A and B
   E. A, B, and C

4. The product formed from the reduction of bilirubin in the small intestine is:
   A. bile
   B. biliverdin
   C. porphyrin
   D. urobilinogen

5. Which of the following may be associated with direct bilirubin?
   A. soluble in water
   B. conjugated with glucuronic acid
   C. cannot be excreted in the urine
   D. both A and B
   E. A, B, and C

6. The enzyme system that catalyzes the conjugation of bilirubin is known as
   A. leucine amino peptidase
   B. glucose-6-phosphate dehydrogenase
   C. UDP-glucuronyl transferase
   D. carbamyl phosphate synthetase

7. Which of the following precautions should be taken when collecting and handling a specimen for bilirubin analysis?
   A. protect specimen from light
   B. avoid hemolysis
   C. freeze specimen immediately
   D. both A and B
   E. both A and C

8. The bilirubin fraction that is covalently attached to albumin and contributes to the conjugated bilirubin value is
   A. delta
   B. direct
   C. indirect
   D. bound

9. Which of the following may be associated with elevated serum bilirubin levels?
   A. increased red cell destruction
   B. decreased conversion of unconjugated to conjugated bilirubin

    C. decreased bilirubin excretion

    D. both B and C

    E. A, B, and C

10. The production of a diazo product is associated with the measurement of

    A. ammonia

    B. bilirubin

    C. creatinine

    D. urea

11. Elevation of only the unconjugated fraction of bilirubin may be associated with

    A. hemolytic anemia

    B. newborns

    C. hepatitis

    D. both A and B

    E. A, B, and C

12. A source of ammonia in the circulation is the

    A. catabolism of creatine

    B. catabolism of creatinine

    C. metabolism of uric acid

    D. hydrolysis of urea

13. The use of a cation exchange resin, an elution step, and reaction of the analyte with phenol-hypochlorite describes the method to quantitate

    A. ammonia

    B. creatinine

    C. urea

    D. uric acid

14. Increased blood ammonia levels may be associated with

    A. hepatic encephalopathy

    B. neurological changes

    C. coma

    D. both B and C

    E. A, B, and C

15. Which of the following enzymes exhibit a decreased serum level in liver disease?

    A. alanine aminotransferase

    B. gamma-glutamyltransferase

    C. lactate dehydrogenase

    D. cholinesterase

    E. all of the above

16. Which of the following enzymes exhibit an increased serum level in liver disease?

    A. alkaline phosphatase

    B. alanine aminotransferase

    C. gamma-glutamyltransferase

    D. lactate dehydrogenase

    E. all of the above

17. Which of the following enzymes is useful in diagnosing hepatobiliary obstructive disorders?

    A. alkaline phosphatase

    B. alanine aminotransferase

    C. gamma-glutamyltransferase

    D. lactate dehydrogenase

18. Which of the following parameters would be detected in blood in the presence of hepatitis B disease?
    A. increased bilirubin
    B. increased ALT
    C. HBsAg
    D. HBeAg
    E. all of the above

19. Which of the following laboratory results characterize Reye's syndrome?
    A. increased plasma ammonia
    B. increased ALT
    C. increased AST
    D. both B and C
    E. A, B, and C

20. Reabsorption of water, sodium, potassium, bicarbonate, glucose, and amino acids is the primary function of the
    A. glomerulus
    B. proximal tubule
    C. loop of Henle
    D. distal tubule

21. An endogenous substance that is commonly quantitated in clearance tests since it is filtered by the glomeruli, not reabsorbed, and only minimally secreted by the tubules is
    A. creatinine
    B. inulin
    C. urea
    D. uric acid

22. The Jaffe reaction, which utilizes picric acid, is employed for the quantitation of
    A. ammonia
    B. creatinine
    C. urea
    D. uric acid

23. Calculate a creatinine clearance based on the following information: urine concentration = 120 mg/dL; plasma concentration = 1.5 mg/dL; urine volume for 24 h = 1520 mL; body surface = 1.60 m^2.
    A. 78 mL/min
    B. 82 mL/min
    C. 84 mL/min
    D. 91 mL/min

24. In the diacetyl monoxime method, diacetyl reacts directly with
    A. ammonia
    B. ammonium ion
    C. urea
    D. uric acid

25. A compound which reacts with phosphotungstic acid causing the reduction of the latter to a tungsten blue complex is
    A. ammonia
    B. creatinine
    C. urea
    D. uric acid

26. In the uric acid ultraviolet procedure, the reaction between uric acid and uricase causes
    A. a destruction of uric acid
    B. the formation of allantoin

    C. a decrease in absorbance proportional to uric acid concentration

    D. both A and B

    E. A, B, and C

27. Elevated serum uric acid levels may be associated with

    A. nephrotic syndrome

    B. hepatitis

    C. gout

    D. both A and B

    E. both A and C

28. Which of the following laboratory findings may be associated with glomerulonephritis?

    A. elevated urine protein

    B. red cells and casts in urine

    C. increased serum creatinine and urea

    D. both A and C

    E. A, B, and C

BIBLIOGRAPHY

Alcoff, J., "Viral hepatitis," *J. Fam. Pract.* 15:141–162, 1982.

Bakerman, S., and Bakerman, P., "Pediatric hyperammonemia," *Lab. Manage.* May, 1986.

Bakerman, S., and Bakerman, P., "Reye's syndrome: Laboratory and clinical features," *Lab. Manage.* August:25–28, 1986.

Bates, H. M., "Hepatitis D. I: Discovery and nature of the pathogen," *Lab. Manage.* April:14–16, 1986.

Bates, H. M., "Hepatitis D. II: Clinical significance and epidemiology," *Lab. Manage.* June:9–13, 1986.

Beck, L. H., "Clinical disorders of uric acid metabolism," *Med. Clin. North Am.* 65:401–411, 1981.

Berlin, N. I., and Berk P. D., "Quantitative aspects of bilirubin metabolism for hematologists," *Blood.* 57:983–999, 1981.

Blumberg, B. S., "Australia antigen and the biology of hepatitis," *Science.* 197:17–25, 1977.

Brenner, B. M., and Humes, H. D., "Mechanics of glomerular ultrafiltration," *New Eng. J. Med.* 297:148–154, 1977.

Burke, M. D., "Liver function," *Human Path.* 6:273–286, 1975.

Centers for Disease Control (Atlanta), "Hepatitis A among drug abusers," *J. Amer. Med. Assoc.* 259:3235–3236, 1988.

Chang, Y-W, "Serologic markers of viral hepatitis," *Diag. Med.* July/August:28–39, 1983.

Cole, T. M., "Viral marker testing for hepatitis B," *Diag. Clin. Test.* May:23–28, 1989.

Czaja, A. J., and Davis, G. L., "Hepatitis non A, non B. Manifestations and implications of acute and chronic disease," *Mayo Clin. Proc.* 57:639–652, 1982.

Dawson, A. M., "Regulation of blood ammonia," *Gut.* 19:504–509, 1978.

Del Greco, F., and Krumlovsky, F. A., "Role of the laboratory in management of acute and chronic renal failure," *Ann. Clin. Lab. Sci.* 11:283–291, 1981.

DeLong, G. R., and Glick, T. H., "Encephalopathy of Reye's syndrome: A review of pathogenetic hypotheses," *Pediatrics.* 69:53–63, 1982.

Diehl, A. M., "Alcoholic liver disease," *Med. Clin. North Am.* 73:815–830, 1989.

Feinstone, S. M., and Hoofnagle, J. H., "Non-A, maybe-B hepatitis," *New Eng. J. Med.* 311:185–189, 1984.

Feld, R. D., "Hepatitis A," *Clin. Chem. News.* July, 1989.

Fields, H. A. et al., "Non A/non B hepatitis detection methodology: a review," *Ligand Quart.* 5:29–33, 1982.

Fields, H. A., "Delta hepatitis: A review," *J. Clin. Immunoassay.* 9:128–142, 1986.

Finkelstein, F. O., and Hayslett, J. P., "Nephrotic syndrome: Etiology, diagnosis, and treatment," *Geriatrics.* 31:39–48, 1976.

Fox, I. H. et al., "Hyperuricemia: A marker for cell energy crisis," *New Eng. J. Med.* 317:111–112, 1987.

Frank, B. B. et al., "Clinical evaluation of jaundice," *J. Amer. Med. Assoc.* 262:3031–3034, 1989.

Fraser, C. L., and Arieff, A. I., "Hepatic encephalopathy," *New Eng. J. Med.* 313:865–873, 1985.

Fried, T. A., "Glomerular dynamics," *Arch. Intern. Med.* 143:787–791, 1983.

Gammal, S. H., and Jones, E. A., "Hepatic encephalopathy," *Med. Clin. North Am.* 73:793–813, 1989.

Gruber, H. E., and Seegmiller, J. E., "The clinical and laboratory significance of uric acid assay," *Lab. Manage.* October:21–28, 1983.

Grundy, S. M., and Vega, G. L., "Rationale and management of hyperlipidemia of the nephrotic syndrome," *Amer. J. Med.* 87:5-3N–5-11N, 1989.

Helzberg, J. H., and Spiro, H. M., "'LFTs' test more than the liver," *J. Amer. Med. Assoc.* 256:3006–3007, 1986.

Hollander, I. J., "Nonhemolytic hyperbilirubinemias," *Ann. Clin. Lab. Sci.* 10:204–208, 1980.

Hoofnagle, J. H., "Chronic type B hepatitis," *Gastroenterol.* 84:422–424, 1983.

Hoofnagle, J. H., "Type D (delta) hepatitis," *J. Amer. Med. Assoc.* 261:1321–1325, 1989.

Jones, A. L., and Schmucker, D. L., "Current concepts of liver structure as related to function," *Gastroenterol.* 73:833–851, 1977.

Jones, S. L. et al., "The architecture of bile secretion. A morphological perspective of physiology," *Dig. Dis. Sci.* 25:609–629, 1980.

Kaplan, M. M., "Primary biliary cirrhosis," *New. Eng. J. Med.* 316:521–528, 1987.

King, J. W., "A clinical approach to hepatitis B," *Arch. Intern. Med.* 142:925–928, 1982.

Klahr, S. et al., "The progression of renal disease," *New Eng. J. Med.* 318:1657–1666, 1988.

Lemon, S. M., "Type A viral hepatitis," *New Eng. J. Med.* 313:1059–1067, 1985.

Lester, R., "Not two, but three bilirubins," *New Eng. J. Med.* 309:183–185, 1983.

Lewis, J. H., and Zimmerman, H. J., "Drug-induced liver disease," *Med. Clin. North Am.* 73:775–792, 1989.

Lewis, S. M., "Pathophysiology of chronic renal failure," *Nurs. Clin. North Am.* 16:501–513, 1981.

Lieber, C. S., "Biochemical and molecular basis of alcohol-induced injury to liver and other tissues," *New Eng. J. Med.* 319:1639–1650, 1988.

Lott, J. A., "New concepts in serum bilirubin measurement," *Lab. Manage.* April:41–48, 1987.

Madaio, M. P., and Harrington, J. T., "The diagnosis of acute glomerulonephritis," *New Eng. J. Med.* 309:1299–1302, 1983.

Moreno-Otero, R. et al., "Primary biliary cirrhosis," *Med. Clin. North Am.* 73:911–929, 1989.

Morrell, G., "Serum bilirubin: An analytical breakthrough," *Lab. Manage.* October:11–18, 1981.

Narayanan, S., and Appleton, H. D., "Creatinine: A review," *Clin. Chem.* 26:1119–1126, 1980.

Pillay, V. K. G., "Clinical testing of renal function," *Med. Clin. North Am.* 55:231–241, 1971.

Ratliff, C. R., and Hall, F. F., "Blood ammonia returns to the laboratory," *Lab. Manage.* August:16–23, 1979.

Rosoff, L., Jr., and Rosoff, L., Sr., "Biochemical tests for hepatobiliary disease," *Surg. Clin. North Am.* 57:257–273, 1977.

Roxe, D. M., "Current status of renal clearance," *Ann. Clin. Lab. Sci.* 11:279–282, 1981.

Rustgi, V. K., "Liver disease in pregnancy," *Med. Clin. North Am.* 73:1041–1046, 1989.

Schiff, L., "Jaundice: Five-and-a-half decades in historic perspective. Selected aspects," *Gastroenterol.* 78:831–836, 1980.

Schmid, R., "Bilirubin metabolism: State of the art," *Gastroenterol.* 74:1307–1312, 1978.

Schrier, R. W., "Acute renal failure," *J. Amer. Med. Assoc.* 247:2518–2525, 1982.

Schumacher, R. T. and Trey, C., "Viral hepatitis types A, B, and non-A/non-B: Current concepts," *Ligand Quart.* 5:12–27, 1982.

Shirey, T. L., "Bilirubin fractions," *Amer. Clin. Prod. Rev.* October:32–35, 1987.

Sloan, R. W., "Hyperuricemia and gout," *J. Fam. Pract.* 14:923–934, 1982.

Soloway, H. B., "Hepatitis panels revisited," *Diag. Med.* April:23–26, 1985.

Steven, M. M., "Pregnancy and liver disease," *Gut.* 22:592–614, 1981.

Thompson, L. S., "Acute viral hepatitis," *Clin. Chem. News.* April, 1984.

Van Lente, F., "A fatal nephrotic syndrome," *Diag. Med.* May:9–14, 1984.

Van Thiel, D. H. et al., "Gastrointestinal and hepatic manifestations of chronic alcoholism," *Gastroenterol.* 81:594–615, 1981.

Venkatachalam, M. A., and Rennke, H. G., "The structural and molecular basis of glomerular filtration," *Circ. Res.* 43:337–347, 1978.

Whitley, K. et al., "Acute glomerulonephritis: A clinical overview," *Med. Clin. North Am.* 68:259–279, 1984.

Yoshida, T., "Substrate metabolism and renal failure," *Pediatr. Clin. North Am.* 23:627–637, 1976.

INTRODUCTION

CLASSIFICATION OF
CARBOHYDRATES
Monosaccharides
Disaccharides
Polysaccharides

DIETARY INTAKE AND
DIGESTION OF
CARBOHYDRATES
Dietary Sources of
Carbohydrates
Digestion of Starch
Digestion of Disaccharides

ABSORPTION AND METABOLISM
OF CARBOHYDRATES
Transport into the
Circulation
Overview of Carbohydrate
Metabolism
Formation of Glycogen from
Glucose
Metabolism of Glucose

Conversion of Galactose to
Glucose

ROLE OF HORMONES IN
CARBOHYDRATE
METABOLISM
Definition of a Hormone
Insulin and Its Action
Glucagon and Its Action
Other Hormones Affecting
Glucose Metabolism

ANALYSIS OF CARBOHYDRATES
General Considerations
Qualitative Determination of
Reducing Substances in
Urine
Qualitative Assessment of
Urine Glucose
Quantitation of Blood
Glucose

CARBOHYDRATE INTOLERANCE
Categories of Intolerance

Lactose Intolerance
Galactosemia

DIABETES MELLITUS
Historical Overview
Types of Diabetes
Oral Glucose Tolerance Test
Glucose/Insulin Relationships
in Diabetes
Ketone Bodies and Diabetes
Mellitus
Insulin Therapy of Diabetes
Mellitus
Home Monitoring of Blood
Glucose
Glycosylated Hemoglobin
Fructosamine

HYPOGLYCEMIA

SUMMARY

FOR REVIEW

CARBOHYDRATE BIOCHEMISTRY

1. Define the following terms:
   A. glycolysis
   B. glycogenolysis
   C. gluconeogenesis
   D. hypoglycemia
   E. diabetes mellitus
   F. gestational diabetes
   G. glucosuria
2. Name the three classes of carbohydrates.
3. Identify the sugar composition of each class of carbohydrates.
4. Explain the processes by which carbohydrates are digested and absorbed into the blood.
5. State the origin and action of each of the glucose regulatory hormones.
6. State the reference ranges for
   A. serum glucose
   B. CSF glucose
   C. glycosylated hemoglobin
7. Explain the relationship between blood and CSF glucose concentrations.
8. State the principles of the copper sulfate and glucose oxidase methods for determining urine sugar.
9. Describe the proper collection and handling of a blood specimen for glucose analysis.
10. Name two chemicals that may be in a blood collection tube that inhibit the glycolytic process.
11. State the principles of the following serum glucose methods:
    A. *o*-toluidine
    B. glucose oxidase
    C. hexokinase
12. State the biochemical defects and the physiological effects associated with lactose intolerance and galactosemia.
13. Explain the purposes of performing the oral glucose and lactose tolerance tests.
14. Explain the major differences between the insulin-dependent and the noninsulin-dependent diabetic.
15. Discuss the clinical implications associated with gestational diabetes.
16. Name the three compounds collectively referred to as ketone bodies.
17. State the principle of the sodium nitroprusside test for detecting acetoacetic acid.
18. Discuss the clinical significance associated with detecting abnormal levels of the following:
    A. urine ketones
    B. glycosylated hemoglobin
    C. serum fructosamine
19. State the specimen requirements and the principles of the column chromatography and electrophoresis methods for measuring glycosylated hemoglobin.

20. State the principles of the colorimetric assays used for quantitating fructosamine.
21. State several causes of hypoglycemia.
22. Explain the use of a 5-h oral glucose tolerance test to facilitate the diagnosis of hypoglycemia.

INTRODUCTION

Providing chemical energy to the body, furnishing part of the structural integrity of the cell membrane, determining your blood type—these are some of the important roles carbohydrates play in the body. These molecules, whether they enter the system by way of our diet or are formed from other biochemical components, are vital constituents of all living organisms. By providing the building blocks for many processes of metabolism, carbohydrates are central ingredients for life.

12.1

CLASSIFICATION OF CARBOHYDRATES

Originally named because they are composed of carbon and the elements of water, **carbohydrates** can be divided into three classes (Table 12–1):

1. monosaccharides—simple sugars
2. disaccharides—composed of two monosaccharides
3. polysaccharides—made up of many monosaccharide units

Table 12–1
CLASSIFICATION OF CARBOHYDRATES

Monosaccharides	Disaccharides	Polysaccharides
Glucose	Sucrose	Starch
Fructose	Lactose	Glycogen
Galactose	Maltose	Cellulose
Ribose		
Deoxyribose		

The polysaccharides can be found in most foods. They are also formed in the body from monosaccharides. Disaccharides are ingested as part of the diet and then broken down to monosaccharides in the small intestine.

Monosaccharides

Though the **monosaccharides** are small molecules, they are extremely vital to the proper functioning of living systems. The most important monosaccharides contain only five or six carbon atoms in the chain. Structures of several important monosaccharides are illustrated in Figure 12–1.

Ribose and **deoxyribose** are **pentoses** (five-carbon sugars). These two molecules are part of the structure of **ribonucleic acid (RNA)** and **deoxyri-**

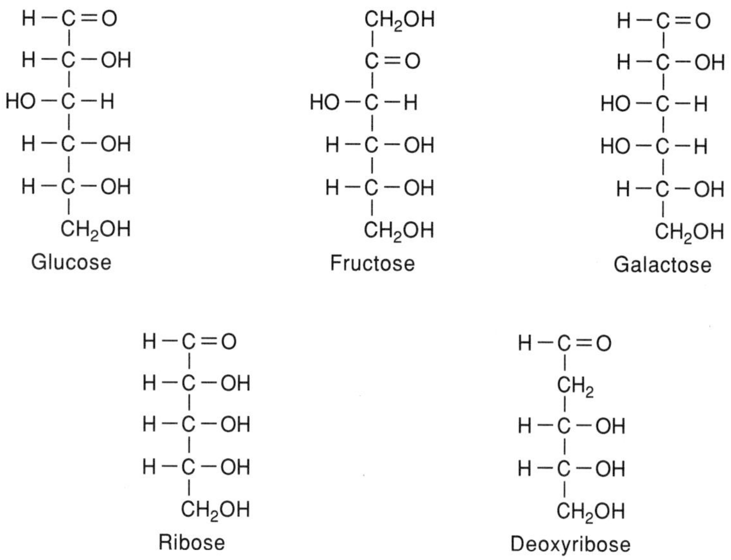

FIGURE 12–1
Structures of important monosaccharides.

bonucleic acid (DNA). Both RNA and DNA are involved with the genetic code, which determines our eye color, blood type, and other inherited characteristics.

There are three **hexoses** (six-carbon sugars) of importance: glucose, fructose, and galactose. **Glucose** is the central, pivotal point of carbohydrate metabolism. In a way, we assess total carbohydrate use by the body when we measure blood glucose levels. **Fructose** is formed from glucose and from the breakdown of sucrose; this molecule is a key intermediate in the utilization of monosaccharides. Of somewhat less significance from a metabolic point of view is **galactose**. This hexose must be converted to glucose before it can be used by the body. Patients with galactosemia have difficulty carrying out this transformation.

Disaccharides

Several **disaccharides** are of importance (Fig. 12-2), both because of the nutrients they provide and in their roles as biochemical markers for certain disorders of carbohydrate malabsorption. **Sucrose** (common table sugar) is the best known of the disaccharides. Obtained from beets, sugar cane, and a variety of other sources, sucrose provides a major portion of the carbohydrate intake for many individuals. **Lactose** ("milk sugar") is found in dairy products such as milk and cheese. Cereals, wheat, and malt products are good sources for the disaccharide **maltose.**

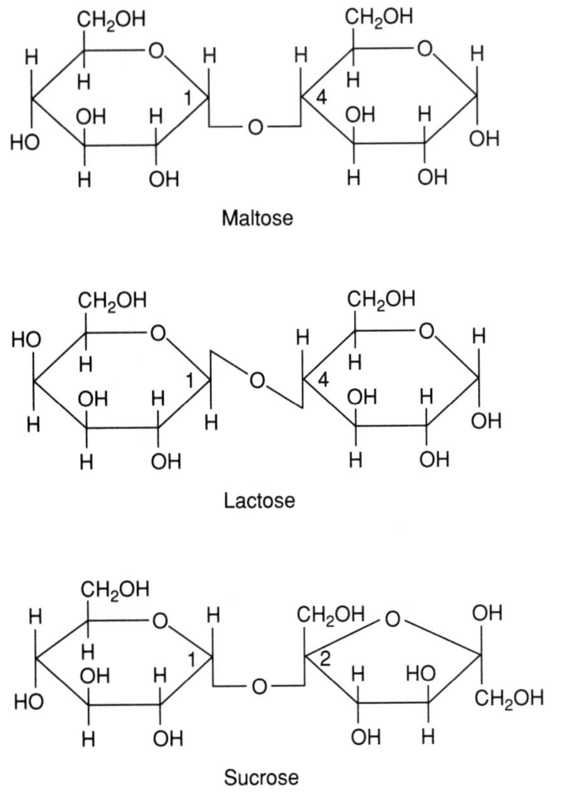

Polysaccharides

Many polysaccharides are known, but only three are important as dietary sources of carbohydrates or for storage of biochemical energy. **Starch** is the primary carbohydrate in the diet and is found in most plants. **Cellulose** is another polysaccharide in plants. Although cellulose is not digested by humans, it does provide bulk for proper intestinal functioning. The storage form of carbohydrate in the body is **glycogen,** formed from glucose by the liver. At those times when the body is not taking in carbohydrate from eating, glycogen breaks down to form glucose to help maintain the proper blood level of this necessary monosaccharide.

Both glycogen and starch are formed from glucose molecules which are joined together in long chains with many branches. Many "smaller" forms of both molecules are found as the digestion of the polysaccharide proceeds, but these intermediate forms are not important for our discussion. Cellulose has a somewhat different structure, which makes it much less digestible by humans.

12.2

DIETARY INTAKE AND DIGESTION OF CARBOHYDRATES

Dietary Sources of Carbohydrates

Three major classes of carbohydrates are consumed in the diet: starch, cellulose, and disaccharides. Since cellulose remains undigested by humans, we will not consider this material as a contributor to carbohydrate intake. Little glycogen remains in animal tissue after death (and is not found in plant tissue), so this polysaccharide does not provide any dietary carbohydrate. Therefore, the only two types of carbohydrates which contribute to human nutrition are starch and the various disaccharides.

Digestion of Starch

Starch is a long, branching chain of glucose molecules which are linked together. Before the body can use this glucose, the starch must be broken down into its individual glucose components. Amylase is the enzyme responsible for the breakdown of starch. Although some amylase is found in saliva, this fraction of the enzyme is not involved to any ex-

FIGURE 12-2 Structures of important disaccharides.

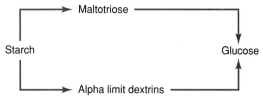

FIGURE 12-3　Conversion of starch to glucose.

tent in the digestion of starch. Whatever salivary amylase is present is quickly inactivated by acid when the food reaches the stomach.

Starch digestion occurs in the small intestine. When the food passes through the stomach and enters the intestinal tract, biochemical signals are sent to the pancreas. This organ then releases a wide variety of materials into the intestine. Pancreatic secretions and bile from the liver neutralize the acid from the stomach and change the pH to a more basic one. A number of enzymes are also released by the pancreas to digest carbohydrates and other materials. Amylase from the pancreas enters the small intestine to break down the starch. Although glucose is the major product of starch digestion, a number of short-chain carbohydrate polymers are also formed (Fig. 12-3). These small polysaccharides are then metabolized further to form monosaccharides by a variety of enzymes.

Digestion of Disaccharides

The disaccharides in the diet are digested by specific enzymes (Table 12-2). Each disaccharide has a specific enzyme responsible for its conversion to the constituent monosaccharides. **Sucrase** catalyzes the conversion of sucrose to fructose plus glucose. The disaccharide lactose is transformed to one molecule each of glucose and galactose by the enzyme **lactase**. **Maltase** forms two molecules of glucose from the disaccharide maltose. Some individuals have a deficiency of one of the enzymes responsible for disaccharide digestion in the intestine. This enzyme lack causes difficulty in the ability to absorb that particular carbohydrate from the intestine. The malabsorption of a disaccharide may produce nausea, vomiting, and/or abdominal cramping. Specific tests for the diagnosis of some of these enzyme deficiencies will be discussed later in this chapter.

Table 12-2
CONVERSION OF DISACCHARIDES TO
MONOSACCHARIDES

Disaccharide	Enzyme	Monosaccharide Products
Sucrose	Sucrase	Glucose + fructose
Lactose	Lactase	Glucose + galactose
Maltose	Maltase	Glucose + glucose

ABSORPTION AND METABOLISM OF CARBOHYDRATES

Transport into the Circulation

Carbohydrates are absorbed across the intestinal wall and into the bloodstream by an active transport process which requires biochemical energy. For the most part, only the monosaccharides are transported (both hexoses and pentoses). Starch, small carbohydrate polymers, and disaccharides are broken down into smaller units before passing into the circulation; they are not transported intact to any appreciable degree.

Overview of Carbohydrate Metabolism

After entering the circulation, the various monosaccharides are brought to the liver for further metabolism. Each monosaccharide must enter the cell before conversion to other biochemical compounds occurs. This process of transport into the cell is complex and regulated by a chemical called insulin. The role of insulin in carbohydrate metabolism will be discussed in detail later in this chapter. Ribose and deoxyribose eventually become incorporated into nucleic acids and thus become a part of the gene structure of the body. Galactose is converted to glucose by a series of enzyme reactions which will be discussed later in this chapter when we talk about the problem of galactosemia. Fructose enters the glycolytic pathway (see below) and is further converted to biochemical energy. Glucose can be channeled in two directions (Fig. 12-4): in one this molecule is converted to glycogen and stored in the liver for "quick energy." In the other glucose is broken down by way of the Embden-Meyerhof glycolytic pathway, resulting ultimately in lipids or other needed materials.

Formation of Glycogen from Glucose

To enter the cell, glucose must first have a phosphate group added to the number 6 carbon, forming glucose-6-phosphate. The enzyme responsible for this conversion is **hexokinase**. The glucose-6-phos-

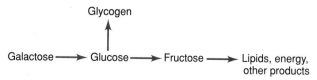

FIGURE 12-4　Interconversions of monosaccharides.

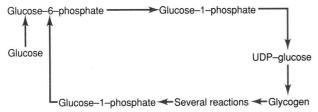

FIGURE 12-5 Interconversion of glucose and glycogen.

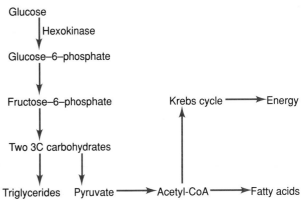

FIGURE 12-6 Overview of glucose metabolism.

phate can then be converted to glucose-1-phosphate (Fig. 12-5), and then to uridine diphosphoglucose (UDP-glucose). We will see later that galactose also can be transformed into UDP-glucose, an important step in the utilization of galactose by the cell. At this point, the UDP-glucose then forms long chains of glucose molecules linked together to form glycogen. (The uridine diphosphate (UDP) is then recycled for further use.) The liver stores glycogen until the body needs a quick source of chemical energy from carbohydrates. When this energy demand is made, glycogen is converted back to glucose-1-phosphate, but not by the same pathway used in its formation. Through a very complex process, this form of glucose is formed and released back into the bloodstream. This sugar is then transported to cells to be used as an energy source.

Metabolism of Glucose

As part of the process of moving glucose into the cell, the carbohydrate has a phosphate group added to become glucose-6-phosphate. This form of the molecule can then be transported into the cell. Once glucose enters the cell, it can be changed into other molecules and used for a number of purposes by the body. Although a portion of the glucose may end up being stored as glycogen, the major route for metabolism is the conversion of glucose to smaller molecules (Fig. 12-6). This set of metabolic transformations is referred to as the **Embden-Meyerhof glycolytic pathway,** named after the two scientists who characterized this area of metabolism. The term **glycolysis** literally means "a splitting of glucose" (*lysis* meaning "splitting") and involves the breaking of glucose into a series of three-carbon molecules.

Glucose-6-phosphate is first converted to fructose-6-phosphate (another six-carbon sugar). The fructose-6-phosphate then adds a second phosphate group to form fructose-1,6-diphosphate. This molecule then breaks into 2 three-carbon compounds which can then be used to form triglycerides (an important class of lipids) or undergo transformation to form pyruvate. The pyruvate then converts to **acetyl coenzyme A** (acetyl-CoA), an important biochemical intermediate. Acetyl-CoA undergoes a complex series of changes which lead to the production of biochemical energy. Another process using acetyl-CoA is the synthesis of fatty acids, another class of lipids. A very intricate and interrelated pattern is woven, with glucose at the center of activity. Changes in one segment of the pattern have far-reaching consequences in other areas of the process. We will see how complex some of these changes can be when we consider some of the biochemical alterations observed in diabetes mellitus.

Conversion of Galactose to Glucose

Galactose is a monosaccharide produced by the digestion of lactose. Galactose is phosphorylated by the enzyme galactokinase and enters the cell as galactose-1-phosphate (Fig. 12-7). In the cell, an exchange reaction takes place between the galactose-

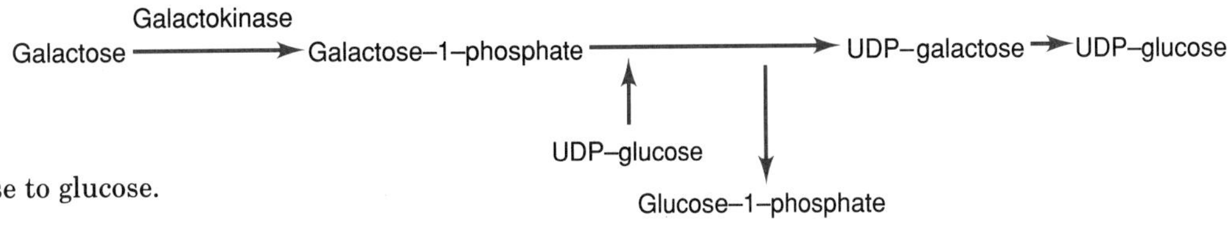

FIGURE 12-7
Conversion of galactose to glucose.

1-phosphate and a molecule of UDP-glucose. The products of the reaction are glucose-1-phosphate and UDP-galactose. The glucose-1-phosphate can then be further converted to glucose-6-phosphate and undergo normal carbohydrate metabolism. The UDP-galactose formed is enzymatically converted to UDP-glucose, providing another molecule of this glucose derivative for reaction either with galactose or for glycogen formation.

12.4

ROLE OF HORMONES IN CARBOHYDRATE METABOLISM

Before glucose can be utilized, it must be moved into the cell. Then the intracellular enzymes have access to the molecule and can convert it to other materials. Two components involved in the transport process are insulin and receptors located on the outside of the cell membrane. These two structures also have a major role to play in determining the factors associated with the rate of glucose metabolism once the carbohydrate enters the cell.

Definition of a Hormone

Many biochemical processes are regulated by chemicals released into the bloodstream in response to changes taking place somewhere in the body. The compounds released are called **hormones** and are produced by the **endocrine system of glands.** These glands are special tissues located in various parts of the body. When the appropriate stimulus is noted by the cells of a particular endocrine gland, a specific hormone is discharged into the circulation. The hormone goes directly into the bloodstream from the endocrine gland and is immediately transported to its site of action in the body. The field of hormone biochemistry (endocrinology) is discussed thoroughly in Chapter 14.

Insulin and Its Action

Insulin is a hormone with a protein structure, composed of two polypeptide chains joined together by sulfur bonds called **disulfide bridges** (Fig. 12-8). This hormone is synthesized and stored in the

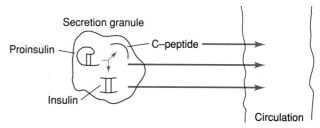

FIGURE 12-9 Release of insulin from beta cells into the bloodstream. Proinsulin cleaves into insulin and C-peptide.

beta cells (in the islets of Langerhans) of the pancreas as **proinsulin,** a single polypeptide chain which has coiled into the correct conformation. As the blood glucose level rises, a signal is sent to the pancreas (the exact nature of which is not known at present). While it is still in the pancreatic beta cell, proinsulin is cleaved by a specific enzyme. The products of this reaction are the active hormone insulin and an inactive portion known as **C-peptide.** The insulin (as well as the C-peptide and a small amount of unreacted proinsulin) are released into the bloodstream. The hormone travels through the circulation to various cells where it exerts its effects to promote glucose entry into the cell. Neither proinsulin or C-peptide have any biochemical activity of their own; they do not stimulate glucose uptake by the cell. Only insulin has the ability to effect glucose transport into the cells and stimulate its metabolism. Figure 12-9 illustrates the process of insulin formation and release into the bloodstream.

The other major part of the glucose transport process is the **cell receptor.** The functions of the receptor are quite complex, involving both the movement of glucose into the cell and stimulation of overall metabolic processes. When insulin acts on a cell, it does so by binding to a specific insulin receptor site on the outside of the cell membrane. Insulin does not enter the cell itself, but causes chemical signals to be sent inside the cell when the hormone attaches to the receptor.

Exactly how the receptor carries out its work is still not very clear, but we have an overall picture of the process (Fig. 12-10). As insulin attaches to the

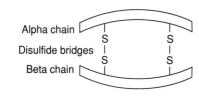

FIGURE 12-8 Structure of human insulin.

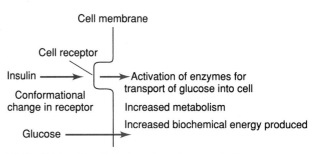

FIGURE 12-10 Effects of insulin on cellular metabolism.

receptor, the shape of the receptor changes. This change of shape causes release of chemicals inside the cell. Some of the chemicals make the movement of glucose into the cell possible by activating the enzyme system involved in the transport process. Other materials released stimulate the enzymes responsible for the conversion of glucose into different materials. The biochemicals activate some enzymes, making them function more efficiently. Other enzymes are synthesized in greater amounts within the cell, giving a higher level of enzyme for more rapid transformation of glucose. The receptor is an integral part of the glucose transport and metabolism process. Changes in receptor structure and activity play a central role in some of the problems associated with glucose utilization in diabetes mellitus.

Glucagon and Its Action

Glucagon is a small polypeptide hormone produced by the alpha cells of the pancreas. Glucagon acts in a manner opposite to that of insulin. When blood glucose levels drop below a certain point, glucagon is released from the pancreatic alpha cell. Whereas insulin affects most cells of the body, glucagon has a specific target tissue—the liver. When this hormone interacts with liver cells, two processes are increased: glycogenolysis and gluconeogenesis.

Glycogenolysis is the process involving the splitting of glycogen (*glycogen* + *lysis*). The enzyme phosphorylase is activated through a complex process, resulting in the breakdown of glycogen to reform glucose-1-phosphate. This carbohydrate is then transformed into glucose-6-phosphate and fed into the Embden-Meyerhof glycolytic pathway. Glucagon causes the breakdown of glycogen to predominate, instead of the reverse process of glycogen synthesis and storage.

Gluconeogenesis is the synthesis of glucose from other compounds (*glucose* + *neo* (new) + *genesis*). In this process, amino acids are converted through a special pathway into glucose, which can then be used in the normal fashion. A small amount of gluconeogenesis is always taking place in the body. When glucagon is released from the pancreas, however, the body is being sent the signal that not enough glucose is available and more needs to be made. So there is an increase in the rate of gluconeogenesis to try to raise the glucose level in the circulation.

A very delicate balance exists between glucose utilization and production. This balance is maintained by the interplay between the hormones insulin and glucagon. Although most research in diabetes mellitus has been concerned with the role of insulin, there is some evidence to suggest that glucagon may be more important in this disease then was first suspected.

Table 12-3
HORMONAL ALTERATION OF GLUCOSE METABOLISM

Hormone	Biochemical Processes Affected
Insulin	Increases cellular uptake, glycolysis
Glucagon	Increases glycogen formation
Cortisol	Stimulates gluconeogenesis
Catecholamines	Stimulate glycogenolysis
Thyroid hormones	Stimulate glycogenolysis
Somatomedins	Act like insulin or inhibit insulin breakdown
Somatostatins	Inhibit glucagon and insulin secretion
Growth hormone	Inhibits insulin action

Other Hormones Affecting Glucose Metabolism

There are a wide variety of other hormones which alter the regulation of glucose utilization by the body (Table 12-3). The effects are complex and usually interrelated. Changes in the level of one hormone often alter levels of one or more other hormones. Frequently, more than one process of glucose utilization or formation is affected simultaneously.

12.5

ANALYSIS OF CARBOHYDRATES

General Considerations

Assessment of the state of carbohydrate production, metabolism, and excretion primarily involves the measurement of glucose in body fluids. The normal blood glucose level is approximately 70–110 mg/dL, depending on the assay method. Variations in glucose metabolism result in levels as low as 20–30 mg/dL (in hypoglycemia) or as high as 800 mg/dL (in diabetes mellitus). Therefore, any method used to quantitate glucose in serum or plasma must be able to measure this material accurately throughout a wide range of values. Levels in CSF are normally about 40–70 mg/dL. Cerebrospinal fluid glucose concentrations are approximately 60–75% of the concentrations found in the blood of a specific individual. When the blood glucose level changes, approximately 2 h is required to restore the CSF/blood glucose equilibrium. Urine glucose is rarely quantitated; a random specimen is usually screened in a qualitative fashion. These screening processes should be able to detect urine glucose at levels of 30 mg/dL or higher.

In addition to the range capabilities of the assay, the issue of specificity must be addressed. There are other carbohydrates besides glucose in blood or urine. Although these sugars generally are present at much lower levels than glucose, they contribute to the value obtained if a nonspecific analytical method is used. For this reason, much effort has been expended over the years to develop tests which measure only glucose and not other carbohydrates. However, a nonspecific method for total carbohydrates is useful when screening urine samples for the presence of sugars. A urine screen may consist of a qualitative test for total sugar in urine (whatever the source) and another qualitative (or semiquantitative) assessment of the amount of glucose present.

Qualitative Determination of Reducing Substances in Urine

There are two major analytical approaches to the measurement of urine sugar. The nonspecific approach (which measures "total reducing substances") uses copper sulfate, generally packaged in the form of a tablet (Fig. 12–11). This tablet is placed in the bottom of a test tube and several drops of urine are carefully layered on top of the tablet. The reaction involves the interaction of the Cu^{2+} ion with the carbohydrate. The sugar is oxidized and the Cu^{2+} ion is reduced to Cu^{1+} (sugars which produce this reaction are called **reducing sugars.** All these compounds contain a carbonyl group which can be oxidized in this reaction). When the copper is in the +2 state, it gives a blue color in solution. The change to Cu^{1+} results in the formation of an orange-red color. The tablet also contains sodium carbonate, citric acid, and sodium hydroxide. When water comes into contact with the tablet, carbon dioxide is released and heat is generated. The CO_2 provides a "blanket" which keeps out oxygen and hinders the reoxidation of the Cu^{1+} ion. The heat produced makes the reaction between the copper and the sugar take place more rapidly.

There are two major limitations to the reducing-sugar approach to urine carbohydrate screening. First, the test is not very sensitive. In the tablet form, the assay may not show positive until a urine sugar level of about 0.25% (250 mg/dL) is reached. Although the problem of false positive test results is minimized by this low degree of sensitivity, the method may not detect those slight increases in urine sugar output which could give an early warning sign for the presence of a clinical abnormality.

In addition, the test lacks specificity. Glucose is not the only carbohydrate which reacts with this system; any sugar which can reduce Cu^{2+} gives a positive test. Although this is a desirable outcome if we are looking for some general sign of excessive car-

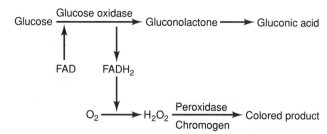

FIGURE 12–12 Glucose analysis with glucose oxidase/peroxidase.

bohydrate excretion, there are drawbacks to the lack of specificity. A wide variety of substances also give a positive reaction if present in the sample. Among these materials are sulfa drugs, vitamin C, high levels of aspirin, and the presence of ketone bodies (formed in more severe cases of diabetes mellitus).

Qualitative Assessment of Urine Glucose

The qualitative (or semiquantitative) measurement of urine glucose relies on a more specific test than that for reducing substances (Fig. 12–12). The enzyme glucose oxidase reacts only with glucose, not with any other carbohydrate. Products of the reaction are gluconic acid and hydrogen peroxide. When a **chromogen** (a material capable of forming a colored product) comes in contact with the hydrogen peroxide, a color change results. The intensity of the color is proportional to the amount of glucose present.

Commercial dipsticks are available which have glucose oxidase and the chromogen incorporated into a reaction package at the end of the stick. After the stick is dipped into a urine sample, the reagents are allowed to react with the glucose in the specimen for 1 min. Then the color is observed and compared with a standard color chart to determine how much glucose is in the sample.

There are three major advantages to the dipstick method compared with the reducing-sugar approach (Table 12–4). First, the reaction is specific for glucose; no other carbohydrate gives a positive test. In addition, the problem of drug interference is elimi-

$$CuSO_4 + Glucose \xrightarrow[\text{Alkali}]{\text{Heat}} Cu_2O + \text{Oxidized}$$
(blue)   (red–orange)   glucose

FIGURE 12–11 Qualitative test for urinary reducing substances.

Table 12–4
COMPARISON OF METHODS FOR ASSAYING URINE CARBOHYDRATE

Method	Reducing Substances	Glucose Oxidase
Specificity	Nonspecific	Glucose only
Lower limit	250 mg/dL	100 mg/dL
Interferences	Many drugs	Negligible
Cost	Very inexpensive	More expensive
Automation	No	Yes

nated. Since the reaction uses a specific enzyme, there are no false positive results. However, the presence of ascorbic acid in the urine can interfere with the peroxidase reaction, producing false negative values. Finally, use of the dipstick allows the detection of lower levels of glucose in urine. Whereas the reducing-sugar test only detects carbohydrates at 250 mg/dL or higher concentration, the glucose oxidase reaction has a lower limit of detection of approximately 100 mg/dL.

One significant problem, however, did become apparent when tests of the accuracy and reproducibility of the dipsticks were carried out several years ago. There was a remarkable variation in the glucose level reported by different laboratory personnel for the same urine sample. Depending on the glucose content of a given specimen, inconsistent readings would occur in 10% to over 50% of the cases. Simply put, many technicians read the color changes on the stick differently from their colleagues. In some instances, there may have been differences in perception of color. Laboratory accreditation standards now require all personnel to be tested for colorblindness. To overcome the drawbacks associated with manual performance of this test, instruments were developed which read the dipstick automatically and print out the results. Some models even do the dipping step without human assistance. Reproducibility and accuracy are significantly improved with the automated system. Laboratories with a large volume of samples for urine glucose screening find these instruments indispensable.

Quantitation of Blood Glucose

The measurement of glucose in blood is one of the most commonly performed laboratory tests today. A wide variety of methodologies and instrumentation have been developed within the last 10–15 years which make it possible to assay glucose accurately in a sample as small as 10 µL (or less) and in a time as short as 1 min. When we look at quantitation of glucose in blood samples, the following factors need to be considered in order to achieve accuracy and reliability:

1. choice of sample
2. choice of method
3. issues related to specificity
4. interferences
5. ease of measurement

Each one of these factors will be addressed separately.

CHOICE OF SAMPLE

When a blood sample is collected for glucose analysis, some of the components involved in the conversion of glucose to other biochemical compounds

are still present. In particular, red blood cells have all the enzymes necessary to metabolize glucose. Therefore, it is important to separate the cells from the liquid portion of the blood as soon as possible. If anticoagulants are used, the sample should be centrifuged as soon as possible (preferably within an hour after collection) and the plasma removed. Serum samples can be obtained after the blood clots; centrifugation again should not be delayed unduly. The use of collection tubes containing a gel-like material as a separation barrier between the cells and the serum minimizes the loss of glucose through cellular metabolism. Fluoride or iodoacetate is sometimes used in sample collection, since these materials inhibit the glycolytic process and prevent most glucose consumption by the erythrocytes.

After separation of either plasma or serum from the cells, glucose remains stable at room temperature for several hours in serum samples; glucose is somewhat less stable in plasma. If samples are not to be assayed soon after collection, glucose stability may be enhanced by cold storage of the samples.

SPECTROPHOTOMETRIC ASSAY OF GLUCOSE

A wide variety of methods for colorimetric measurement of glucose have been used over the years with varying degrees of success. Early methods involved an adaptation of the copper reduction technique previously described for qualitative determination of urine glucose. Both manual and automated procedures employed measurement of the color change from blue to orange-red as copper was reduced. However, as we pointed out, these methods were not very specific and were interfered with by many medications. As a result, other methods (both colorimetric and enzymatic) grew in popularity while the quantitative copper reduction methods slowly died out.

Once colorimetric method used today involves the reaction of glucose in strong acid with *o*-toluidine (Fig. 12–13). When heated at 100°C in glacial acetic acid, glucose forms a bluish-green derivative with *o*-toluidine, with an absorbance maximum of approx-

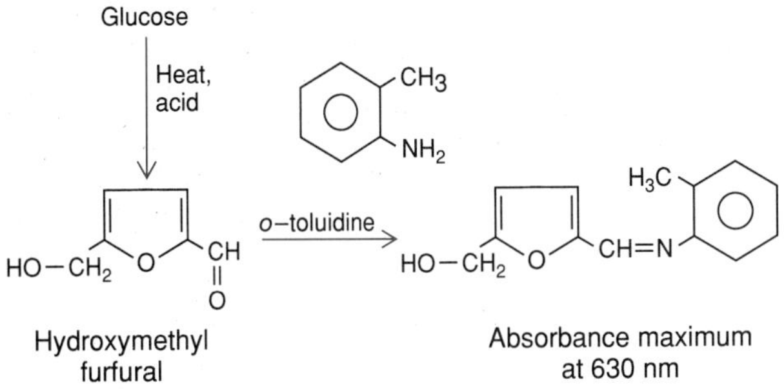

FIGURE 12–13 Colorimetric assay of glucose using *o*-toluidine.

imately 620 nm. The reaction is reasonably specific for glucose. Other carbohydrates which give this color are galactose, mannose, and lactose, all of which are present in extremely low concentrations in blood. No significant drug interferences are known. The major drawbacks to the o-toluidine method are the harsh chemicals and strong reaction conditions required. Heat and concentrated acid produce hazardous conditions and are very destructive to instrumentation over the long term.

ENZYMATIC ASSAYS

A higher degree of specificity can be achieved using an enzymatic method. Because of this improvement in accuracy, enzymatic methods for glucose analysis are more popular than colorimetric assays. In addition, the hazardous conditions and chemicals used in colorimetric procedures can be avoided. One such approach, the **glucose oxidase assay,** has already been discussed briefly in our consideration of urine glucose measurements. For serum or plasma, a coupled assay involving both glucose oxidase and peroxidase is frequently employed. The reaction principle is the same one used for urine glucose. As mentioned previously, glucose oxidase specifically converts glucose to another derivative, forming hydrogen peroxide as a second product in the reaction. The hydrogen peroxide then reacts further with the enzyme peroxidase to regenerate oxygen. Interaction of the oxygen with a chromogen produces a colored compound which can be detected spectrophotometrically, allowing quantitation of glucose. Some of the chromogens which have been employed in this reaction are o-anisidine, o-tolidine, and diethyl-p-phenylenediamine.

Glucose in solution exists in two forms (approximately 35% alpha and 65% beta) determined by the orientation of the hydroxyl group on carbon-1. Glucose oxidase is specific for the beta form of the molecule and does not react with the alpha form. Some commercially available assays include the enzyme mutarotase in the system to convert any of the alpha form present to the reactive beta configuration.

A more recent variant of the glucose oxidase theme involves some interesting technology developed commercially by at least two different companies (Fig. 12–14). In the **glucose oxidase method,** for each molecule of glucose reacted there is consumption of one molecule of oxygen (Fig. 12–14A). This disappearance of oxygen can be measured quite readily. The key ingredient is the use of an electrode which assesses the concentration of oxygen in solution in much the same way as a pH electrode determines hydrogen ion concentration. By measuring the rate of oxygen consumption, the glucose concentration can be determined in 1 min without requiring all the glucose in the sample to be consumed.

Glucose is converted by glucose oxidase in the

$$\text{Glucose} + O_2 + H_2O \xrightarrow{\text{Glucose oxidase}} \text{Gluconic acid} + H_2O_2$$

A

$$H_2O_2 + \text{Ethanol} \xrightarrow{\text{Catalase}} \text{Acetaldehyde} + \text{Water}$$

B

$$H_2O_2 + 2H^+ + 2I^- \xrightarrow{\text{Molybdate}} I_2 + \text{Water}$$

C

FIGURE 12–14   Oxygen rate method for glucose analysis. A. For each molecule of glucose that reacts, one molecule of oxygen is consumed. B. Hydrogen peroxide reacts with ethanol to produce acetaldehyde. C. Hydrogen peroxide reacts with iodide to form molecular iodine.

usual fashion. The oxygen concentration decreases as the glucose reacts with the enzyme. The rate of decrease is proportional to the concentration of glucose present (refer back to Chapter 8 to review substrate concentration/reaction rate principles). If left to itself, the hydrogen peroxide breaks down again, regenerating oxygen. This would obviously create a problem in measuring the rate of oxygen utilization. Therefore, two other reactions are incorporated into the system to take care of this difficulty. Hydrogen peroxide is either reacted with ethanol to form acetaldehyde (Fig. 12–14B) or with iodide to form molecular iodine (Fig. 12–14C). In either case, the hydrogen peroxide is destroyed and cannot regenerate oxygen.

Although the glucose oxidase method is reasonably free of interferences, it appears that acetaminophen and a few related compounds interfere with the reaction, probably by interacting with the enzyme system themselves. Other medications do not seem to present a problem with the assay. An alternative enzymatic approach is even more specific and

$$\text{Glucose} + \text{ATP} \xrightarrow{\text{Hexokinase}} \text{Glucose–6–phosphate} + \text{ADP}$$

A

$$\begin{array}{c}\text{Glucose–6–}\\\text{phosphate}\\+\\\text{NAD (or NADP)}\end{array} \xrightarrow{\text{Glucose–6–phosphate dehydrogenase}} \begin{array}{c}\text{6–phospho–}\\\text{gluconolactone}\\+\\\text{NADH (or NADPH)}\end{array}$$

B

FIGURE 12–15   Hexokinase assay of glucose. A. Hexokinase catalyzes the conversion of glucose to glucose-6-phosphate. B. Glucose-6-phosphate dehydrogenase catalyzes glucose-6-phosphate to 6-phosphate-gluconolactone.

involves the use of hexokinase in a coupled enzyme assay system. Hexokinase catalyzes the conversion of glucose to glucose-6-phosphate (Fig. 12–15A). A second enzyme (glucose-6-phosphate dehydrogenase) then carries the transformation one more step (Fig. 12–15B). In the reaction with the dehydrogenase enzyme, a molecule of NAD (or NADP, depending on the source of the dehydrogenase) is reduced to NADH (or NADPH). This product can be measured directly at 340 nm or it can be coupled to a reaction which yields a colored compound absorbing at another wavelength. For each molecule of glucose reacting in the system, one molecule of NADH (or NADPH) or resulting chromogen is formed.

The specificity of this system does not lie in the use of hexokinase, since this enzyme promotes the phosphorylation of several hexoses. It is the second enzyme (glucose-6-phosphate dehydrogenase) which makes the reaction specific for glucose; no other phosphate derivative is converted by this enzyme.

A number of commercial adaptations of the hexokinase system have been marketed. Most of them are colorimetric and rely on a further reaction of NADH (or NADPH) with a chromogen to form the final product. The colorimetric step usually involves some variation of the same type of reaction used to visualize lactate dehydrogenase (LD) isoenzymes or to assay for enzyme activity in systems where the reduced cofactor (NADH or NADPH) is formed. Phenazine methosulfate coupled to a final chromogen such as a nitrophenyl tetrazolium derivative is commonly employed in this type of assay. Interferences by other carbohydrates, uric acid, urea, ascorbic acid, or creatinine are usually negligible or nonexistent.

12.6

CARBOHYDRATE INTOLERANCE

Categories of Intolerance

Although the processes of digestion and absorption of carbohydrates usually proceed without any hindrance, there are situations where the movement of a given sugar from the intestinal tract into the circulation is impaired. Other clinical situations arise where there is a defect in the conversion of one carbohydrate to another. In both syndromes, the basic problem is caused by a lack of some specific enzyme. Since the amount of the enzyme is much lower than normal, the particular biochemical process cannot take place at the normal rate. This decrease in the ability of the body to transform one material into another results in the accumulation of the unconverted carbohydrate either in the intestine, in the circulation, or in the cell. The consequences of this accumulation may range from physical discomfort to cellular damage, mental retardation, and early death. We want to consider briefly two such medical problems: lactose intolerance, due to the inability of the small intestine to break down lactose; and galactosemia, where galactose accumulates because the body cannot convert it to glucose.

Lactose Intolerance

BIOCHEMICAL DEFECT

Lactose intolerance is a remarkably common disorder, affecting large percentages of several ethnic groups. The biochemical defect is characterized by a marked decrease in the level of lactase in the small intestine. Because this enzyme is at very low levels in the lactase-deficient patient, whatever lactose enters the system cannot be readily broken down into glucose and galactose. The disaccharide accumulates in the intestine and is metabolized by bacteria, causing a great deal of gastrointestinal discomfort, cramping, and diarrhea. Elimination of lactose from the diet (by discontinuing or greatly decreasing the consumption of dairy products) usually diminishes the symptoms significantly.

LACTOSE TOLERANCE TEST

If necessary, this disorder can be fairly easily diagnosed with a **lactose tolerance test.** The patient fasts overnight, then a fasting blood glucose level is determined. An oral lactose load is administered (for adults, use 50 g of lactose in 200–300 mL of water), and blood samples are taken every 30 min for the next 2 h. If the glucose level (changed because of the lactose load) rises 25 mg/dL or more above the fasting level, the patient has normal lactose tolerance (Table 12–5). An increase in blood glucose of less than 20 mg/dL is presumptive evidence of an inability to convert this disaccharide. Quite often, the individual also experiences abdominal bloating, and diarrhea during the test, providing additional evidence for the presence of the intolerance. As a practical matter, each sample is analyzed as soon as it is collected. If the glucose level in any single sample rises 25 mg/dL or more above the baseline, lactose absorption is normal and the test may be dis-

Table 12–5
LACTOSE TOLERANCE TEST DATA
(GLUCOSE mg/dL)

Patient Intervals	Normal	Lactose-Intolerant	Ambiguous
Fasting	89	85	93
30 min	104	96	107
60 min	121	92	115
90 min	*	90	104
120 min	*	88	97

* No further testing needed; change is greater than 25 mg/dL

continued. This approach reduces the number of venipunctures and assays necessary and the expense in some instances.

CAUSES OF LACTOSE INTOLERANCE

Lactose intolerance may be congenital (present from birth) or acquired later in life. If it is acquired, quite often the problem arises from some environmental cause. Even in this situation, there still may be a genetic component, making the person more susceptible to the disease. Populations vary markedly in their incidence of lactose tolerance. Whereas Caucasians generally exhibit a 5–20% incidence, some Chinese and Indian groups may have between 50 and 100% of their populations affected. Other Oriental ethnic groups and a large number of African populations also show high frequencies of lactose intolerance.

Galactosemia

DESCRIPTION

Although lactose intolerance is rarely life-threatening and usually creates only inconvenience and discomfort for those afflicted, **galactosemia** is a much more serious disorder. This disease is the result of the lack of an enzyme necessary for the conversion of galactose to glucose after the monosaccharides have been absorbed into the bloodstream and have entered the cells. Because galactose accumulates in this disease, the patient develops cataracts and mental retardation at a very early age. In most cases, the individual lives for only a few years after birth. The incidence of galactosemia is quite low, affecting about 1 in 62,000 newborns.

BIOCHEMICAL DEFECT

In most cases of galactosemia, the enzyme galactose-1-phosphate uridyltransferase is deficient, often being present at a level of less than 0.1% of normal. This enzyme catalyzes the exchange reaction between galactose-1-phosphate and UDP-glucose (Fig. 12–16). Galactose-1-phosphate cannot be converted to its glucose counterpart and accumulates in tissues. In addition, the formation of galactose-1-phosphate decreases, leading to high concentrations of unphosphorylated galactose in the

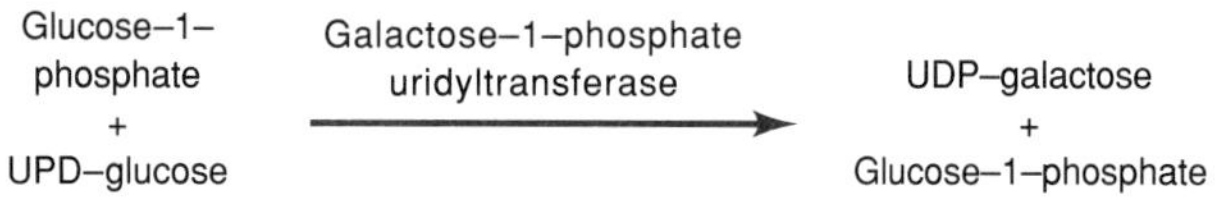

FIGURE 12–17   Conversion of galactose to galactilol.

circulation. A side reaction then becomes important: the formation of galactilol catalyzed by the enzyme aldose reductase (Fig. 12–17). It is believed that the presence of excessive amounts of galactilol in tissues is a primary contributor to the cataracts and other problems associated with galactosemia. A similar concern is seen with aldose reductase in some of the problems of diabetes mellitus.

LABORATORY DIAGNOSIS

Laboratory testing for galactosemia can follow several routes (Table 12–6). Serum levels of liver enzymes are elevated because of the liver damage so frequently seen in this disorder (probably caused by galactilol). Elevated bilirubin is also a common manifestation. Urine testing shows the presence of elevated amounts of reducing substances, although urine glucose is normal (as is serum glucose). If specific tests for urine galactose are available (usually through an outside reference laboratory), the results are positive. Galactose can also be quantitated in serum, where an elevated level is seen. The definitive test for this disorder is the quantitation of galactose-1-phosphate uridyltransferase in red blood cells. A marked decrease in this enzyme is seen in patients with galactosemia. This assay is available only at a few specialized reference laboratories and is rather expensive to perform.

FIGURE 12–16   Enzyme deficiency in galactosemia.

Table 12–6
LABORATORY DATA IN GALACTOSEMIA

Urine	Serum	Red Blood Cells
Positive reducing substances	Elevated bilirubin	Greatly decreased levels of galactose-1-phosphate uridyltransferase
Positive galactose	Elevated alkaline phosphatase	
Negative glucose	Elevated AST	
Positive protein	Elevated ALT	
Elevated galactilol	Normal glucose	
	Elevated galactose	

12.7

DIABETES MELLITUS

Historical Overview

Named because of its "sweet urine" characteristic, **diabetes mellitus** is one of the major diseases seen in this country today. More frequently referred to simply as *diabetes,* this disorder has been known from ancient times. There are several descriptions of the clinical signs of diabetes in the Egyptian, Greek, and Roman literature. During the Middle Ages, diabetes was diagnosed by tasting the urine to detect the large amount of glucose excreted. Fortunately, we have more reliable (and less distasteful) methods for detecting glucose loss today.

Types of Diabetes

The primary metabolic problem in diabetes mellitus is an inability of the body to utilize glucose. Absorption of this carbohydrate across the intestinal wall is unimpaired, and glucose enters the bloodstream unhindered. However, the process of glucose transfer into the cell for further metabolism is impaired. At the heart of the issue is insulin, the hormone responsible for cellular glucose uptake. In some situations, the synthesis and release of insulin by the pancreas is greatly decreased. Therefore, less insulin is available to exert its metabolic effect. In other categories of diabetes, however, the amount of insulin is adequate (and often more than adequate), but its biochemical effect is significantly diminished owing to changes in the number and type of cell receptors responsible for glucose uptake.

Diagnosis of diabetes mellitus from clinical symptoms is often difficult because of the various ways the disease manifests itself. In one type of the disorder, the clinical picture is one of weight loss, thirst, and loss of glucose in the urine (**glucosuria**). Frequently we also see the presence of other substances in the urine known as ketone bodies. In another category is the patient who is overweight, physically inactive, and has glucosuria. A third sit-

uation may be that of a woman who shows transient abnormalities of glucose metabolism only during pregnancy. The picture is complicated and sometimes very confusing.

A brief comparison of the characteristics of the two major types of diabetes mellitus (Table 12–7) is helpful (the transient stage in pregnancy is omitted from this discussion). **Insulin-dependent diabetes** (type I) usually manifests itself at an early age (generally below 20 years old) and is seen in less than 10% of all diabetics. Over 90% of those with diabetes have the **noninsulin-dependent** variety (type II), which appears much later in life. In the insulin-dependent form, there are decreased numbers of beta cells in the pancreas and little production of insulin. Frequently, inflammation of the pancreas is also seen. The noninsulin-dependent variety is generally characterized by excessive insulin production, with little or no pancreatic damage until much later in the course of the disease. Whereas the insulin-dependent diabetic often experiences significant weight loss during the onset of the disease, the patient with noninsulin-dependent diabetes is usually obese, a major contributing factor for the disease. Even the therapies are implied by the nomenclature. Insulin injection is the major form of treatment for the insulin-dependent diabetic, but the noninsulin-dependent patient is usually treated by diet control and weight loss. These latter patients do not normally require insulin administration to regulate blood glucose levels.

Oral Glucose Tolerance Test

To facilitate the diagnosis of diabetes, several laboratory tests are needed. The initial clue usually comes from the routine urinalysis performed during an office or clinic out-patient visit. The urine glucose level is elevated, sometimes quite markedly. A fasting blood glucose would then be ordered for follow-up. If this test shows a blood glucose level greater than 140 mg/dL, the patient is presumed to have diabetes mellitus. If the fasting glucose is elevated, but not above 140 mg/dL, further testing is necessary and an **oral glucose tolerance test (OGTT)** is ordered.

Unlike its counterpart (the lactose tolerance test), the OGTT is designed to determine how well the body utilizes glucose after it has been absorbed into the circulation. Since blood glucose levels are already high and urine glucose is also elevated, there is obviously no problem with absorption. The American Diabetes Association (ADA) published criteria for the performance of the glucose tolerance test (Table 12–8). The directions must be followed closely for the results to be meaningful.

For 3 days prior to the test, the patient is asked to consume a diet consisting of at least 150 g of carbohydrate per day. The patient is told to fast overnight, to avoid excessive physical activity, and to re-

Table 12–7
CHARACTERISTICS OF TYPES OF
DIABETES MELLITUS

Feature	Insulin-Dependent (Type I)	Noninsulin-Dependent (Type II)
Age of onset	Usually <20	Usually >40
Percentage of diabetics	<10%	>90%
Ketone bodies	Usually	Rarely
Obesity at onset	Rare	Common
Serum insulin	Very low	Normal or high

Table 12–8
GUIDELINES FOR ADMINISTRATION OF
ORAL GLUCOSE TOLERANCE TEST

1. Patient is asked to consume 150 g carbohydrate a day for 3 days prior to test.
2. Patient is to fast overnight and to avoid excessive physical activity.
3. Measure fasting glucose level before giving glucose.
4. If fasting glucose > 140 mg/dL, terminate test.
5. If fasting glucose < 140 mg/dL, administer glucose:
   adults: 75 g
   pregnant women: 100 g
   children: 1.75 g/kg body wt., up to 75 g
6. Collect samples every 30 min for 2 h.

Table 12–9
CRITERIA FOR DIAGNOSIS OF DIABETES
USING GLUCOSE TOLERANCE TEST

1. Diabetes mellitus is considered to be present if:
   Fasting glucose > 140 mg/dL (no GTT required)
   Two-hour value and one other value both > 200 mg/dL
2. Impaired glucose tolerance is considered to be present if:
   Two-hour value 140–200 mg/dL and one other value > 200 mg/dL
3. Gestational diabetes is considered present if two or more values exceed:
   105 mg/dL fasting
   190 mg/dL at 1 h
   165 mg/dL at 2 h
   145 mg/dL at 3 h

port to the laboratory early the next morning. Because of the length of the test, an early start eases the patient's strain and discomfort somewhat. A fasting blood glucose sample is collected and analyzed before the OGTT is begun. It is important to have this fasting value before administering the glucose load. If the fasting level is above 140 mg/dL, the test does not need to be run and could be dangerous to the patient in some circumstances. However, always check with the ordering physician first before discontinuing the test.

If the full glucose tolerance test is to be performed, get information from the patient about what medications he or she is currently using. You cannot have the patient discontinue these medications, but the physician may need this information in order to interpret the test results completely. Explain that the patient is not to smoke, eat, or drink during the test (although a few sips of water occasionally are permitted). The patient should be physically inactive, either sitting quietly or lying down. A great deal of physical activity noticeably alters the results. The ADA guidelines state that the OGTT should be administered only to out-patients. Hospitalized individuals have other medical problems which may result in misleading information being obtained from the tolerance test.

A glucose load is given to the patient to drink and the time noted. For adults, a solution containing 75 g of glucose is the standard amount administered, although pregnant women should receive 100 g. If the test is being administered to a child, use a load based on body weight. Give 1.75 g of glucose per kilogram of body weight up to a maximum of 75 g of glucose. There are a number of commercial preparations on the market which combine the glucose with flavored soft-drink material to make the solution more palatable. Even so, a patient may occasionally vomit some of the drink. If this happens, the test should be discontinued and rescheduled for a later date. Both the stress of the situation and the loss of administered glucose (however little it may be) affect the results.

After giving the glucose load, blood samples are collected every 30 min for 2 h. The diagnosis of di-

abetes mellitus is made on the basis of criteria published in 1979 by an international committee of scientists and physicians (Table 12–9). If the glucose level of the 2-h sample exceeds 200 mg/dL and one other sample also has a glucose value greater than 200 mg/dL, the diagnosis of diabetes is confirmed and no further testing is needed. A 2-h value for blood glucose between 140 and 200 mg/dL and one other value over 200 mg/dL suggests a situation of impaired glucose tolerance. Although diabetes mellitus is not clearly present, the situation warrants follow-up and further testing at some later date. Keep in mind that a fasting glucose level of greater than 140 mg/dL is diagnostic for the disease. No glucose tolerance test is required in this situation.

Some pregnant women may manifest symptoms of diabetes, but the regular diagnostic criteria do not show themselves clearly. This condition, known as **gestational diabetes,** is characterized by two or more glucose values above the following limits:

fasting > 105 mg/dL
at 1 h > 190 mg/dL
at 2 h > 165 mg/dL
at 3 h > 145 mg/dL

Gestational diabetes apparently arises when the extra metabolic demands of the fetus put stress on the ability of the mother to produce adequate amounts of insulin for proper glucose utilization. Although it is not a real problem during pregnancy, this pattern can be an early warning sign, since a large percentage of these patients develop diabetes within 5 to 10 years. Obviously, the diagnosis of gestational diabetes would not apply to a known diabetic who later becomes pregnant.

Glucose/Insulin Relationships in Diabetes

To understand in more detail the biochemical changes which take place during the absorption and

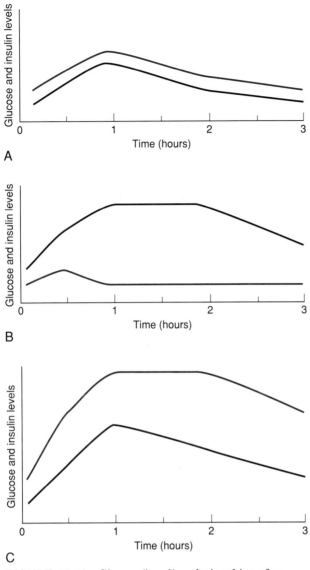

A

B

C

FIGURE 12-18 Glucose/insulin relationships after glucose load. A. Normal response. B. Insulin-dependent response. C. Noninsulin-dependent response.

utilization of glucose and to see more clearly the metabolic changes in diabetes mellitus, let us look at a detailed glucose tolerance curve (Fig. 12–18). If we measure the glucose levels for 3 or more hours (as was the earlier practice in carrying out an OGTT), we observe some characteristic changes in the values for both normal and diabetic individuals. Figure 12–18 also shows insulin levels for each situation measured using the same blood samples from which the glucose values were obtained. Studies of these data help us see how the current classification of diabetes mellitus was established.

After administration of a glucose load, the normal individual (part A) shows a rapid rise in blood glucose which peaks between 30 and 60 min later. The glucose level then begins to drop, generally returning to somewhere in the range of the initial fasting

level within 3 h. If we were to run the test for a longer period, we might see a slight dip in the glucose concentration at 4 h, with a rebound to the fasting level by 5 h. The peak glucose value does not greatly exceed 150–160 mg/dL. The insulin levels rise and fall in parallel with the glucose values. As the blood glucose concentration increases, the insulin concentration also increases to facilitate glucose uptake and metabolism by the cell. When the blood glucose concentration drops, less insulin is needed by the body and insulin production decreases. Insulin levels normally do not exceed 100 μU/mL.

In the individual who is insulin-deficient, the blood glucose rises markedly and continues to rise after 30 min (part B). Peak glucose levels may not be seen until 2 or 3 h after the glucose load was given. These glucose concentrations are considerably higher than normal. However, there is essentially no change in the concentration of insulin in the blood. Because of damage to the pancreas (the exact causes of which are still unknown), this organ cannot manufacture much insulin at all. Therefore, little or no insulin is released into the bloodstream to be used by the cells in metabolism. There is no way to help the glucose enter the cell. As a result, blood glucose levels remain high for a very long time. This type of diabetes was called **juvenile-onset diabetes** (because the disease usually appears at a fairly early age), but is currently termed **insulin-dependent diabetes.**

In contrast, a very different pattern is seen in many adults, particularly those who are overweight (part C). In the initial stages, the blood glucose pattern shows an increase above what is normally expected at 30 min and 1 h, followed by a return to fasting glucose levels by 3 h. Insulin levels rise to concentrations much higher than normal and then gradually drop also. In more advanced cases, glucose levels are even higher, but there is still a significant increase in blood insulin concentrations. This paradoxical behavior may be explained (at least in part) when we look at the body fat content of individuals with this glucose/insulin pattern. There is a much higher amount of body fat present; the vast majority of these patients are overweight. The fatty (adipose) tissue has a lower number of insulin receptors on each cell and is therefore less responsive to the same amount of insulin released. This type of diabetes was referred to as **maturity-onset diabetes** (seen mainly in patients over 40 years old), but is currently categorized as **noninsulin-dependent diabetes.**

Ketone Bodies and Diabetes Mellitus

The major abnormality seen in diabetes mellitus is the impairment of carbohydrate metabolism. This biochemical alteration has some far-reaching con-

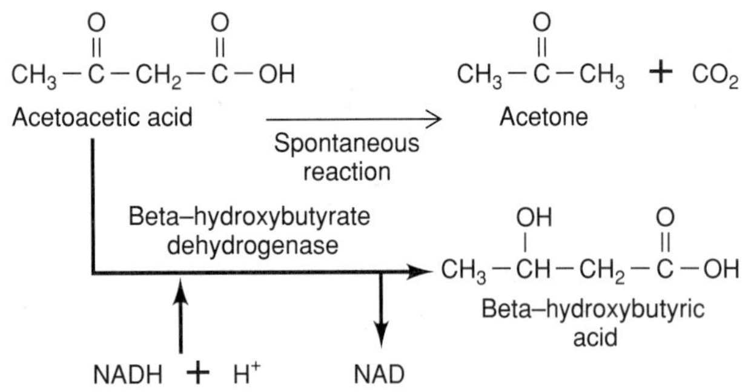

$$CH_3-\overset{\overset{\displaystyle O}{\|}}{C}-CH_2-\overset{\overset{\displaystyle O}{\|}}{C}-OH \qquad CH_3-\overset{\overset{\displaystyle O}{\|}}{C}-CH_3 + CO_2$$

Acetoacetic acid $\xrightarrow{\text{Spontaneous reaction}}$ Acetone

Beta–hydroxybutyrate dehydrogenase

$$CH_3-\overset{\overset{\displaystyle OH}{|}}{CH}-CH_2-\overset{\overset{\displaystyle O}{\|}}{C}-OH$$

Beta–hydroxybutyric acid

NADH + H$^+$ NAD

FIGURE 12–19   Formation of ketone bodies.

sequences, not only for glucose utilization, but also for metabolism of fats by the body. When biochemical energy is not produced from the conversion of glucose to other materials, metabolism shifts to the breakdown of lipids to provide the chemicals needed. As a result of this shift, certain end products of lipid metabolism begin to appear in the urine of the diabetic (primarily the insulin-dependent patient). These compounds, called **ketone bodies,** are either ketones or structurally related to them (Fig. 12–19).

Adipose tissue is made up primarily of **triglycerides,** which are esters composed of one molecule of glycerol and three molecules of fatty acids. When biochemical energy demands of the body cannot be met by glucose, the triglycerides break down to form glycerol and fatty acids. The fatty acids undergo further metabolism by losing two carbons at a time, with these two-carbon units forming acetyl-CoA. If excessive conversion of fatty acids is taking place (as in the case of insulin-dependent diabetes), there is an increase in the amount of acetoacetic acid. This ketoacid can then either break down to form acetone (a ketone) and carbon dioxide or it forms beta-hydroxybutyric acid. All three of these products of excessive lipid metabolism are excreted in the urine. Simple chemical tests permit the easy detection of these materials.

When testing for ketone bodies, detection of any one of the three compounds should be sufficient evidence for the presence of altered metabolism. Since acetoacetic acid constitutes about 20% of the total of the three derivatives, most rapid screening procedures focus on the qualitative detection of this compound. Addition of sodium nitroprusside to urine produces a purple color in basic solution if acetoacetic acid is present. This reagent does not react with either acetone or beta-hydroxybutyric acid. Commercially marketed dipstick procedures for the assay are available from a variety of sources. When dipped into the urine sample, the pad on the stick turns a shade of purple. The deeper the color, the higher the concentration of acetoacetic acid. Quantitation is not required since only a rough indication of the level is needed.

Caution must be taken when interpreting this test. If the patient is using aspirin, the presence of salicylate metabolites in the urine give a false positive result. Certain antihypertensive medications (such as methyldopa) also give false positive results, as do phenothiazine derivatives. If patients do their own urine testing at home (as many diabetics do), they must fully understand the problems associated with drug interferences in the screen for ketone bodies.

Two major problems result from excessive production of ketone bodies: acidosis and electrolyte loss. The large amounts of acetoacetic acid and beta-hydroxybutyric acid that form release high quantities of H$^+$ ions because of dissociation of protons from the carboxyl groups. This increase in H$^+$ ions produces a decrease in blood pH, called **acidosis.** In addition, the ionized acids (now containing negatively charged carboxylate anions) are excreted in the urine. As these compounds pass out of the body, they carry positive ions (Na$^+$ and K$^+$) with them. This excessive **electrolyte loss** can produce adverse changes in body biochemistry. In severe cases, the patient may lapse into a coma. Prompt diagnosis of this situation is imperative so that corrective therapy can be administered quickly.

Insulin Therapy of Diabetes Mellitus

For the insulin-dependent diabetic, regulation of blood glucose is accomplished by daily administration of insulin. Currently, much of the insulin used is obtained either from cows or from pigs. Although the structures of human, pork, and beef insulins are similar, there are some differences (Table 12–10). Pork insulin has one amino acid different from human insulin; beef has three. These seemingly insignificant structural changes produce some major effects over a long period. Because of the difference in amino acid sequence in the protein chain(s), the human body sees animal insulin as "foreign" and eventually begins to generate antibodies to these alien molecules. When the insulin antibodies bind to insulin, the biochemical activity of the hormone is lost. The molecule can no longer bind to a cell receptor site to promote glucose transport and metabolism. Therefore, more insulin is required to produce the same amount of glucose utilization. However, as the insulin dosage is increased, the for-

Table 12–10
COMPARATIVE STRUCTURES OF MAMMALIAN INSULINS

Insulin	Alpha Chain Position 8	Alpha Chain Position 10	Beta Chain Position 30
Human	Threonine	Isoleucine	Threonine
Pork	Threonine	Isoleucine	Alanine
Beef	Alanine	Valine	Alanine

mation of insulin antibodies also increase. In cases of severe diabetes, regulation of insulin administration can become complex.

One of the goals of the biotechnology industry is to be able to manufacture human insulin in large amounts. Human genetic material coding for the insulin molecule can be incorporated into a bacterium (*E. coli* is commonly used for this process). The bacterium "reads" the human gene and begin to synthesize the molecules specified in the human genetic code. After some time, the bacteria are isolated, lysed, and the human insulin is separated from other materials and further purified. By producing a molecule with exactly the right structure, formation of insulin antibodies in the circulation can be avoided.

Home Monitoring of Blood Glucose

The laboratory contribution to diagnosis and management of diabetes mellitus does not stop with the initial testing. For the insulin-dependent diabetic, glucose levels in the blood must be monitored on a regular basis in order to adjust insulin dosage. Although more and more patients self-monitor their glucose (either through checking urine glucose with a dipstick or by measurement of blood glucose with dipstick and reading device), the laboratory must maintain some level of involvement. With the recent shift toward an emphasis on out-patient services, laboratory personnel may assume an educational role in addition to the analytical services already provided.

As improved instrumentation becomes available, we will see an increased emphasis on home testing of blood glucose. Automatic lancet devices are available that do fingersticks for those persons who are hesitant to stick themselves. Small hand-held dipstick readers for quantitative blood glucose are advertised in our daily newspapers. These home testing devices are quite accurate when directions are followed and materials are properly stored. Many hospitals and clinics are using these instruments for bedside and other types of testing. This approach to monitoring of a disease state may become more popular as technology advances and the cost of medical care continues to increase.

Glycosylated Hemoglobin

A major drawback to the monitoring of urine or serum glucose levels is that the value obtained provides information about the metabolic status only at that moment. Until recently, there has been no way to assess the long-term control of glucose use by the body. It would be very helpful to know how well the patient has maintained dietary and medication regimens over the last week or month, particularly on

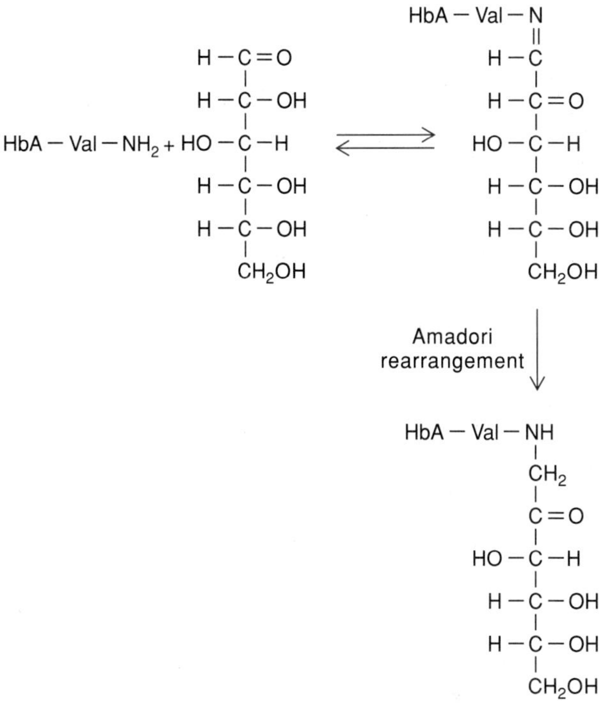

FIGURE 12-20   Formation of glycosylated hemoglobin.

an out-patient basis. Information concerning glucose control can now be obtained with relative ease by measuring the amount of a particular hemoglobin fraction present in red blood cells.

Glucose enters the erythrocyte and binds to hemoglobin to a very small degree (Fig. 12-20). Part of this interaction is reversible; the glucose molecule can dissociate from the hemoglobin because a covalent bond has not been formed. However, some of the glucose reacts chemically, forming a bond with a particular amino acid in the hemoglobin protein. When this glucose/hemoglobin complex is formed, the chemical properties of the hemoglobin are changed. This form of hemoglobin can now be separated easily from other hemoglobin by several techniques.

The important point to remember is that the concentration of this glucose/hemoglobin derivative (called **glycosylated hemoglobin**) increases as the blood level of glucose increases. Since formation of the complex is a slow process, the blood level of glucose must stay elevated for a long time in order to produce a higher than normal amount of glycosylated hemoglobin. Therefore, measurement of this hemoglobin fraction tells us whether the blood glucose level has been normal or elevated (on the average) over the last several weeks. From this we can infer the faithfulness with which the patient has carried out the instructions from the health-care provider and the effectiveness of the prescribed treatment.

Why use red blood cells? One obvious response is: where else do you get hemoglobin? Secondly, the erythrocytes are easily isolated by centrifugation after a blood sample is collected. Finally, the red cell has a life span of approximately 120 days. We can therefore expect the extent of glycosylation to reflect average blood glucose levels over a period of several months prior to the time of sampling. Many other proteins in blood are also glycosylated. Several current research studies are exploring the use of these other protein/glucose complexes in monitoring treatment effectiveness in diabetes.

Glycosylated hemoglobin is named in accordance with international conventions regarding hemoglobin classification. Since the hemoglobin is "normal" with regard to its primary structure, it is designated as **hemoglobin A.** On the basis of electrophoretic mobility, it would be broadly categorized as **hemoglobin A₁.** Since this particular hemoglobin has a slightly different mobility than the usual hemoglobin A_1, it is designated as **hemoglobin A₁c.** Technically speaking, glycosylated hemoglobin encompasses more than one fraction and the term *hemoglobin A_{1c}* may be slightly inaccurate. A newer term is **glycated,** which has been recommended by an international commission on biochemical nomenclature. You will encounter any or all three of the terms in much of the literature you may read.

The two major approaches to the measurement of glycosylated hemoglobin are column chromatography and electrophoresis. Both require a preliminary lysing step before sample analysis. Whole blood (with edetic acid as an anticoagulant) is treated with a lysing agent to break down the red-cell membrane. This reagent contains a dilute aqueous solution of a detergent which disrupts the red cells and keeps the membrane somewhat in solution. Centrifugation of the lysed sample removes some of the cellular debris and minimizes some interference with the test.

For column chromatography, a measured amount of the lysate is placed on an ion-exchange column. Glycosylated hemoglobin and "regular" (nonglycosylated) hemoglobin both bind to the resin, but to different degrees. Addition of a phosphate buffer (pH about 6.7) causes the glycosylated portion to be eluted from the column, while the other hemoglobin fractions remain. Since hemoglobin has a strong absorbance peak at 415 nm, measurement of the absorbance of the eluate using a spectrophotometer and comparing it to the absorbance of the total hemolysate solution allows for calculation of the percentage of glycosylated hemoglobin in the sample. Abnormal hemoglobins, if present, can interfere with the test. False elevations of glycosylated hemoglobin arise from the fact that the abnormal molecules may be glycosylated themselves or they may elute from the column with the normal glycosylated hemoglobin. The reference range is somewhat method-dependent and should be based on data supplied by the manufacturer. Normal values are somewhere in the range of 4.5 to 8.5%.

Electrophoresis has not gained as wide a following for quantitation of glycosylated hemoglobin. Nevertheless, separation on the basis of electric charge has some advantages over column technology. Most ion-exchange column methods are sensitive to small changes in temperature and pH, producing markedly altered results with a 2°C change in temperature or a 0.02-unit change in pH. Electrophoresis, on the other hand, provides control over all these variables because all the fractions are separated and observed. In addition, if hemoglobins S or C are present, they migrate to a different area of the electrophoresis film and do not interfere with the fractionation of glycosylated hemoglobin. The major drawback to electrophoresis is that it is less accurate in the normal and slightly elevated range relative to column methodology. Normal values are approximately 5.5–7.0% and are method-dependent.

By using the results from glycosylated hemoglobin measurements, a reliable assessment of diabetic control by the patient is obtained. If the values are elevated, obviously the blood glucose levels have been high for the past few weeks. This suggests either that the therapy is not working or that the patient is not complying with the prescribed program (unfortunately, the latter is usually the case). As carbohydrate metabolism comes under closer regulation (by diet, insulin therapy, or other means), the glycosylated fraction of hemoglobin decreases over time. If the patient becomes less stringent in complying with the therapy, this fraction again increases. Therefore, long-term compliance can be reliably determined, and the data can be used to monitor both the effectiveness of the treatment and the cooperation of the patient.

Fructosamine

Hemoglobin is not the only protein which reacts with glucose to form a carbohydrate/protein derivative. Serum albumin and a wide variety of other proteins in blood and tissues also undergo this reaction. Some of the degenerative effects of long-term diabetes mellitus are thought to be due to this process taking place in nerves and other body sites. Glycosylation of these proteins occurs predominantly at the epsilon-amino side chain of the amino acid lysine. The term **fructosamine** refers to glycosylated albumin and other proteins.

A number of analytical procedures have been developed for the quantitative estimation of fructosamine. Some methods involve lengthy hydrolysis steps followed by column chromatography to isolate the fructosamine derivatives of interest. These procedures are too long and cumbersome for routine clinical work. Affinity chromatography methods have been developed and are commercially available. The glycosylated derivatives selectively bind to side chains of the column chromatographic material, allowing all other proteins to pass through the column. Elution and analysis for fructosamine is easily carried out using approaches similar to those for the measurement of glycosylated hemoglobin.

Colorimetric assays have received significant attention in recent years. If fructosamines are reacted with strong acid, a cyclic derivative known as 5-hydroxymethylfurfuraldehyde forms. This process is similar to the reaction employed for the colorimetric determination of glucose. The furfuraldehyde derivative can then react with thiobarbituric acid, forming a product which absorbs light in the visible region. Spectrophotometric analysis permits quantitation of this product.

In alkaline solution, fructosamine reduces nitro-blue tetrazolium. The reaction is complex, but yields a product that absorbs at 530 nm. Glucose does not react under these conditions. This assay approach has been adapted to a wide variety of automated chemistry systems.

Whereas glycosylated hemoglobin measurements provide an indication of glucose control over the last 3 months or so, measurement of fructosamines gives a clear picture of more short-term glucose levels. Serum albumin has a half-life of 2–3 weeks (compared with the 2-month half-life of hemoglobin). Other proteins measured in this system possess even shorter half-lives, some as low as 2–3 days. The more rapid turnover of proteins allows examination of a shorter period, permitting more rapid adjustment of treatment to maintain blood glucose concentrations at the proper level.

12.8

HYPOGLYCEMIA

On occasion, a patient may complain of dizziness and fainting, which they may attribute to "low blood sugar," or **hypoglycemia.** Although there has been a great deal published in the popular literature (much of it inaccurate), there are several clinical causes of low blood glucose levels. Many of these causes are secondary to other illnesses and resolve themselves when the primary disorder is treated.

Patients with various types of cancer often show hypoglycemic symptoms. Many tumors secrete proteins which alter a variety of metabolic processes, including glucose metabolism. Perhaps the best known tumor is an insulinoma, which produces massive amounts of insulin, resulting in a striking decrease in glucose levels. Various liver disorders (hepatitis, cirrhosis, and others) lead to hypoglycemia following the loss of liver tissue responsible for glucose formation from glycogen. Many types of gastrointestinal disorders and surgery give rise to hypoglycemia due to changes in absorption of carbohydrates from the intestine. A few emotionally disturbed individuals induce marked hypoglycemia with surreptitious insulin injections in order to gain the attention they feel they need.

Diagnosing hypoglycemia requires closer investigation than simply monitoring the glucose level once during the day. On occasion, a 5-h oral glucose tolerance test may be requested. The clue for possible hypoglycemia comes in looking at glucose levels from hours 4 and 5. The values drop strikingly below the fasting level and may or may not rebound sufficiently. Perhaps the most clear-cut diagnostic test is the 72-h fast. Healthy individuals can maintain glucose values at a normal level during this period, but the true hypoglycemic experiences a noticeable drop in blood glucose over the 3 days of the fast. This test is not requested very often and is certainly not well liked by the patient.

SUMMARY

Carbohydrates, for which monosaccharides are the fundamental units, are important energy sources for the body. Disaccharides are composed of two monosaccharides, and polysaccharides represent long chains of monosaccharides (usually glucose). Starch (the primary dietary polysaccharide) is digested in the small intestine by amylase, forming smaller units and free glucose. Each disaccharide has a specific enzyme located in the walls of the small intestine which hydrolyzes it to form two monosaccharides.

The monosaccharides enter the circulation, followed by phosphorylation before entry into the cell. Galactose is first converted to glucose, but fructose can be phosphorylated and participate in intermediary metabolism more directly. Glucose is the focal point for carbohydrate metabolism and its regulation, either being converted to glycogen and stored in the liver or by being channeled through the glycolytic pathway and into acetyl-CoA for energy production or lipid synthesis. Formation of glycogen is regulated by the hormone glucagon, whereas the phosphorylation, cell entry, and glycolytic metabolism of glucose is under the control of insulin.

Carbohydrate analysis focuses on the analysis of glucose in serum, plasma, or urine. For urine screening, either total reducing substances (all monosaccharides excreted) or urine glucose can be assessed by colorimetric or coupled enzyme assays. Although some colorimetric methods are still in use for the quantitative measurement of glucose in serum or plasma, the preferred methodologies involve conversion of glucose with the formation of hydrogen peroxide or NAD(P)H, both of which can be assayed by a wide variety of techniques.

Two major types of carbohydrate intolerance (other than glucose) can be identified through laboratory studies. Lactose intolerance is a defect in the digestion of lactose in the small intestine. By measuring the increase in glucose after a lactose load, this condition can be readily identified. Galactosemia results from an inability to convert galactose to glucose. This disorder produces poor growth and development, and can result in mental retardation if not detected during early infancy. The definitive diagnosis involves measurement of galactose-1-phosphate uridyltransferase.

The major clinical condition resulting from impaired glucose metabolism is diabetes mellitus. Two major forms of the disease exist. In type I (insulin-dependent) diabetes mellitus, the major defect appears to be a lack of insulin production by the pancreas. The body cannot utilize glucose as efficiently, since the carbohydrate cannot enter the cells and be metabolized. The type II (noninsulin-dependent) form of the disorder is characterized by high circulating levels of insulin, but impaired response by the cells to insulin. Again, glucose does not enter the cells as efficiently. A glucose tolerance test permits the diagnosis of the disease, but does not differentiate between the type I and type II categories. Assessment of urine ketone-body excretion indicates the severity of the situation. Measurement of glycosylated hemoglobin or fructosamine are means of monitoring long-term control of glucose metabolism.

Hypoglycemia is a condition characterized by low levels of blood glucose. In rare instances, this disorder is caused by a tumor of the pancreas which synthesizes high amounts of insulin and is not under feedback regulation. More frequently, the disorder is secondary to other disease states or is a result of excessive insulin administration (either accidental or purposeful).

FOR REVIEW

Directions: For each question, choose the best response.

1. The term *gluconeogenesis* refers to the
   A. conversion of hexoses into lactate or pyruvate
   B. conversion of glucose to glycogen
   C. breakdown of glycogen to form glucose
   D. synthesis of glucose from noncarbohydrate sources

2. All of the following are classified as disaccharides *except*
   A. fructose
   B. lactose
   C. maltose
   D. sucrose

3. Which of the following enzymes is involved in the digestion of starch?
   A. amylase
   B. lactase
   C. lipase
   D. sucrase

4. Insulin is synthesized and stored in the
   A. adrenal cortex
   B. adrenal medulla
   C. alpha cells of the pancreas
   D. beta cells of the pancreas

5. In normal cerebrospinal fluid, the glucose concentration is approximately
   _____ of the plasma glucose level.
   A. 30–45%
   B. 60–75%
   C. 40–70%
   D. equal to that

6. Which of the following methods estimates the urine sugar concentration by
   measuring total reducing substances?
   A. copper sulfate
   B. glucose oxidase
   C. hexokinase
   D. ferricyanide

7. Which of the following chemicals is used in blood collection tubes to help
   preserve the integrity of the blood glucose concentration?
   A. citrate
   B. EDTA
   C. fluoride
   D. oxalate

8. In which of the following methods is the glucose concentration linked to a
   change in the oxygen concentration in the reaction mixture?
   A. copper sulfate
   B. glucose oxidase
   C. hexokinase
   D. ferricyanide

9. Due to an enzyme deficiency, which disaccharide accumulates in the intes-
   tine causing significant gastrointestinal discomfort?
   A. glucose
   B. galactose
   C. lactose
   D. maltose

10. Which form of diabetes manifests itself later in life and is associated with
    obesity, physical inactivity, and glucosuria?
    A. congenital
    B. gestational
    C. insulin-dependent
    D. noninsulin-dependent

11. The term *ketone bodies* refers to all of the following compounds *except*
    A. acetic acid
    B. acetoacetic acid
    C. acetone
    D. beta-hydroxybutyric acid

12. A patient with a severe, uncontrolled case of diabetes mellitus exhibits all
    of the following conditions *except*
    A. glycosuria
    B. hyperglycemia
    C. ketosis
    D. metabolic alkalosis

13. Which of the following methods provides information about glucose metabolism over a 3-month period?
   A. fructosamine
   B. glucose oxidase
   C. glycosylated hemoglobin
   D. oral glucose tolerance

14. A patient with insulinoma may exhibit dizziness and fainting attributable to
   A. acidosis
   B. ketosis
   C. hypoglycemia
   D. hyperglycemia

15. Which hormone functions to increase the cellular uptake of glucose?
   A. catecholamines
   B. cortisol
   C. glucagon
   D. insulin

BIBLIOGRAPHY

American Diabetes Association, "Statement on hypoglycemia," *Diabetes Care.* 5:72–73, 1982.

Armbruster, D. A., "Fructosamine: Structure, analysis, and clinical usefulness," *Clin. Chem.* 33:2153–2163, 1987.

Asbury, A. K.,"Understanding diabetic neuropathy," *New Eng. J. Med.* 319:577–578, 1988.

Brownlee, M., and Cerami, A., "The biochemistry of the complications of diabetes mellitus," *Ann. Rev. Biochem.* 50:385–432, 1981.

Bunn, H. F., "Evaluation of glycosylated hemoglobin in diabetic patients," *Diabetes.* 30:613–617, 1981.

Bunn, H. F., "Nonenzymatic glycosylation of protein: Relevance to diabetes," *Amer. J. Med.* 70:325–330, 1981.

Burke, M. D.,"Diabetes mellitus: Test strategies for diagnosis and management," *Postgrad. Med.* 66:213–217, 1979.

Cahill, G. F. et al., "Insulin-dependent diabetes mellitus: The initial lesion," *New Eng. J. Med.* 304:1454–1465, 1981.

Cahill, Jr., G. F., "Beta-cell deficiency, insulin resistance, or both?" *New Eng. J. Med.* 318:1268–1270, 1988.

Clements, Jr., R. S.,"Diabetic neuropathy—New concepts of its etiology," *Diabetes.* 28:604–611, 1979.

Cryer, P. E., "Glucose counterregulation in man," *Diabetes.* 30:261–264, 1981.

Czech, M. P., "Insulin action and the regulation of hexose transport," *Diabetes.* 29:399–409, 1980.

Czech, M. P., "The nature and regulation of the insulin receptor: Structure and function," *Ann. Rev. Physiol.* 47:357–381, 1985.

DeFronzo, R. A., "Glucose intolerance and aging," *Diabetes Care.* 4:493–501, 1981.

deShazo, R. D., "Insulin allergy and insulin resistance," *Postgrad. Med.* 63:85–92, 1978.

The Diabetes Control and Complications Trial Research Group, "Are continuing studies of metabolic control and microvascular complications in insulin-dependent diabetes mellitus justified?" *New Eng. J. Med.* 318:246–250, 1988.

Eisenbarth, G. S., "Type I diabetes mellitus, A chronic autoimmune disease," *New Eng. J. Med.* 314:1360–1368, 1986.

Exton, J. H., "Gluconeogenesis," *Metabolism.* 21:945–990, 1972.

Foster, D. W., and McGarry, J. D., "The metabolic derangements and treatment of diabetic ketoacidosis," *New Eng. J. Med.* 309:159–169, 1983.

Frank, P. M., "Monitoring the diabetic patient with glycosylated hemoglobin measurements," *Amer. Clin. Prod. Rev.* April:29–33, 1988.

Free, H. M., and Free, A. H., "Analytical chemistry in the conquest of diabetes," *Anal. Chem.* 56:664A–684A, 1984.

Gabbay, K. H., "The sorbitol pathway and the complications of diabetes," *New Eng. J. Med.* 288:831–836, 1973.

Gabbe, S. G., "Diabetes in pregnancy: Clinical controversies," *Clin. Obstet. Gynecol.* 21:443–453, 1978.

Gabbe, S. G., "Diabetes mellitus in pregnancy: Have all the problems been solved?" *Amer. J. Med.* 70:613–618, 1981.

Goldstein, D. E., "Is glycosylated hemoglobin clinically useful?" *New Eng. J. Med.* 310:384–385, 1984.

Gossain, V. V., and Rovner, D. R., "Pancreatic glucagon. Possible implications of the hyperglycemic hormone in diabetes control," *Postgrad. Med.* 72:87–96, 1982.

Gray, G. M., "Carbohydrate digestion and absorption. Role of the small intestine," *New Eng. J. Med.* 292:1225–1230, 1975.

Greene, D. A. et al., "Sorbitol, phosphoinositides, and sodium-potassium-ATPase in the pathogenesis of diabetic complications," *New Eng. J. Med.* 316:599–606, 1987.

Hadorn, B. et al., "Biochemical mechanisms in congenital enzyme deficiencies of the small intestine," *Clin. Gastroenterol.* 10:671–690, 1981.

Haunz, E. A., "Pitfalls in the early diagnosis of diabetes," *Geriatrics.* October, 1979.

Hers, H. G., and Hue, L., "Gluconeogenesis and related aspects of glycolysis," *Ann. Rev. Biochem.* 52:617–653, 1983.

Hodgkins, D. C., "The structure of insulin," *Diabetes.* 21:1131–1150, 1972.

Holst, J. J., "Gut glucagon, enteroglucagon, gut glucagonlike immunoreactivity, glicentin—Current status," *Gastroenterology.* 84:1602–1613, 1983.

Johnson, K. E. et al., "Islet amyloid, islet-cell polypeptide, and diabetes mellitus," *New Eng. J. Med.* 321:513–518, 1989.

Jovanovic, L., and Peterson, C. M., "The clinical utility of glycosylated hemoglobin," *Amer. J. Med.* 70:331–338, 1981.

Kahn, C. R., "Insulin resistance: A common feature of diabetes mellitus," *New Eng. J. Med.* 315:252–254, 1986.

Knowles, Jr., H. C., "Evaluation of a positive urinary sugar test," *J. Amer. Med. Assoc.* 234:961–963, 1975.

Krowlewski, A. S. et al., "Epidemiologic approach to the etiology of type I diabetes mellitus and its complications," *New Eng. J. Med.* 317:1390–1398, 1987.

Kwentus, J. A. et al., "Hypoglycemia. Etiologic and psychosomatic aspects of diagnosis," *Postgrad. Med.* 71:99–104, 1982.

Leggett, J., and Favazzi, A. F., "Hypoglycemia: An overview," *J. Clin. Psychiatry.* 39:51–57, 1978.

Nanji, A. A., "Hypoglycemia: When it's the real thing," *Diagn. Med.* October:60–70, 1983.

National Diabetes Data Group, "Classification and diagnosis of diabetes mellitus and other categories of glucose intolerance," *Diabetes.* 28:1039–1057, 1979.

Orci, L. et al., "The insulin factory," *Sci. Amer.* September: 85–94, 1988.

Passey, R. B. et al., "Evaluation and comparison of 10 glucose methods and the reference method recommended in the proposed product class standard (1974)," *Clin. Chem.* 23:131–139, 1977.

Pfeiffer, M. A. et al., "Insulin secretion in diabetes mellitus," *Amer. J. Med.* 70:579–588, 1981.

Reaven, G. M., "Insulin-independent diabetes mellitus: Metabolic characteristics," *Metabolism.* 29:445–454, 1980.

Redmond, G. P., "Pulling the patient through diabetic ketoacidosis," *Diagn. Med.* November/December:41–46, 1983.

Rosen, O. M., "After insulin binds," *Science.* 237:1452–1458, 1987.

Rosenbloom, A. L. et al., "Classification and diagnosis of diabetes mellitus in children and adolescents," *J. Pediat.* 98:320–323, 1981.

Schade, D. S., and Eaton, R. P., "Insulin delivery: How, when, and where?" *New Eng. J. Med.* 312:1120–1121, 1985.

Schwartz, M. L., and Brenner, W. E., "The need for adequate and consistent diagnostic classifications for diabetes mellitus diagnosed during pregnancy," *Amer. J. Obstet. Gynecol.* 143:119–124, 1982.

Sharon, N., "Carbohydrates," *Sci. Amer.* November:90–116, 1980.

Sherwin, R., and Felig, P., "Pathophysiology of diabetes mellitus," *Med. Clin. North Am.* 62:695–711, 1978.

Sperling, M. A., "Diabetes mellitus," *Pediat. Clin. North Am.* 26:149–169, 1979.

Trundle, D. S., and Weizenecker, R. A., "Capillary glucose testing: A cost-saving bedside system," *Lab. Manage.* May:59–62, 1986.

Turner, R. C. et al., "Relative contributions of insulin deficiency and insulin resistance in maturity-onset diabetes," *Lancet,* pp. 596–598, March 13, 1982.

von Schenk, H., "Glucagon—Biochemistry, physiology, and pathophysiology," *Acta Med. Scand.* 209:145–148, 1981.

Watts, N. B., "Oral glucose tolerance test may be done too fast," *Lab World.* June:68–72, 1981.

Wheeler, T. J., and Hinkle, P. C., "The glucose transporter of mammalian cells," *Ann. Rev. Physiol.* 47:503–517, 1985.

Winegrad, A. I., "Does a common mechanism induce the diverse complications of diabetes?" *Diabetes.* 36:396–406, 1987.

• •

LIPIDS AND HEART DISEASE

1. Define the term *lipids*.
2. State the major functions of fatty acids, cholesterol, and triglycerides.
3. Explain the role of phospholipids in respiration.
4. List the three major categories of lipids in the diet.
5. Describe the process of lipid digestion and absorption into the blood.
6. Explain why lipids require transport proteins for circulation in the blood.
7. List the four classes of lipoproteins.
8. Name the two lipoprotein classes that contain the greatest amounts of cholesterol and the two classes that play major roles in triglyceride transport.
9. Name the two forms of cholesterol which together make up total cholesterol.
10. Discuss the role of HDL in removing cholesterol from cells.
11. Explain why proper patient preparation is critical for accurate lipid testing.
12. State the principle of the enzymatic method for quantitating triglycerides.
13. State the principles of the Liebermann-Burchardt and enzymatic methods for quantitating cholesterol.
14. State the principle of lipoprotein electrophoresis.
15. Name the dye used to visualize the fractions in lipoprotein electrophoresis.
16. List the electrophoretic migration order of the lipoprotein fractions starting at the origin.
17. Discuss how the results of an HDL cholesterol assay are used to assess the risk of developing coronary heart disease.
18. State the two basic steps required to measure HDL cholesterol.
19. List three precipitating agents that have been used to remove VLDL and LDL components when performing the HDL cholesterol analysis.
20. Explain the clinical significance associated with measuring apolipoproteins A and B and their general functions.
21. List seven factors that have been shown to affect total cholesterol and HDL cholesterol levels.
22. Discuss the clinical significance associated with abnormal cholesterol, triglyceride, and lipoprotein levels.
23. Discuss the controversy associated with the establishment of a reference range for cholesterol.

INTRODUCTION

Lipids play a variety of roles in our bodies, from providing structure to furnishing biochemical energy. This interesting class of compounds is characterized by solubility, not by structure. **Lipids** are defined as compounds which dissolve in organic solvents (such as diethyl ether), but not in water. This unifying definition encompasses the long hydrocarbon chains of the fatty acids, the phosphate linkages in phospholipids, and the four-ring structure characteristic of cholesterol and the various steroid hormones. Since lipids are not water-soluble, they perform the important function of forming membranes to separate components from one another.

13.1

LIPIDS: STRUCTURES AND FUNCTIONS

Fatty Acids

The simplest lipid class (in terms of structure) are the **fatty acids.** There are two types of fatty acids: saturated and unsaturated. The **saturated fatty acids** are medium- to long-chain carboxylic acids of the general formula

$$CH_3(CH_2)_x\overset{O}{\overset{\|}{C}}-OH$$

where x is usually 10 to 16. The **unsaturated fatty acids** are similar in structure, with carbon-carbon double bonds as part of the long chain. The presence of the double bonds makes the unsaturated fatty acids more liquid at room temperature. This property may play an important role in decreasing lipid storage and lowering the tendency to block blood flow in arteries.

The fatty acids serve two major functions:

1. building blocks for triglycerides and phospholipids
2. sources of metabolic energy

These important molecules undergo a great deal of synthesis and breakdown in response to the ever-changing energy needs of the body.

Triglycerides

The triple esters formed between glycerol and fatty acids are known as **triglycerides,** or triacylglycerols. The relevant structures are

$$
\begin{array}{ll}
CH_2-OH & \\
| & \\
HO-CH & + \text{3 fatty acids} \rightarrow \\
| & \\
CH_2-OH & \\
\text{Glycerol} &
\end{array}
\qquad
\begin{array}{l}
\quad\quad\quad O \\
\quad\quad\quad \| \\
CH_2-O-C-(CH_2)_n-CH_3 \\
| \quad\quad\quad\; O \\
| \quad\quad\quad\; \| \\
CH-O-C-(CH_2)_x-CH_3 \\
| \\
CH_2-O-C-(CH_2)_y-CH_3 \\
\quad\quad\quad \| \\
\quad\quad\quad O \\
\text{Triglyceride}
\end{array}
$$

The three fatty acids incorporated into the triglyceride structure do not all necessarily have the same structure. In the example above, each of the three fatty acids has a different structure. In any given triglyceride, there may be a mixture of saturated and unsaturated fatty acids.

Triglycerides serve important functions as a part of the cell membrane and as storage forms of lipids. Because of the structural variety allowed in the fatty acid portion of the molecule, the presence of saturated and unsaturated acids provides a number of different conformational possibilities in the molecule (due to the double bonds in the unsaturated fatty acids). Some very specific "tailoring" of compounds for the membrane results from this specificity.

Phospholipids

Phospholipids are structurally similar to triglycerides. The glycerol backbone and two fatty acid esters are present, but the third ester is a group containing phosphate

$$
\begin{array}{l}
CH_2-O-\text{fatty acid} \\
| \\
\text{fatty acid}-O-CH \quad\quad O\bar{V} \\
| \quad\quad\quad\quad\quad\quad | \\
CH_2-O-P-O-\text{charged group} \\
\quad\quad\quad\quad \| \\
\quad\quad\quad\quad O
\end{array}
$$

The charged groups vary from one class of phospholipids to another. Phospholipids are very important constituents of cell membranes because they contain both polar and nonpolar sections. These differing groups allow the phospholipid to interact both with water (charged portion) and organic compounds (noncharged portion). Important constituents of the lung are the phospholipids which make up the lung surfactant material. Without these phospholipids, the inner lining of the lung would adhere to itself whenever air was exhaled, and respiration would be extremely difficult.

Cholesterol

One of the more complicated lipid structures is **cholesterol,** illustrated as follows:

Cholesterol

This molecule plays an important role in providing structure for the cell membrane and is the starting material for the synthesis of all the various steroid hormones and for vitamin D. Another important role for cholesterol is in the formation of bile salts, utilized for the digestion of lipids in the small intestine. These compounds contain the cholesterol ring, but have a polar side chain which allows them to interact with an aqueous system and help dissolve lipids. Cholesterol is a waxy material which forms platelike crystals. This molecule plays an important role in plaque formation in the blood vessels, blocking the flow of blood and subsequently leading to heart damage. Cholesterol is perhaps one of the most studied molecules we know about; a dozen or more Nobel Prizes have been awarded for research in the chemistry and biochemistry of cholesterol structure, function, synthesis, and regulation in the body.

Other Lipids

Based on our operational definition of lipids, several other classes of compounds can also be included in this category. The **steroid hormones** all derive from cholesterol and are considered lipids. Even though these molecules all have the same ring structure, the substituents on the ring and the differences in the side chain create wide variations in the biochemical effects they produce.

An interesting group of long-chain fatty acid derivatives are the **prostaglandins.** These molecules function as modulators of hormone action in the body and exert a variety of different influences. One important area of research involves the role of prostaglandins in blood clotting.

Terpenes are long-chain unsaturated lipids that, among other functions, act as some of the intermediates in the synthesis of cholesterol. The yellow material in carrots is a terpene called beta carotene. In the body, this molecule can be converted to vitamin A.

13.2

DIGESTION AND ABSORPTION OF LIPIDS

The major categories of lipids in the diet are fatty acids, triglycerides, and cholesterol. Other lipids are present in smaller amounts and are usually not considered in a discussion of lipid digestion. Triglycerides and cholesterol are found primarily in animal fat, but vegetable sources usually do not contain cholesterol. However, vegetable oil may contain other sterol molecules similar to cholesterol in structure. The health implications of consuming large amounts of these vegetable sterols is presently under debate.

Little digestion of lipids takes place in the stomach. There are no enzymes specific for lipid breakdown and the acid conditions do little for the hydrolysis of triglycerides. The major site of lipid digestion and absorption is the small intestine. Pancreatic fluids render the pH of the food mixture more basic. Bile salts (synthesized by the liver and stored in the gallbladder) form complexes with the water-insoluble lipids so they can be acted upon by enzymes (which function best in an aqueous medium). Specific enzymes are released by the pancreas into the small intestine to hydrolyze lipids, breaking them down into smaller particles.

Free fatty acids are simply absorbed across the intestinal wall, but the triglycerides must first be metabolized to lower molecular-weight components. The hydrolysis of triglycerides is accomplished by the pancreatic enzyme lipase, which acts on the ester linkage between the glycerol and the fatty acid portions of the molecule to form an alcohol (the glycerol portion) and a carboxylic acid (the fatty acid portion).

The entire triglyceride molecule is not broken down before being absorbed; usually only one or two fatty acids are removed and at least one fatty acid remains attached to the glycerol.

After this partial breakdown of the triglyceride, the fatty acids and the acylglycerol remainder are transported across the intestinal wall. These components are then reassembled back into triglycerides before entering the bloodstream. Here they form complexes with specific proteins which transport the triglycerides through the circulation. All lipids are water-insoluble and must become associated with certain proteins (called apolipoproteins) before they can move through the circulation. These initial triglyceride-rich lipoprotein complexes are called **chylomicrons.** After transport to their destination, the triglycerides may be broken down to provide metabolic energy or they are stored in adipose tissue cells to become available at a later time.

Dietary cholesterol does not undergo any significant metabolism in the small intestine. This molecule binds to bile salts and is transported across the intestinal wall into the circulation. After becoming associated with lipoproteins which are specific for the transport of cholesterol, the complex is taken to the liver for further distribution or conversion to bile salts. Fifteen percent or less of the cholesterol in our bodies comes from diet; the remainder is synthesized by various tissues.

13.3

SYNTHESIS AND UTILIZATION OF FATTY ACIDS AND TRIGLYCERIDES

A complex interchange exists between glucose metabolism and triglyceride formation (Fig. 13-1). Shortly after the six-carbon carbohydrate fructose is split into two molecules of three-carbon carbohydrates, one of those three-carbon units is partially diverted for the formation of the glycerol backbone of the triglyceride molecule. Pyruvate, the end product of glucose metabolism, is converted to acetyl CoA. This derivative can then undergo further metabolism to provide biochemical energy or to be used as a building block for fatty acids. Beginning with acetyl CoA, fatty acids are built up in a cyclic manner (two carbons at a time) into a chain which can contain as many as 16 carbons. Longer carbon

chains and incorporation of double bonds (to synthesize unsaturated fatty acids) are furnished through other biochemical processes. After synthesis of the fatty acids, an ester reaction unites these molecules with the glycerol backbone to form the final triglyceride. This product is then transported by specific proteins to adipose tissue where it is stored in the adipose cells until needed for other purposes.

When biochemical energy is required and is not readily available from carbohydrates, the breakdown of triglyceride begins. One at a time, the fatty acids are hydrolyzed from the molecule and undergo stepwise degradation. Again, they are broken down two carbons at a time in a cyclic process. Each step of the process produces one molecule of acetyl CoA to be used for energy. Excessive conversion of fatty acids to acetyl CoA can result in the formation of some acetoacetyl CoA. This molecule then further metabolizes to acetoacetate, which produces acetone and beta-hydroxybutyrate. The glycerol derivative formed recycles back through the glycolytic pathway or is reused for further triglyceride synthesis.

Although pancreatic lipase is responsible for digestion of triglycerides in the intestinal tract, this enzyme does not play a noticeable role in metabolism of circulating triglycerides. Once the lipids are absorbed across the intestinal wall, they are resynthesized and circulate as triglycerides bound to specific transport proteins (lipoproteins). Metabolism of these circulating lipids is primarily carried out by another lipase known as lipoprotein lipase. This enzyme is also referred to as **postheparin lipase** since it is activated in the circulation by the administration of heparin.

Lipoprotein lipase is apparently bound to the inside wall of the blood vessel. This enzyme is also present in high amounts in adipose tissue, muscle, mammary tissue, and lung. The triglycerides which serve as substrates must be bound to specific lipoproteins for full activity to be seen. If the lipoproteins are not present, the enzyme metabolizes the triglycerides at low rates of activity.

Analysis of lipoprotein lipase is somewhat complicated. Steps must be taken to inactivate or inhibit the "regular" lipase which may be present in the serum sample. Selective inhibition by antibodies or by addition of protamine sulfate has been used to differentiate between the two activities. Most assays published have been used for research studies and are not suitable for routine clinical work.

Lipoprotein lipase activity is low in patients with an inherited disorder of hypertriglyceridemia. The presence of decreased levels of this enzyme results in a reduced ability to metabolize circulating triglycerides, and elevated serum levels are seen. In diabetics who are insulin-resistant, the elevated triglyceride levels seen as a part of this disease are due to impairment of the lipoprotein lipase activity. In this case, activity is affected in some manner by the

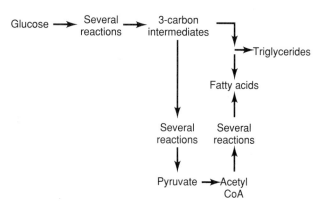

FIGURE 13-1 Carbohydrate/lipid interrelationships.

abnormal levels of circulating carbohydrates. Elevated levels of lipoprotein lipase have been reported in alcoholics, especially after a prolonged period of drinking.

Obviously, further study is necessary to understand the role of lipoprotein lipase in the regulation of lipid metabolism. As our knowledge of control factors for lipid and carbohydrate utilization improves, we may be able to see more clearly how this enzyme interacts with other processes to control fat storage and use by the body. Useful clinical tests may result from this research.

13.4

FORMATION AND UTLIZATION OF CHOLESTEROL

Dietary Sources

Approximately 15% of the cholesterol in our bodies is provided by diet; the remainder is synthesized by the liver. The major sources of dietary cholesterol are meats, eggs, some seafoods, and dairy products (Table 13–1). Vegetable material is generally cholesterol-free. However, other steroid molecules are contained in plants which may play a role in either supplying cholesterol precursors or slowing down the absorption of cholesterol from the small intestine into the bloodstream.

Biosynthesis

The majority of the cholesterol molecules in our body are synthesized. The basic building block for cholesterol is the small molecule acetyl CoA. This precursor goes through a series of reactions to form chains of increasing length by a condensation process. Two short chains join together to form a longer chain; two of these longer chains connect and the process is repeated. The end result is a 30-carbon chain which then undergoes cyclization to form the ring compound **lanosterol**. From lanosterol a complex series of reactions (approximately 20 in number) take place to rearrange double bonds and form **cholesterol** as the final product.

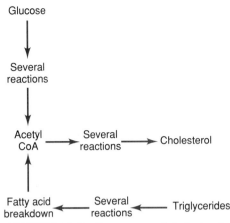

FIGURE 13–2   Precursors of cholesterol.

Since the initial molecule in the pathway leading to cholesterol synthesis is acetyl CoA, we can easily see the interrelationships between dietary intake of lipids and carbohydrate and the formation of cholesterol from these molecules (Fig. 13–2). Since acetyl CoA is formed from the conversion of glucose to smaller molecules, excess production of this intermediate can result in more being shifted to cholesterol synthesis. In a similar manner, the breakdown of fatty acids by the body results in production of acetyl CoA, with the possibility that some might be used for the synthesis of more cholesterol.

Utilization

Once cholesterol is formed, further transformation is possible to provide specialized molecules. Cholesterol is the initial steroid in the complex pathway leading to the formation of all the steroid hormones. Cortisol (which controls many metabolic processes), testosterone (the male sex hormone), the estrogens (female sex hormones), aldosterone (involved in sodium retention and regulation of blood pressure)—all these and other steroids are formed by the further metabolism of the cholesterol ring.

A significant portion of the cholesterol molecule is converted to cholesteryl esters by a specific enzyme. These esters constitute a major portion of the lipoprotein molecules and are involved in the process of cholesterol transport and utilization by the high-density lipoproteins. Interconversion of cholesterol and cholesteryl esters allows the body to eliminate some of the excess cholesterol present.

The other major conversion of cholesterol is to bile salts. With the addition of a polar group on the cholesterol side chain, the molecule is now able to dissolve partially in water, but can still interact with lipids. The bile salts are important in the digestion of lipids in the small intestine. They bind lipids and bring them into contact with lipase (which acts in an aqueous system). By making the

Table 13–1
DIETARY SOURCES OF CHOLESTEROL

Food Source	Serving	Cholesterol (mg)
Egg	1 large	213
Beef	3.5 oz chuck roast	101
Lamb	3.5 oz roast loin	87
Pork	3.5 oz broiled loin	98
Chicken	3 oz white, no skin	65
Oysters	7	55

Adapted from *U.S. News & World Report*, June 5, 1989, p. 61.

lipids water-soluble, metabolism of these compounds can take place more readily.

Cholesterol can also be incorporated into the cell membrane. This large planar molecule interacts with other lipids and with specific proteins to form part of the cell membrane which maintains a barrier between the material inside the cell and the exterior.

13.5

LIPID TRANSPORT AND STORAGE

Purpose of Lipoproteins

Because they are not soluble in water, lipids cannot circulate through the bloodstream by themselves. Some sort of transport mechanism is required to assist these materials in their travels. A number of proteins are synthesized by the liver or intestine which are specifically responsible for moving lipid classes throughout the aqueous environment of the blood. These proteins are designated as **lipoproteins** because of their function.

There are a number of classes of lipoproteins and their specificities for different lipid categories seem to overlap greatly. In each case, the protein provides a coat for the lipid mixture. Cholesterol and other lipids aggregate, forming a large complex. The protein serves as an outer layer and partially encases this complex of lipids, making them water-soluble. Many hundreds of lipid molecules may be contained within a single lipoprotein complex with a thin protein outer layer coating this aggregation.

Apolipoproteins serve four major functions in the body. These proteins allow the transport of water-insoluble lipid molecules throughout the circulation. Furthermore, they play an important role in regulating the synthesis and breakdown of the various lipoprotein complexes as they travel throughout the body. Some apolipoproteins function as activators or inhibitors of the specific lipase enzymes involved in the synthesis or breakdown of fatty acid esters (with either triglyceride or cholesterol) during lipoprotein interconversions. Lipid exchange from one lipoprotein fraction to another is often aided by a specific apolipoprotein. The protein portion of these lipid-protein complexes provides a dynamic function in affecting the rates of several reactions and does much more than serve as a passive carrier of triglycerides, cholesterol, and other lipids.

Classes of Lipoproteins

CLASSIFICATION BY DENSITY

Several specific lipoprotein classes exist, determined by both the lipid composition and the type of proteins present in the complex. These different lipoprotein groups were first studied by ultracentrifugation, a high-speed centrifugal method for separating compounds. The groups separated according to density and thus were designated as **high-density** or **low-density** materials. The ultracentrifuge is still a valuable research tool for the study of lipoproteins, but is not used in the routine clinical chemistry laboratory.

CLASSIFICATION BY LIPID COMPOSITION

Probably the most useful categorization of lipoproteins is according to lipid composition (which affects the density). The different classes of lipoproteins vary in the amount of protein, triglycerides, cholesterol (both free and esterified), and phospholipids present in the molecule. These fractions (Table 13-2) are categorized as **chylomicrons, very low-density lipoproteins (VLDL), low-density lipoproteins (LDL),** and **high-density lipoproteins (HDL).** Another class of lipoproteins (**IDL, or intermediate-density lipoproteins**) exists, but has not been characterized as well as the previous four.

From Table 13-2, we see that the focus on triglyceride transport lies with the chylomicrons and the VLDL fractions. Cholesterol is contained primarily in the LDL and HDL components. These latter two lipoprotein groups are of major importance in the discussion of the relationship between cholesterol and heart disease. Although three of the fractions listed above contain significant amounts of phos-

Table 13-2
LIPOPROTEIN CLASSES IN NORMAL SUBJECTS

Composition	Chylomicrons	VLDL	LDL	HDL
Density	<0.94	0.94–1.006	1.006–1.063	1.063–1.21
Protein (%)	1–2	6–10	18–22	45–55
Triglyceride (%)	85–95	50–65	4–8	2–7
Cholesterol (%)				
free	1–3	4–8	6–8	3–5
esterified	2–4	16–22	45–50	15–20
Phospholipid (%)	3–6	15–20	18–24	26–32

Adapted from Schaefer, E. J., and Levy, R. I., "Pathogenesis and management of lipoprotein disorders," *New Eng. J. Med.* 312:1300–1310, 1985.

pholipids, this lipid component had not received a major research emphasis in comparison with the studies carried out on cholesterol.

PROTEIN COMPOSITION OF LIPOPROTEIN COMPLEXES

Just as the lipid composition of the various lipoprotein fractions varies from one group to another, so also does the apolipoprotein composition (Table 13–3). We currently know of over 15 different specific proteins which are constituents of the lipoproteins. These various proteins play a variety of roles, both structurally and functionally. Some serve as sites of recognition by cells, others affect the activity of specific enzymes, and still others provide a role in maintaining structure of the complex.

The protein portion is very important to the metabolism of the lipids. Along with furnishing some structure to the complex, the various proteins serve as activators and inhibitors of lipases which convert triglycerides to glycerol and fatty acids. **Lecithin-cholesterol acyltransferase (LCAT)** is an important enzyme for the formation of cholesteryl esters from free cholesterol. Both the B and E apolipoproteins have received intensive study because of their role in binding the lipoprotein fractions to the cell membrane for intake and further metabolism. Several other apolipoproteins have been discovered, but we do not have a clear picture of their functions yet.

Lipoproteins and Triglyceride Utilization

When triglycerides are digested in the small intestine, they are hydrolyzed by pancreatic lipase. After the fatty acids and acylglycerol fragments pass across the intestinal wall, they are resynthesized into triglycerides. At this point, the re-formed lipids become associated with the lipoproteins B-48 and

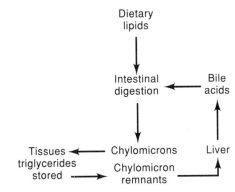

FIGURE 13-3 Exogenous pathway of lipid metabolism.

C-II, both of which are synthesized in the intestine. The resulting complex (the chylomicron) is mainly lipid with only about 2–3% protein associated. The major component is triglyceride, with small amounts of cholesterol and cholesteryl esters also present.

The chylomicrons enter the circulation and travel to the various tissues. The lipoprotein lipase in the target tissue is activated by the apolipoprotein C-II in the chylomicron. Triglyceride digestion occurs and free fatty acids become available for the cell to utilize as energy or as building blocks for other biochemicals.

The resulting lipoprotein particle is now much smaller and much denser, having lost most of the triglyceride it contained. These chylomicron remnants now contain mainly cholesterol and proteins. The remnants then go to the liver where cholesterol is taken up by the liver cells. Alternatively, the cholesterol and phospholipid portion of the remnant are transferred to HDL, which then moves to the liver for further metabolism.

The process described above is referred to as the **exogenous pathway for lipid metabolism,** since it deals with those lipids entering the system through diet (Fig. 13–3). A connected route, the **endogenous pathway,** is involved with transport and disposal of triglycerides and cholesterol synthesized within the organism. The endogenous pathway will be described later in this section.

A number of other apolipoproteins are synthesized in the liver. These protein fractions are then combined with triglycerides and cholesterol to form lipoprotein complexes designated as the VLDL fraction. Release of VLDL into the circulation is followed by activation of lipoprotein lipase with the release of glycerol and free fatty acids. The remainder of the complex becomes the LDL fraction.

Lipoproteins and Cholesterol Utilization

Dietary cholesterol enters the small intestine where it is transported across the intestinal wall.

Table 13–3
FUNCTIONS OF APOLIPOPROTEINS

Protein Class	Lipoprotein Category	Function
A-I	Chylo, HDL	Activates LCAT Structural role in HDL
A-II	Chylo, HDL	Activates hepatic lipase Structural role in HDL
B-100	VLDL, LDL	Binds cell receptor Structural role
B-48	Chylomicrons	Structural role
C-I	Chylo, VLDL, HDL	Activates LCAT
C-II	Chylo, VLDL, HDL	Activates lipoprotein lipase
C-III	Chylo, VLDL, HDL	Inhibits lipoprotein lipase
E (several)	Chylo, VLDL, HDL	Binds cell receptor

Adapted from Schaefer, E. J., and Levy, R. I., "Pathogenesis and management of lipoprotein disorders," *New Eng. J. Med.* 312:1300–1310, 1985.

With triglycerides and the appropriate lipoproteins, cholesterol then forms chylomicrons, which undergo metabolism as described above to form either HDL or chylomicron remnants. Both entities enter the liver for further disposal. One route for cholesterol elimination by the liver is the formation of bile acids, which are excreted into the small intestine to participate in digestive processes.

The endogenous pathway for cholesterol (and triglyceride) utilization (Fig. 13–4) facilitates transport and removal of those lipids formed through intracellular metabolism or not dealt with by way of the exogenous cycle. In the liver, triglycerides, cholesterol, and cholesteryl esters interact with specific apolipoproteins to form VLDL. Following entry into the circulation, the triglyceride portion of VLDL is partially hydrolyzed by lipoprotein lipase in capillary tissue. The resulting IDL is less lipid-rich and is further converted in two ways. Some of the IDL interacts with cell receptors. The lipid material is taken into the cell and the lipoprotein portion is broken down. The major route of IDL processing involves further lipase hydrolysis of triglycerides with loss of more lipid, leading to the formation of the cholesterol-rich LDL fraction.

Our knowledge of the further processing of cholesterol by the LDL fraction (and the role of HDL in cholesterol removal) represents one of the most significant advances in understanding the transport and metabolism of cholesterol since the discovery of this molecule in 1784. Discoveries during the 1970s and early 1980s contributed greatly to both basic science and clinical medicine. Michael Brown and Joseph Goldstein were awarded the Nobel Prize in Medicine in 1985 for discoveries dealing with the role of LDL in cholesterol utilization by the body.

The low-density lipoprotein complex contains mainly the apolipoprotein B, cholesterol, and cholesterol esters. The protein portion of LDL interacts with a specific receptor on the cell membrane. This LDL/receptor complex then enters the cell and is broken down by enzymes within the lysosome. The LDL protein is hydrolyzed into its consituent amino acids, cholesterol is esterified with fatty acids by the enzyme acyl CoA: cholesterol acyltransferase, and the receptor moves back to the cell surface for another intake cycle.

Once inside the cell, the cholesterol can be stored or utilized, depending on cellular needs and cell type. Conversion to steriod hormones may take place, followed by hormone release into the circulation. Cell membranes contain significant amounts of cholesterol (free and esterified) which must be obtained from intracellular sources. In addition, the amount of cholesterol within the cell exerts significant control over the synthesis of new cholesterol. As the cholesterol concentration increases, the activity of 3-hydroxy-3-methylglutaryl (HMG) CoA reductase diminishes, lowering the rate of cholesterol production.

Not all the cholesterol enters the cell. Excess amounts of this molecule are esterified by the enzyme lecithin-cholesterol acyltransferase (LCAT) and attach to an apolipoprotein to form the HDL fraction. In addition, there appears to be a process referred to as **reversed cholesterol transport** in which HDL helps remove excess free (unesterified) cholesterol from cells through interaction with a specific cell receptor. This free cholesterol is also esterified and transported for further use elsewhere. One role for HDL is thought to be the movement of cholesterol to the liver, where it can be converted to bile acids and then excreted. High-density lipoprotein also appears to play a redistribution role in moving cholesterol to cells where it is needed. This function is accomplished by transfer of cholesterol molecules from HDL to either VLDL or LDL for recycling within the system. The biochemical role of HDL in cholesterol transport and metabolism has not yet been worked out in great detail.

The Purpose of Studying Lipoproteins

Although lipoprotein levels may change because of a variety of disease states, the major interest in studying cholesterol and lipoprotein metabolism lies in their contribution to coronary heart disease. Understanding the relationship between the concentrations of different lipids and lipoproteins allows more reliable prediction of the amount of risk a given individual has for a heart attack. Intervention through diet, exercise, and medication (if necessary) can lower that risk.

13.6

ANALYSIS OF LIPIDS IN BODY FLUIDS

Sampling

Patient preparation and sampling are critical factors in obtaining reliable data on lipid status. For

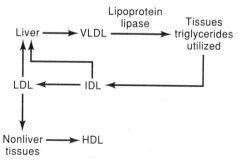

FIGURE 13–4   Endogenous pathway of lipoprotein metabolism. *Note.* Cholesterol esters are transferred from HDL to IDL with lecithin-cholesterol acyltransferase.

optimum results, the patient should remain on a regular diet for at least 3 weeks prior to the blood sample being taken. There should be no significant weight gain or loss. The sample must be collected after a fast of at least 12 h. Water is permitted, but no food should be consumed. A fasting specimen is particularly important for any assessment involving triglycerides. The blood level of these lipids rises markedly within 2 h after a meal and may peak several hours later. High levels of chylomicrons are seen in these specimens, and interpretation of the data can be difficult. Ethanol consumption should be restricted or eliminated at least 2 days prior to a blood sample being taken since this substance causes some increase in serum triglyceride levels.

Although cholesterol levels are not significantly different in fasting or nonfasting specimens, a fasting sample is still recommended. Further testing may be warranted and recollection can be avoided if a fasting specimen was initially collected. High levels of chylomicrons in a nonfasting sample may produce interference with some testing procedures. Some medications (oral contraceptives, antihypertensives, and some vitamins) may alter lipid values; a medication history is of value when interpreting the laboratory data.

Triglycerides

Measurement of triglycerides (whether by chemical or enzymatic means) involves the following general processes:

1. hydrolysis of triglyceride to form glycerol plus free fatty acids
2. measurement of glycerol present (either as glycerol or after conversion to another product)

Although the most accurate results are obtained by treatment of the sample to separate triglycerides from any free glycerol present, this fractionation procedure is cumbersome and time-consuming. Most laboratories take the expedient route of simply assaying glycerol after triglyceride hydrolysis. The results obtained may be 5–10% higher than if free glycerol were eliminated prior to analysis. Alternatively, a sample blank can be employed to correct for free glycerol. The blank does not undergo hydrolysis. Any glycerol detected in this specimen is due to free glycerol present.

CHEMICAL METHODS OF TRIGLYCERIDE ANALYSIS

Analysis of triglycerides by chemical methods requires pretreatment of the sample to remove the lipids from the proteins and other constituents. Organic solvents such as alcohols or chloroform are usually employed for this step to extract the lipids, leaving other materials in the aqueous phase. Di-

ethyl ether would be the solvent of choice, but safety factors dictate employing somewhat less effective (and less flammable) alternatives. In addition, the use of an alcohol results in only one solvent layer being formed, requiring less manipulation. Treatment with a solid absorbing material such as Florisil or zeolite then removes interfering substances such as carbohydrates which would react with the color reagent employed later in the analysis.

Hydrolysis is accomplished in basic solution at elevated temperature. Potassium hydroxide in ethanol is the preferred reagent for breakdown of the ester linkage between the glycerol and the fatty acid. The hydroxide ion promotes the hydrolysis, and the ethanol allows the material to remain in solution. An excess of ethanol also minimizes any reforming of the triglyceride. Any free fatty acid present could react with the ethanol to form a new ester, leaving the glycerol molecules (present in much lower concentration) available for further conversion. Since ethanol has a lower boiling point than water, the heating step does not need to be carried out at a very high temperature, but the increase in temperature markedly enhances the rate of hydrolysis.

Following hydrolysis, all the glycerol present is oxidized by periodate anion, forming formaldehyde and formic acid. This step needs some careful control since it is only the formaldehyde that is assayed by color reagent. Formaldehyde can undergo further oxidation under the proper conditions to produce formic acid, which does not react in this system.

The formaldehyde produced combines with a variety of reagents to form a colored product. Some of the materials employed give products whose absorption maxima are in the 500–600 nm range (a good area for spectrophotometric assays since interferences by bilirubin and hemoglobin can be minimized). However, these reagents (chromotropic acid, phenylhydrazine) have not found wide application in the clinical analysis of triglycerides.

For chemical analysis, the reactant of choice appears to be acetylacetone. In the process known as the **Hantzsch reaction,** this compound forms a cyclic derivative with formaldehyde. The product exhibits a strong absorption maximum at 412 nm and also has good fluorescence, giving two methods of detection for the quantitation of triglycerides.

A summary of the steps in the chemical analysis of triglycerides is presented in Figure 13–5.

ENZYMATIC METHODS OF TRIGLYCERIDE ANALYSIS

The use of enzymes for the quantitation of triglycerides allows a one-step procedure which eliminates the extraction step. The hydrolysis of triglycerides with strong base is not necessary since this step can be accomplished enzymatically. Glycerol formed from the breakdown of triglycerides then reacts directly with an indicator system in a series of

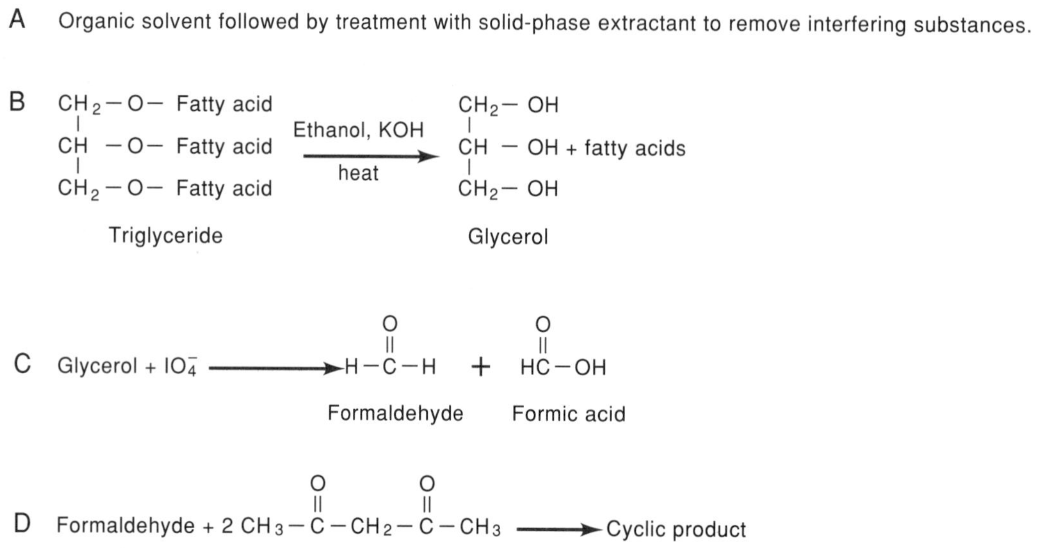

A Organic solvent followed by treatment with solid-phase extractant to remove interfering substances.

B

$$
\begin{array}{ccc}
CH_2-O- \text{Fatty acid} & & CH_2-OH \\
| & \xrightarrow[\text{heat}]{\text{Ethanol, KOH}} & | \\
CH\ \ -O- \text{Fatty acid} & & CH\ -OH + \text{fatty acids} \\
| & & | \\
CH_2-O- \text{Fatty acid} & & CH_2-OH
\end{array}
$$

Triglyceride   Glycerol

C Glycerol + IO_4^- $\longrightarrow$ $\underset{\text{Formaldehyde}}{H-\overset{\displaystyle O}{\overset{\|}{C}}-H}$ $+$ $\underset{\text{Formic acid}}{HC-OH}\ \overset{\displaystyle O}{\overset{\|}{}}$

D Formaldehyde + 2 $CH_3-\overset{\displaystyle O}{\overset{\|}{C}}-CH_2-\overset{\displaystyle O}{\overset{\|}{C}}-CH_3$ $\longrightarrow$ Cyclic product

FIGURE 13-5   Steps in the measurement of triglycerides. A. Extraction of triglycerides. B. Hydrolysis of triglycerides. C. Oxidation of glycerol. D. Reaction with acetylacetone.

enzyme reactions. The actual product measured is either NAD^+ or NADH, depending on the specific system used.

Lipase is employed as the enzyme for hydrolysis of triglycerides, producing glycerol plus free fatty acids. An excess of enzyme must be employed to achieve complete breakdown of all ester linkages. In some assay systems, a proteolytic enzyme such as chymotrypsin is employed as part of the hydrolysis step. Although the exact role of this second enzyme has not been well defined, cleavage of the protein portion of any lipoprotein complex containing triglycerides may be necessary for complete reaction of triglycerides.

Hydrolysis with lipase and formation of glycerol-3-phosphate from glycerol (part A) are common steps in all enzymatic methods. Two approaches can be taken for subsequent analysis. In the Bucolo and David method (Fig. 13-6), the ADP undergoes further reaction, producing pyruvate. NADH then reacts with pyruvate (catalyzed by lactate dehydrogenase) to form lactate and NAD^+, with the decrease in absorbance at 340 nm indicating the extent of the reaction. An alternative approach assesses glycerol-3-phosphate synthesis in the initial reaction. Conversion of glycerol-3-phosphate involves reaction with NAD^+ to form either dihydroxyacetone or dihydroxyacetone phosphate, depending on the enzyme system and cofactors employed. The other product of the reaction, NADH, can be quantitated directly at 340 nm. More commonly, the NADH formed then reacts with a reagent, forming either a colored product with strong light absorbance in the visible region or a highly fluorescent product.

With enzymatic methods, correction for free glycerol is easily accomplished by running a second sample through the analysis, but eliminating the lipase hydrolysis step. This process is rather time-consuming and the reagents are expensive, so most laboratories do not make the effort to correct for free glycerol. Failure to correct for the presence of free glycerol, however, leads to an overestimation of triglyceride values.

STANDARDIZATION OF TRIGLYCERIDE MEASUREMENTS

One of the major problems in the quantitation of triglycerides is the lack of reliable reference materials. In the serum, the triglycerides are a mixture of compounds with differing fatty acids attached to the glycerol backbone. The standards that are employed have reflected this heterogeneity of composition of the natural materials. In addition, there is wide variability in the amount of free glycerol contained in different standard materials. Various government science agencies are addressing the issue of standards with some agreement and success. Any material employed in a clinical laboratory ideally should be referenced against Centers for Disease Control materials or controls from some similar recognized organization.

1. Glycerol + ATP $\xrightarrow{\text{Glycerol kinase}}$ Glycerol-3-phosphate + ADP

2. ADP + Phosphoenolpyruvate $\xrightarrow[\text{kinase}]{\text{Pyruvate}}$ ATP + Pyruvate

3. Pyruvate + NADH + H^+ $\xrightarrow[\text{dehydrogenase}]{\text{Lactate}}$ Lactate + NAD^+

FIGURE 13-6   Enzymatic assay for triglycerides.

Cholesterol

As is the case with triglyceride measurements, analysis of cholesterol can take two major routes: chemical and enzymatic. The chemical methods suffer in terms of accuracy and specificity. The materials used are also rather hazardous. Although enzymatic methods demonstrate greater reliability, accuracy, and safety, these approaches are more costly. Most laboratories today are employing some sort of an enzymatic method for the quantitation of serum cholesterol.

The analysis of total cholesterol is complicated by the presence of two forms of this molecule. Cholesterol circulates both as the free steroid and in the esterified form. The esterified forms represent approximately two-thirds of the total cholesterol present. In colorimetric reactions, cholesterol esters give a greater color intensity than the same amount of free cholesterol. Other sterols present may also contribute to the amount of color formed.

CHEMICAL METHODS OF CHOLESTEROL ANALYSIS

Colorimetric procedures for the quantitation of serum cholesterol follow a rather simple format, although a wide variety of specific methods have been published. The simplest approach is the one-step direct method, known as the **Liebermann-Burchardt (L-B) procedure.** Cholesterol (in both free and ester forms) reacts with a mixture of sulfuric acid and acetic anhydride to form oxidation products which absorb strongly in the region of 410 nm. The strong acid combination removes the hydroxyl group and introduces another double bond into the steroid ring, thereby increasing the color intensity.

Other serum constituents such as hemoglobin and bilirubin absorb strongly in this region and may produce falsely elevated values. Turbidity produced by high triglyceride levels may also contribute to the measured absorbance. Appropriate blanks minimize these interferences, but a simple reagent blank is not sufficient to correct for these other influences on the measured absorbance reading.

The one-step assay does not resolve the problems associated with the presence of cholesterol esters and other sterols. Some type of preliminary treatment is necessary to overcome difficulties produced by these two entities. An initial extraction with zeolite removes all forms of cholesterol from the system, leaving behind most other sterols. After redissolving the cholesterol, the esters are hydrolyzed and the total cholesterol measured. This process forms the basis of the current reference method for cholesterol, known as the **Abell-Kendall method.** Precipitation of cholesterol esters after extraction separates esterified from free cholesterol, permitting measurement of only the free fraction. The clinical utility of data obtained by this approach has not been clearly demonstrated.

Although the one-step chemical method has some severe limitations, it has long been the standard method of cholesterol measurement (until the advent of enzymatic methods). Because samples need to be pretreated, this practice has not been widely adopted. Considerations of cost, complexity, and time dictated the compromises made in the analysis of cholesterol, which has become one of the most routinely requested analyses in initial health work-ups. Large numbers of analyses are performed every day on automated instruments, and preextraction or ester hydrolysis by chemical means are simply not feasible in this setting. The more elaborate procedures have been reserved for reference methods, whereas the day-to-day analyses were carried out using somewhat less accurate (but more practical and cost-effective) methods.

ENZYMATIC METHODS OF CHOLESTEROL ANALYSIS

To diminish the problems associated with esters and other sterols, enzymatic methods were developed for the analysis of cholesterol (Fig. 13–7). A common theme of these procedures is the initial hydrolysis of cholesterol esters followed by an enzymatic conversion of cholesterol. A product of the second reaction is then monitored enzymatically. The specificity of this type of analysis allows us to eliminate extraction processes and avoids the need to separate cholesterol esters from other components. The one-step enzymatic procedure for quantitation of serum cholesterol is perhaps the most widely used method today.

The initial step in the process involves the hydrolysis of cholesterol esters by the enzyme cholesterol esterase (Fig. 13–7 A). At the end of this process, all cholesterol present in the sample is in the form of free cholesterol. Other sterol esters should not be affected by this procedure because of the specificity of the enzyme.

Following hydrolysis, cholesterol oxidase converts all the cholesterol present into cholest-4-ene-3-one,

A Cholesterol esters $\xrightarrow{\text{Cholesterol esterase}}$ Cholesterol + Fatty acids

B Cholesterol + O_2 $\xrightarrow{\text{Cholestrol oxidase}}$ Cholest-4-ene-3-one + H_2O

C H_2O_2 + Indicator dye (reduced) $\xrightarrow{\text{Peroxidase}}$ Indicator dye (oxidized) + H_2O

FIGURE 13–7 Enzymatic determination of cholesterol.

an oxidation form of the molecule in which the hydroxyl group on carbon 3 of the steroid ring is converted to a ketone. The other product of the reaction is hydrogen peroxide, which is measured as the indicator for the amount of cholesterol originally present (Fig. 13–7 B).

Assay for hydrogen peroxide is enzymatic, involving the enzyme peroxidase (Fig. 13–7 C). The peroxide molecule is reduced to water, while an indicator (present in the reaction mixture) is oxidized. Most indicators employed in this method are colorless under reduced conditions and form a colored material when oxidized. The colored compound can then be detected spectrophotometrically. In some instances, a fluorometric assay may be possible. By careful selection of indicator dye, absorbance can be at a wavelength where interference from hemoglobin or bilirubin is minimal.

There is some question about the completeness of ester hydrolysis if an enzymatic method is used. A few assay approaches employ a chemical hydrolysis with saponification under alkaline conditions. More complete hydrolysis may occur in this situation, but specificity is lost since other sterol esters will also react with the hydrolysis reagent. Enzymatic hydrolysis is more specific, although both the specificity and completeness of reaction depend to some extent on the source of the esterase enzyme employed in the assay.

In some systems, an oxygen electrode is employed to measure oxygen consumption directly. This variation allows some simplification of the assay, but has some drawbacks of its own. In other analyses involving oxygen electrodes, the rate of O_2 consumption is determined (as in the case of glucose), and the concentration of the analyte is determined from this data. With cholesterol, however, the more reliable parameter appears to be the total amount of oxygen consumed in the reaction. The problem with this approach is the length of time necessary for the reaction to go to completion (several minutes), which lowers sample processing rates in batch mode.

The most common system for quantitating end points in the enzymatic cholesterol assay involves some variation on the theme of measuring the hydrogen peroxide formed. The most popular indicator is 4-aminophenazone (or some structurally similar compound), which forms a strongly colored derivative when oxidized. Other methods have included the oxidation of methanol to formaldehyde by the peroxide present, followed by reaction of the formaldehyde with acetylacetone (using the same principle as the triglyceride assay).

Major concerns with enzymatic methods have to do with interferences by bilirubin and ascorbate. High concentrations of these materials may give falsely lowered values for cholesterol measurement due to interferences with the peroxidase reaction.

OUT-PATIENT TESTING FOR CHOLESTEROL

Because of the tremendous emphasis in recent years on the relationship between cholesterol levels and coronary heart disease, a striking increase in the number of cholesterol assays being offered to outpatients is being seen. "Consumer clinics," pharmacies in drug-stores, drop-in minor emergency clinics, and physicians' offices are all in the cholesterol screening business to some extent or another. To respond to this increased and varied demand, manufacturers have developed a number of compact, single-assay systems for cholesterol measurement.

The assay methodologies usually involve some variant of the enzyme-based quantitation of cholesterol. In some systems, a dipstick methodology is employed, similar to the techniques for urinalysis and for home monitoring of blood glucose. Other methods include a cartridge system for separating plasma from cells, followed by a measurement of plasma cholesterol. The Kodak system employs a dry-reagent slide similar to those used in Kodak's larger pieces of equipment. Most devices are designed to require minimal user interaction, making problems associated with technical error less likely.

In general, when assays are performed according to the manufacturers' stated instructions by trained personnel, the data meet all criteria for reliable measurement of cholesterol. There have been documented problems in accuracy and reproducibility if the individuals running the assays do not have a solid laboratory background. Although we will see a rise in this type of out-patient testing, it is good practice to have results confirmed by a reliable, accredited laboratory.

ACCURACY, PRECISION, AND STANDARDIZATION OF CHOLESTEROL ASSAYS

How reliable are cholesterol measurements using current technology? During the last several years, significant improvement has been seen in the accuracy and reproducibility of measurements. Most laboratories are able to meet the recommendations of expert boards established by the National Institutes of Health and the Centers for Disease Control. These analytical goals include being able to measure serum total cholesterol with an accuracy of $\pm 5\%$ or less of the target value (as established by the isotope dilution/mass spectrometry reference method). Ideal goals for accuracy are $\pm 3\%$ of the target value, already achieved by approximately 70% of the U.S. clinical laboratories participating in survey programs. Day-to-day reproducibility should be at least $\pm 3\%$; many laboratories can achieve even better statistics with some care.

The major problems associated with accuracy in

cholesterol measurements have to do with the standard used, the matrix employed for the controls and calibrators, and the presence of interferences (notably bilirubin, hemoglobin, vitamin C, or turbidity in the sample due to lipemia). Although these variables may cancel each other out in some instances, they can also provide additive effects that complicate the situation even further.

The standard employed for calibration and reference materials must be carefully characterized. Differences in the amount of cholesterol esters produce significant differences in the final value, particularly if less specific methods are used. Matrix differences between standards and patient samples create enormous problems in accuracy and reproducibility. All reference materials should be serum-based. Changes in the matrix produced by lyophilization and other procedures may produce errors in the final measurement of cholesterol and must be examined carefully. Obviously, standards and controls should not have interfering substances present, a potential problem when using a serum-based control containing a wide variety of constituents.

A number of patient-related factors affect results obtained for cholesterol and other lipids. Fasting is recommended, but not required, if only total cholesterol is being measured. There are measurable differences in the cholesterol value if large amounts of fluids have been consumed a few hours prior to the test or if the patient is lying down rather than standing or sitting. If an anticoagulant is used, EDTA is preferred. Other anticoagulants can introduce significant errors in the plasma measurement versus the serum values because of variable effects on the water distribution between cells and plasma. There is a striking amount of intraperson variability in the cholesterol value measured. An individual may have results which vary as much as 5–10% in 1 month even under carefully controlled conditions. Reliable determination of a clinical problem with cholesterol requires careful study over several months.

Lipoproteins

Measurement of lipoproteins as a diagnostic tool has undergone several major changes within the last 20 years. Our growing knowledge of the relationship between lipids, lipoproteins, and coronary heart disease has led to several revisions in the protocols used for lipid studies in clinical situations. Whereas the focus in the 1970s was on lipid profiles involving electrophoresis, current emphasis is more on cholesterol levels and exploration of specific lipoproteins involved in cholesterol transport and metabolism. Details of the quest for biochemical markers which will allow us to predict coronary heart disease will no doubt change as much in the future as they have in the past.

LIPOPROTEIN ELECTROPHORESIS

The fractionation of lipoprotein classes can be fairly easily accomplished by electrophoresis. A variety of separating media (including paper, cellulose acetate, agarose, and polyacrylamide) have been employed; agarose gel appears to be the method of choice for routine clinical work. Commercially available materials allow rapid set-up and fractionation using the same equipment and gel employed for routine protein electrophoresis.

The serum sample is loaded on the gel and separated into components in the same manner previously described for proteins. To locate the lipid fractions, the gel is stained with Fat Red 7B (or similar) dye, which binds to lipids in the system. Following destaining and drying, the film is examined and the fractions are qualitatively assessed for location and concentration. Because of the variability of composition and concentration of the various lipids in each fraction, quantitation by scanning densitometry provides no further useful information.

Four components are usually detected by lipoprotein electrophoresis. The alpha lipoproteins migrate the furthest, followed by prebeta, beta, and chylomicron fractions. The relationships among lipoprotein fractions, major lipoprotein contained in each fraction, and mobility relative to serum protein fractions are illustrated in Table 13–4.

The classification used above is not always operable in a given patient. In some instances, multiple bands are seen in the prebeta lipoprotein region. When the chylomicron fraction is extremely ele-

Table 13-4
RELATIVE MOBILITIES AND COMPOSITIONS OF LIPOPROTEIN FRACTIONS

Serum Protein Class	Lipoprotein Class	Major Lipoprotein Component
Albumin (migrates furthest)	—	—
Alpha₁ globulin	Alpha	HDL
Alpha₂ globulin	—	—
—	Prebeta	VLDL
Beta globulins	Beta	LDL
Gamma globulins	Chylomicrons (at origin)	VLDL

Table 13-5
FREDRICKSON CLASSIFICATION OF LIPOPROTEIN PATTERNS

Lipid Parameter	Lipoprotein Electrophoresis Type				
	I	II*	III	IV	V
Cholesterol	Normal	Increase	Increase	Mild increase	Mild increase
Triglycerides	Marked increase	Normal, increase	Increase	Increase	Marked increase
Chylomicrons	Marked increase	Absent	Absent or slight amount	Absent	Present
Beta	Decrease	Increase	Merge	Normal, decrease	Decrease
Prebeta	Normal, slight increase	Normal, increase	Merge	Increase	Increase
Alpha	Normal	Normal	Normal	Decrease, normal	Decrease

* Type II can be further subdivided into IIa and IIb.

vated, it can extend into the prebeta lipoprotein region. Although gamma globulins move away from the origin toward the cathode (negative pole), chylomicrons do not.

The ability to separate lipoprotein classes by electrophoresis led to the development of a lipoprotein phenotyping classification by D. S. Fredrickson and colleagues. A combination of lipid analysis and visual examination of the electrophoresis strip allows the categorization of patterns into five (sometimes six) abnormal types, numbered I–V (Table 13-5).

These lipoprotein electrophoresis patterns can be seen in a variety of disorders, either produced by primary deficiencies in specific lipoproteins or secondary to other diseases. If primary, the defect lies in a deficiency of a specific lipoprotein required for the transport and elimination of cholesterol. With the protein carrier absent or present in only low amounts, cholesterol accumulates in the tissues, often forming deposits (called *xanthomas*) under the skin. Individuals with these lipoprotein defects manifest lipid abnormalities early in life. The incidence of coronary heart disease is quite high in patients with primary lipid disorders of these types, and onset of severe cardiac problems is quite often in the 20s or early 30s.

Disorders which produce hyperlipoproteinemias secondary to the primary disease state include diabetes mellitus, pancreatitis, alcoholism, thyroid disorders, and liver disease. In these situations, the primary disorder produces alterations in the metabolism, transport, and/or excretion of lipids. Treatment of the primary disease frequently leads to elimination of (or at least a marked reduction in) the lipid abnormalities.

Although the classification of patients by lipoprotein electrophoresis permitted progress in our understanding of lipid metabolism and its relationship to disease state, it became apparent that a preliminary categorization based on triglyceride and total cholesterol measurements, accompanied by visual inspection of the sample to detect the presence of chylomicrons, provided as much clinically useful information as the electrophoresis pattern. With the advent of specific methods for the quantitation of cholesterol fractions and apolipoproteins, electrophoretic fractionation now appears to provide little useful clinical information.

HDL CHOLESTEROL ASSAYS

As clinical chemistry combined forces with cardiology in the late 1970s, some significant discoveries were made regarding the relationship between specific cholesterol fractions and coronary heart disease. The famous Framingham study (so named because one major location for the research was Framingham, Massachusetts) illustrated the predictive value of HDL cholesterol in assessing the risk of coronary heart disease in a given individual. This study clearly demonstrated an inverse relationship between the level of cholesterol in the HDL fraction and the probability of having a heart attack; the higher the HDL cholesterol, the lower the chance of a heart attack.

With the publication of this data came the realization that the methodology for measurement of HDL cholesterol required a great deal of study and improvement. Although we could fairly accurately measure serum total cholesterol in the 150–350 mg/dL range, the values for HDL cholesterol were more likely to fall between 25 and 75 mg/dL. Improvements in the fractionation procedure were also needed so that HDL cholesterol could be separated from other fractions easily and reliably. Currently, several different assay approaches are available for the quantitation of the cholesterol fraction bound only to the HDL class of lipoproteins.

The measurement of HDL cholesterol involves two basic steps. First, other lipoprotein classes must be separated from the HDL proteins. Second, cholesterol must be quantitated. Both of these steps represent some significant analytical problems which have not yet been completely resolved.

The separation process has been accomplished primarily with the use of various precipitating agents to remove VLDL and LDL components of the mixture. The initial precipitant of choice was heparin/Mn^{2+} to remove the unwanted proteins, followed by quantitation of the cholesterol in the HDL fraction using a colorimetric assay. In recent years, the method selected by many laboratories employs a phosphotungstate/Mg^{2+} precipitant, with the HDL cholesterol measured by an enzymatic method. A number of laboratories have found dextran sulfate to be useful for removing unwanted lipoprotein fractions. At least one company developed an electrophoretic method for separation of fractions, followed by enzymatic detection of HDL on the agarose film. The method of choice for quantitation of HDL in all these separation schemes appears to be the enzyme system.

A major source of error when using precipitation methods is the failure to achieve clear separation of fractions. Visual inspection of the tube after addition of reagents and centrifugation usually indicates whether this goal has been achieved. The supernatant should be completely clear, or the subsequent cholesterol measurement does not accurately reflect the HDL concentration. High triglyceride concentrations often create a problem in this step. Samples with triglyceride levels above about 400 mg/dL have a high probability of error when the cholesterol in the supernatant is assayed.

Although operationally simpler, the electrophoresis approach has not gained a wide following. The main problem appears to be a significant overestimation of the amount of HDL cholesterol present when compared with reference methods. Errors greater than 20% have been seen in the estimation of HDL cholesterol using electrophoresis followed by color reaction (enzymatic) and densitometry to assess the amount of HDL cholesterol in the samples.

There are few clear-cut advantages to any of the precipitation methods. Each of them has some bias in the actual data obtained and all show significant coefficients of variation (from 11–15%, depending on specific methodology). If a sample has a target value of 37 mg/dL, this translates into a range of values for any sample of approximately 27–47 mg/dL. As we will see later, this range seriously interferes with any attempt to reliably predict coronary heart disease.

The measurement of cholesterol is affected to some extent by the precipitant used. Various combinations of heparin plus metal cation or phosphotungstate plus metal cation have been shown to affect the amount of color obtained with a set amount of cholesterol using an enzyme assay system. Standardization of results must take into consideration the total system and not just the cholesterol.

APOLIPOPROTEIN MEASUREMENTS

We are becoming more aware of how the concentrations of certain transport proteins affect the cholesterol levels. Much research in recent years has focused on the usefulness of quantitating the serum or plasma levels of these proteins in an attempt to predict more reliably the possibility of a heart attack in a given patient.

A wide variety of techniques have been employed for the analysis of specific lipoproteins. Most approaches have been immunochemical in nature, including RIA, ELISA, nephelometric, and rocket immunoelectrophoretic methods. These assays have all the advantages and shortcomings of similar techniques for measurement of proteins in other clinical conditions. A number of companies now offer commercial kits containing all the assay and standardization materials needed for measurement of a specific protein.

In spite of the wealth of analytical procedures currently available, there is no clear consensus about the clinical utility of these methods. A great deal of attention has been focused on quantitation of apolipoproteins A and B. The apo-A components are important constituents of the HDL fraction, whereas apo-B proteins are found in relatively high amounts in the LDL and VLDL lipoproteins. Although the data are inconclusive at present, we may see the measurement of these specific proteins become a routine part of our lipid profile in the future.

13.7

REFERENCE RANGES FOR LIPIDS

For both males and females, there is a gradual increase in serum cholesterol values from childhood on (Table 13–6). Serum cholesterol levels for men appear to peak at about age 45–50, but the levels for women do not appear to peak until the mid- to late 50s. After these peaks, the cholesterol values decline somewhat in both men and women. The serum values for women are approximately 10% lower than levels for men during early and midadulthood. In the mid-40s (at the onset of menopause) the values are roughly equivalent. After that time, serum cholesterol levels for women are much higher than those for men.

Expected values for plasma triglycerides show a somewhat less complicated picture (Table 13–7). Concentrations for both men and women are lower as children than they are as adults. For males, values rise during adulthood and peak in the late 40s–mid-50s, followed by a noticeable drop in later years. Serum triglyceride levels in women show a gradual rise in adulthood to a plateau in the late 50s. In adult years the values for women are generally lower than for men.

When we look at the commonly accepted definition of *reference range*, the assumption is that this span of values defines what is normal for a given population. The range theoretically defines those values seen in presumably healthy individuals. For

Table 13–6
OBSERVED CHOLESTEROL VALUES ACCORDING TO AGE
AND SEX

Age Range (Years)	Males (mg/dL)	Females (mg/dL)
0–4	125–186	120–189
10–14	127–190	131–190
20–24	130–204	130–203
30–34	148–239	139–213
40–44	163–250	154–235
50–54	169–261	172–268
60–64	171–259	186–280
>70	162–252	180–278

All values represent the 10–90 percentile range.
Adapted from Schaefer, E. J., and Levy, R. I., "Pathogenesis and management of lipoprotein disorders," *New Eng. J. Med.* 312:1300–1310, 1985.

cholesterol (and possibly for triglycerides also), this assumption is not valid. Although the range of values for serum cholesterol goes up to 270–280 mg/dL, this does not represent the healthy situation, but merely reflects the actual cholesterol levels observed in a given population. Current data indicate that serum cholesterol levels in excess of 190–200 mg/dL are related to an increased incidence of coronary heart disease. Similar data will be offered in the next section for reference values for HDL and LDL cholesterol. Triglyceride reference ranges in the literature often run up to 300 mg/dL or higher, although more recent evidence suggests that values above 250 mg/dL are associated with some added risk of heart attack. As we look at serum levels of various lipids, we will not be focusing on reference ranges, but on cutoff points and their association with risk of coronary heart disease. This approach creates some complications in the interpretation of data and eliminates the "quick answer" we all like, but the result is a much more informed diagnosis and improved effectiveness in developing a treatment program.

Table 13–7
OBSERVED TRIGLYCERIDE VALUES
ACCORDING TO AGE AND SEX

Age (Years)	Males (mg/dL)	Females (mg/dL)
0–4	33–84	38–96
10–14	37–102	44–114
20–24	50–165	41–112
30–34	58–213	44–123
40–44	64–248	51–155
50–54	68–250	59–186
60–64	68–235	64–202
>70	67–212	69–204

All values represent the 10–90 percentile range.
Adapted from Schaefer, E. J., and Levy, R. I., "Pathogenesis and management of lipoprotein disorders," *New Eng. J. Med.* 312:1300–1310, 1985.

13.8

CHOLESTEROL AND HEART DISEASE

Formation of Atherosclerotic Plaques

The formation of the deposits on the inside of major blood vessels begins early in childhood. As the deposits grow, they gradually evolve into plaques, semisolid material composed of lipids and other biochemical components. The result is the blockage of blood flow through the tissues, followed by tissue death (due to lack of oxygen and nutrients). Clinical signs include sudden death due to cardiac failure, myocardial infarction, stroke, or other complications, depending on the tissue affected.

In childhood, the coronary arteries have a different structure than they do in adults. Over a period of years, the arteries become thicker and lose some of their flexibility. In addition, fatty streaks begin to appear on the inside of the arteries. Cholesterol and cholesterol esters are the primary lipids seen in these streaks. During early adulthood, fibrous plaques begin to form, presumably using the fatty streaks as a foundation for their growth. In addition to a high concentration of cholesterol esters, the plaque accumulates cellular debris, platelets, calcium, and other materials. Smooth muscle and connective tissue grow around the plaque, increasing its size and rigidity. The blood vessel blocks, and hemorrhage and small blood clots occur, leading to cardiac damage or stroke.

Total Cholesterol and Risk of Heart Attack

The relationship between the serum total cholesterol concentration and the probability of coronary

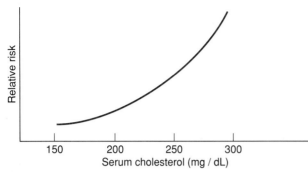

FIGURE 13–8   Relationship between serum total cholesterol and relative risk of coronary heart disease.

heart disease has been well established. All the available data suggest that the likelihood of a heart attack increases as the serum total cholesterol level goes above about 200 mg/dL (Fig. 13–8). Whereas there is only a slight increase in the probability of a heart attack if the cholesterol value increases from 150 mg/dL to 200 mg/dL, the probability jumps sharply as the cholesterol rises higher. We realize that other factors such as blood pressure and cigarette smoking also play important roles in the pathogenesis of this medical problem. In addition, as age increases, the probability becomes greater (at any given serum cholesterol level) that a heart attack might occur.

A panel of experts convened by the National Institutes of Health in 1984 recommended cutoff points for serum cholesterol which might be used to identify patients at moderate or high risk for developing coronary heart disease (Table 13–8).

Lipoprotein Cholesterol Fractions

Although serum total cholesterol can provide an early warning of risk, other parameters allow for somewhat more reliable prediction. With the development of methods for the measurement of HDL cholesterol, a significant tool in developing predictive values became available. A number of clinical studies have shown that increased HDL cholesterol levels lead to decreased risk of coronary heart dis-

Table 13–8

RECOMMENDED CUTOFF POINTS FOR SERUM CHOLESTEROL

Age (Years)	Moderate Risk	High Risk
2–19	>170 mg/dL	>185 mg/dL
20–29	>200 mg/dL	>220 mg/dL
30–39	>220 mg/dL	>240 mg/dL
40 and over	>240 mg/dL	>260 mg/dL

Adapted from Rosenfeld, L., "Atherosclerosis and the cholesterol connection. Evolution of a clinical application," *Clin. Chem.* 35:521–531, 1989.

ease. The problem has been in defining exactly what these HDL concentrations should be.

The highest risk occurs in adults with HDL levels below 40 mg/dL. At these values, even patients with total cholesterol levels of less than 200 mg/dL have significantly higher probability of a heart attack. High-density lipoprotein cholesterol levels between 40 and 49 mg/dL markedly lower the risk for those individuals with total cholesterol below 230 mg/dL. As the HDL concentration increases further, the risk factor is lowered for patients with even higher total cholesterol levels. There are no simple answers, but the combination of total cholestrol, HDL cholesterol, and assessment of other risk factors (particularly blood pressure and smoking) allows the physician to make a somewhat reliable determination of an individual's relative risk.

More recently, scientists have developed techniques which allow the study of HDL subclasses. There appear to be at least three groups of HDL lipoproteins (HDL_1, HDL_2, and HDL_3), with differing properties and possibilities for serving as predictive markers for coronary heart disease. Some early data indicate that HDL_3 may be a better marker than total HDL, but much more research needs to be carried out before measurement of this particular protein is employed in the routine clinical setting.

Similarly, a great deal of interest has been shown in the measurement of apo-A, found in the HDL class of lipoproteins. In this case, current research has produced conflicting data about the relationship between the blood levels of this protein and heart disease, although much work is still being done in this area.

To a great extent, women have a much lower incidence of coronary heart disease than men, even if total cholesterol levels are the same. In women, this disorder is rarely seen in patients whose total cholesterol is less than 260 mg/dL. The chief explanation at present for this difference appears to be the higher HDL cholesterol levels in women (approximately 39–80 mg/dL in contrast to approximately 30–60 mg/dL for men). The presence of estrogens appears to enhance the synthesis of this class of lipoproteins, offering a protective effect particularly to premenopausal women.

A number of studies have also explored the relationship between LDL cholesterol values and risk of heart attack. Here the data are less clear, but a trend is still evident. As the LDL cholesterol level decreases, the relative risk also decreases. Low-density lipoprotein quantitation is often included in the assessment of the risk of coronary heart disease.

Factors Affecting HDL Cholesterol Levels

A variety of factors have been shown to affect total cholesterol and HDL cholesterol levels in humans.

These factors include

1. heredity
2. diet
3. obesity
4. alcohol consumption
5. smoking
6. exercise
7. estrogens and related hormones

There is a strong genetic component to the levels of HDL and LDL in an individual. Patients can be heterozygous or homozygous for the gene coding for the production of LDL receptors on the cell. A decrease in the amount of receptors leads to a lowering of the quantity of cholesterol taken up by the cell, keeping more cholesterol in the circulation. The levels of apo-A proteins (which contribute greatly to the HDL class) vary greatly from one individual to another in a genetically linked fashion. There does not appear to be a sex-linked process; both men and women show these genetic variants. However, only about 2% of the population in this country manifest a major gene defect. Other factors play a much more important role, but may interact with the genetic component in ways we do not yet understand.

Since a certain amount of cholesterol is consumed in the diet, it seems obvious that dietary reduction of cholesterol intake should be accompanied by a decrease in serum cholesterol. To some extent, this is true, but the issue is complicated by the changes in cholesterol synthesis by the body in response to lowered cholesterol intake. Cholesterol restriction appears to be more effective for men than women, although both groups see some decrease in total serum cholesterol after a decrease in dietary consumption of saturated fats and cholesterol. Lowered amounts of both LDL and HDL cholesterol are also observed, with a more marked decrease in the LDL fraction. A diet with a high content of polyunsaturated fatty acids appears to have some beneficial effect on the lowering of cholesterol, particularly in those patients with elevated lipid values.

Obesity is strongly related to total cholesterol levels in both men and women. Patients with higher body mass have somewhat higher total cholesterol levels and a somewhat more elevated LDL cholesterol fraction than others. The HDL cholesterol level in obese individuals is some 10% or so less than in those with lower body weights. After a weight loss program, HDL levels are often found to increase. It is not clear how much of this effect is due ultimately to diet and how much is due to changes in metabolism determined by different fat-to-muscle ratios in the body.

Exercise is generally thought to have a beneficial effect on cholesterol metabolism through lowering total cholesterol and LDL cholesterol levels while increasing HDL cholesterol. The extent of these changes is debatable, as is the exact levels and duration of exercise needed to bring about benefits. Men appear to show more significant changes in the

various cholesterol and lipoprotein parameters than women.

Moderate alcohol consumption has long been believed to produce some increase in the level of HDL. There are more data available on males than on females, and men appear to have greater increases (15–21%) than do women (6–18%). There is also some increase seen in total cholesterol in moderate drinkers. The effect on the LDL fraction is unclear. Some data suggest that the LDL level increases after drinking, whereas other studies indicate there is either no measurable change or a slight decrease in the concentration of this lipoprotein component.

Smoking adversely affects cholesterol levels. The HDL fraction is lowered by smoking, and the LDL fraction is increased. Decreases in HDL of from 5 to 14% have been reported, with greater lowering associated with heavier smoking. This decrease may be seen more in women than men, although much of the data are inconclusive. Complicating the picture is the fact that many smokers also consume more alcohol than do nonsmokers. We do not know at present exactly how smoking exerts its effects on cholesterol metabolism.

A major hormonal effect which appears to contribute to the higher amount of HDL cholesterol in women than in men is the presence of estrogens. A variety of data suggest that premenopausal women have higher HDL and lower LDL and total cholesterol that do women in menopause. Simple estrogen therapy during menopause has been shown to produce an increase in HDL cholesterol values. Changes in lipoprotein concentrations do not appear to take place to any great extent during the various phases of the menstrual cycle. During pregnancy, as serum estrogen increases, there also appears to be an increase in HDL cholesterol. These data are inconclusive at present and require further study to define the exact contribution of endogenous estrogens and other hormones to cholesterol metabolism.

Oral contraceptives have been shown to adversely affect cholesterol parameters. Current medications of this type are usually a combination of synthetic estrogen derivatives and progestins (usually derived from testosterone). Studies strongly indicate that use of these materials results in a decrease in HDL levels and an elevation of LDL and total cholesterol. The exact amount of change differs, depending on the specific composition of the oral contraceptive.

Approaches to Therapy

Two major therapies have evolved to deal with elevated cholesterol levels: diet and drugs. Restricting the intake of meat, eggs, and other high-cholesterol foods is certainly a good place to start in lowering serum cholesterol. However, only a small fraction of the cholesterol in our bodies comes from dietary sources, making it difficult to see significant reductions by this approach. In addition, if intake of cholesterol decreases, the body somehow compensates in part by manufacturing more cholesterol.

Another dietary approach has been to alter the ratio of saturated to unsaturated fatty acids in the diet. Although we do not at present understand the mechanism, it is known that an increase in the amount of unsaturated fatty acids ingested (and some decrease in our intake of fatty acids) lowers blood cholesterol levels. Considerable interest has been shown in a certain class of unsaturated fatty acids called the **omega-3 fatty acids** (the name indicates the location of the double bond). These compounds apparently improve cholesterol levels dramatically if included as part of the diet (and accompanied by reduction of saturated fats). A great deal of research is being carried out to explore the mechanism of this process.

Pharmacological intervention is still in its infancy. An ion-exchange resin (cholestyramine) has been used for years to help decrease cholesterol levels somewhat. Taken orally, the resin binds bile salts in the intestine and removes them from the system. The body then needs to synthesize more bile salts, which can then in turn be removed by the resin, helping to lower blood cholesterol. There are some side effects with this medication, and it cannot be used successfully by all patients, but it does represent an effective treatment for many people.

In recent years, a number of drugs have been developed which decrease the synthesis of cholesterol directly. The key enzyme controlling the manufacture of cholesterol is HMG CoA reductase. Several pharmaceutical agents are available which inhibit this enzyme and lower the amount of synthesis. This may represent a significant therapeutic approach in the future.

13.9

TRIGLYCERIDES AND HEART DISEASE

Often lost in the clamor over cholesterol levels and coronary heart disease is the contribution of another important lipid class, the triglycerides. Plasma triglyceride concentrations are high in a variety of disorders. The cause of the elevation may be primary, in hereditary diseases such as familial hypertriglyceridemia or familial combined hyperlipoproteinemia. Elevation of triglycerides in the circu-

lation is frequently secondary to diseases such as diabetes mellitus, a variety of kidney and liver disorders, and hypothyroidism. In these situations, the elevated lipid is produced as a result of other metabolic changes within the system.

Triglycerides make up the primary component of the chylomicrons and the VLDL classes of lipoproteins. In the normal individual, chylomicrons and VLDL are progressively reduced to LDL by enzymatic hydrolysis of the ester portion of the triglyceride molecule. The result is the cholesterol-rich LDL particle which then interacts with a cell membrane receptor, allowing cholesterol to enter the cell. High-density lipoprotein also contains a high amount of cholesterol fated for disposal.

In the circulation, a great deal of lipid exchange takes place among the various lipoprotein fractions. This transfer of lipids from one lipoprotein class to another is facilitated by several enzymes and lipid transfer proteins. In patients with hypertriglyceridemia, the VLDL class loses a great deal of triglyceride, as is expected. However, more cholesterol than normal is transferred to this lipoprotein component. As a result, VLDL metabolism slows considerably in high-triglyceride situations, less cholesterol is transferred to LDL (and HDL), and more cholesterol remains available in the circulation for potential incorporation into atherosclerotic plaques.

Quantitation of triglyceride levels has long been a part of most lipid profiles, but the data relating triglyceride levels to heart disease are not clear. A 1984 report from the National Institutes of Health set guidelines for interpreting these values and indicated their clinical utility and limitations as a predictor of heart attack. Patients with triglyceride values of greater than 500 mg/dL are at severe risk for pancreatitis and should undergo treatment to alleviate that problem. Individuals who have triglyceride values of 250 mg/dL or less are considered to have no added risk for cardiovascular disease. Those patients with values between 250 and 500 mg/dL may have an approximately two-fold increase in their probability of having a heart attack if there are other risk factors present, even when the plasma cholesterol level is normal. Patients with normal cholesterol values, triglycerides in the 250–500 mg/dL range, and no other risk factors do not seem to have an increased incidence of coronary heart disease. These are general guidelines, and a great deal of research is still needed in order to understand better any relationships between triglyceride levels and coronary heart disease.

SUMMARY

Lipids comprise a heterogeneous category made up of compounds with quite different chemical structures. Major classes of lipids include fatty acids (both saturated and unsaturated), triglycerides, cholesterol, and phospholipids. Because of their relative insolubility in water, lipids require transport proteins to carry them throughout the circulation and to allow movement into the cell.

Lipids initially enter the circulation as chylomicrons, rich in triglycerides. As enzymes gradually hydrolyze the triglycerides, the resulting lipoprotein particles become denser and have a higher percentage of cholesterol as a part of their structure. The usual breakdown pathway proceeds from chylomicrons to VLDL to IDL to LDL. The LDL then reacts with cell membrane receptors that move cholesterol into the cell for storage or conversion to other compounds. The HDL fraction facilitates excretion of cholesterol from the body, although the mechanisms for this process are still not clear.

Measurements for triglycerides, serum total cholesterol, and HDL cholesterol are valuable in the prediction of risk for coronary heart disease. These lipids can be assayed either chemically or enzymatically, with most laboratories preferring enzyme-based approaches. Although specificity of measurement is good, work still needs to be done on accuracy and precision. Standardization presents a major problem in the development of reliable assay methods.

Lipoprotein quantitation has followed several directions. Early work with lipoprotein electrophoresis allowed the development of the Fredrickson classification of hyperlipidemic states. More recent studies have shown the value of quantitating the HDL cholesterol fraction. Precipitation of other lipoproteins followed by measurement of cholesterol remaining in the supernatant gives a fairly reliable estimation of HDL cholesterol. Quantitation of specific proteins using immunochemical techniques is possible, but the clinical utility of these types of measurements has not yet been well established.

A great deal of research has been done on the relationship between cholesterol levels and coronary heart disease. As the cholesterol level in the blood increases, local deposits begin to form in arteries. After a time, these deposits thicken and begin to incorporate other material besides cholesterol. Finally, the blood vessel is occluded, decreasing blood flow, and resulting in tissue necrosis and a heart attack.

At present, it appears that cholesterol levels in adults need to be maintained below 200 mg/dL to minimize the probability of a myocardial infarction. Higher levels give rise to a much higher risk for this disorder. An increase in HDL cholesterol diminishes the risk of coronary heart disease, presumably by making it easier for the body to dispose of excess cholesterol. Factors which affect HDL levels include genetics, diet, alcohol consumption, smoking, exercise, and estrogens. Triglyceride levels below 250 mg/dL are not associated with increased risk. A variety of dietary and pharmacological interventions exist for the treatment of elevated blood cholesterol.

FOR REVIEW

Directions: For each question, choose the best response.

1. Which of the following serves as the building block for triglyceride and phospholipid synthesis?
   A. acetate
   B. cholesterol
   C. amino acids
   D. fatty acids

2. Which of the following lipids functions as a surfactant in the lungs?
   A. cholesterol
   B. fatty acid
   C. phospholipid
   D. triglyceride

3. The secretion of pancreatic lipase into the small intestine is necessary for the hydrolysis of
   A. cholesterol
   B. fatty acids
   C. phospholipids
   D. triglycerides

4. Lipids are transported in the blood by
   A. albumin
   B. amino acids
   C. gamma globulins
   D. lipoproteins

5. The lipoprotein class involved in the transport of triglycerides from the small intestine through the circulation to various tissues is
   A. chylomicron
   B. VLDL
   C. LDL
   D. HDL

6. The lipoprotein class associated with the removal of cholesterol from the body is
   A. chylomicron
   B. VLDL
   C. LDL
   D. HDL

7. The lipid component whose measurement would be the most severely affected by analyzing a nonfasting specimen is
   A. apolipoprotein
   B. cholesterol
   C. triglyceride
   D. fatty acid

8. In general, the quantitation of triglyceride is based on measuring
   A. chylomicron
   B. fatty acid
   C. glycerol
   D. triglyceride directly

9. In what form must cholesterol be for it to react with cholesterol oxidase in the enzymatic methods?
   A. esterified
   B. free
   C. bound to lipoproteins
   D. bound to albumin

10. What dye is used to stain the lipid fractions in lipoprotein electrophoresis?
    A. Amido black
    B. Fat Red 7B
    C. Ponceau S
    D. Sudan black

11. When performing lipoprotein electrophoresis using agarose gel, the lipoprotein fraction that migrates the farthest toward the anode is
    A. chylomicron
    B. alpha

    C. beta

    D. prebeta

12. Precipitating agents used to remove VLDL and LDL from serum so that only the HDL cholesterol remains for measurement include all of the following *except*

    A. citrate/fluoride

    B. heparin/manganese

    C. phosphotungstate/magnesium

    D. dextran sulfate

13. Apolipoprotein A is the primary protein component of

    A. HDL

    B. IDL

    C. LDL

    D. VLDL

14. Laboratory tests that aid in predicting the risk of coronary heart disease include all of the following *except*

    A. fatty acid

    B. HDL cholesterol

    C. total cholesterol

    D. triglyceride

15. Which phenotype in the Frederickson classification of hyperlipoproteinemias is characterized by an increase in cholesterol, an increase in the beta lipoproteins, normal triglycerides, and the absence of chylomicrons?

    A. I

    B. II

    C. III

    D. IV

BIBLIOGRAPHY

Albers, J. J., and Warnick, G. R., "Lipoprotein measurement in the clinical laboratory," *Lab. Manage.* February:31–38, 1981.

Bates, H. M., "Hypercholesterolemia and heart disease: The atherogenic index," *Lab. Manag.* October:46–49, 1987.

Brown, M. S. et al., "Regulation of plasma cholesterol by lipoprotein receptors," *Science.* 212:628–635, 1981.

Brown, M. S., and Goldstein, J. L., "Lipoprotein metabolism in the macrophage: Implications for cholesterol deposition in atherosclerosis," *Ann. Rev. Biochem.* 52:223–261, 1983.

Brown, M. S., and Goldstein, J. L., "How LDL receptors influence cholesterol and atherosclerosis," *Sci. Am.* November:58–66, 1984.

Brown, M. S., and Goldstein, J. L., "A receptor-mediated pathway for cholesterol homeostasis," *Science.* 232:34–47, 1986.

Bush, T. L. et al., "Cholesterol, lipoproteins and coronary heart disease in women," *Clin. Chem.* 34:B60–B70, 1988.

Calvert, G. D., and Abbey, M., "Plasma lipoproteins, apolipoproteins, and proteins concerned with lipid metabolsim," *Adv. Clin. Chem.* 24:217–298, 1985.

Consensus Development Panel (National Institutes of Health), "Treatment of hypertriglyceridemia," *J. Amer. Med. Assoc.* 25:1196–1200, 1984.

Cooper, G. R. et al., "Standardization of lipid, lipoprotein and apolipoprotein measurements," *Clin. Chem.* 34(8B):B95–B105, 1988.

Dujovne, C. A., and Harris, W. S., "The pharmacological treatment of dyslipidemia," *Ann. Rev. Pharmacol. Toxicol.* 29:265–288, 1989.

Eisenberg, S., "Lipoprotein abnormalities in hypertriglyceridemia: Significance in atherosclerosis," *Amer. Heart J.* 13:555–561, 1987.

Fosslein, E., "Current strategies in lipid analysis," *Diag. Med.* April:43–47, 1985.

Gibson, J. C., and Brown, W. V., "The human plasma apolipoproteins: Assay methods. Parts I and II," *Lab. Manage.* March:19–27, and April:27–37, 1983.

Hulley, S. B., and Lo, B., "Choice and use of blood lipid tests. An epidemiologic perspective," *Arch. Intern. Med.* 143:667–673, 1983.

Kannel, W. B., "Cholesterol and risk of coronary heart disease and mortality in men," *Clin. Chem.* 34:B53–B59, 1988.

Le, N. -A., and Brown, W. V., "Apolipoprotein immunoassays in screening for coronary risk," *Lab. Manage.* November:55–61, 1987.

Levy, R. I., "Cholesterol, lipoproteins, apoproteins, and heart disease: Present status and future prospects," *Clin. Chem.* 27:653–662, 1981.

Mahley, R., "Apolipoprotein E: Cholesterol transport protein with expanding role in cell biology," *Science.* 240:622–630, 1988.

Merz, B., "Is it time to include lipoprotein analysis in cholesterol screening?" *J. Amer. Med. Assoc.* 261:497–498, 1989.

Motulsky, A. G., "The 1985 Nobel Prize in physiology or medicine," *Science.* 231:126–129, 1986.

Naito, H. K., "The clinical significance of apolipoprotein measurements," *J. Clin. Immunoassay.* 9:11–20, 1986.

Naito, H. K., "Reliability of lipid, lipoprotein, and apolipoprotein measurements," *Clin. Chem.* 34(8B):B84–B94, 1988.

Rosenfeld, L., "Atherosclerosis and the cholesterol connection: Evolution of a clinical application," *Clin. Chem.* 35:521–531, 1989.

Schaefer, E. J., and Levy, R. I., "Pathogenesis and management of lipoprotein disorders," *New Eng. J. Med.* 312:1300–1310, 1985.

Scott, J., "Unravelling atherosclerosis," *Nature.* 338:118–119, 1989.

Slickers, K. A., "Enzyme-linked assays for cholesterol," *CRC Crit. Rev. Clin. Lab. Sci.* 8:193–212, 1977.

Slutztsky, G. M., and Inbar, M., "Lipoproteins, apolipoproteins, and cholesterol in cardiovascular disease," *Amer. Clin. Prod. Rev.* October:18–23, 1987.

Statland, B. E., "Fasting and lipid values." *Med. Lab. Obser.* May, 1986.

Steinberg, D., "Plasma lipid levels: How much is too much?" *Lab. Manage.* May:31–41, 1986.

Stoecklein, L., "Lipid metabolism diagnostics: Electrophoretic determination of apolipoproteins," *Amer. Clin. Prod. Rev.* April:16–21, 1986.

Zak, B., "Cholesterol methodologies: A review," *Clin. Chem.* 23:1201–1214, 1977.

Chapter Fourteen

CLINICAL ENDOCRINOLOGY I: GENERAL CONCEPTS AND THYROID HORMONES

UPON COMPLETION OF THIS CHAPTER, THE STUDENT WILL BE ABLE TO

1. Define the following terms:
   A. hormone
   B. hypothyroidism
   C. hyperthyroidism
2. State three characteristics of hormones.
3. Name the hormone(s) produced by each of the following tissues:
   A. hypothalamus   F. thyroid
   B. anterior pituitary   G. parathyroid
   C. posterior pituitary   H. pancreas
   D. adrenal medulla   I. ovaries
   E. adrenal cortex   J. testes
4. List examples of hormones classified on the basis of the following chemical structures:
   A. peptide
   B. steroid
   C. amino acids
   D. fatty acids
5. Explain the role of receptor molecules in facilitating the action of hormones.
6. Discuss the differences between hormones that bind at the cell surface and those that require internal binding.
7. Explain why a hormone bound to protein is biochemically inactive.
8. Name the two hormones collectively referred to as the thyroid hormones.
9. Name the three proteins that T3 and T4 bind to in the circulation for transport to target tissues, noting which functions as the major carrier protein.
10. Describe the roles of TRH, TSH, and thyroglobulin in thyroid hormone synthesis.
11. List three immunoassay methods used for the quantitation of total thyroxine and two methods used to quantitate TSH levels.
12. Explain why it is necessary to perform a separation step in total thyroxine immunoassay methods.
13. Describe the solid-phase and the analog methods for quantitating free thyroxine.
14. Name the method used to determine the concentration of thyroxine-binding globulin indirectly and two methods used to determine TBG concentration directly.
15. Discuss the clinical significance associated with abnormal thyroid function test values.
16. Discuss the importance of performing neonatal thyroid screening.
17. Describe the collection of blood samples for neonatal thyroid screening.

INTRODUCTION

Clinical endocrinology is one of the fastest growing fields in clinical chemistry. As a result of the development of radioimmunoassay (RIA) and other immunoassay procedures, we are now able to quantitate the minute amounts of polypeptides, steroids, and other compounds which regulate so many of the biochemical processes in the body. Fundamental questions about growth and development can now be addressed by studying the changing hormone patterns. Disruption of the endocrine processes in a variety of disease states has profound implications for other biochemical systems. Increasingly, we are seeing an emphasis on hormonal changes as explanations for some of the problems of mental illness, particularly in severe depression. Some hormone measurements (to investigate pregnancy or deal with infertility) can now be carried out with kits purchased from the local drugstore. Endocrinology has moved from an esoteric science to a place of significance in the clinical laboratory.

14.1

DEFINITION OF A HORMONE

Hormones serve as chemical messengers in the body. Coming from a Greek term meaning "to spur on," hormones interact with other components of the organism to produce biochemical changes, usually to increase the level of activity of a process or series of reactions. There are three characteristics of hormones:

1. Each hormone is produced by a specific tissue (gland).
2. Hormones are released directly from the tissue into the bloodstream and carried to the site of action.
3. Each hormone acts at a specific site or sites (target cells) to induce certain characteristic biochemical changes.

The collection of hormones, carrier proteins, and other components of these processes is called the **endocrine system** ("internal secretion").

14.2

CLASSES OF HORMONES

Hormones can be classified by either their structure (to explore biochemical similarities) or their tissue of origin (to examine the effect of damage or removal of a given endocrine gland). Hormones with similar structure have closely related mechanisms of action, even though the specific biochemical changes produced might be quite different. The pro-

Table 14–1
TISSUES AND THE HORMONES THEY PRODUCE

Tissue of Origin	Hormone(s) Produced
Hypothalamus	TRH, CRF, GnRH, others
Anterior pituitary	TSH, ACTH, FSH, LH, prolactin, growth hormone
Posterior pituitary	Vasopressin, oxytocin
Adrenal medulla	Epinephrine, norepinephrine
Adrenal cortex	Cortisol, 11-deoxycortisol, aldosterone
Thyroid	T3, T4, calcitonin
Parathyroids	Parathyroid hormone (PTH)
Pancreas	Insulin, glucagon
Ovaries	Estrogens
Testes	Testosterone, other androgens
?	Prostaglandins

cess which causes the changes is similar to the cellular level. Interestingly, many different hormones may be synthesized by a specific tissue (such as the adrenal gland), but these hormones produce a wide range of very different end results.

Classification by Tissue of Origin

Each hormone is formed by a specific tissue of origin (Table 14–1). Some tissues produce a variety of different hormones which may or may not have related functions. Included in the table are releasing factors, which are generated by one tissue and stimulate the release of a certain hormone by another tissue. There are a number of other hormones which are not listed since our information about their sites of production and/or biochemical function is rather incomplete.

Classification by Structure

The categorization of hormones on the basis of structure has proven to be very useful (Table 14–2).

Table 14–2
HORMONE CLASSIFICATION BY STRUCTURE

Structure	Examples of Hormones
Peptide	Insulin, parathyroid hormone, LH, FSH, TSH, TRH, ACTH, prolactin, growth hormone, calcitonin, glucagon
Steroid	Cortisol, progesterone, estrone, estradiol, testosterone, aldosterone
Amino acids	Epinephrine, norepinephrine, T4, T3
Fatty acids	Prostaglandins

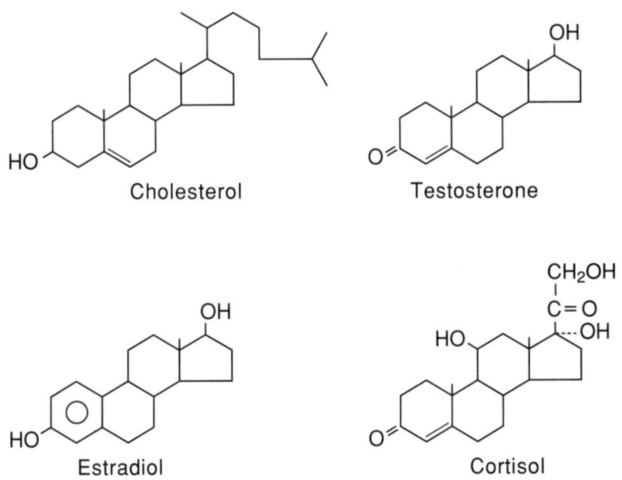

FIGURE 14-1   Structures of typical steroid hormones.

Most hormones of the same basic structure appear to produce the same fundamental biochemical changes at the cellular level, although the specific reactions altered are quite different. However, it is becoming apparent that there are a wide variety of subtle differences in the manner of expression of even similar hormones. Our simple earlier picture must frequently give way to the growing complications we now observe as research reveals more details of how each different hormone acts.

PEPTIDE

Probably about two-thirds of the hormones we know about today are proteins or peptides. All the known releasing factors which stimulate hormone production by another tissue are peptides of some sort. One of the exciting topics of current research is **neurohormones,** compounds which play important roles in development and function of the nervous system. To date, these materials all appear to be peptide in nature. Because of their structures, the peptide hormones are water-soluble and do not re-

quire transport proteins to move through the circulation.

STEROID

The **steroid hormones** are primarily involved in regulation of sexual development and characteristics. These materials are all synthesized from cholesterol as the basic molecule (Fig. 14–1). Differences in the side chain at carbon-17 and variations in substituents on the A and B rings of the steroid nucleus determine the structure of the steroid hormones. **Testosterone** and the **estrogens** are steroid hormones which affect development of sexual characteristics. **Cortisol** and derivatives constitute another major class of steroid hormones which are involved in some complex regulation of overall metabolism. **Aldosterone** is a steroid hormone partially responsible for fluid and electrolyte balance.

As is the case with cholesterol, the steroid hormones are not water-soluble. Therefore, each of these hormones requires a transport protein in order to travel through the bloodstream. Some steroids have specific transporters, and others share a common carrier. The concentration of the transport protein is an important determinant of steroid action and plays a role in regulation of the secretion rate for the particular hormone.

AMINO ACID DERIVATIVES

Two important pairs of hormones are synthesized from aromatic amino acids (Fig. 14–2). Tyrosine serves as the precursor for the thyroid hormones **thyroxine (T4)** and **triiodothyronine (T3),** formed through a very complex process. This amino acid can also be more directly converted to the adrenal hormones epinephrine and norepinephrine, which help control blood pressure and heart rate in addition to some complex effects upon general metabolic rate. Both T3 and T4 require carrier proteins for

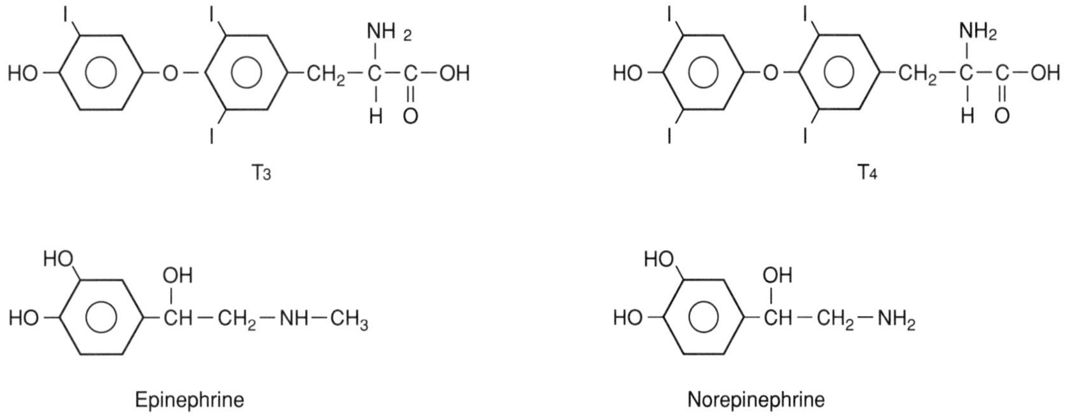

FIGURE 14-2   Hormones derived from amino acids.

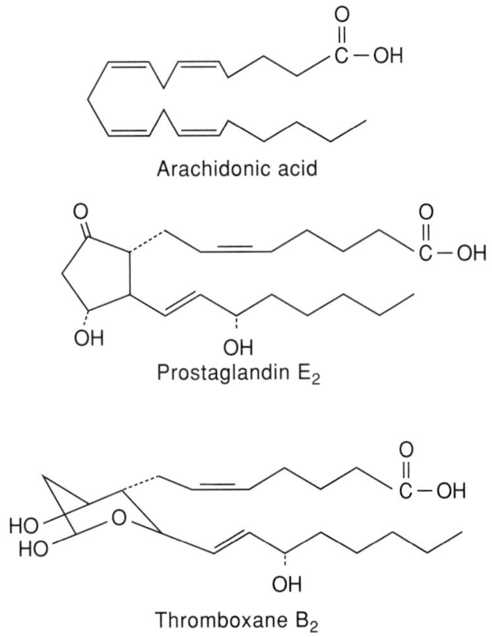

Arachidonic acid

Prostaglandin E_2

Thromboxane B_2

FIGURE 14–3   Prostaglandins.

transport, whereas epinephrine and norepinephrine can easily circulate unbound.

DERIVATIVES OF FATTY ACIDS

The **prostaglandins** are a class of hormones formed from arachidonic acid (Fig. 14–3). This hormone precursor is a 20-carbon fatty acid containing four double bonds. **Arachidonic acid** is an essential fatty acid, required as part of a healthy diet. Since the prostaglandins are synthesized in a variety of tissues and behave in ways different from the "classic" hormones, it is unclear exactly how to categorize these compounds. There is increasing interest in the clinical aspects of prostaglandins, both as biochemical regulatory molecules and as pharmacological agents.

14.3

MECHANISMS OF HORMONE ACTION

How do hormones produce their biochemical effects? This question has been the focus of research for a number of years. Early results indicated different mechanisms for steroid hormones, peptide hormones, and thyroid hormones. More recent studies indicate a multitude of processes taking place when peptide hormones interact with cells, with a single hormone sometimes using more than one process to accomplish its purpose. Only a brief summary of current thinking will be presented to help you see this intricate set of biochemical interactions in proper perspective.

Hormone/Receptor Interaction

The heart of hormone action is the binding of hormone to a specific receptor molecule. The receptor may be located on the surface of the cell or in the cell interior. Each receptor has a specific structure, with sites on the complex that recognize the structure of the hormone and bind tightly to that molecule. An individual cell may have hundreds of thousands of receptors for a specific molecule. As is the case with enzymes, matching complementary portions of receptor and hormone molecules provides specificity of interaction to promote the desired biochemical end result.

Binding at Cell Surface

For most polypeptide hormones, the site of binding to receptor is the surface of the cell membrane (Fig. 14–4). Part of the receptor complex resides on the outer portion of the cell. Attachment of hormone to receptor produces a conformational change in the receptor complex. This shift in three-dimensional configuration is transmitted to that segment of the

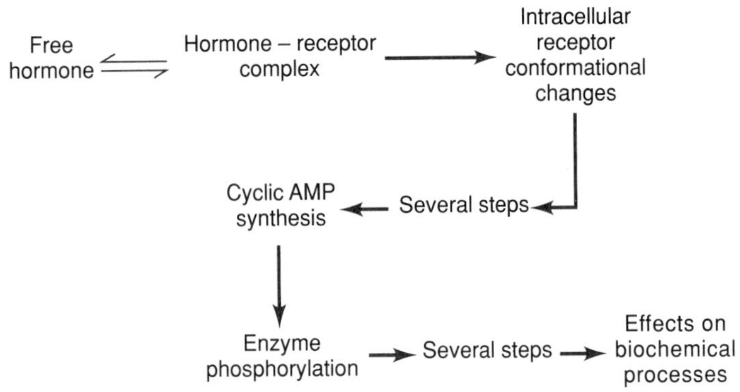

FIGURE 14–4
Peptide hormone/receptor interactions.

receptor which lies on the inside surface of the cell membrane, triggering further biochemical changes.

Inside the cell, the activation of the receptor leads to the release of a protein subunit attached to the receptor complex. This free subunit binds to the enzyme adenyl cyclase (also attached to the cell membrane), causing the formation of cyclic AMP (adenosine monophosphate) from ATP (adenosine triphosphate). Cyclic AMP serves as a "second messenger," transmitting to the cell interior the signal initially generated by the hormone/receptor interaction outside the cell. As a result of cyclic AMP formation, certain enzymes are phosphorylated and exert regulatory control on specific intracellular biochemical reactions.

Some hormones do not trigger a cyclic AMP cascade, but produce their effects through another pathway. The release of inositol triphosphate (a carbohydrate derivative) and diacylglycerol can be promoted through interaction of a hormone such as serotonin with its specific cell receptor. The details are different, but the end result is the same—the formation of phosphorylated enzymes involved in particular metabolic sequences.

Insulin is a polypeptide hormone with a complex method of producing a wide variety of biochemical effects. Binding of insulin to receptor on the outside of the cell leads to direct phosphorylation of proteins in the cell interior without the need for a second messenger such as cyclic AMP. In addition, there is strong evidence to indicate that insulin is also taken up by the cell. At least some of the insulin molecules bind to the exterior of the cell and are transported into the cell before exerting a biochemical effect, probably directly at the DNA level through the regulation of protein synthesis.

Internal Binding

In contrast to the primarily extracellular interactions of peptide hormones and their receptors, the steroid hormones exert their regulatory control at the intracellular level (Fig. 14-5). Transported by its carrier protein to the cell, a steroid molecule diffuses into the cell before it attaches to a specific receptor in the cytoplasm. A given type of cell usually has receptors for only one class of steroids, although some cells have multiple steroid receptors. The steroid/receptor complex then migrates to the DNA in the nucleus. After binding to the DNA, the complex exerts control over the type and amount of DNA produced, which, in turn, determines what kinds of proteins are synthesized. Steroids produce their effects through direct involvement with the process of protein synthesis.

Thyroid hormones function in a very similar way. The only major difference in the mechanism is the site of interaction with the receptor. Entrance of the hormone into the cell is by diffusion, but there is no attachment to a cytoplasmic receptor. Apparently, the thyroid hormone molecule binds directly to a nuclear receptor already attached to DNA. Again, the primary effect is on the control of protein synthesis.

14.4

CONTROL OF HORMONE RELEASE

Regulation by Releasing Factors

What causes an endocrine system to release a hormone? In some instances, the process is relatively simple and direct. In Chapter 12 we discussed briefly the release of insulin. As the glucose level in the blood increases, this rise is detected by a specific system in the pancreas. Insulin from the beta cells is released into the bloodstream to promote glucose utilization by the cells. When the blood glucose level decreases again, the pancreas cuts back on insulin secretion. This process is an example of direct feedback control.

In other situations the regulation is much more complex. Release of cortisol from the adrenal cortex is regulated by a complex system of releasing factors produced by other tissues (Fig. 14-6). The hypothalamus, located in the base of the brain, produces

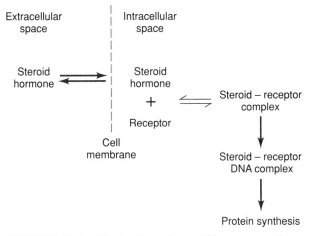

FIGURE 14-5 Mechanism of steroid hormone action.

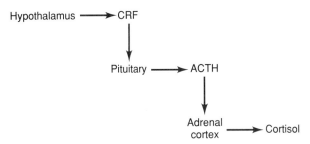

FIGURE 14-6 Regulation of cortisol production by releasing factors.

minute amounts of a substance called **corticotropin-releasing factor (CRF)** in response to stimuli. This material then travels to the pituitary (found near the hypothalamus) and stimulates the formation of **adrenocorticotrophic hormone (ACTH).** After its secretion into the circulation, ACTH arrives at the adrenal gland, located on top of the kidneys. Interaction of ACTH with receptors on the adrenal cortex leads to the synthesis and release of the steroid hormone cortisol. This process of multiple levels of control is also seen in the production of thyroid hormones and in several other endocrine systems.

Prohormones

Some peptide hormones are initially secreted as prohormones, which do not exert any biochemical activity of their own. Only after a portion of the peptide chain is hydrolyzed do we see an active hormone capable of binding to receptor to produce the appropriate changes in the system.

Insulin is one example of a hormone with a prohormone precursor. The initial form of this molecule produced by the pancreatic beta cell is proinsulin, a single polypeptide chain. Proinsulin does not demonstrate any noticeable activity until it has a peptide portion (the C-peptide) removed. The resulting insulin molecule is then biochemically active. Other hormones with prohormone forms are parathyroid hormone, vasopressin, melanocyte-stimulating hormone, and the endorphins (not yet universally considered to be a hormone system). In some instances, the laboratory is concerned with analysis of both prohormone and active material. Appropriate precautions need to be taken in sample collection and storage to minimize conversion of inactive precursor to active hormone.

Feedback Control

The released hormone can also regulate prior steps in the process. In the above example, cortisol was the end product of a series of reactions leading to its release. As the concentration of cortisol increases, it interacts with receptors which slow down reactions that lead to its synthesis. Changes in cortisol blood levels affect the production and release of both CRF and ACTH. This negative feedback process is a common one in endocrine systems.

14.5

HORMONE TRANSPORT PROTEINS

Types of Transport Proteins

A major factor affecting both the biochemically effective concentration of a hormone and its net influence on the system is the transport protein for that hormone. Although most peptide hormones circulate essentially unbound, steroid hormones and the thyroid hormone family have a large fraction of the total amount of the compound interacting with a specific protein or proteins. The steroid hormone cortisol circulates attached to cortisol-binding globulin. Two different proteins are involved in the transport of the thyroid hormones T3 and T4. In each case, attachment of hormone to protein is through noncovalent linkages. The hormone is capable of reversible association/dissociation interaction with the carrier protein, although the bound form is the predominant one in most systems.

Transport Protein Level and Hormone Action

The relative amounts of bound and free hormone determine the degree of stimulus provided by the hormone in a biochemical system. Hormone and transport protein exist in an equilibrium

Hormone + Protein ⇌ Hormone/protein complex

The relative amounts of bound and free hormone differ from one system to another. For example, only about 0.05% of the total concentration of thyroid hormone is unbound in the circulation, and it is that small fraction which is biochemically active. Hormone bound to protein cannot exert any activity at the cell receptor site. The hormone is hindered from attachment to the site because of its interaction with the carrier protein.

Since only the free (unbound) fraction exhibits activity, any situation which affects the protein level or the degree of binding has an impact on the concentration of that unbound portion. A number of drugs bind to specific transport proteins and displace the hormone, increasing its free concentration and (in essence) raising the active concentration of the hormone. Even though the total concentration has not changed, the bound/free ratio has been altered and the response of the system changes accordingly.

Frequently we see changes in the level of the transport protein, either an increase or a decrease. Again, alterations in this part of the equation affect the bound/free hormone ratio. An increase in the concentration of the transport protein shifts the equilibrium in the direction of the bound fraction. The relative amount of free hormone decreases, lowering the observed biochemical effect. Conversely, if the synthesis of the transport protein is impaired, the level of that protein will be lower than usual. The bound/free ratio moves in the direction of more free hormone, enhancing the observed response. In some instances, measurement of the transport protein concentration is an integral part of the assay for a specific hormone system.

14.6

INTRODUCTION TO THE THYROID HORMONES

The **thyroid gland,** located near the larynx, exerts significant control over the rate of metabolism in humans. This organ, two lobes connected by a thin piece of tissue, produces the hormones triiodothyronine (T3) and thyroxine (T4). Other specialized cells in the thyroid synthesize **calcitonin,** involved in the regulation of calcium levels and bone formation. The **parathyroids** (imbedded in the tissue of the thyroid gland) produce **parathyroid hormone,** also responsible for Ca^{2+} utilization and bone metabolism.

The thyroid hormones stimulate metabolism throughout the body. Their major effect seems to be increased consumption of oxygen, probably through hormone involvement with mitochondria. Most metabolic processes involving utilization of materials appear to be enhanced by thyroid hormone, including carbohydrate metabolism and lipid utilization. Because of the nature of the thyroid hormone/receptor interaction with DNA, we expect protein synthesis to be stimulated. The hormone also produces an increase in the strength of cardiac contraction and overall enhancement of growth.

Overview of Thyroid Hormone Production

The process of thyroid hormone synthesis begins at the hypothalamus where thyrotropin-releasing hormone (TRH) is produced (Fig. 14–7). This tripeptide is released in response to specific biochemical stimuli, including decreases in the levels of the thyroid hormones. TRH acts on the pituitary, causing the synthesis and release of thyroid-stimulating hormone (TSH). The thyroid glands interact with TSH to initiate the process of thyroid hormone syn-

thesis. T3 and T4 released from the glands bind to thyroxine-binding globulin and thyroxine-binding prealbumin in the circulation for transport to target tissues. Production of TRH and TSH is regulated through **negative feedback control** by T3 and T4; that is, as the levels of T3 and T4 increase, production of TRH and TSH decreases.

Thyrotropin-Releasing Hormone

TRH is a small peptide composed of only three amino acids. This molecule is synthesized primarily in the hypothalamus, but has been identified in other tissues. Apart from its role in stimulating TSH production, TRH appears to be an effective trigger for the release of prolactin by the pituitary. Administration of TRH to humans leads to a peak in the serum TSH levels in about 15–30 min. Thyroid hormone production after a single dose of TRH peaks somewhere between 90 and 150 min. Response to TRH seems to be greater in women than men and declines with age.

Thyroid-Stimulating Hormone

Interaction of the pituitary gland with TRH results in the synthesis and release of TSH. This glycoprotein has two subunits with several attached carbohydrate groups. The beta subunit is specific for TSH and is the entity recognized by the receptor. The alpha subunit has the same structure as the alpha subunits of three other dipeptide hormones—luteinizing hormone, follicle-stimulating hormone, and chorionic gonadotropin. This identity of structure can create some problems in cross-reactivity when immunoassay procedures are employed for quantitation of TSH.

TSH interacts with receptors on the membrane of cells in the thyroid gland to stimulate production and release of T3 and T4. The amount of TSH released by the pituitary is regulated through negative feedback control by the thyroid hormones.

Triiodothyronine and Thyroxine

The formation of the thyroid hormones is a complex process, involving protein synthesis, iodine uptake and incorporation, intramolecular rearrangements, protein hydrolysis, and release of hormones into the circulation (Fig. 14–8). All the steps in the process are regulated in some way or another by TSH.

Thyroglobulin is synthesized in the initial stages of this process. This protein is a large-molecular-weight material, with carbohydrate groups constituting about 10% of the total structure. Approximately 2% of the amino acids in thyroglobulin are tyrosines, which form the backbone for the thyroid

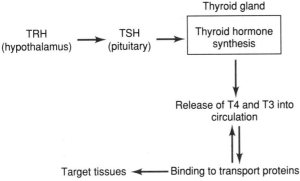

FIGURE 14–7   Overview of thyroid hormone production.

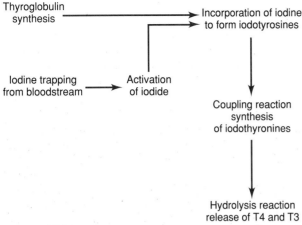

FIGURE 14-8   Synthesis of T3 and T4.

hormone molecules. Since the hormones are synthesized on the protein before being released into the circulation, thyroglobulin could be considered a prohormone.

Iodine plays an important role in the structure and function of the thyroid hormones, since the only difference between T3 and T4 is one iodine atom attached to the tyrosine ring. Iodine from the diet undergoes a complex series of reactions before being incorporated into the amino acid structure. The anion is first filtered from the plasma and concentrated in a section of the thyroid gland. It then undergoes an *activation* reaction (details of this process are not yet clear) before becoming a part of the tyrosine ring. Either one or two iodine molecules can react with a single tyrosine to form monoiodotyrosine or diiodotyrosine. In some parts of the United States, the diet is low in iodide, leading to an increased incidence of thyroid problems. One common way to provide supplementary iodine for thyroid hormone synthesis is with iodized salt.

After iodide is incorporated into the tyrosine ring, two iodinated tyrosines undergo a coupling reaction while still a part of the thyroglobulin protein. The combination of mono- and diiodotyrosines forms **3,5,3′-triiodothyronine (T3).** Two diiodotyrosine groups produce one molecule of **3,5,3′,5′-tetraiodothyronine (thyroxine, T4).** Enzymatic hydrolysis of thyroglobulin (still within the cells of the thyroid gland) releases T3 and T4 into the circulation. A small but measurable amount of thyroglobulin also enters the bloodstream.

When they reach the circulation, the thyroid hormones immediately bind to three proteins: **thyroxine-binding globulin (TBG), thyroxine-binding prealbumin (TBPA),** and serum albumin. The majority of the hormones attach to TBG, with lesser amounts interacting with the other two carriers. Binding of T4 is essentially complete, with only about 0.03–0.05% of the hormone existing in the free state. The attachment of T3 is somewhat weaker; approximately 0.5% of T3 circulates unbound.

Once in the circulation, the hormones can interact with cell membranes to produce appropriate biochemical changes in the system. Both T3 and T4 undergo metabolic inactivation in the liver, where iodine atoms are removed to be recycled for further thryoid hormone synthesis. Amino groups are cleaved and esters can be formed with other compounds. Little urine excretion of thyroid hormones takes place under normal circumstances.

There is still some debate about which hormone is the *primary* thyroid hormone. T4 (present in highest concentration) was long thought to be the major contributor to the effects produced by proper thyroid function, even though T3 has a much more potent biochemical effect, molecule for molecule. In more recent years, we have learned that significant conversion of T4 to T3 takes place in the peripheral circulation, suggesting a more important role for T3 than previously believed. T4 could then be considered a type of prohormone. The true situation is likely to be much more complex and intricate.

Thyroxine-Binding Globulin

The major transport protein for T3 and T4 is TBG. This material is a low-molecular-weight protein synthesized by the liver. Evidence suggests that the rate of TBG formation is regulated in part by the serum T4 concentration. TBG binds one molecule of T4 or T3 per molecule of protein. Some TBG can be detected in the urine since this small protein (molecular weight approximately 13,000) can easily be filtered out of the bloodstream at the glomerulus.

14.7

MEASUREMENT OF THYROID HORMONE LEVELS

Thyroxine and Triiodothyronine

TOTAL THYROXINE

The backbone of thyroid testing is the assay for total thyroxine in serum. The earlier chemical methods involved a lengthy extraction and destruction of serum proteins, followed by chemical conversion of "organic iodide" (attached to the tyrosine ring) to a form which could undergo an oxidation reaction and produce a color change. This chemical procedure was time-consuming and very much subject to interference by compounds containing iodine.

The next stage in the development of assays for T4 was the application of **competitive protein-binding procedures** to the study of thyroid hormones. These methods, developed in the 1960s, were based on competition between patient T4 and radiolabeled T4 (usually using ^{125}I) for binding to TBG added to the reaction mixture. The initial step in the procedure involved separation of bound T4 from patient TBG, with subsequent removal of that TBG from the reaction mixture. A standard amount of TBG and labeled T4 was then added to all tubes. The degree of binding of labeled T4 was inversely proportional to the amount of patient T4 present.

A major difficulty associated with the competitive protein-binding approach was interference with the binding of T4 to the added TBG. Fatty acids competed with T4 for sites on the protein molecule, often causing serious distortions in the assay values. If T4 were displaced on TBG by fatty acids (or various drugs), more T4 would appear in the free fraction. Detection of a higher amount of radioactive T4 in this fraction led to the erroneous conclusion that the patient T4 level was higher than it actually was. Recognition of this problem pushed the development of alternative methods for T4 quantitation. The competitive binding assay is rarely employed today for measurement of T4.

A wide variety of immunoassay methods are currently available for the assay of total T4 in serum. Although RIA still appears to be the most prevalent approach, enzyme immunoassays of several types and fluoroimmunoassays are gaining in popularity. For most of the procedures, a separation step is still required to remove bound T4 from TBG so it can be analyzed (Fig. 14-9). In many cases, a blocking agent is added to the system, essentially completely dissociating T4 from the protein.

Serum is the preferred sample for total T4 assays. Thyroxine remains stable for several days after sample collection, although storage at low temperatures is recommended. Hemolysis or lipemia usually does not create a problem in this measurement. If the patient has been given radioisotopes up to a few days prior to sample collection, a check for radioactive contamination needs to be made if an RIA procedure is employed.

FREE THYROXINE

When a request for T4 assay is received by the laboratory, it is usually assumed that the total T4 value is desired. However, there may be occasions in which assessment of the minute amount of unbound T4 is needed. The measurement of *free T4* is still a somewhat controversial subject, both in terms of

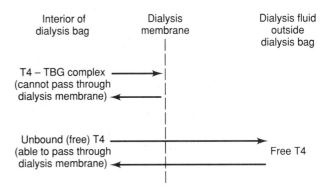

FIGURE 14-10   Equilibrium dialysis assay for free T4. Analysis is by the appropriate immunoassay technique.

techniques employed and the clinical utility of the results.

The research method for measuring free T4 utilizes equilibrium dialysis (Fig. 14-10). Patient serum is placed in buffer in a dialysis bag and T4 is allowed to diffuse through the pores of the bag into the surrounding dialysis medium. Since the T4/TBG complex is too large to pass through the pores, only the unbound T4 appears outside the bag. By measurement of total T4 in the sample and the amount of T4 detected in the dialysis medium, a reasonable estimate of free T4 can be obtained.

This method is obviously cumbersome and time-consuming, not something the average clinical chemistry laboratory likes to perform. In attempts to simplify the process, several companies have produced materials for analysis of free T4 by indirect methods. One approach involves a timed binding of T4 to a solid-phase antibody. The sample is not pretreated to dissociate T4, and therefore, the only thyroid hormone binding is that due to free T4 in the sample. The incubation is followed by separation of the sample from the solid phase at a time assumed to be short enough not to distort appreciably the equilibrium between the native bound and free T4.

A second strategy uses a labeled thyroxine analog, designed so that it does not bind to TBG or TBPA (Fig. 14-11). Incubation of the untreated sample, labeled analog, and antibody specific for T4 provides the conditions for a classic RIA technique. Equilibrium between T4 bound to antibody and free T4 in the sample is quickly established, as is competition between labeled analog and patient T4. After a short incubation, the antibody/bound fraction is separated and radioactivity is measured.

A major criticism of the analog methods is their degree of analog binding to albumin, particularly in cases where albumin concentrations are altered or in samples in which there may be abnormal proteins present for some reason. In these situations, some degree of binding to albumin or other protein has been detected, distorting the free T4 results. More research needs to be performed before analog methods for free T4 measurement are widely accepted.

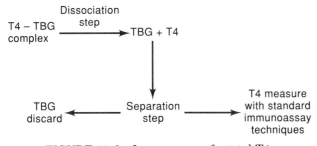

FIGURE 14-9   Immunoassay for total T4.

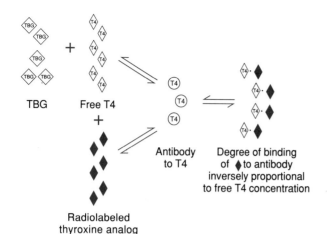

FIGURE 14–11 Analog free T4 assay. ⬡ represents thyroxine-binding globulin, ◈ represents T4 protein, ⬡ is the antibody to T4, and ◆ is the radiolabeled thyroxine analog. The degree of binding of ◆ to ⬡ is inversely proportional to free ◈ concentration.

DIRECT TRIIODOTHYRONINE

On occasion, an assay for the total amount of T3 in serum may be requested. Methodology is very similar to that for T4 by immunoassay. Addition of a dissociation reagent such as salicylate or thimerosal removes T3 bound to proteins, rendering this fraction of the total concentration available for binding to antibody. Commercial kits are available from several companies for RIA measurement of direct T3. The use of highly specific antibodies eliminates almost all cross-reactivity with T4; interferences from thyroxine are usually less than 0.5%.

Triiodothyronine Uptake and Thyroxine-Binding Globulin

A useful method of indirect measurement of the concentration of thyroxine-binding globulin is the T3 uptake (Fig. 14–12). The term *T3 uptake* is a misnomer, however; the test does not measure T3 concentration and it does not assess uptake by TBG. Patient serum is incubated with ^{125}I-labeled T3 for a time. A binder which adsorbs any T3 not already bound to TBG is then added. The binder is then separated from the sample and the amount of radiolabel on the binder is determined. Results are expressed as a percentage of the total label originally added. Therefore, a high-percentage T3 uptake indicates a greater-than-normal attachment of label to the binder, which means that less label than normal attached to the TBG in the patient serum. This finding is interpreted to suggest a low serum TBG level in that patient. A low-percentage T3 uptake implies a high concentration of TBG.

Binders can take many forms, from ion-exchange resins to solid-phase antibodies. The primary requirement is that the binder have a high affinity for T3 with no interference produced by any T4 present. Labeled T3 is used since it has little effect on T4 binding to TBG and does not displace any T4 from the carrier protein. A number of drugs (salicylate, phenytoin, and heparin) may compete with T3 for binding sites on TBG and falsely elevated T3 uptake.

Direct measurement of thyroxine-binding globulin can be accomplished by RIA or nephelometric analysis using specific antibodies to TBG. No prior displacement of thyroid hormones is required and there is little or no interference by materials which affect the T3 uptake value. Although the quantitative TBG assay provides more specific useful information than the T3 uptake, problems of cost and the greater familiarity of most laboratory and medical staff with the older test have limited the applications for TBG quantitation.

Thyroid-Stimulating Hormone

Quantitation of serum TSH levels has been shown to be extremely useful in the diagnosis of hypothyroid disorders, where TSH values are usually elevated. Both RIA and immunoradiometric assay (IRMA) methods have been successfully employed. The IRMA method has the advantage of giving a di-

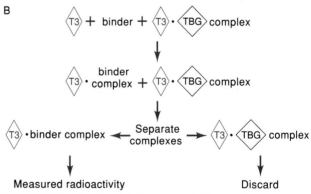

FIGURE 14–12 T3 uptake assay. A. Serum is incubated with ^{125}I-labeled T3. B. Binder is added to adsorb any unbound T3. ⬡ represents the thyroxine-binding globulin, and ◈ is the radiolabeled T3 protein.

rect response in the assay. The amount of label bound to antibody is directly proportional to the quantity of TSH present. Some studies have attempted to document subnormal production of TSH, but have been unsuccessful. The low end of the normal range is usually undetectable with most TSH assays.

14.8

CLINICAL APPLICATIONS OF THYROID HORMONE MEASUREMENTS

Reference ranges for thyroid function tests are listed in Table 14-3. These values should be considered as approximate, since they are somewhat method-dependent. There are no appreciable differences in ranges between men and women. T4 values for newborns are considerably higher than for adults, but decline to near-adult levels by age 10.

Hypothyroidism

Disorders caused by low production of thyroid hormone can be classified into two types: congenital and acquired. The incidence of thyroid deficiency in neonates is approximately 1 per 4000 live births, suggesting the need for screening programs to detect this medical problem. The usual form of a **congenital thyroid defect** is a failure of the thyroid gland to develop fully. During pregnancy, sufficient hormone apparently can be obtained by way of the fetal/maternal circulation. After birth, this external supply disappears and a state of hypothyroidism sets in. Lethargy, weakness, failure to develop, and mental retardation are the almost inevitable outcomes if the disorder is not treated promptly.

Acquired hypothyroidism in later childhood is usually due to the development of autoantibodies to thyroid tissue components. This disorder, known as **Hashimoto's thyroiditis,** can be seen in approximately 1% of the school-aged population. Although many patients may be euthyroid (at least initially), the physical manifestations are apparent: failure to

gain height, some obesity, lethargy, weakness, and some impairment of school performance.

Many of the same symptoms are seen in hypothyroidism which develops in adulthood. The major initial clinical signs are weakness, tiredness, and obvious weight gain with no change in the diet. Laboratory examination (Table 14-4) reveals a low T4, decreased T3 uptake in many cases (if the hypothyroidism is due to impaired T4 production), and an elevation in TSH. If the primary defect is pituitary damage, the TSH level decreases along with the T4 and T3 uptake. In this situation, the main problem is TSH deficiency, leading to understimulation of the thyroid gland and diminished production of thyroid hormones.

Hyperthyroidism

The major cause of increased thyroid output **(hyperthyroidism)** is apparently some type of antibody reaction which stimulates hormone production. The patient experiences weight loss, sleeplessness, fatigue, sweating, a sense of warmth, and increased cardiac rhythm. Quite often the thyroid gland is enlarged. Both T4 and T3 uptake are elevated (Table 14-5), while TSH values are either low-normal or undetectable. Treatment involves use of medications to decrease thyroid hormone production or surgical removal of the thyroid. If surgery is required, postoperative thyroid replacement therapy is initiated and must be monitored by the laboratory until a stable pharmacological treatment schedule is established.

Table 14-4
LABORATORY VALUES IN
HYPOTHYROIDISM

Test	Value
T4, total	Decreased
T4, free	Decreased
T3, direct	Decreased
T3 uptake	Decreased
TBG	Normal
TSH	Increased

Table 14-3
REFERENCE RANGES FOR THYROID
FUNCTION TESTS

Test	Value
T4, total	4.5–12.0 μg/dL
T4, free	0.7–1.9 ng/dL
T3 uptake	25–35%
T3, direct	105–195 ng/dL
TSH	Up to 5.5 μIU/mL

Table 14-5
LABORATORY VALUES IN
HYPERTHYROIDISM

Test	Value
T4, total	Increased
T4, free	Increased
T3, direct	Increased
T3 uptake	Increased
TBG	Normal or decreased
TSH	Low-normal or undetectable

Table 14–6
THYROID HORMONE LEVELS IN
PREGNANCY AND ESTROGEN USE

Test	Value
T4, total	Increased
T4, free	Normal
T3, direct	Increased
T3 uptake	Decreased
TBG	Increased

Thyroid Function and Pregnancy

The values for various thyroid tests shift markedly during pregnancy or in a patient on oral contraceptives or estrogen replacement therapy (Table 14–6). Laboratory values commonly seen include an elevated total T4 (with free T4 normal) and a greatly decreased T3 uptake. If we examine TBG concentrations, we see an increase during the course of the pregnancy, often doubling by the end of the term. The elevated total T4 value can then be explained on the basis of increased binding to the higher amount of TBG, requiring more total T4 to maintain the correct level of free T4 in the circulation. TBG levels are affected in the same manner in patients using estrogens.

Neonatal Thyroid Screening

Approximately 1 in every 4000 live births in the United States produces a child with a thyroid gland problem leading to diminished production of T4 and T3. There are a variety of causes including inappropriate diet during the pregnancy, stress of birth, or various antibodies which may be present in the maternal circulation. The result of this neonatal hypothyroidism is poor physical development and mental retardation in many cases. The high incidence of this disorder (approximately four times as prevalent as phenylketonuria [PKU]) has led to the development of screening programs in many states.

To facilitate sample collection and keep the system compatible with that for PKU screening, whole-blood samples are collected as spots on filter paper. Care must be taken to get good saturation of the paper; failure to reproduce test values is a problem if this precaution is not taken. The spot is punched with a paper punch and placed in a test tube for elu-

tion of T4 and TSH. Assays for these materials then follow standard protocols.

In many instances the T4 value may not reveal the thyroid deficiency. There are situations where the TBG level may be low, producing a low total T4 value as a side effect. The free T4 result is within the normal range. The preferred test is the quantitation of TSH. If this result is increased, follow-up testing is warranted and the suspicion of neonatal hypothyroidism is quite high. States with screening programs have the protocols spelled out by legislation.

Thyroid Function in the Elderly

Elderly patients often present with thyroid problems which are difficult to diagnose on the basis of physical findings. Laboratory values can play an important role in identifying these individuals. As a person ages, the production of T4 by the thyroid decreases, but this slowdown is compensated for by a decline in the rate of breakdown of T4. The serum levels of this hormone remain essentially unchanged from those of younger patients. However, there is diminished conversion of T4 to T3, lowering the serum T3 concentration and leading to less metabolic stimulation by this thyroid hormone component. TSH production appears to be unchanged in the elderly, unless some other disease state interferes with this process.

Other Disorders

Patients with a number of nonthyroidal medical problems often manifest clinical symptoms of thyroid disease, usually hypothyroidism. Laboratory studies may not always be helpful, because the primary disease state frequently alters the thyroid profile. Surgery can exert a profound effect on thyroid hormone levels, often lowering all parameters for several days postoperatively. Much of this effect is thought to be due to stress and the changes in cortisol levels produced by the operation. Kidney disease, particularly if there is significant protein loss, may produce decreases in thyroid hormone levels because of loss of TBG in the urine. The effects produced by liver disease are variable and can encompass changes in the rate of deiodination of thyroid hormones as well as alterations in TBG levels.

SUMMARY

Hormones are biochemical substances which are synthesized at one site and travel through the bloodstream to produce their effects at another site. Production of a hormone is often stimulated by the formation of a releasing factor produced in another endocrine gland. Feedback control is exerted by the hormone to regulate its level in the blood. Since many hormones are carried by transport proteins, any factors affecting the level of the protein have an impact on the circulating level and availability of the hormone.

Hormones stimulate cellular biochemical activity through interactions with receptors. Many peptide hormones bind to receptors on the cell wall, producing a series of biochemical reactions inside the cell which initiate metabolic changes. Some peptide hormones are also incorporated into the cell before these signals are activated. Steroid hormones first diffuse into the cell, where they interact with cytoplasmic receptors. The hormone/receptor complex then binds to DNA in the nucleus, generating the synthesis of specific proteins. Thyroid hormones bind directly to specific receptors in the nucleus to create their metabolic effects.

Synthesis of T4 and T3 occurs in the thyroid gland through a complex process regulated by TSH from the pituitary. TRH formed in the hypothalamus serves as a control of TSH output, as does negative feedback regulation by T4 and T3. Iodine from the diet is incorporated into tyrosine molecules, which are a part of the protein thyroglobulin. Although still attached to the protein, two iodinated tyrosines undergo a coupling reaction to form T4 and T3. These thyroid hormones are released from the protein and enter into the circulation, where they are transported by thyroid-binding globulin. The major effect of the thyroid hormones is to produce an overall increase in the metabolic rate.

A variety of analytical techniques are available for quantitation of thyroid hormones. When measuring either T4 or T3, an initial separation step must take place to dissociate the hormone from TBG, allowing the total hormone concentration to be determined. The predominant methodology for T4 assays is RIA, with some enzyme immunoassays becoming more widely used in recent years. T3 total concentration is also measured by RIA, although this test is not requested with anywhere near the same frequency as the one for T4. Measurement of free T4 is preferentially done by equilibrium dialysis in the rare instances when this test is requested. Analog binding methods for free T4 currently appear to have some methodological shortcomings. Determination of T3 uptake by measuring the degree of binding of labeled T3 to TBG provides a fairly reliable indirect estimation of the TBG level. Quantitation of TSH by RIA or IRMA permits the detection of elevated values in patients who are hypothyroid.

In primary hypothyroidism, laboratory results show decreases in total T4, free T4, total T3, and T3 uptake. TSH levels increase in this disorder. A patient who is hyperthyroid has increases in total T4, T3, and T3 uptake, with a low-normal or undetectable TSH value. Pregnant women and those on oral contraceptives or estrogen replacement therapy show an anomalous pattern consisting of an increase in total T4 and a decreased T3 uptake. These values are a result of the increase in production of TBG seen in these situations. Patients with other systemic illnesses often manifest altered laboratory values for thyroid hormones, although they may be clinically euthyroid.

FOR REVIEW

Directions: For each question, choose the best response.

1. The anterior pituitary produces all of the following hormones *except*
   A. ACTH
   B. FSH
   C. PTH
   D. TSH

2. The thyroid gland produces all of the following hormones *except*
   A. aldosterone
   B. calcitonin
   C. thyroxine
   D. triiodothyronine

3. All of the following hormones are classified as steroids *except*
   A. cortisol
   B. epinephrine
   C. estrogen
   D. testosterone

4. Thyroid hormones function by binding to
   A. cell membrane receptors
   B. cytoplasmic receptors
   C. nuclear receptors
   D. cyclic AMP

5. For a hormone to be biochemically active and able to bind at its receptor site, the hormone must be
   A. bound to glucose
   B. bound to lipoprotein
   C. bound to protein
   D. free, not bound to protein

6. Decreased blood levels of the T3 and T4 hormones stimulate the hypothalamus to produce
   A. growth hormone
   B. thyroid-stimulating hormone
   C. thyrotropin-releasing hormone
   D. thyroxine-binding globulin

7. Normally in the circulation, the thyroid hormones are bound to all of the following proteins *except*
   A. albumin
   B. thyroglobulin
   C. thyroxine-binding globulin
   D. thyroxine-binding prealbumin

8. Which amino acid is directly involved in thyroid hormone synthesis?
   A. alanine
   B. glutamine
   C. threonine
   D. tyrosine

9. The major carrier protein of T3 and T4 in the circulation is
   A. albumin
   B. thyroglobulin
   C. thyroxine-binding globulin
   D. thyroxine-binding prealbumin

10. The T3 uptake test is used to quantitate the concentration of
    A. free T3
    B. protein-bound T3
    C. total T3
    D. TBG

11. When measuring total T4, the purpose of the separation step is to remove
    A. bound T4 from TBG
    B. bound T3 from TBG
    C. total T3 from the mixture
    D. TBG from the mixture

12. In hypothyroidism, one would expect the total T4 level to be _____, the T3 uptake to be _____, and the TSH level to be _____.
    A. increased, increased, decreased
    B. decreased, decreased, increased
    C. decreased, increased, increased
    D. decreased, decreased, decreased

13. All of the following statements pertain to neonatal hypothyroidism *except*
    A. it occurs more frequently than phenylketonuria
    B. it causes mental retardation
    C. it results in poor physical development
    D. it can be tested for only by using 0.5 mL of serum

BIBLIOGRAPHY

Burroughs, V., and Shenkman, L., "Thyroid function in the elderly," *Amer. J. Med. Sci.* 283:8–17, 1982.

Doss, R. C., and Green, B. J., "Thyroid-stimulating hormone," *Clin. Chem. News.* February 1986.

Evans, R. M., "The steroid and thyroid hormone receptor superfamily," *Science* 240:889–895, 1988.

Gershengorn, M. C., "Mechanism of thyrotropin-releasing hormone stimulation of pituitary hormone secretion," *Ann. Rev. Physiol.* 48:515–526, 1986.

Jackson, I. M. D., "Thyrotropin-releasing hormone," *New Eng. J. Med.* 306:145–155, 1982.

Levey, G. S., and Robinson, A. G., "Introduction to the general principles of hormone-receptor interactions," *Metabolism* 31:639–645, 1981.

Ray, R. A., and Howanitz, P., "RIA in thyroid function testing," *Diag. Med.,* May: 55–70, 1984.

Schwartz, H. L., and Oppenheimer, J. H., "Nuclear receptors in thyroid function," *Diag. Med.* October: 11–20, 1982.

Thompson, L. S., "Hyperthyroidism," *Clin. Chem. News.* April 1983.

Van Herle, A. J. et al., "Control of thyroglobulin synthesis and secretion. Parts I and II," *New. Eng. J. Med.* 301:239–249, 307–314, 1979.

Walfish, P. A., "The best way to screen for neonatal hypothyroidism," *Diag. Med.* February: 67–75, 1984.

Wartofsky, L., and Burman, K. D., "Alterations in thyroid function in patients with systemic illness: the 'euthyroid sick syndrome'," *Endocr. Rev.* 3:164–217, 1982.

Wilke, T. J., "Estimation of free thyroid hormone concentrations in the clinical laboratory," *Clin. Chem.* 32:585–592, 1986.

Wilson, B. E. et al., "Congenital hypothyroidism and transient thyrotropin excess: Differential diagnosis of abnormal newborn thyroid screening," *Ann. Clin. Lab. Sci.* 12:223–233, 1982.

Yamamoto, K. R., "Steroid receptor regulated transcription of specific genes and gene networks," *Ann. Rev. Genet.* 19:209–252, 1985.

Zack, B. G., "Hypothyroidism in childhood," *Postgrad. Med.* 70:177–184, 1981.

Chapter Fifteen

CLINICAL ENDOCRINOLOGY II: STEROID HORMONES

• • • • • • • • • • • • • • • • • • •

UPON COMPLETION OF THIS CHAPTER, THE STUDENT WILL BE ABLE TO

1. Identify the three categories of steroid hormones and the general function of each category.
2. Identify the most clinically significant mineralocorticoid, glucocorticoid, and sex hormones and the functions associated with each.
3. Name the three major estrogens, noting the estrogen present in the greatest concentration during pregnancy.
4. Name the three major androgens synthesized by the testes.
5. State the primary purpose for measuring urine 17-ketosteroids.
6. Describe the procedure for measuring urine 17-ketosteroids.
7. Explain the purpose for using the Allen correction.
8. Identify the steroid hormones measured in the 17-hydroxycorticosteroid and 17-ketogenic steroid assays, respectively.
9. Identify the type of steroids that react in the Porter-Silber and Zimmermann reactions, respectively.
10. Explain the purpose for measuring urine estrogens during pregnancy.
11. Identify the colorimetric assay for quantitating urine estrogens.
12. List three techniques, in addition to the colorimetric methods, that may be employed to quantitate the steroids.
13. Describe several effects that the ingestion of drugs may have on steroid metabolism.
14. Describe metyrapone's major effect on steroid synthesis.
15. Name the glands responsible for the production of cortisol and ACTH, respectively.
16. Describe the regulation of cortisol synthesis.
17. Explain the relationship between the diurnal rhythm associated with ACTH and cortisol synthesis.
18. Explain the purpose for evaluating plasma cortisol samples both in the morning and evening.
19. List the major physiological characteristics associated with Cushing's syndrome and Addison's disease and the laboratory tests that are useful in diagnosing these disorders.
20. Describe the expected laboratory results for each of the following conditions associated with Cushing's syndrome:
   A. pituitary tumor
   B. ectopic ACTH production
   C. adrenal tumors
21. Explain the clinical significance associated with performing the dexamethasone and metyrapone tests.
22. Discuss the use of the dexamethasone suppression test as a diagnostic tool for identifying biochemical depression.

INTRODUCTION

The steroids constitute one of the more complex groups of hormones. Although these compounds are structurally similar, the biochemical effects produced by various steroids differ markedly. Testosterone, a male sex hormone, differs by only two double bonds in the A ring and a ketone group from estradiol, a female sex hormone. Other steroids are responsible for regulation of metabolic rate and a variety of metabolic conversions. Aldosterone plays a primary role in regulating electrolyte and fluid balances, and blood pressure. The complexity of the steroids lies not in their structure, but in the multiplicity of biochemical effects these cyclic compounds produce.

15.1

CLASSES OF STEROID HORMONES

Steroid hormones may be broadly classified into three categories according to function (Table 15-1):

1. **mineralocorticoids**—responsible for fluid and electrolyte balance
2. **glucocorticoids**—regulate glucose production and protein metabolism
3. **sex steroids**—regulate sexual development and control many aspects of pregnancy

In each case, the major effect of the steroid is at the cellular level, where it interacts with receptors in the cytoplasm to produce the appropriate biochemical effect.

Mineralocorticoids

Aldosterone, the major mineralocorticoid of clinical significance, is synthesized primarily by the adrenal cortex. Approximately 30% of the total aldosterone in plasma circulates bound to cortisol-binding globulin, with another 42% interacting with albumin. The high fraction of unbound aldosterone permits rapid metabolism and inactivation, mainly through the formation of a glucuronide derivative in the liver.

The chief effect of aldosterone is the promotion of sodium ion reabsorption by the kidney to maintain an appropriate sodium/potassium/H^+ balance. This reabsorption process also affects water retention by the body. Aldosterone production is regulated to a great extent by the renin/angiotensin system (to be discussed in Chapter 17) and to a much lesser extent by ACTH concentrations.

Table 15-1
FUNCTIONS OF STEROID HORMONES

Steroid Type	Major Tissue of Origin	Plasma Concentration
Mineralocorticoids		
Aldosterone	Adrenal cortex	Female: 5–30 ng/dL
		Male: 6–22 ng/dL
Glucocorticoids		
Cortisol	Adrenal cortex	8 A.M.: 5–23 µg/dL
		8 P.M.: ~ one half the A.M. value
Cortisone	Adrenal cortex	0.13–2.3 µg/dL
11-Deoxycortisol	Adrenal cortex	<1.0 µg/dL
Sex Steroids		
Androgens		
Testosterone	Testes, ovaries, adrenal cortex	Male: 300–840 ng/dL
		Female: 17–57 ng/dL
Estrogens	Ovaries, placenta, adrenal cortex	
Estradiol		Male: 8–36 pg/mL
		*Female: 10–500 pg/mL
Estriol		Male: <2 ng/mL
		Female: <2 ng/mL
Estrone		Male: 3–8 ng/dL
		*Female: 1–31 µg/dL
Progesterone	Adrenal cortex, ovary	Male: 0.12–0.3 ng/mL
		*Female: 0.02–30 ng/mL

* Actual value depends on phase of menstrual cycle.
Note: Values for plasma estrogen levels in women change markedly during the course of pregnancy.
Adapted from Tietz, N. W., ed., *Textbook of Clinical Chemistry.* Philadelphia: W. B. Saunders Co., 1986, pp. 1810–1850.

Glucocorticoids

Cortisol is the glucocorticoid that is most significant physiologically, whereas cortisone plays a less important role in affecting metabolism. The average daily synthesis of cortisol is 10–40 mg, in contrast to the 2–4 mg/day rate for cortisone. Both steroids are converted to inactive tetrahydro derivatives through reduction reactions. Conjugation with glucuronic acid or sulfate yields water-soluble derivatives which can be excreted in the urine.

Cortisol circulates primarily bound to plasma proteins. Approximately 83% of cortisol is attached to cortisol-binding globulin (CBG), with another 12% bound to serum albumin; only 5% of the hormone moves through the bloodstream in the free state.

The major role for cortisol in metabolic control is the enhancement of glucose production from proteins and amino acids. In this function, cortisol works in a way opposite that of insulin. Protein breakdown is enhanced by high cortisol levels. Some indirect effects on water retention by the kidney can be exerted by this hormone. Increased amounts of cortisol diminish the synthesis of antibodies and can produce immunosuppression. The antiinflammatory properties of cortisol may be due to its effect on protein synthesis.

Sex Steroids

A wide variety of hormones are produced which affect one or more aspects of sexual development and function. Brief descriptions of the major hormones are provided here.

PROGESTERONE

Some **progesterone** is produced by the adrenal cortex, but its primary role at this site is to furnish a precursor for the formation of other steroids. The major tissue responsible for progesterone synthesis is the ovary. Very low amounts of this hormone are produced during the early stages of the menstrual cycle, but the rate of progesterone synthesis rises markedly after the midcycle surge of luteinizing hormone and subsequent ovulation. The entire time of enhanced progesterone synthesis covers about 12 days, with peak production occupying only 4–5 days. By the end of the menstrual cycle, progesterone production has once again declined to its low baseline level.

Progesterone circulates attached to CBG if sites on that protein are available. However, since cortisol occupies the vast majority of the binding sites, little room is left for progesterone to attach. As a result, most progesterone in the circulation exists unbound.

The major route for progesterone inactivation and excretion is through the formation of conjugates, rendering the molecule more water-soluble. Progesterone serves mainly to promote the growth of the endometrial cells, which must be present before the fertilized ovum can be implanted and pregnancy develop. The changing levels of this compound also exert negative feedback control on the synthesis and release of luteinizing hormone and follicle-stimulating hormone.

ESTROGENS

There are three major estrogens: estradiol, estriol, and estrone. Both **estriol** and **estrone** are of clinical significance only during pregnancy, and estradiol plays an important role in the regulation of the menstrual cycle.

The primary source of **estradiol** is the ovary, where it is produced in a cyclic fashion throughout the 28-day menstrual period. Two peaks in the blood level for estradiol are seen. The first peak occurs at the same time as the midcycle peak for luteinizing hormone (approximately day 13). A decline in production then occurs, followed by a second peak seen at the same time as the progesterone peak (approximately day 21).

A significant amount (some 38%) of the estradiol in the circulation is bound to the sex-hormone-binding globulin, with the majority of the hormone (60%) circulating attached to serum albumin. Only 2–3% of estradiol exists unbound. Conjugation is the major form of inactivation.

Estradiol assists in the regulation of follicle-stimulating hormone and luteinizing hormone release by the pituitary. This steroid also plays a role in the development of the endometrium in conjunction with progesterone. As the level of estradiol decreases near the end of the cycle, the drop in its concentration triggers the process of menstruation.

Estriol and estrone are present in very low concentrations in the nonpregnant female, but estriol production rises markedly during the course of pregnancy. This hormone is produced as a result of interactions between the fetus and the mother, and can be used to monitor the course of fetal growth and development.

ANDROGENS

The major **androgens** synthesized by the testes are testosterone, androstenedione, and dehydroepiandrosterone. Small amounts of these steroids are also manufactured by the adrenal cortex. Production of testosterone appears to be cyclic, with a peak around 7 A.M. and a trough at approximately 8 P.M. Only low amounts of androstenedione and dehydroepiandrosterone are synthesized and both of these steroids have weak biochemical activity. The vast majority (greater than 98%) of testosterone circulates bound either to serum albumin or to sex-hormone-binding globulin.

The primary function of **testosterone** is to facilitate development of secondary male sexual characteristics. Protein synthesis is enhanced, leading to growth in both skeletal muscle and bone. This phenomenon is particularly significant during puberty.

15.2

BIOSYNTHESIS OF STEROID HORMONES

We can examine steroid synthesis in two different ways. One approach involves looking at the interconversions of steroid molecules produced by a series of enzymatic reactions. Starting with the parent molecule (cholesterol), a complex chain of ring sub-

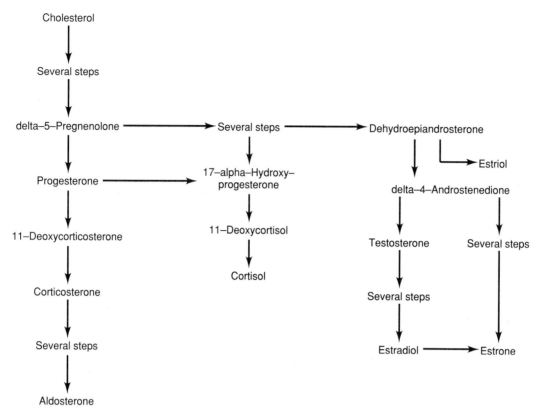

FIGURE 15-1 Biosynthetic pathways of adrenal steroids.

stitution and conversion reactions lead to the formation of the various steroid hormones, one deriving from another (Fig. 15-1). Other tissues demonstrate some of the same reactions. Note that only the transformations themselves are shown. This pathway does not indicate what triggers the conversions and produces the varying levels of the different steroids in the circulation.

To explore the control processes involved in steroid synthesis, we need to move to the specific tissue (adrenal cortex, testes, or ovaries) in which the steroid is produced. At this level, we see an often complex series of events leading to the formation of a particular compound. A brief overview of such a process was offered for cortisol in Chapter 14. Hormones produced by other tissues stimulate the production of cortisol through their interaction with specific receptors on cell membranes of tissues responsible for cortisol synthesis. The amount of cortisol formed is regulated in part by the circulating level of this hormone and its feedback control of the hormones which originally stimulated its synthesis. Different specific regulatory pathways exist for other steroid hormones and will be described when appropriate. Knowledge of the means of regulation of synthesis allows us to predict the effects of a given medication or disease state on the circulating levels of a given hormone.

15.3

STEROID HORMONE TRANSPORT

Like all lipids, steroids are relatively insoluble in water. As is the case with cholesterol, the transport of the various steroid hormones requires the presence of some sort of carrier protein. In most instances, the vast majority of the molecules are loosely bound to this carrier protein in a noncovalent fashion. The free portion is the only biochemically active fraction of the hormone, capable of diffusing into the cell and interacting with intracellular receptors.

There are apparently three major proteins responsible for transport of steroid hormones within the circulation (Table 15-2). Albumin, the protein present in the highest concentration, is a relatively non-

Table 15-2
STEROID TRANSPORT PROTEINS

Protein	Steroid Specificity
Albumin	Nonspecific; carries many steroids
CBG	Cortisol and derivatives; progesterone
SHBG	Testosterone; estradiol

selective binder. This large molecule can bind to a variety of different steroids and can transport a considerable total number of molecules. Although a single albumin molecule may only bind one molecule of a certain steroid, this limitation apparently does not interfere greatly with albumin's ability to bind another steroid molecule of a different type on the same protein.

Cortisol-binding globulin and **sex-hormone-binding globulin (SHBG)** are the two major proteins other than albumin responsible for steroid transport in the bloodstream. Both are glycoproteins, containing significant amounts of sialic acid and other carbohydrates, which may contribute to their electrophoretic heterogeneity (but not to their steroid-binding capabilities). CBG displays the electrophoretic mobility of an alpha$_1$ globulin, whereas SHBG migrates in the beta-globulin region. The liver is the site of synthesis for both proteins; changes in hepatic function can lead to altered levels of one or both transport proteins.

15.4

STEROID HORMONE ANALYSIS

A variety of analytical approaches have been developed over the years for the quantitation of one or more steroid hormones in body fluids. Techniques encompass procedures for measurement of a single steroid, as well as methods for analysis of classes of closely related steroids. Measurement of steroid mixtures is performed in situations where it is not necessary or feasible to quantitate only a single component. Reference ranges for urine steroids are found in Table 15–3.

Colorimetric

URINE 17-KETOSTEROIDS

The **17-ketosteroids** are compounds containing a carbonyl oxygen group at the number 17 carbon on the D ring of the steroid. These materials are metabolites of several precursors to cortisol, and include androsterone, dehydroepiandrosterone and some related androgens. The primary purpose for measurement of 17-ketosteroids is to assess androgen production by the adrenal glands.

An aliquot of a 24-h urine collection is first hydrolyzed in strong acid (a mixture of HCl and glacial acetic acid) to remove the glucuronic acid and sulfate conjugates present. This step renders the steroids less water-soluble and more soluble in an organic solvent. The steroids are then extracted with dichloroethane, followed by washing with sodium hydroxide to remove interfering materials. Color development (known as the Zimmermann reaction) takes place when the steroid reacts with m-dinitrobenzene in alcoholic KOH solution (Fig. 15–2). A purple color forms with an absorbance maximum at 520 nm.

Proper hydrolysis is crucial for the success of the assay. Incomplete destruction of conjugate linkages leads to low yields, since a portion of the steroids does not transfer into the organic layer. Prolonged acid hydrolysis causes some breakdown of the molecules being analyzed. Treatment with enzymes has been employed to diminish these difficulties, but enzymatic hydrolysis has not been completely successful.

A frequent problem in 17-ketosteroid assays is the formation of nonspecific color due to contaminants in the reagents or created by side reactions in the assay system. This difficulty can be minimized by use of the **Allen correction** (Fig. 15–3), making absorbance measurements at wavelengths equidistant on either side of the 520-nm maximum. The exact wavelengths employed are determined by the width of the absorption peak. The corrected absorbance can be calculated using the following formula:

$$A_{corr} = A_{520} - \frac{(A_x + A_y)}{2}$$

This formula can best be applied after the absorption curve in the 520-nm region has been deter-

---

Table 15–3
REFERENCE RANGES FOR URINE STEROIDS

Steroid Fraction	Males (mg/24 h)	Females (mg/24 h)
17-Ketosteroids	8–20	6–15
17-hydroxycorticosteroids	3–10	2–8
17-Ketogenic steroids	5–23	3–15
Estrogens (total)	5–25 (μg/24 h)	Preovulation: 5–25 μg/24 h Ovulation: 28–100 μg/24 h Luteal peak: 22–80 μg/24 h Postmenopausal: <10 μg/24 h

Note: All female reference ranges are for the nonpregnant state.
Adapted from Tietz, N. W., ed., *Textbook of Clinical Chemistry.* Philadelphia: W. B. Saunders Co., 1986, pp. 1810–1850.

FIGURE 15-2
Zimmermann reaction for 17-ketosteroids.

17-Ketosteroid + m-Dinitrobenzene → (Alcoholic alkali) → Purple compounds

mined. The Allen correction (using appropriate wavelengths) has application in a variety of spectrophotometric measurements.

A number of drugs cause interferences with the 17-ketosteroid assay (Table 15–4). Falsely low values are produced by materials which interfere with the color formation in the Zimmermann reaction. Substances giving this effect include chlordiazepoxide, digoxin, estrogens, propoxyphene, secobarbital, and spironolactone. Synthesis of 17-ketosteroids may be impaired by aminoglutethimide, ampicillin, cortisone or corticosteroids, phenytoin, meperidine, morphine, oral contraceptives, phenothiazines, propoxyphene, and spironolactone.

Increases in the formation of 17-ketosteroids can be induced by ampicillin and chlorpromazine. Muscular exercise and pregnancy also enhance production of these steroids. Materials affecting the Zimmermann reaction and leading to false elevations include acetophenone, chlorpromazine, dexamethasone, meprobamate, morphine, several penicillin derivatives, phenothiazines, quinidine, secobarbital, and spironolactone.

URINE 17-HYDROXYCORTICOSTEROIDS

Two different assays are available for measurement of steroids containing a hydroxyl group on carbon 17 of the D ring (Fig. 15–4). The "classic" 17-hydroxycorticosteroid technique (Porter-Silber method) measures only cortisol, cortisone, 11-deoxycortisol, and some of their reduced derivatives. The Zimmermann method allows quantitation of an expanded group of these 17-hydroxy compounds which includes not only the cortisol derivatives, but also other cortisol metabolites, pregnanetriol and metab-

olites, 17-hydroxyprogesterone, and 17-hydroxypregnenolone. These groups of compounds are called **17-ketogenic steroids** since chemical treatment can convert each specific steroid into a 17-ketosteroid.

Porter-Silber Reaction

The **Porter-Silber assay** for urine cortisol and derivatives capitalizes on a reaction long employed for carbohydrate analysis (Fig. 15–5). The dihydroxyacetone side chain (carbon-20 and carbon-21) of the steroid reacts with 2,4-dinitrophenylhydrazine to form a yellow derivative having an absorbance maximum of 410 nm. Compounds which lack this specific side-chain structure do not participate in the reaction.

Initial steps in the procedure include enzymatic hydrolysis (these steroids are extremely susceptible to damage if acid is used) and extraction into an organic solvent, followed by washing with sodium hydroxide. This process removes impurities which might react with the 2,4-dinitrophenylhydrazine and decreases the concentrations of nonspecific colored compounds (such as bilirubin derivatives)

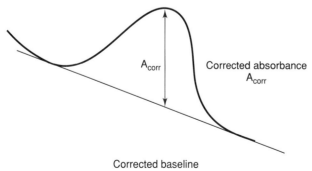

FIGURE 15–3   Allen correction of absorbance measurement.

Table 15–4
FACTORS AFFECTING 17-KETOSTEROID RESULTS

False Low Results	False High Results
Chlordiazepoxide	Acetophenone
Digoxin	Chlorpromazine
Estrogens	Dexamethasone
Propoxyphene	Meprobamate
Secobarbital	Morphine
Spironolactone	Penicillin derivatives
	Phenothiazines
	Quinidine
	Secobarbital
	Spironolactone
Impairment of Synthesis	**Enhancement of Synthesis**
Aminoglutethimide	Ampicillin
Ampicillin	Chlorpromazine
Cortisone and derivatives	Muscular exercise
Phenytoin	Pregnancy
Meperidine	
Morphine	
Oral contraceptives	
Phenothiazines	
Propoxyphene	
Spironolactone	

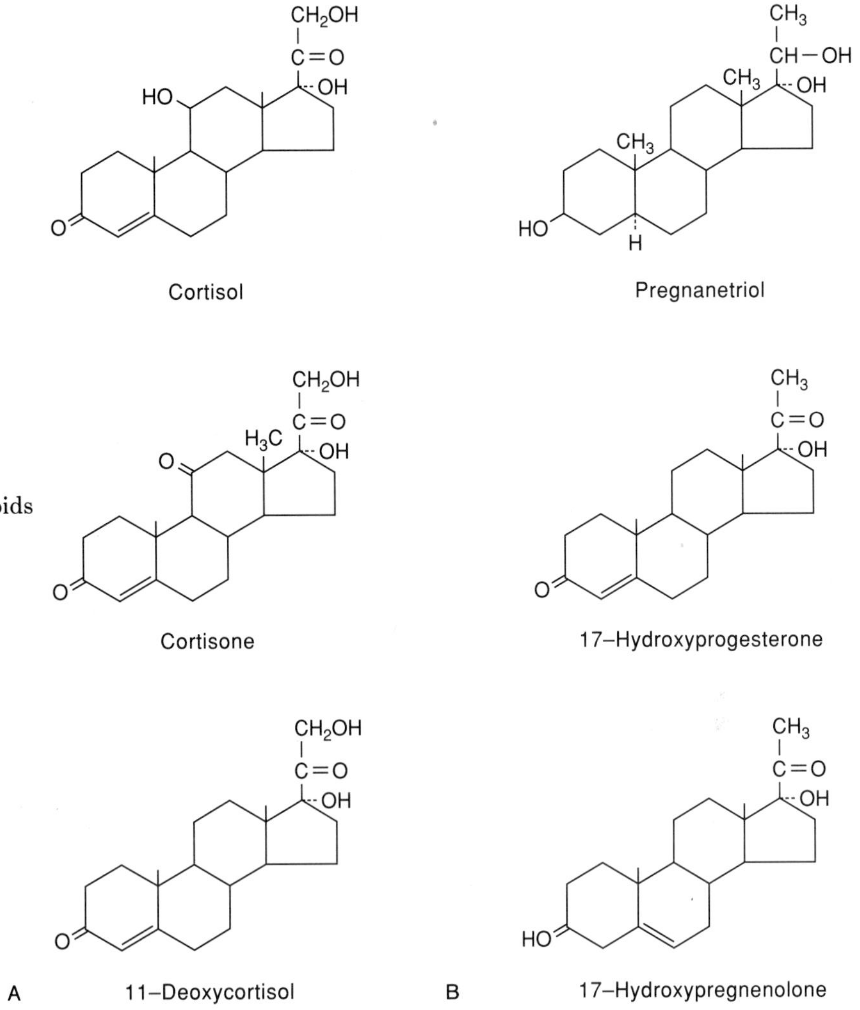

FIGURE 15-4
Structures for 17-hydroxycorticosteroids (A) and 17-ketogenic steroids (B).

Cortisol

Pregnanetriol

Cortisone

17–Hydroxyprogesterone

A   11–Deoxycortisol

B   17–Hydroxypregnenolone

which could absorb at 410 nm. Addition of 2,4-dinitrophenylhydrazine and heating produce the final yellow 2,4-dinitrophenylhydrazine derivative.

A number of compounds form colored derivatives in this reaction, including acetone, fructose, and some forms of vitamin C. Positive interference is also produced by a variety of medications such as phenothiazines, sulfa compounds, spironolactone, and propoxyphene. Caffeine consumption should be restricted or eliminated for several days prior to specimen collection to minimize interference from this material.

Zimmermann Reaction

An expanded array of steroids can be analyzed with the **Zimmermann procedure** (Fig. 15–6). Any steroid with an oxygen-containing carbon-20 (either carbonyl or hydroxyl groups) and a group on carbon-21 (oxygen-containing or methyl) undergoes reaction to form a 17-ketosteroid (part A). The 17-ketosteroids naturally present in the urine sample are converted to 17-hydroxysteroids in the initial stages of the assay and do not react with the final Zimmermann reagent (part B).

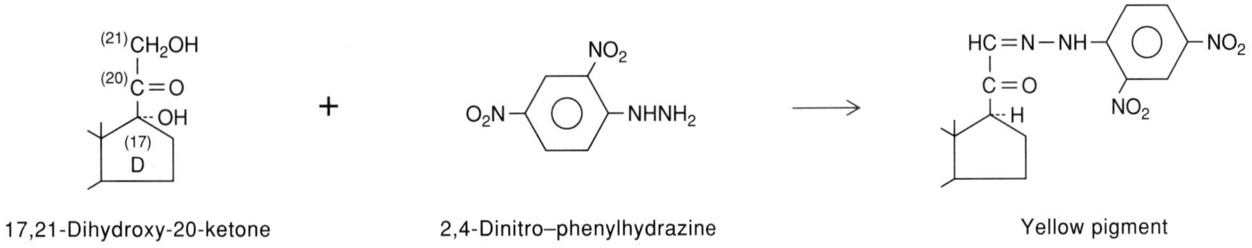

17,21-Dihydroxy-20-ketone

2,4-Dinitro–phenylhydrazine

Yellow pigment

FIGURE 15-5   Porter-Silber reaction.

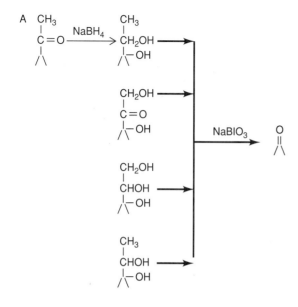

FIGURE 15–6   Zimmermann assay of 17-ketogenic steroids. A. Steroid with oxygen-containing carbon-20 and carbon-21 reacts to form 17-ketosteroid. B. 17-Ketosteroid converts to 17-hydroxysteroid, but does not react with final Zimmermann reagent. ∧ represents bonds at carbon-17.

The first step in the procedure involves assessment of glucose concentration using a dipstick method, since high glucose levels interfere with further reactions. If the glucose concentration is below 500 mg/dL, the assay may continue. Higher glucose concentrations require a preextraction with an ethanol-ether solvent system to isolate the steroids and eliminate the glucose interference.

Reduction of carbonyl groups is necessary before conversion of the steroids to the 17-keto derivatives. Sodium borohydride ($NABH_4$) is added to convert all $C=O$ structures to the corresponding hydroxyl groups. Addition of sodium bismuthate ($NaBiO_3$) removes all side chains at C-17 and oxidizes the 17-hydroxy group, leaving a ketone at that site. Sodium bismuthate does not reoxidize any of the original 17-ketosteroids which had been reduced. Destruction of excess sodium bismuthate is accomplished by mixing sodium bisulfite with the sample. Acid hydrolysis is then employed to destroy any conjugates. Extraction of steroids into dichloroethane, followed by sodium hydroxide pellet treatment and solvent evaporation completes the sample preparation process. The Zimmermann reaction can then be performed as previously described.

A variety of medications create problems for accurate assessment of 17-ketogenic steroid production. Any compound producing an interference in the Zimmermann reaction affects the value for this group of steroids. Reductions in the output of 17-ketogenic steroids can be caused by aminoglutethimide, ampicillin, chlorpromazine, phenytoin, meperidine, oral contraceptives, phenothiazines, and propoxyphene. The values obtained in these situations reflect the actual steroid production, which is impaired by the medication. Interferences in the color reaction which result in false low values can be seen in patients using estrogen replacement therapy, meprobamate, propoxyphene, and secobarbital.

URINE ESTROGENS

The measurement of urine estrogens is employed primarily to monitor development of the unborn child during pregnancy. As the pregnancy progresses, increases in the 24-h urine output of estrogens is seen. The rise over the course of 9 months is due primarily to increases in estriol formation. Since the overall change in total estrogen levels provides sufficient clinical information, there is usually no further value in fractionating the components of the system.

The classic assay employed for urine estrogen determination is the **Kober reaction.** This process, still poorly characterized as to chemical details, involves heating a urine sample in a strong aqueous sulfuric acid solution containing hydroquinone. After cooling and dilution, the absorbance of the resulting reddish-brown color is measured and the total estrogen concentration determined.

Prior to analysis, the sample must be checked for glucose concentration since high levels of any compound forming acetaldehyde produce false negative results by reacting with the hydroquinone. Hydrolysis of estrogen conjugates is then accomplished, either with HCl and heat or enzymatically. Extraction into ethyl acetate separates the steroids from other materials. After drying to remove the organic solvent, a mixture of hydroquinone and sulfuric acid is added to the tube containing the estrogens. The tubes are heated to produce the color. Absorbance is determined at 472, 512, and 556 nm, followed by an Allen correction to calculate the estrogen concentration using standard solutions for calibration.

The colorimetric assay has sufficient sensitivity to quantitate total urine estrogens only during the middle and latter stages of pregnancy. Slight modifications of the procedure permit fluorometric assays with good sensitivity and accuracy for specimens from nonpregnant women.

Significant decrease in urine estrogen output occurs in patients taking ampicillin or neomycin. These antibiotics reduce bacterial levels in the intestine, causing a diminished hydrolysis of estriol conjugates. The re-uptake of estriol into the circulation becomes impaired, as is the urine excretion of estrogens. Hydrochlorothiazide produces falsely lowered results due to its destruction of estrogens during the acid hydrolysis step. Meprobamate, L-dopa, and phenolphthalein all give reactions with the Kober reagent, falsely increasing the results for urine total estrogen output.

Immunoassay Techniques

A wide variety of immunoassay techniques have been explored over the years for the analysis of specific steroids. These techniques are particularly attractive for measurement of plasma levels since they can detect much lower quantities than the formerly employed colorimetric or fluorometric methods. In addition, immunoassay approaches are much more specific due to improvements in antibody-generating technology. Although radioimmunoassay has been the primary mode of detection, a number of other techniques are being employed for measurement of a particular steroid component. Because of their simplicity and ease of automation, enzyme immunoassay methods are gaining wide popularity. Significant work is also being done using a combination of antibody and luminescent methodologies, although there is little commercial progress in this area at present.

One major hindrance to development of simple immunoassay techniques is the protein-binding capabilities of the steroids. In the circulation, a large majority (if not essentially all) of the steroid molecules are transported by carrier proteins. Since immunoassay techniques rely on competition for binding sites on an antibody, accurate assay of the total concentration of a given steroid requires that it first be dissociated from its transport protein before analysis takes place. Although the dissociation step is usually fairly easy to perform (often involving only an extraction into an organic solvent), the process does require extra steps and extra time.

Specificity of antibody is also a major concern in steroid analysis. The various steroids have quite similar structures, calling for creativity in the techniques employed for antibody generation. The site of labeling by radioisotope or other signal is crucial for success in this instance. Synthetic steroids employed as pharmacological agents may cross-react with the antibody if proper precautions are not taken. The assay must be checked with samples from patients in a variety of clinical situations to ascertain whether a specific steroid is present in amounts much higher than normal. The metabolic intermediate, 11-deoxycortisol, has been shown to interfere with some immunoassays for plasma cortisol when present in elevated amounts, but has no significant effect on the assay at normal concentrations.

Gas Chromatography

Although not a routine clinical chemistry procedure at present, gas-chromatographic fractionation and quantitation of steroids represents the single most powerful tool for analysis of steroid mixtures today. When coupled with mass spectrometry, the steroid is usually unequivocally identified and quantitated.

Extraction of the steroids into an organic solvent is a prerequisite for GC analysis. The protein portion is dissociated from the steroid and discarded; otherwise, severe damage to the column occurs. Evaporation of the solvent allows concentration of the steroids in the mixture. If a 10-mL urine sample is extracted and concentrated to a volume of 0.1 mL, a 100-fold increase in the effective concentration occurs.

Although some steroids can be injected directly onto the GC column for separation, significant improvement in analysis occurs when a volatile derivative is first formed. Reaction with acetic anhydride often forms useful derivatives with little effort. Perhaps the most widely used reagents are those which yield tetramethylsilyl ethers. Reaction occurs with a hydroxyl group on either the ring or a side chain to yield a nonpolar compound. Ketones may be derivatized using methoxamine hydrochloride.

There are several advantages to the GC analysis of derivatized steroids instead of the free compounds. Most steroids are not very heat-stable and tend to decompose at the temperatures employed for GC fractionation. Derivatives provide much better heat stability, allowing good separations with much less artefact formation. Irreversible adsorption onto the column is markedly decreased by derivatization, giving more reliable estimates of actual steroid content and composition. Volatility of derivatives is usually much greater than that of the unreacted steroid, providing faster throughput and more rapid analyses. Often, the separating capability of a particular column is enhanced when derivatives are used, again providing more accurate measurements and improved information obtained from the study.

High-Performance Liquid Chromatography

Development of procedures for HPLC analysis of steroids has been hindered by the lack of sensitive derivatization reagents. Although some steroids (such as cortisol) are present in plasma in concentrations which allow detection by measurement of absorbance at 254 nm, other steroids exist in much lower amounts and require more rigorous approaches. Postcolumn detection systems would seem to have the greatest applicability, since they allow column fractionation on the basis of ring and side-chain substituents before performing a chemical reaction which alters that specificity. As is the case with GC, HPLC column techniques require an initial extraction step to dissociate the steroid bound to protein.

15.5

DRUGS AFFECTING STEROID HORMONE METABOLISM

Steroid hormone levels can be significantly altered through the effects of various pharmaceutical agents. These changes may affect the levels of enzymes (within the tissue) concerned with the processes of steroid production or metabolism. Steroid hormone output can also be altered by affecting the hypothalamic/pituitary regulatory functions. Changes in the output of either of these glands, affecting either CRF or ACTH release (or both), directly affects steroidogenesis.

Effects on Peripheral Metabolism of Steroids

A number of drugs stimulate an increase in the amount of certain liver enzymes responsible for drug metabolism. These enzymes have broad specificities, acting on a wide variety of similarly structured compounds. Significant enhancement of the reaction leading to the introduction of a hydroxyl group on the A ring is commonly seen.

PHENYTOIN

Phenytoin (an anticonvulsant) affects steroid metabolism through the cytochrome P_{450} metabolic sequence in the liver. By increasing the amount of this enzyme system, phenytoin causes an elevation in the amount of 6-betahydroxycortisol. Although serum cortisol levels are not lowered (due to increased production stimulated by this metabolic shift), the levels of the urine total 17-hydroxysteroids appear to be decreased. Since 6-betahydroxycortisol is not detected by the methods used in this assay, an underestimation of urine steroid output can occur.

In addition, prednisone and dexamethasone also undergo increased degradation in the presence of phenytoin. Among other applications, prednisone is used after an organ transplant to minimize tissue rejection. A lowering of the effective level of this drug can adversely affect the success of the transplant. Dexamethasone is commonly employed as a suppressor of cortisol production and is given to patients to assess the ability of the adrenal gland to function independently. Enhanced metabolism of this compound can lead to misdiagnosis of adrenal cortex malfunction if the amounts available for gland suppression are too low to be effective.

BARBITURATES

In a somewhat similar fashion, administration of barbiturates affects hepatic enzymes involved in the metabolism of several steroids. Increased conversion of estrogens, testosterone, cortisol, and progesterone to metabolites has been reported. One problematic effect of prolonged barbiturate treatment is the enhanced metabolism of oral contraceptives. Decreases in the effective concentrations of these compounds can have a number of significant influences.

ETHANOL

The major pharmacological effect of ethanol on cortisol metabolism is inhibitory. The cytochrome P_{450} system in the liver responsible for cortisol degradation is blocked by ethanol, leading to elevations in plasma cortisol levels. Another effect of ethanol appears to be an enhancement of ACTH production. Prolonged ethanol use leads to higher than normal serum ACTH levels with a loss in the normal diurnal variation of ACTH release.

Testosterone concentrations in plasma decrease after ethanol intake, even in normal individuals. This effect appears to be due to enhanced production of reducing enzymes in the liver responsible for the degradation of testosterone. There is no compensatory increase in the synthesis of testosterone, so the overall level declines. Ethanol apparently interferes in some manner with the regulation of testosterone production by the hypothalamus.

Inhibition of Steroid Synthesis

Those drugs which decrease the synthesis of specific steroids are employed in clinical situations where there is an overproduction of one or more of these hormones. The site of inhibition varies from one medication to another, and multiple sites may be affected along the metabolic pathway. The analyte steroid may be measured for diagnostic purposes (to determine the exact biochemical problem) or to assess therapy. The health-care team is aware of the direct impact of the pharmacological intervention, but sometimes may not be familiar with the drug's effects on the production of other steroid families.

AMINOGLUTETHIMIDE

The metabolic impact of aminoglutethimide on steroid synthesis is complex. This medication is used with patients who would not benefit from surgical treatment for elevated cortisol production. Aminoglutethimide's inhibitory effect on cortisol synthesis occurs at two or more sites, involving blockage of reduction and dehydrogenase reactions. Inhibition leads to diminished production of cortisol and several of its metabolites. The effects of aminoglutethimide are best monitored through measurement of urine free cortisol. After treatment for several months, this inhibition decreases and cortisol levels begin to rise again.

Aminoglutethimide has been shown to decrease blood pressure as a result of its blockage of the synthesis of aldosterone, a steroid intimately involved with electrolyte and water balance. A complex effect of this drug on estrogen production leads to the decrease in the synthesis of most fractions. The mechanism of this effect is not known at present.

METYRAPONE

The major effect of metyrapone is its blockage of synthesis of 11-hydroxysteroids, including cortisol. This inhibition leads to an increase in the amount of deoxycortisol derivatives in the urine. As a result of the decrease in cortisol synthesis, ACTH levels in the blood rise after metyrapone treatment. This medication also apparently increases the rate at which cortisol is cleared from the circulation, enhancing its inhibitory effects on cortisol production.

Estrogen synthesis is altered by metyrapone in a complex manner. Urine total estrogen levels may either decline or remain unchanged, but the specific estrogen fractions vary greatly after administration of metyrapone. The major effect is a marked decline in the amount of urine estradiol, with variable changes in estriol and estrone. Alterations of progesterone output are seen, with stimulation of the production of this steroid being common.

SPIRONOLACTONE

The major therapeutic application for this steroid analog is in the treatment of hypertension caused by elevated aldosterone levels. The effects of spironolactone on steroid production appear to be two-fold. The analog inhibits at least one of the enzymes involved in the synthesis of cortisol and testosterone, perhaps at the level of progesterone. This lowering of enzyme activity leads to a decrease in the production of aldosterone. Spironolactone also seems to directly affect the cell membrane receptor. By attaching to receptor sites for a testosterone derivative, spironolactone diminishes the biochemical effect of this steroid hormone as well as increasing the hormone's effective concentration in the circulation. Elevation of the concentration leads to a negative feedback, producing a decrease in synthesis of testosterone. For some individuals, increases in plasma progesterone concentrations may also occur after administration of spironolactone.

15.6

CLINICAL CHEMISTRY OF ACTH AND CORTISOL

Because of its central role in the regulation of several aspects of metabolism, disorders of cortisol production furnish important insights into steroid bio-

chemistry. The hypothalamic/pituitary/adrenal system tightly controls the synthesis of cortisol by the adrenal cortex. Failure of hormonal regulatory processes or of enzymes involved in the biosynthesis of cortisol leads to a variety of clinical problems which are amenable to study using laboratory techniques.

Regulation of Cortisol Synthesis

The production of the steroid hormone cortisol is regulated at many levels in the body (Fig. 15–7). The initial site for the stimulation or reduction of cortisol manufacture is the hypothalamus, located at the base of the brain. This tissue produces a small peptide hormone called **corticotropin-releasing factor (CRF)**. After traveling a very short distance through the bloodstream to the pituitary (located right below the hypothalamus), CRF interacts with pituitary receptors, causing the synthesis of **adrenocorticotropic hormone (ACTH)**. This larger polypeptide then passes through the circulation until it reaches the adrenal gland, located on top of the kidneys. ACTH attaches to specific receptors in the zona fasciculata (the middle portion of the adrenal cortex) to activate the enzyme system responsible for cortisol formation. Both CRF and ACTH production are regulated through negative feedback control by cortisol.

CORTICOTROPIN-RELEASING FACTOR

Corticotropin-releasing factor is a 41-amino acid peptide hormone found primarily in the hypothalamus, although some CRF activity has been detected throughout the central nervous system. There is increasing evidence to suggest that one of the major stimuli to the production and release of CRF is activity of the cerebellum (a portion of the brain). Stress and pain appear to trigger an increase in the synthesis of this peptide. CRF production decreases when the blood levels of either ACTH or cortisol rise.

Recent research suggests that CRF may stimulate the production of hormones other than ACTH. In-

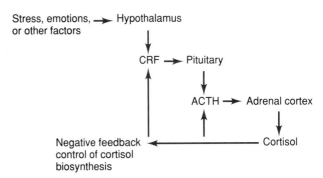

FIGURE 15–7 Hormonal regulation of cortisol biosynthesis.

creases in blood pressure after CRF release strongly suggest that this peptide exerts some influence over the production of catecholamines. CRF has been shown to increase respiration rates in humans, again implying a wider range of influences than previously thought.

Measurement of CRF in plasma is best carried out by RIA. The technique is not widely employed in clinical laboratories, but has been used in a number of research applications. As our knowledge of cortisol production and metabolism increases, more clinical applications of CRF measurements will become apparent.

Reference ranges for CRF levels in both men and women are the same: approximately 2–28 pg/mL. There do not appear to be any diurnal variations in CRF production, as there are for ACTH and cortisol. During pregnancy there is a substantial rise in plasma CRF levels. Concentrations of CRF near term are several hundred times as high as the values seen in a nonpregnant woman. Much of this increase seems to be a result of CRF production by the placenta.

ADRENOCORTICOTROPIC HORMONE

In recent years our knowledge of ACTH biosynthesis has increased tremendously. With this newly gained information, however, has come an awareness of the complexities of the regulatory system and a large number of unanswered questions.

ACTH synthesis is much more complicated than was previously thought. Earlier concepts envisioned the formation of a single polypeptide, possibly as a prohormone. Cleavage of a small number of amino acids was believed to yield the final ACTH molecule. This simple picture has had to be replaced by a complex process in which a number of hormones are produced simultaneously, some of which have very poorly defined (or unknown) functions.

The anterior lobe of the pituitary produces a long polypeptide chain (proopiomelanocortin) which is the precursor for several hormones, including ACTH (Fig. 15–8). Enzymatic cleavage of this pro-

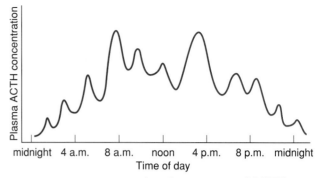

FIGURE 15–9 Twenty-four-hour pattern of ACTH release.

tein yields ACTH (or "pro"-ACTH) as well as lipotropins (the function of which is currently unknown), melanocyte-stimulating hormones (which are involved in pigmentation), and some of the endorphins. ACTH can be further broken down to form a melanocyte-stimulating hormone (MSH) and corticotropin-like intermediate lobe peptide (CLIP). The function of this complicated process is unknown at present.

ACTH acts upon the adrenal gland to promote the synthesis of cortisol and other steroid hormones. The major stimulus is on the enzyme pathway leading from cholesterol to a cortisol precursor called delta-5-pregnenolone. Further hydroxylation of this derivative results in the formation of cortisol (approximately 20 mg/day produced in the normal adult) and corticosterone (production rate of some 2 mg/day). Stimulus of androgen synthesis, particularly dehydroepiandrosterone, is also facilitated by ACTH. Aldosterone production is regulated to some extent by ACTH levels in the circulation.

Release of ACTH into the bloodstream is not a smooth, continuous process, but occurs in small spurts or pulses (Fig. 15–9). This intermittent secretion takes place throughout the day and night, with greater clusters of pulsed release around the time of awakening. There is also a diurnal rhythm associated with ACTH synthesis, apparently regulated by the hypothalamus (but not by CRF). The rise begins in the early hours of the morning and peaks shortly after awakening. Following this maximum production, the ACTH level in the circulation gradually declines throughout the day and reaches its lowest (trough) level in late evening. If the sleep/wake cycle is altered, the diurnal variation in ACTH production shifts in accordance with the new sleeping pattern.

CORTISOL METABOLISM AND EXCRETION

Once cortisol is synthesized by the adrenal gland, it is released into the circulation. Transport is achieved primarily through noncovalent binding

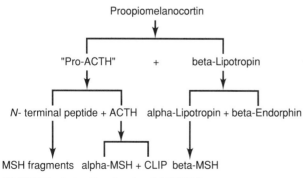

FIGURE 15–8 Synthesis of ACTH from proopiomelanocortin.

with CBG, often referred to as **transcortin.** A smaller amount of cortisol binds to serum albumin. Approximately 10% or less of the cortisol in the circulation is free; that is, it is not bound to plasma transport proteins. This free fraction can diffuse into cells to exert metabolic effects. Unbound cortisol is also filtered at the glomerulus. A portion is reabsorbed by the kidney, but a small fraction is excreted into the urine. Quantitation of this "urine free cortisol" can give an accurate assessment of the level of cortisol production over time.

Conversion of cortisol to biochemically less active products occurs mainly through reduction of double bonds on the steroid ring system and conversion of keto side groups to hydroxyl derivatives. Attachment of glucuronic acid to the hydroxyl moiety converts the insoluble steroids to water-soluble derivatives which can be readily excreted in the urine. These metabolic products are measured as part of the urine 17-hydroxycorticosteroid group.

DIURNAL RHYTHM

As is the case with ACTH, cortisol is produced in short pulses throughout the day and exhibits a diurnal rhythm (Fig. 15–10). The peak for cortisol production is in the morning, usually around 8–9 A.M. The trough values are seen at approximately 10–11 P.M. As expected, there sometimes appears to be a slight lag between ACTH peaks and cortisol peaks in this circadian cycle of production. The absence of the diurnal rhythm can be an important diagnostic clue. For this reason, laboratory requests for plasma cortisol assays should include evaluation of both morning and late evening samples to document the presence or absence of the cyclic production of this steroid. Patients undergoing significant stress often lose the rhythmic production and exhibit elevated cortisol levels with no discernible cycle of synthesis.

This regular rise and fall of cortisol production during the day and night can be disrupted by meals. After eating, there is a prompt and noticeable rise in the plasma cortisol level. The plasma cortisol concentration may not return to the appropriate base-line value for 2 h or so. This phenomenon often has not been closely examined when a patient is being evaluated for possible alterations of cortisol production. Caffeine also stimulates a measureable and prolonged rise in plasma cortisol concentrations, a factor which could confuse laboratory results.

FEEDBACK CONTROL OF CORTISOL SYNTHESIS

Since cortisol exerts a series of metabolic effects resulting in significant changes in carbohydrate and protein utilization, an internal mechanism must therefore exist to regulate the output of this hormone. The body employs the process of negative feedback control as a means of altering the rate of cortisol manufacture. As the cortisol level increases, there is a rise in the circulating level of both total and unbound hormone. Some of the cortisol molecules not attached to transport proteins interact with receptors in both the hypothalamus and the pituitary. These receptors then bind to nuclear DNA within these two tissues and diminish the process of protein synthesis leading to the manufacture and release of the peptides CRF and ACTH. As less of these signal peptides are formed and released into the circulation, there is less stimulus to the adrenal gland to produce cortisol molecules. The plasma cortisol level drops, leading to diminished feedback inhibition of the hypothalamus and pituitary. Synthesis of CRF and ACTH increases once more, causing enhanced formation of cortisol, and the cycle begins again. This feedback inhibition is presumably responsible for the minute-by-minute fluctuations in the production of ACTH and cortisol which can be measured. The processes controlling the diurnal rhythm are under separate regulation and are not clearly understood at present.

OTHER FACTORS

A major contributor to higher than normal cortisol levels is stress, either physical or emotional. Although moderate exercise has not been shown to increase cortisol output, extreme physical stress can elevate blood cortisol values. Postoperative stress generally stimulates a marked elevation of plasma cortisol, which may last for 24 h. Emotional stress (acute anxiety, depression) also clearly increases cortisol output.

Pregnancy leads to an elevation of plasma cortisol, mainly in response to changes in the concentration of transport proteins. Although many plasma proteins decrease during the course of pregnancy, the concentration of cortisol-binding globulin increases. Since there is more binding of cortisol by this increased amount of transport protein, the body must then synthesize more cortisol in order to maintain the appropriate amount of free hormone in the

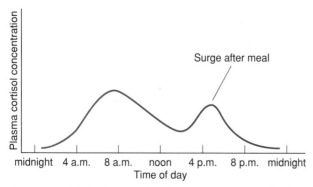

FIGURE 15–10   Twenty-four-hour pattern of cortisol production.

Table 15-5
LABORATORY VALUES IN CUSHING'S SYNDROME

Clinical Situation	Plasma Cortisol	Plasma ACTH	Urinary 17-OHCS
Pituitary tumor or hyperplasia	Normal/high	Normal/high	Normal/high
Ectopic ACTH production	High	High	High
Adrenal tumors	High	Low	High

circulation. By the end of 9 months, the plasma cortisol levels may be double what they were prior to pregnancy.

Cushing's Syndrome

There are a number of clinical situations in which an increase in plasma and urine cortisol and metabolites can be seen. These disorders are collectively referred to as **Cushing's syndrome,** first described by Harvey Cushing in the early 1930s. The syndrome can be subdivided into two categories:

1. Cushing's disease: increase in cortisol production caused by excessive development and activity of the pituitary gland.
2. Cushing's syndrome: increase in cortisol production as a result of tumors which produce either excessive ACTH or cortisol.

The various possibilities can be studied through a combination of plasma and urine steroid measurements (Table 15-5), both before and after stimulation and suppression tests.

Excessive growth (hyperplasia) of the pituitary results in an increase in the production of ACTH, causing elevations of plasma cortisol. Alternatively, a pituitary tumor develops which stimulates increased production of ACTH by the pituitary gland itself. In both instances, the primary site of disruption is the pituitary: rises in steroid production and excretion are primary laboratory indicators of the disorder. Plasma ACTH values also increase.

A number of tumors in the lung and pancreas synthesize ACTH as part of their biochemical output. This ectopic production of hormone is not under feedback regulation and does not diminish when the plasma cortisol levels rise. Since ACTH levels also increase, other procedures are necessary to distinguish this situation from one due to pituitary malfunction.

Adrenal carcinoma and adrenal adenoma are two cancers of the adrenal gland in which the tumors synthesize large amounts of cortisol and other adrenal hormones. Often the imbalance enhances male secondary sexual characteristics in women or increases feminization in men, frequently the first clinical sign of the disease. Plasma ACTH levels are low because of feedback suppression by the cortisol produced by the tumor.

Suppression of ACTH output by dexamethasone is often used as a diagnostic tool in establishing the root cause of hypercortisolism (Table 15-6). This synthetic steroid acts in the same manner as cortisol to induce feedback inhibition at the level of the pituitary. Use of high-dose dexamethasone (8 mg/day) is usually necessary for suppression. If the elevated cortisol is due to pituitary problems, approximately 50% or more suppression of cortisol synthesis relative to base-line values occurs. Ectopic ACTH production is not suppressed, so cortisol levels remain unchanged after dexamethasone treatment in these situations. Production of cortisol by adrenal adenoma or carcinoma is also not suppressed by dexamethasone since pituitary ACTH does not appear to regulate steroid synthesis in these cases.

The metyrapone inhibition test, although not as frequently employed as the dexamethasone suppression test, can be useful in distinguishing pituitary malfunction from cancers in other tissues as the reason for elevated cortisol synthesis (Table 15-6). Metyrapone inhibits the conversion of 11-deoxycortisol to cortisol. Since less cortisol is formed, the feedback inhibiting ACTH production is diminished. The ACTH level rises, causing the adrenal gland to form more steroids. In a normal individual, administration of metyrapone results in a decrease in plasma cortisol with a rise in the blood level of the 11-deoxy precursor. The total 17-hydroxycorticosteroid excretion also increases while the amount of urine free cortisol declines. No change in cortisol output after metyrapone is seen in cases of adrenal carcinoma or adenoma. The response to metyrapone in patients with ectopic ACTH synthesis is variable,

Table 15-6
INHIBITION/SUPPRESSION OF CORTISOL SYNTHESIS

Clinical Situation	Dexamethasone	Metyrapone
Pituitary tumor or hyperplasia	50% suppression	Increased 17-OHCS
Ectopic ACTH production	No suppression	Variable response
Adrenal tumors	No suppression	No response

with some individuals showing changes in cortisol values after administration of this drug.

Addison's Disease

A marked decrease in the production of cortisol and other steroids is seen either in cases of Addison's disease, where the defect is in the synthesis of hormones by the adrenal gland, or in cases where ACTH production is low due to damage to the hypothalamus or pituitary. In both cases, plasma cortisol values, urine free cortisol, and 17-hydroxycorticosteroid excretion is diminished.

In **Addison's disease,** the disorder is primarily one in which adrenal damage has occurred through a variety of processes. This tissue destruction impairs the gland's ability to manufacture steroids. Since cortisol output is diminished, there is less feedback control on the pituitary. As a result, plasma ACTH levels in Addison's disease are usually quite elevated. Electrolyte imbalances due to a decrease in aldosterone synthesis are also common.

Adrenal insufficiency secondary to impaired ACTH production is another cause for decreased cortisol production. Any damage to the hypothalamus or pituitary results in diminished ACTH output and a decline in stimulus of the adrenal cortex. Frequently, low ACTH levels are observed during or after treatment with glucocorticoids, which result in transient declines in cortisol production.

Administration of synthetic ACTH analogs allows us to distinguish between the two causes for hypocortisolism. If the decrease in plasma cortisol is due to the inability of the adrenal cortex to synthesize steroids, neither plasma nor urine steroid levels are enhanced after ACTH analog treatment. However, if the decrease in cortisol is secondary to a decline in ACTH release, the plasma and urine cortisol levels after treatment increase over baseline values.

Cortisol and Depression

There has been a long-standing interest in identifying biochemical markers for mental illness. The traditional forms of diagnosis rely on the patient accurately reporting symptoms and responses to situations. When individuals are emotionally disturbed, their ability to describe correctly what they are experiencing is seriously impaired. A laboratory test which provides a reasonably clear distinction between emotionally healthy individuals and those with some mental illness would be of great value in psychiatric diagnosis.

Measuring plasma cortisol and relating cortisol rhythms to emotional state has been carried out for years. A number of studies have shown that a significant percentage of patients with Cushing's syndrome experience severe depression. This situation usually disappears as soon as the physical problem is corrected that originally caused the excessive production of cortisol. Other research has shown that patients whose cortisol diurnal rhythm is disturbed for any reason are much more prone to depressive illness.

In patients whose primary problem is severe depression with no discernible accompanying physical illness, the plasma cortisol levels may be higher than normal. In some instances, a diurnal rhythm is observed, but it is much less pronounced than the cycle seen in a healthy individual. Although ACTH and CRF concentrations have been measured in some cases, it is still very unclear what produces the altered cortisol metabolism. Current theories suggest that stress or some other emotional state induces changes in the rate of synthesis and release of CRF which result in abnormal levels of plasma cortisol in depressed patients.

The **dexamethasone suppression test (DST)** has been employed as a means of assessing cortisol production in depressed patients and as a potential diagnostic tool for identifying the possible cause of depression. The standard protocol involves administration of 1.0 mg dexamethasone orally at 11 P.M. Samples for plasma cortisol measurement are collected the following day at 4 P.M. and 11 P.M. There are no special patient preparations, although individuals who are pregnant or have a major physical illness often have altered cortisol production due to their physical condition and should not have the DST administered.

If either the 4 P.M. or the 11 P.M. sample shows a total cortisol level greater than 5.0 μg/dL, the test is presumed positive for the presence of endogenous depression. A rise in the 4 P.M. value above the 5.0-μg/dL cutoff is termed **early escape,** implying a strong degree of resistance to suppression of cortisol synthesis by dexamethasone.

Since a fairly low level of cortisol is being measured, it is very important that the laboratory assure that the assay procedure employed will reliably differentiate between values above and below the cutoff point. Linearity and reproducibility need to be assessed carefully in the 1.0–5.0-μg/dL range so that borderline results can be properly interpreted.

A variety of situations (Table 15–7) produce false positive DST results (apparent failure to suppress below 5.0 μg/dL or early escape). In addition to the possible major physical illness or pregnancy mentioned above, individuals who are dehydrated or nauseated will fail to suppress after DST. Patients with Cushing's disease or unstable diabetes reflect their abnormal hormonal production in nonsuppression by DST. Individuals who are malnourished or anorexic have diminished capability to suppress since the undernourished state alters production of several hormones. Alcoholics (either currently abusing ethanol or undergoing withdrawal) do not suppress completely. A number of medications, including anticonvulsants, tranquiliz-

Table 15-7
FACTORS AFFECTING THE DST

False Positive Results	False Negative Results
Major physical illness	Addison's disease
Pregnancy	Steroid therapy
Dehydration, nausea	Pituitary damage
Cushing's syndrome	Benzodiazepines
Unstable diabetes mellitus	
Malnutrition, anorexia	
Alcohol abuse	
Tranquilizers	
Anticonvulsants	
Barbiturates	

ers, and barbiturates, also produce false positive tests.

False negative tests are seen in patients with Addison's disease, those on steroid therapy (either corticosteroids or estrogen replacement), patients with pituitary damage, and those individuals on high levels of benzodiazepines. The diminished cortisol production is a result of the underlying physical condition or pharmacological treatment, not a reflection of the presence or absence of the depressive state.

The DST is not as reliable as was once thought. For in-patients (using both the 4 P.M. and 11 P.M. cortisol measurements), the true positive rate is approximately 65%, whereas the true negative rate is much better, some 95%. The test is apparently very good at excluding patients whose depression is not due primarily to elevated plasma cortisol levels, but is not highly successful in identifying those individuals who do have nonsuppressible cortisol. When out-patients are tested, the 4 P.M. sample is the only one collected. In this group, the true positive rate drops to 50%. Although a positive DST suggests better success in treatment with antidepressants and electroconvulsive therapy, the test cannot be used as a primary diagnostic technique to identify endogenous depression.

SUMMARY

The steroid hormones are all composed of the four-ring skeleton characteristic of cholesterol. Variations in substituents C-3 and C-17 result in the formation of different steroid structures. There are three major classes of steroids: (1) the mineralocorticoids (primarily aldosterone) responsible for fluid and electrolyte balance; (2) the glucocorticoids (cortisol and derivatives), which regulate some aspects of carbohydrate and protein metabolism; and (3) the sex steroids (testosterone, estrogens, progesterone), involved with the development of secondary sexual characteristics and reproduction.

A complex network of reactions link the different steroid classes. All molecules are formed initially from cholesterol, with metabolic interconversions leading to synthesis of aldosterone and cortisol through separate pathways by way of progesterone and 17-hydroxyprogesterone. Androgens (testosterone) and estrogens derive from dehydroepinandrosterone, which is synthesized from delta-5-pregnenolone. After synthesis in the adrenal gland, testes, ovaries, or a combination of these tissues, the steroids enter the circulation and are carried through the body by specific transport proteins or by serum albumin.

A wide variety of analytical approaches are available for measurement of steroids in plasma or urine. Classes of urine steroids can be measured colorimetrically. The 17-ketosteroids are determined with the Zimmermann reaction and represent metabolites of dehydroepiandrosterone. The Porter-Silber reaction using 2,4-dinitrophenylhydrazine measures 17-hydroxycorticosteroids (cortisol and derivatives). A two-step process (reduction of keto side chains with $NaBH_4$ followed by oxidation with $NaBiO_3$) allows the assay of 17-ketogenic steroids (a wide variety of cortisol, pregnanetriol, and progesterone derivatives). During the course of pregnancy urine estrogens can be monitored with the Kober reaction, in which the estrogens form a colored complex with hydroquinone in aqueous sulfuric acid. Specific steroids can be quantitated with a number of immunoassay procedures. Gas chromatography (alone or coupled with mass spectrometry) provides a sensitive and versatile method for examining mixtures of steroids. HPLC has been explored to some extent, but currently is not a widely used analytical approach for steroid determination.

In addition to interferences produced by drugs in the chemical reactions used for steroid analysis, several pharmacological agents alter steroid synthesis. Phenytoin, barbiturates, and ethanol stimulate the peripheral degradation of steroids, leading to lower levels being detected in plasma. Aminoglutethimide, metyrapone, and spironolactone impair the synthesis of steroids, making these medications useful in treating clinical problems characterized by excessive steroid production.

The hypothalamus initiates the process of cortisol production through release of corticotropin-releasing factor; stress and emotional factors, in part, appear to regulate CRF release. The pituitary is stimulated by CRF and synthesizes adrenocorticotropic hormone (ACTH), which then interacts with receptors on the adrenal cortex to cause the synthesis and release of cortisol. Binding of cortisol to intracellular receptors changes protein synthesis and enhances the formation of glucose from amino acids. Conversion of cortisol to metabolites and water-soluble conjugates occurs in the liver, followed by excretion in the urine.

Cortisol production follows a diurnal rhythm, with a peak in plasma levels about 8 A.M. and a trough at approximately 10 P.M. Slight fluctuations in plasma levels are seen at all times, since the production of cortisol is pulsed. Both CRF and ACTH are under constant negative feedback control by cortisol.

Cushing's disease is a clinical situation characterized by pituitary abnormalities leading to increased cortisol production. This disorder is distinguished from tumors which cause high cortisol output by ectopic release of ACTH. Adrenal adenomas and carcinomas stimulate excessive production of cortisol in the face of suppressed ACTH synthesis. Use of dexamethasone and metyrapone to alter the metabolism of cortisol enables us to distinguish among these possibilities.

Addison's disease is caused by a variety of situations leading to adrenal gland damage. The primary laboratory finding is a low rate of cortisol synthesis. The hypothalamic/pituitary axis is not impaired, and ACTH production is increased since there is little feedback regulation by cortisol.

Measurement of plasma cortisol levels after dexamethasone suppression has proven helpful in the diagnosis of endogenous depression. Individuals with no physical or psychiatric problem have plasma cortisol levels less than 5.0 μg/dL for 24 h after administration of 1.0 mg dexamethasone. The failure to suppress below 5.0 μg/dL or the presence of early escape (a rise above the cutoff point within 24 h after suppression) suggests the presence of a depression amenable to treatment with antidepressant medications or electroconvulsive therapy.

FOR REVIEW

Directions: For each question, choose the best response.

1. All of the following are categories of steroid hormones *except*
   A. catecholamines
   B. glucocorticoids
   C. mineralocorticoids
   D. sex hormones

2. The glucocorticoid hormones function in the regulation of
   A. glucose production
   B. lipid synthesis
   C. protein metabolism
   D. both A and B
   E. both A and C

3. The hormone that functions to promote sodium ion reabsorption by the kidney and thus water retention is
   A. aldosterone
   B. cortisol
   C. estrogen
   D. progesterone

4. The principal estrogen produced during pregnancy is
   A. epiestriol
   B. estradiol
   C. estriol
   D. estrone

5. Which hormone is considered to be the primary glucocorticoid?
   A. progesterone
   B. cortisol
   C. cortisone
   D. corticosterone

6. The androgen that is the most biologically active is
   A. androstenedione
   B. androsterone
   C. dehydroepiandrosterone
   D. testosterone

7. The test used to assess androgen production by the adrenal glands is
   A. 17-ketogenic steroids
   B. 17-ketosteroids
   C. 17-hydroxycorticosteroids
   D. total 17-hydroxycorticosteroids

8. Which group of steroid hormones is quantitated by the 17-hydroxycorticosteroid and the 17-ketogenic steroid assays?
   A. androgens
   B. glucocorticoids
   C. mineralocorticoids
   D. both A and B
   E. both A and C

9. Which hormone is measured in the mother's urine during pregnancy to assess the health status of the fetus?
   A. aldosterone
   B. androgens
   C. cortisol
   D. estrogens

10. The secretion of cortisol is such that blood specimens should be drawn
    A. in early morning
    B. in late evening
    C. anytime of the day or night
    D. both A and B

11. Which hormone interacts with receptors on the adrenal cortex to stimulate cortisol synthesis?
    A. ACTH
    B. CRF

C. TRH

D. TSH

12. Which of the following characterize Cushing's syndrome caused by an adrenal tumor?
    A. plasma ACTH decreased
    B. plasma cortisol increased
    C. urine 17-hydroxycorticosteroids increased
    D. all of the above

13. Which of the following plasma hormone levels characterize Addison's disease when the primary defect is in the adrenal gland?
    A. decreased cortisol
    B. increased aldosterone
    C. increased ACTH
    D. both A and B
    E. both A and C

14. Which of the following tests may be used in the differential diagnosis of depression?
    A. ACTH stimulation test
    B. dexamethasone suppression test
    C. metopirone inhibition test
    D. metyrapone inhibition test

15. Which of the following results would be expected when the metyrapone inhibition test is performed on a patient who has a pituitary tumor?
    A. decreased plasma cortisol
    B. increased plasma ACTH
    C. increased urine total 17-hydroxycorticosteroids
    D. all of the above

BIBLIOGRAPHY

Abraham, G. E., "Radioimmunoassay of steroids in biological fluids," *J. Steroid Biochem.* 6:261–270, 1975.

Carroll, B. J., "Use of the dexamethasone suppression test in depression," *J. Clin. Psychiatry.* 43:44–48, 1982.

Dluhy, R. G., and Williams, G. H., "Cushing's syndrome and the changing times," *Ann. Intern. Med.* 97:131–133, 1982.

Elias, A. N., and Gwinup, G., "Effects of some clinically encountered drugs on steroid synthesis and degradation," *Metabolism.* 29:582–595, 1980.

Ellyin, F., and Singh, S. P., "Cushing's disease," *Postgrad. Med.* 70:131–143, 1981.

Garcia, M. M., "Immunoassay measurement of corticotropin (ACTH) and corticotropin-releasing factor (CRF)," *J. Clin. Immunoassay.* 6:291–295, 1983.

Goeroeg, S. (ed.), *Advances in Steroid Analysis.* New York: Elsevier Publishing Co., 1982.

Goeroeg, S. (ed.), *Quantitative Analysis of Steroids.* New York: Elsevier Publishing Co., 1983.

Health and Public Policy Committee, American College of Physicians, "The dexamethasone suppression test for the detection, diagnosis, and management of depression," *Ann. Intern. Med.* 100:307–308, (1984).

Hsu, T-H., "The pituitary-adrenal axis: Clinical considerations," *J. Clin. Immunoassay,* 6:277–287, 1983.

Imura, H., "ACTH and related peptides: Molecular biology, biochemistry, and regulation of secretion," *Clin. Endocrinol. Metab.* 14:845–866, 1985.

James, V. H. T., and Few, J. D., "Adrenocorticosteroids: Chemistry, synthesis, and disturbances in disease," *Clin. Endocrinol. Metab.* 14:867–892, 1985.

Malvano, R. et al., "Recent trends in methodological simplification of steroid radioimmunoassay," *Hormone Res.* 9:422–439, 1978.

Meltzer, H. Y., and Fang, V. S., "Cortisol determination and the dexamethasone suppression test. A review." *Arch. Gen Psychiatry.* 40:501–505, 1983.

Orth, D. N., "The old and the new in Cushing's syndrome," *New Eng. J. Med.* 310:649–651, 1984.

Rivier, C. L., and Plotsky, P. M., "Mediation by corticotropin-releasing factor (CRF) of adenohypophyseal hormone secretion," *Ann. Rev. Physiol.* 48:475–494, 1986.

Taylor, A. L., and Fishman, L. M., "Corticotropin-releasing hormone," *New Eng. J. Med.* 319:213–222, 1988.

van Weemen, B. K. et al., "Enzyme-immunoassay of steroids: Possibilities and pitfalls," *J. Steroid Biochem.* 11:147–151, 1979.

CHAPTER OUTLINE

Chapter Sixteen

CLINICAL ENDOCRINOLOGY III: CATECHOLAMINES

• • • • • • • • • • • • • • • • • • • •

Upon completion of this chapter, the student will be able to

1. Name the amino acid that acts as the immediate precursor of catecholamine synthesis.
2. Name the two best known catecholamines and two end products of catecholamine metabolism.
3. Identify the relationship between epinephrine and adrenaline and between norepinephrine and noradrenaline.
4. Identify the primary site of catecholamine synthesis and storage.
5. Describe the neuronal control of catecholamine release.
6. List five metabolic effects of the catecholamines.
7. Describe the function of dopamine in catecholamine synthesis and storage.
8. Describe specimen collection and proper handling for quantitation of plasma and urine catecholamines.
9. Describe the fluorometric assay for the quantitation of urine catecholamines.
10. Identify the significance of normetanephrine and metanephrine in catecholamine metabolism.
11. Describe the colorimetric assays for quantitating urine total metanephrines and vanillylmandelic acid (VMA).
12. Identify dietary causes of false elevations of VMA.
13. Describe the radiometric and HPLC methods for quantitating plasma catecholamines.
14. Define the terms *pheochromocytoma* and *neuroblastoma.*
15. State the major physiological characteristics associated with pheochromocytoma and neuroblastoma.
16. Identify urine screening tests employed for the detection of pheochromocytoma and neuroblastoma.
17. Identify clinical conditions that manifest abnormal catecholamine levels.

INTRODUCTION

Study of catecholamine production and excretion has been a complex field since the discovery of these compounds in the late 1890s. A major area of interest is the effects of these materials on blood pressure and their contribution to the problem of hypertension. There is increasing recognition of the role played by catecholamines in certain aspects of other common disease states such as pituitary malfunction, diabetes mellitus, and depression. Extensive study is being undertaken to elucidate the contribution of impaired catecholamine synthesis to the problem of Parkinson's disease. Catecholamine synthesis and metabolism are important components of a variety of biochemical control processes.

16.1

BIOCHEMISTRY OF CATECHOLAMINES

Types of Compounds

All the **catecholamines** are formed by conversions of **tyrosine,** an amino acid containing a substituted benzene ring (Fig. 16–1). Tyrosine itself derives from phenylalanine which undergoes a ring hydroxylation reaction to yield this important catecholamine precursor. The name *catecholamine* derives from the common catechol structure (1,2-dihydroxybenzene) possessed by all these compounds. The best known catecholamines are **epinephrine (adrenaline)** and **norepinephrine (noradrenaline),** responsible in part for the regulation of blood pressure. End products of catecholamine metabo-

lism are homovanillic acid and vanillylmandelic acid.

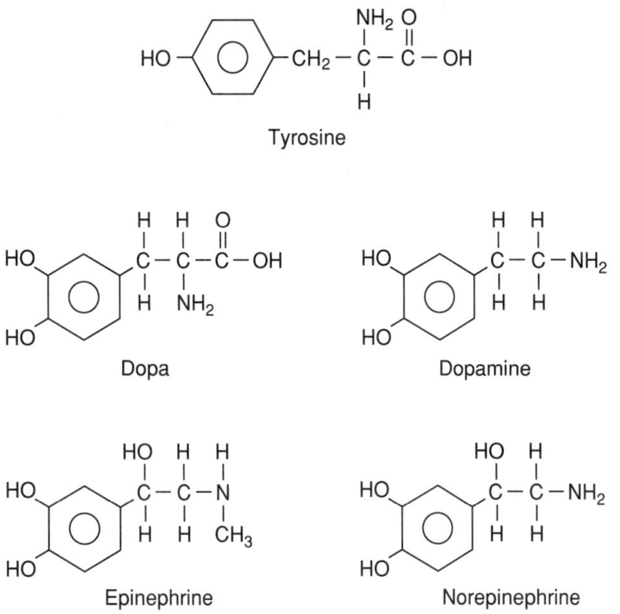

Sources

The catecholamines are primarily synthesized and stored in vesicles of the chromaffin cells in the adrenal medulla until released into the circulation. The **chromaffin cells** are located at the top of the kidneys and are part of the adrenal gland, which comprises both the adrenal cortex and the adrenal medulla. The adrenal glands in healthy adults weigh about 4 g each. The other site of catecholamine formation is in the sympathetic neuron, part of the central nervous system.

Synthesis and Storage in Chromaffin Cell

Tyrosine resulting from protein metabolism or formed by conversion of phenylalanine enters the chromaffin cell and undergoes ring hydroxylation (catalyzed by tyrosine hydroxylase) to form 3,4-dihydroxyphenylalanine (dopa). Subsequent reactions then involve only side-chain modifications; the benzene ring does not undergo further change (Fig. 16–2). Removal of the carboxyl group from dopa by the enzyme dopa decarboxylase produces dopamine. **Dopamine** then enters the chromaffin vesicle inside the cell to undergo further conversion. Reaction with dopamine-beta-hydroxylase adds an $-OH$ to the side chain, producing norepinephrine. The norepinephrine then leaks out of the granule and is N-methylated to yield epinephrine (catalyzed by phenylethanolamine-N-methyltransferase). Re-uptake of epinephrine in the chromaffin vesicle completes the synthesis and storage process. In the vesicle mainly norepinephrine (51%) and epinephrine (43%) are found with a small amount of dopamine (6%).

16.2

CATECHOLAMINE RELEASE

Neuronal Control of Release

The splanchnic nerves of the central nervous system are connected to the adrenal medulla, impulses from which cause the nerve endings to release acetylcholine. This neurotransmitter binds to receptor sites on the cells of the adrenal medulla and causes changes in the cell membrane potential which allow calcium ions (among others) to enter the cell. The influx of calcium is a trigger for the release of catecholamines from the granules into the bloodstream.

The process by which catecholamines leave the cell is called **exocytosis.** The vesicles migrate toward the interior of the cell membrane and fuse to

FIGURE 16–1 Important catecholamines and precursors.

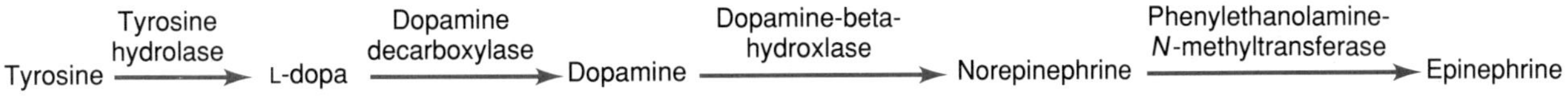

Tyrosine $\xrightarrow{\text{Tyrosine hydrolase}}$ L-dopa $\xrightarrow{\text{Dopamine decarboxylase}}$ Dopamine $\xrightarrow{\text{Dopamine-beta-hydroxlase}}$ Norepinephrine $\xrightarrow{\text{Phenylethanolamine-N-methyltransferase}}$ Epinephrine

FIGURE 16-2   Synthesis of catecholamines from tyrosine.

the membrane. After fusion, a portion of the exterior membrane opens, releasing the catecholamines into the circulation. This process is energy-efficient in that it requires no metabolism of ATP and no active transport system.

Effect of Adrenal Cortex Hormones

The sympathetic nervous system has primary control over catecholamine production and release, but the adrenal cortex also exerts an influence over production of these neurotransmitters. These two segments of the adrenal gland are connected by blood vessels which carry cortisol and other hormones from the cortex portion through the medulla. Conversion of norepinephrine to epinephrine is profoundly affected by cortisol. Decreases in the concentration of this steroid hormone enhance breakdown of the enzyme phenylethanolamine-*N*-methyltransferase, leading to a decrease in the amount of epinephrine synthesized.

Inactivation and Excretion of Catecholamines

Unlike many other hormones, the catecholamines circulate free, not bound to any plasma proteins. As a result, uptake by tissues is rapid, whether the tissue is a receptor or simply a vehicle for inactivating the circulating hormone. Catecholamines released at neuronal sites primarily undergo re-uptake at the neuron, followed either by storage and reutilization or by metabolic inactivation. In the circulation, platelets play an important role in the uptake and removal of catecholamines. The liver and other tissues also contain enzymes involved in catecholamine inactivation.

The initial step in the metabolism of catecholamines involves either the conversion of the side chain $-NH_2$ to a carboxyl group or the *O*-methylation of a ring hydroxyl (Fig. 16–3). The oxidation reaction is catalyzed by **monoamine oxidase (MAO),** found in high concentrations in platelets, liver, and brain. **Catechol-*O*-methyltransferase (COMT)** is the enzyme responsible for the methylation reaction. Alcohol dehydrogenase and aldehyde reductase play important roles in the conversion of intermediates.

The end product of dopamine metabolism is **homovanillic acid (HVA)** (Fig. 16–4), regardless of which enzyme (MAO or COMT) catalyzes the initial reaction. If epinephrine and norepinephrine are first acted upon by COMT, metanephrines are formed as metabolites. These compounds are frequently measured in urine to assess altered catecholamine production in various clinical states. Further oxidation by MAO produces **vanillylmandelic acid (VMA),** an end product of catecholamine metabolism. A number of sulfated derivatives are formed through the action of phenolsulfotransferase, but the clinical significance of these derivatives has not yet been established. Another metabolite is 3-methoxy-4-hydroxyphenylglycol, of interest to those studying the biochemistry of depression.

Role in Transmission of Nerve Impulses

As neurotransmitters, the catecholamines play important roles in the propagation of the nerve im-

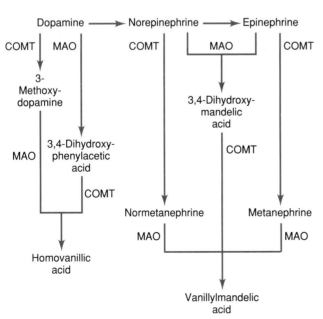

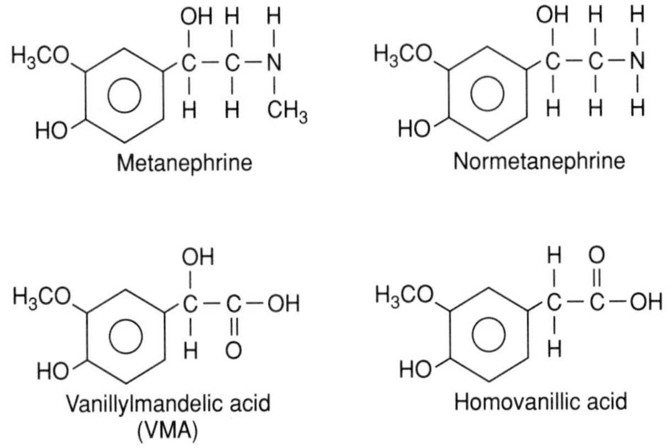

FIGURE 16-3   Metabolism of catecholamines.

FIGURE 16-4   Structures of catecholamine metabolites.

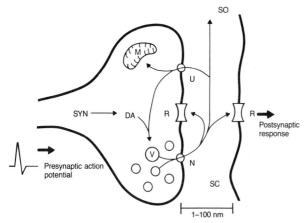

FIGURE 16–5 The terminal (or synaptic) region coupling two cells. SYN = synthesis, DA = dopamine, V = dopamine storage vesicles, N = neuronal release, SC = synaptic cleft, R = pre- and postsynaptic dopamine receptors, U = dopamine uptake, SO = synaptic overflow, and M = metabolic enzymes. (From Wightman, M. R., May, L. J., and Michael, A. C., "Detection of Dopamine Dynamics in the Brain," *Anal. Chem.* 60:770A, 1988. Copyright 1988 American Chemical Society. Reprinted with permission.)

pulse (Fig. 16–5). Synthesis and storage of catecholamines in vesicles takes place as previously described. An electric current passing down the nerve stimulates the release of catecholamine by exocytosis at the presynaptic membrane. Specific neurons release either dopamine, epinephrine, or norepinephrine. After the neurotransmitter passes across the synaptic cleft, it attaches to receptors at the postsynaptic membrane, triggering an impulse at that segment of the nerve. A number of different types of receptors exist, each specific for a separate catecholamine and each producing somewhat different effects. After binding to the receptor and assisting in the further propagation of the nerve impulse, the catecholamine dissociates from the receptor. Inactivation then occurs through metabolism by COMT, or the catecholamine can be taken back up again by the presynaptic site. Some of the catecholamine is then inactivated by monoamine oxidase. Small amounts of both active catecholamines and metabolites can enter the circulation and contribute to the total amount of these materials in the bloodstream. Norepinephrine is the major fraction contributed from nervous tissue.

Metabolic and Physiological Effects of Catecholamines

Catecholamines profoundly affect metabolism (Table 16–1). When catecholamine levels increase

Table 16–1
METABOLIC EFFECTS OF CATECHOLAMINES

1. Increased breakdown of triglycerides
2. Enhanced synthesis of glucose from amino acids
3. Enhanced breakdown of liver glycogen
4. Decrease in protein synthesis
5. Increase in blood glucose levels

in the circulation, breakdown of triglycerides is enhanced with a corresponding rise in the availability of free fatty acids. Less pronounced is the effect on protein metabolism: apparently there is some decrease in the rate of protein breakdown in the presence of increased catecholamines.

Perhaps the most striking effects are seen in relation to glucose metabolism. A rise in catecholamine release triggers increases in blood glucose levels. This shift comes (in part) from an enhanced rate of glycogen breakdown in the liver, with glucose as the end product of that conversion. Gluconeogenesis (formation of glucose from amino acids) also increases. Glucose concentrations in plasma remain higher than expected since the release of insulin is inhibited by the elevated amounts of catecholamines.

There are some complex relationships between catecholamine levels and carbohydrate release. Hypoglycemia stimulates the release of catecholamines into the circulation. Blood levels of hormones such as glucagon (which promotes glycogen breakdown), cortisol, and growth hormone also increase during a hypoglycemic episode. The regulatory effects of the catecholamines and other hormones then serve to stimulate metabolism and restore the euglycemic state. Biochemical energy is provided more through fatty acid metabolism and ketogenesis than through glucose utilization, resulting in a lowered demand for available glucose until the hypoglycemia is corrected.

The physiological effects produced by epinephrine and norepinephrine are mediated by two types of receptors located in smooth muscle and other tissues. Interaction between the neurotransmitter and alpha adrenoreceptors usually leads to muscle contraction, whereas stimulation of the beta adrenoreceptor induces relaxation. There are, however, several exceptions to this rule. The role of dopamine in affecting peripheral tissue is unclear. This compound binds to both the alpha and beta adrenergic receptors, but also has specific receptors of its own.

Both epinephrine and norepinephrine stimulate cardiac muscle. The more significant contributor to increased heart rate and enhanced strength of contraction is epinephrine. Although norepinephrine also has some effect on heart rate, its primary role is to promote vasoconstriction. Thus, the effects of these neurotransmitters on heart rate and decreasing size of vessel openings furnish a complex and sensitive means of regulating blood pressure. The renin/angiotensin/aldosterone system involved in electrolyte balance (which contributes to changes in blood pressure) may also be regulated in part by changes in catecholamine output.

16.3

ANALYSIS OF PLASMA AND URINE CATECHOLAMINES

A variety of methods are available for the quantitative estimation of catecholamines in plasma and urine. Approaches allow measurement of total catecholamine output or the fractionation of specific components of this endocrine system. In each case, attention should be given to proper sample collection and storage, particularly in the case of plasma samples. Platelets in blood bind readily to catecholamines and metabolize epinephrine and norepinephrine through both the MAO and COMT enzyme systems. Selection of the analytical method depends on the information desired and the equipment available. Assays vary from reasonably simple colorimetric approaches to sophisticated enzyme assays involving radioisotopes. With modern technology, fractionation and quantitation of the various catecholamine fractions in plasma can be readily accomplished.

Sample Collection and Stability

If plasma catecholamines are to be determined, care must be taken to avoid raising the plasma levels of these materials during the collection process. Since standing has been shown to increase norepinephrine levels markedly, the patient should be lying down for approximately 15–30 min prior to sample collection (some diagnostic protocols may require collection of samples both supine and standing). Sitting also causes increases, but to a lesser extent. To avoid elevations related to the stress of venipuncture, the needle or catheter should be inserted 15 min or so prior to sample collection and kept open with a heparin lock or intravenous drip.

Sample collection should employ an anticoagulant appropriate for the specific analytical method. Since the vast majority of plasma catecholamine assays are carried out by reference laboratories, a collection protocol should be provided by that laboratory. Immediately upon collection, the blood sample should be cooled and centrifuged to separate plasma from cells. Any prolonged contact with platelets and other cellular components decreases the catecholamine content. Plasma samples must be divided into aliquots and frozen in case duplicate assays are needed. Repeated thawing and refreezing of the sample leads to significant loss of catecholamines. An antioxidant, such as reduced glutathione, stabilizes the catecholamines somewhat against oxidative degradation.

Urine specimens for assessment of 24-h output of

catecholamines should be collected on ice in a container with 10 mL added HCl. During the period of sample collection, the urine should be refrigerated to minimize degradation of materials. After the collection is complete, the pH should be adjusted to somewhere between pH 2 and 5 with 6M HCl. Aliquots can be frozen indefinitely, but repeated thawing and refreezing should be avoided.

Drug interferences present a real difficulty with catecholamine measurements, especially with assays for urine constituents. Most individuals having their catecholamine status assessed have high blood pressure and may be taking antihypertensive medication. A number of these medications interfere with the chemical assays by giving false high results. It is not always practical to have the patient stop medications for the several days required to clear them completely from the system. Coordination with the health-care team is necessary to provide the best possible analytical specimen without jeopardizing the health of the individual.

Urine Catecholamines: Fluorometric Assay

Epinephrine and norepinephrine excretion in urine can be measured in a variety of ways. These compounds are excreted both in the free (unconjugated) and conjugated states. Most clinical work has been done using data on free catecholamines. Inclusion of an initial acid hydrolysis step allows the assay of total epinephrine and norepinephrine using the same methods employed for the quantitation of the free fractions.

The initial step in the fluorometric assay of free catecholamines is to (1) adjust the urine pH to the alkaline range (Fig. 16-6). Both epinephrine and norepinephrine exist as neutral compounds at pH values above 8 since the amine groups on these molecules are not protonated. These forms of the molecules then bind to alumina or resin ion-exchange columns (2). After washing to remove other materials, the catecholamines are eluted with dilute acid (3). Oxidation with ferricyanide (4) converts the catecholamines to cyclic derivatives. Addition of an antioxidant (ascorbic acid) and adjustment of pH to basic converts these trihydroxyindole derivatives to fluorescent forms (5).

The major drawbacks to this assay are incomplete adsorption on alumina, high blank values, and interferences with antihypertensive medications which react in the same manner as the native catecholamines. Proper adjustment of pH prior to alumina extraction is critical; pH values below about 8.5 lead to lowered extraction efficiency. Prolonged refrigeration sometimes discolors the sample, causing increased blank values. Medications such as alpha methyldopa (which serve to inhibit catecholamine synthesis and reduce hypertension) are structurally similar to epinephrine and norepinephrine. These drugs undergo the same reactions and form

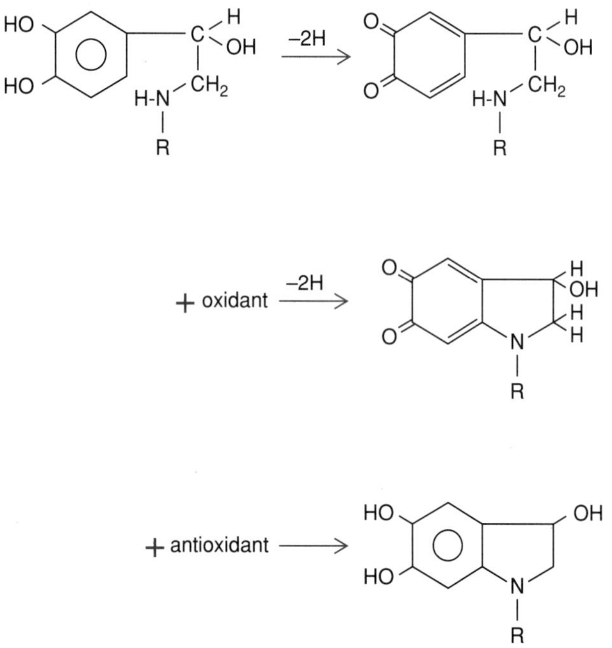

FIGURE 16-6 Trihydroxyindole method for plasma catecholamine determination. In epinephrine, R represents CH_3; in norepinephrine, R stands for H. I. Epinephrine (norepinephrine). II. Epinephrine-quinone (norepinephrine-quinone). III. Epinephochrome (norepinephochrome). IV. Epinepholutine (norepinepholutine)—fluorescent.

similar fluorescence derivatives, creating false positive results when present.

Urine Metanephrines: Colorimetric Assay

Total metanephrines can be analyzed by the **Pisano method,** which involves extraction followed by colorimetric reaction (Fig. 16-7). This assay does not distinguish between metanephrine and normetanephrine but gives the total of the two components.

Hydrolysis with HCl breaks down all conjugates, allowing more uniform extraction of all fractions. The solution containing the metanephrine mixture is poured onto an ion-exchange column, which loosely binds the metanephrines but allows other components to pass through and be removed from the reaction mixture. Elution with ammonium hydroxide washes the metanephrines off the column. Conversion to vanillin (absorbance maximum 360 nm) is accomplished through periodate oxidation.

Very few false negative results are obtained with this assay, but a number of compounds do give false positive values. Interferences leading to artefactual increases are caused by chlorpromazine, imipramine, phenothiazines, methyldopa, and some tetracycline derivatives. A number of medications alter

1. Hydrolysis with HCl

2. Column absorption of metanephrines

3. Elution of metanephrines with NH_4OH

4. Conversion to vanillin

Vanillin

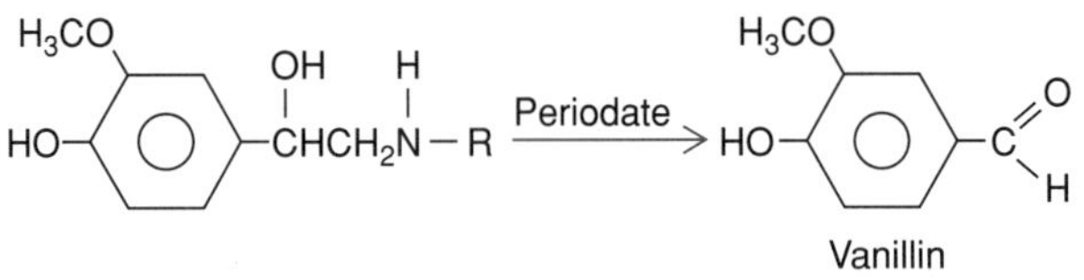

FIGURE 16-7 Steps in the assay of urine metanephrines. For normetanephrine R stands for H; for metanephrine R stands for CH_3.

the metabolism of catecholamines; the elevated metanephrine values seen in these situations are real but are caused by the pharmacological intervention and not necessarily by the disease state. Hydrazine derivatives and MAO inhibitors both interfere with the metabolic processes leading to the conversion of metanephrines to other derivatives.

Urine Vanillylmandelic Acid: Colorimetric Assay

Measurement of urine VMA involves the same final chemical reaction used in the assay for metanephrines but employs a different separation procedure (Fig. 16-8). Hydrolysis with HCl to destroy conjugates is the initial step in the process. Solvent extraction with ethyl acetate separates the VMA into the organic layer. Further purification of the mixture is accomplished by reextracting the VMA into an aqueous potassium carbonate solution. Treatment with sodium metaperiodate oxidizes VMA to vanillin, which can be quantitated spectrophotometrically. Often, vanillin is extracted into an organic solvent and reextracted into an aqueous one to separate the vanillin from other materials which might interfere in the absorbance measurement.

A variety of materials produce false decreases in the value for urine VMA. If the urine has a pH above 3, the VMA decomposes, lowering the amount avail-

1. Hydrolysis to remove conjugates

2. Extraction of VMA into ethyl acetate

3. Reextraction into aqueous potassium carbonate

4. Oxidation to vanillin

FIGURE 16-8 Steps in the assay of urine VMA.

able for assay. Compounds which react with the periodate and lower the amount of this reagent available for reaction with VMA include salicylates, clofibrate, and L-dopa. Any catecholamine metabolite (such as homovanillic acid) may also react with the periodate to decrease its effective concentration. A variety of medications alter catecholamine output and impair the formation of VMA. These agents include clonidine, disulfiram, hydrazine derivatives, imipramine, MAO inhibitors, morphine, phenothiazines, and some radiographic agents.

The major false elevations of VMA excretion are usually due to diet. Ingestion of bananas, coffee, tea, vanilla, or chocolate results in materials in the urine which react with the metaperiodate to produce colored derivatives. A wide variety of medications produce chemical interference with the assay leading to higher than expected values. These materials include salicylates, disulfiram, nalidixic acid, oxytetracycline, and various phthalein dyes. Elevations due to altered metabolism of catecholamines may be seen in patients on insulin, isoproterenol, L-dopa, lithium, and reserpine.

Plasma Catecholamines: Radiometric Analysis

Although catecholamine levels in plasma can be determined by modifying the column extraction flu-orescent approach used for free urine catecholamines, the method is somewhat lengthy and not as sensitive as other approaches. In addition, the fluorescent technique requires several milliliters of plasma and 2 days to perform. Two other types of analyses have become widely used for the measurement of plasma catecholamines: HPLC fractionation and the radiometric estimation. HPLC techniques will be discussed in a subsequent section.

Catecholamines can be assayed in plasma without the need for deproteinization or extraction (Fig. 16–9). Incubation of plasma with a mixture of catechol-O-methyltransferase and tritiated S-adenosylmethionine transfers radioactivity to the catecholamine to form a radiolabeled metanephrine derivative. Extraction of the labeled derivatives into an organic solvent separates this label from the excess S-adenosylmethionine containing ^{3}H. The various catecholamine fractions are then separated by thin-layer chromatography and isolated from the TLC plate. Fractions corresponding to the original epinephrine and norepinephrine are oxidized to vanillin with sodium metaperiodate and reextracted before determining the amount of radioactivity present. The dopamine-labeled fraction has its radioactivity measured without oxidation and reextraction.

Some patients have an endogenous inhibitor of the COMT reaction in their sample. This material (identity unknown at present) decreases the activity of the enzyme and reduces the amount of radiolabel

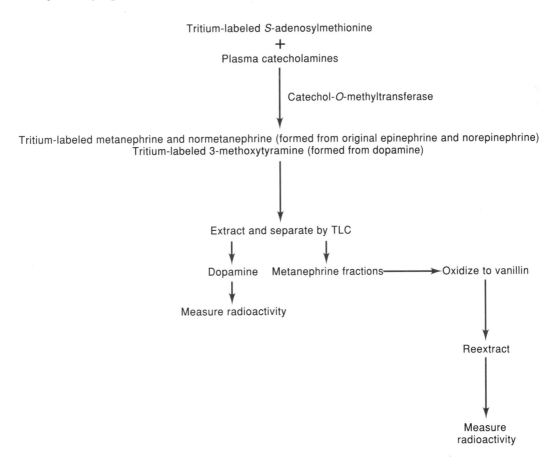

transferred to the catecholamines in the sample. This problem can be detected by adding to the sample an internal standard solution containing known amounts of catecholamines. The final results show the labeling of patient catecholamines alone in one sample and the labeling of the sum of patient material plus added catecholamines in the other sample. The difference between the two tubes is the amount of radiolabeled methyl group attached to the added known concentration of catecholamine. If this value is lower than that obtained under conditions where no inhibitor was known to be present, interference with the reaction is demonstrated. Correct concentrations of catecholamines in these samples can be calculated using the data provided by the internal standard.

High-Performance Liquid Chromatographic Determination of Catecholamines

The high resolution power of HPLC has been successfully adapted to the quantitative measurement of catecholamines in a variety of situations. Methods are available for fractionation of both plasma and urine catecholamines in the clinical laboratory. Research applications include the study of catecholamine levels in various tissues, providing invaluable information about the details of catecholamine metabolism and the changes seen in various pathological conditions.

Three major steps are involved in the HPLC analysis of catecholamines. First, the catecholamines are isolated from other components of the sample through preliminary column chromatography. Next, the catecholamines are column fractionated in a mixture of water and an organic solvent. Finally, after the individual fractions elute from the column, each catecholamine is detected through measurement of the weak natural fluorescence of the compounds or by forming quinone derivatives with the following electrochemical process:

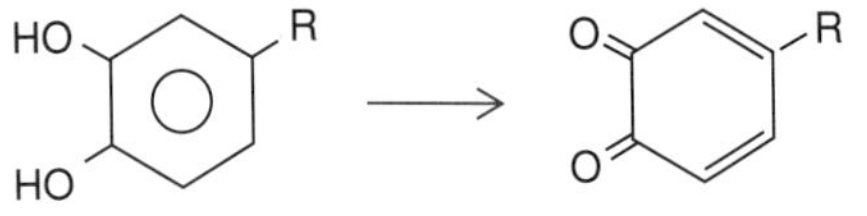

The change in electric current flowing through the detector produced by the oxidation reaction is related to the concentration of the specific catecholamine eluting from the column at that time.

Drug Interferences with Catecholamine Measurements

A wide variety of medications influence the levels of various catecholamines in the body. Since a single urine or plasma specimen may be employed for several different assays, a thorough medication history is necessary before sample collection and analysis.

Table 16–2
MEDICATIONS AFFECTING ANALYSIS OF URINE CATECHOLAMINE FRACTIONS

False Elevations	False Decreases
Salicylates (C, V)	Chlorpromazine (C)
Chloral hydrate (C)	Salicylates (V)
Chlorpromazine (M, C)	Clofibrate (V)
Imipramine (M)	
Tetracyclines (C, M, V)	
Erythromycin (C)	
Isoproterenol (C)	
Methyldopa (C, M, V)	
Nalidixic acid (V)	
Niacin (C)	
Phenothiazines (M)	
Phthalein derivatives (V)	
Quinidine (C)	
Riboflavin (C)	

C = catecholamines (epinephrine + norepinephrine)
M = metanephrines
V = vanillylmandelic acid

A summary of medications that interfere with one or more assays for urine catecholamines is presented in Table 16–2. Those medications that actually change the level of one or more catecholamines are listed in Table 16–3. Values obtained under these circumstances reflect the actual levels of catecholamines in the system and are not artefacts of the measuring process. No attempt will be made to describe interferences with HPLC assays since these interferences are very method-dependent.

In some instances, the same medication may either increase or decrease the catecholamine value (real or artefactual). The exact influence and the degree of change depend on a variety of factors, including the specific modification of an analytical method (for a chemical interference), the dosage level of the medication, or how long the medication has been used (for an actual change in levels). Catecholamine production and excretion vary depending on the amount and time span of medication use. Initially,

Table 16–3
MEDICATIONS AFFECTING LEVELS OF URINE CATECHOLAMINE FRACTIONS

Increases	Decreases
Isoproterenol (C, V)	Clonidine (C, V)
Nitroglycerin (C)	Methyldopa (C, V)
Reserpine (C, V)	Radiographic agents (C)
Theophylline (C)	Reserpine (C, V)
MAO inhibitors (M, V)	L-Dopa (M, V)
Glucagon (V)	Disulfiram (V)
Insulin (V)	Imipramine (V)
	MAO inhibitors (V)
	Phenothiazines (V)

C = catecholamines (epinephrine + norepinephrine)
M = metanephrines
V = vanillylmandelic acid

a given medication may stimulate increased output of catecholamines. After a time, the synthesis of these materials may decline markedly due to depletion in the supply of precursors. Tables 16–2 and 16–3 give only approximate guidelines and do not represent an exhaustive summary of medication effects on catecholamine assay or synthesis.

16.4

PHYSIOLOGICAL FACTORS AFFECTING CATECHOLAMINE LEVELS

A number of factors affect catecholamine production and excretion in the healthy individual. Awareness of the changes that can take place under normal conditions facilitates the establishment of appropriate testing conditions and permits more reliable interpretation of data in various situations. Reference ranges for patients who are relaxed and resting are shown in Table 16–4.

Catecholamine synthesis and release is intermittent and varies according to the time of day. Release into the circulation is greatest during the waking hours and is three to five times higher than production during the night. There does not seem to be a discernible diurnal cycle, as is the case for cortisol. Rather, the increase in output during the day is more likely a response to the increases in physical activity (including the simple task of standing) and emotional stress which accompany our waking hours.

Epinephrine production increases markedly during times of mental stress, pain, loud noises, or other situations producing stress. Norepinephrine levels also increase somewhat, but not to the extent seen for epinephrine. This fact has profound implications for the environment and technique used for collecting plasma samples for catecholamine studies. Stressful situations should be minimized even when collecting a 24-h urine specimen so that catecholamine excretion is not influenced unduly by external factors.

There is a marked increase in catecholamine output (particularly norepinephrine) upon standing. This change is associated with altered blood pressure and the need to stimulate vasoconstriction to maintain appropriate blood flow while in the upright position. Alterations in the heart rate seen when shifting from the reclining to the upright position also reflect the shift in catecholamine production rate.

Increases in physical activity involving heavy muscular work provoke a marked elevation in norepinephrine output. This increase in hormone production may reflect a complex system of changes which include vasodilation in the muscle circulatory system as well as responses to the changing metabolic needs of the body.

Catecholamine biosynthesis and metabolism are partially under genetic control Studies in humans have focused primarily on changes in the enzyme levels of dopamine-beta-hydroxylase, catechol-*O*-methyltransferase, and monoamine oxidase. In each case, synthesis of enzyme (and, therefore, amount of enzyme present in tissues) has been shown to have a strong genetic component. These data have been extensively studied in patients with mental illness (depression and schizophrenia, in particular) in an attempt to identify potential biochemical markers for these disorders. The impact of these genetic studies on other clinical situations is unclear at present.

16.5

PATHOLOGICAL FACTORS AFFECTING CATECHOLAMINE LEVELS

A number of clinical conditions are associated with altered catecholamine output (Table 16–5). Until fairly recently, the major focus was on disorders relating to blood pressure. Increasingly, we see catecholamine changes as part of a wider spectrum of clinical disorders, particularly involving changes in production of other hormones. As we unravel the intricate interworkings of the endocrine system and clarify the relationships of effects produced by various hormones, a clearer picture of several aspects of a number of disease states will unfold.

Pheochromocytoma

Although **pheochromocytoma** (a tumor of neuroectodermal origin) is responsible for hypertension

Table 16–4
REFERENCE RANGES* FOR
CATECHOLAMINE FRACTIONS

Component Measured	Reference Range
Urine Free Catecholamines	
Epinephrine	0–15 µg/24 h
Norepinephrine	0–100 µg/24 h
Dopamine	65–400 µg/24 h
Urine Total Metanephrines	
Metanephrine	74–297 µg/24 h
Normetanephrine	105–354 µg/24 h
Homovanillic acid	66–222 µg/24 h
Urine VMA	2–7 mg/24 h
Plasma Catecholamines (random)	
Epinephrine	<88 pg/mL
Norepinephrine	104–548 pg/mL
Dopamine	<136 pg/mL

* All values are for adults.
Adapted from Tietz, N. B., ed., *Textbook of Clinical Chemistry*. Philadelphia: W. B. Saunders Co., 1986.

Table 16-5
CATECHOLAMINE LEVELS IN CLINICAL CONDITIONS

Increased Catecholamines	Decreased Catecholamines
Pheochromocytoma	Hyperthyroidism
Neuroblastorna	Diabetes: long-term
Essential hypertension (?)	
Hypothyroidism	
Diabetic acidosis	
Cardiac disease	
Burns	
Septicemia	
Depression	

in only about 0.1–0.7% of the population with high blood pressure, there is a great deal of interest in diagnosing the presence of this tumor. If pheochromocytoma is the cause of the elevation in blood pressure, surgical removal has a success rate of greater than 90%. In addition, these tumors can also be found in patients with other endocrine disorders, complicating both the diagnostic picture and the treatment regimen in these patients.

Pheochromocytomas are usually found on the adrenal gland itself. Occasionally an extraadrenal tumor is detected, either instead of or in addition to the adrenal tumor. After diagnosis, precise location of the tumor site is important in order to minimize the amount of surgery needed and the hazards associated with the operation.

The major clinical signs which point toward the presence of this tumor are very severe headaches, palpitations of the heart with possible increase in heart rate, and excessive sweating. These symptoms are a result of excessive catecholamine production by the tumor.

Testing strategies for the detection of a pheochromocytoma vary. Most experts recommend a 24-h urine metanephrine and/or VMA assay as the initial screening procedure. If these values are markedly elevated (and interferences can be ruled out), there is a high probability the tumor is present. Borderline high results should be followed up with an assay for urine free catecholamines. Collection of a 24-h urine specimen on more than one occasion may be necessary, since production of catecholamines by these tumors is often intermittent.

There is currently a great deal of debate about the value of plasma catecholamine assays in the diagnosis of a pheochromocytoma. Sampling is easier than the collection of a 24-h urine specimen, even though somewhat elaborate precautions need to be taken in obtaining the sample. Since the catecholamine release by the tumor is not steady, a single plasma sample may not reflect the actual catecholamine status over time.

The **clonidine suppression test** is being seen as a useful adjunct to plasma catecholamine measurements in the diagnosis of pheochromocytoma. Plasma epinephrine and norepinephrine levels are elevated in both primary hypertension (not associated with a tumor) and pheochromocytoma. If clonidine is administered, patients with primary hypertension show decreases in plasma catecholamine levels to (or below) the reference range. Individuals with pheochromocytomas show some decline in the values, but the levels invariably still remain above the reference range. Clonidine functions as a suppressor of catecholamine release from the nervous system. Since the pheochromocytoma producing catecholamines is under no regulatory control, clonidine has no effect on secretion by the tumor. Any decline seen in these cases is due only to the central nervous system catecholamine production normally seen and not to suppression of synthesis by the tumor.

Neuroblastoma

Among the other catecholamine-secreting tumors, neuroblastomas and associated tumors are more widely spread throughout the body wherever sympathetic nervous tissue is found. These tumors also differ from pheochromocytomas in their lack of the enzyme responsible for the conversion of norepinephrine to epinephrine. In these patients, analysis of urine free catecholamines appears to be the most reliable biochemical indicator for the presence of the tumor. Metanephrine and VMA excretion are variable and frequently within the normal range. Dopamine output is invariably elevated, but testing for this metabolite is not commonly performed.

Essential Hypertension

Primary (essential) hypertension is a poorly understood disorder of high blood pressure of unknown cause(s). In contrast to the pheochromocytoma, there is at present no clearly understood mechanism for the production of hypertension in these cases. Apparently, some disorder of the sympathetic nervous system may trigger vasoconstriction, leading to elevations in blood pressure.

Measurement of plasma catecholamines has been somewhat helpful in exploring this phenomenon, but the data are contradictory. A number of studies suggest that plasma norepinephrine increases in this disorder, whereas little or no change in epinephrine levels is seen. There are still questions about the extent of the changes since the reference range for plasma catecholamines has not been reliably established. The values apparently rise somewhat with age, and this factor has not always been considered carefully in all studies.

Other Disorders

Plasma catecholamine values and urine output of catecholamines are altered in a variety of clinical situations. Disorders of the thyroid affect catecholamine production and excretion, with hyperthyroid individuals having decreases in these parameters and hypothyroid patients manifesting elevated levels. Individuals with diabetic acidosis have increased plasma and urine catecholamines, although long-term diabetics show lowered levels, possibly due to development of diabetic neuropathy and diminished peripheral nerve function. Any stressful situation (such as cardiac disease, burns, or septi-cemia) elevates catecholamines in both plasma and urine. Patients with manic-depressive disorder or depressive neurosis have been reported to demonstrate increases in catecholamine output, although urine catecholamine excretion may diminish in some patients. It is not known at present whether these alterations in catecholamine metabolism represent actual changes which cause the disorders or merely reflect the psychological stress experienced by these patients.

SUMMARY

Catecholamines are formed from the amino acid tyrosine through a series of hydroxylation and decarboxylation reactions. Epinephrine and norepinephrine are stored in chromaffin granules within the adrenal medulla and released into the circulation. Unlike many other hormones, the catecholamines circulate free, not bound to a transport protein. Further conversion to metanephrines and to vanillylmandelic acid (the final product of metabolism) takes place in the liver and other tissues. Conjugation of catecholamines and metabolites is an important means of inactivating and excreting these compounds.

Catecholamines are also stored in presynaptic membranes of nerve tissue. When a nerve impulse is propagated, the particular catecholamine involved in that system is released. The compound passes across the synapse to interact with specific receptors on the postsynaptic membrane, continuing the process of impulse propagation. Release from the receptor is followed either by inactivation (catalyzed by catechol-O-methyltransferase) or re-uptake by the presynaptic membrane. Inactivation at this site is under the control of monoamine oxidase.

Dopamine, epinephrine, and norepinephrine play important roles in regulating metabolism and nerve function. They are involved in control of heart rate, blood vessel tone, and some aspects of kidney function. In addition, there are complex interactions with some aspects of metabolism, including glucose synthesis, lipid breakdown, and protein turnover.

Catecholamine assay is a sensitive and complex process. Most current methodologies involve measurement of catecholamines or metabolites excreted in the urine. Proper sample collection is crucial to success in obtaining accurate measurements. Urine free (unconjugated) catecholamines can be measured after extraction on alumina and conversion to fluorescent derivatives. Ion-exchange chromatography and oxidation to vanillin serves as the method of choice for the colorimetric analysis of total metanephrines. Solvent extraction allows VMA to be separated from other reacting materials before oxidation to vanillin for measurement of this end product of catecholamine metabolism. Plasma catecholamines are determined using a radioenzymatic method which adds a tritiated methyl group to the 3-hydroxy portion of the catecholamine. Epinephrine and norepinephrine can be fractionated with this technique. Significant interest is being shown in the measurement of catecholamine fractions by HPLC, with electrochemical techniques being employed for detection and quantitation of each fraction.

Catecholamine production and release can be affected by a variety of physiological factors. Extreme physical activity, pain, stress, or upright posture increase plasma and urine catecholamine levels. Some medications simply interfere with assays for catecholamines, but a wide variety of pharmaceutical agents stimulate catecholamine output, raising the levels in body fluids.

Increased catecholamine production is seen in patients with pheochromocytoma, a tumor of the adrenal gland. Although this disease is responsible for only 0.1–0.7% of those patients with hypertension, detection is important since surgical treatment is usually quite successful. Measurement of urine metanephrines and VMA usually provides sufficient diagnostic information, although follow-up testing for urine free catecholamines is often warranted. A related tumor (neuroblastoma) can be detected using urine free catecholamine assays. This tumor does not produce consistently elevated levels of metanephrines or VMA.

The laboratory diagnosis of essential hypertension is still difficult. In some instances, plasma catecholamine levels are elevated and can be suppressed by clonidine. At present, there appear to be many factors besides catecholamines involved in this disorder.

Catecholamine levels change in a variety of other disease states. Patients with hypothyroidism usually demonstrate elevated levels of catecholamines, but hyperthyroid individuals have decreased values. Although patients with diabetic ketoacidosis have increases in catecholamine production, those with long-term diabetes usually show a lowering of catecholamine activity. Cardiac patients and those with burns or septicemia exhibit elevations in catecholamines. Patients with depressive illness have variable catecholamine levels, but the values are frequently increased.

FOR REVIEW

Directions: For each question, choose the best response.

1. The amino acid that is the immediate precursor for the biogenesis of the catecholamines is
   A. tryptophane
   B. threonine
   C. tyrosine
   D. phenylalanine

2. Which of the following may be classified as an end product of catecholamine metabolism?
   A. dopamine
   B. metanephrine
   C. normetanephrine
   D. vanillylmandelic acid

3. The primary site for catecholamine synthesis is the
   A. adrenal cortex
   B. adrenal medulla
   C. anterior pituitary
   D. pancreas

4. Which of the following may be classified as a catecholamine hormone?
   A. dopamine
   B. epinephrine
   C. norepinephrine
   D. all of the above

5. Homovanillic acid is the principal urinary metabolite of
   A. dopamine
   B. epinephrine
   C. metanephrine
   D. vanillylmandelic acid

6. All of the following represent metabolic effects caused by the action of the catecholamines *except*
   A. increased breakdown of triglycerides
   B. increased breakdown of liver glycogen
   C. decreased protein synthesis
   D. decreased blood glucose level

7. Which urine analysis may be performed initially to aid in the diagnosis of pheochromocytoma?
   A. HVA
   B. VMA
   C. metanephrine
   D. both A and B
   E. both B and C

8. In the test for VMA, false elevations may occur due to the presence of
   A. salicylates in the urine
   B. clofibrate in the urine
   C. an alkaline urine pH in the specimen container during collection
   D. metabolites of bananas and coffee in the urine

9. Which of the following methods may be used to quantitate plasma catecholamines?
   A. direct spectrophotometry
   B. HPLC fractionation
   C. radiometric assay
   D. both A and B
   E. both B and C

10. All of the following clinical conditions may be associated with increased catecholamine synthesis *except*
    A. hypertension
    B. hyperthyroidism
    C. pheochromocytoma
    D. neuroblastoma

11. Which urine analysis is considered the most reliable biochemical indicator of the presence of neuroblastoma?
    A. free catecholamines
    B. metanephrine
    C. HVA
    D. VMA

BIBLIOGRAPHY

Anderson, G. M., and Young, J. G., "Determination of neurochemically important compounds in physiological samples using HPLC," *Schizophr. Bull.* 8:333-348, 1982.

Axelrod, J., "Noradrenaline: Fate and control of its biosynthesis," *Science.* 173:598-606, 1971.

Bakes-Martin, R. C., "Urine catecholamine assays in suspected pheochromocytoma: A practical approach," *Lab Manage.* June:47-52, 1987.

Binder, S. R., and Sivorinovsky, G., "Analysis of urinary metanephrines by HPLC," *Amer. Clin. Prod. Rev.* June:34-41, 1986.

Bravo, E., "The clinical value of plasma catecholamine measurement," *Lab Manage.* June:53-69, 1982.

Bravo, E. L., and Gifford, Jr., R. W., "Pheochromocytoma: Diagnosis, localization, and management," *New Eng. J. Med.* 311:1298-1303, 1984.

Carmichael, S. W., and Winkler, H., "The adrenal chromaffin cell," *Sci. Amer.* August:40-49, 1985.

Christensen, N. J., "The role of catecholamines in clinical medicine," Acta Med. Scand, suppl. 624:9-18, 1979.

Falterman, C. J., and Kreisberg, R., "Pheochromocytoma: Clinical diagnosis and management," *South. Med. J.* 75:321-328, 1982.

Johnson, G. A. et al., "Single isotope derivative (radioenzymatic) methods in the measurement of catecholamines," *Metabolism.* 29 (suppl. 1):1106-1113, 1980.

Koch, D. D., "Urinary catecholamines," *Clin. Chem. News.* August, 1985.

Laverty, R., "Catecholamines: Role in health and disease," *Drugs.* 16:418-440, 1978.

McCann, D. S., and Huber-Smith, M. J., "Plasma catecholamines," *J. Clin. Immunoassay.* 6:308-312, 1983.

Molinoff, P. B., and Axelrod, J., "Biochemistry of catecholamines," *Ann. Rev. Biochem.* 40:465-500, 1971.

Nanji, A. A., and Campbell, D. J., "Pheochromocytoma: Which tests are best?" *Diag. Med.* Sept./Oct.:52-56, 1982.

Tracy, R. P., "Testing for the important catecholamines," *Clin. Chem. News,* February, 1980.

Waraska, J., "HPCL analysis of catecholamines," *J. Anal. Purif.* March:61-65, 1986.

Weinshilboum, R. M., "Biochemical genetics of catecholamines in humans," *Mayo Clin. Proc.* 58:319-330, 1983.

Young, D. S. et al., "Effects of drugs on clinical laboratory tests," *Clin. Chem.* 21(no. 5): April, 1975.

Chapter Seventeen

CLINICAL CHEMISTRY OF PREGNANCY AND BIRTH

• • • • • • • • • • • • • • • • • • • •

UPON COMPLETION OF THIS CHAPTER, THE STUDENT WILL BE ABLE TO

1. Describe the chemical composition of the protein hormone human chorionic gonadotropin.
2. Describe the purpose for measuring human chorionic gonadotropin (HCG) or the beta subunit of HCG to determine pregnancy.
3. Identify the structural similarities and differences among HCG, LH, FSH, and TSH.
4. State the principle of the urine agglutination test for detecting the presence of HCG.
5. Describe the use of radioreceptor, RIA, EIA, and IRMA techniques for measuring HCG in serum.
6. Contrast the progression of serum HCG levels seen in a healthy pregnancy with those seen in an ectopic pregnancy.
7. Describe the biochemical changes that occur during pregnancy.
8. Define the following terms:
   A. respiratory distress syndrome
   B. amniotic fluid
   C. jaundice
   D. kernicterus
9. Discuss the use of amniotic fluid to assess prenatal biochemistry.
10. Describe the collection and handling of amniotic fluid specimens.
11. Discuss the use of measuring estrogens to assess fetoplacental well-being.
12. Describe how surfactants aid in the respiration process, noting the major surfactant in the lung.
13. Describe the chromatography method for measuring the lecithin/sphingomyelin (L/S) ratio.
14. Describe the foam stability test for assessing the amount of surfactant in amniotic fluid.
15. Discuss the use of L/S ratio in assessing fetal lung maturity.
16. Describe why the excess accumulation of bilirubin in the blood of a newborn creates a major health risk.
17. Identify the enzyme that may be inactive in the newborn which is necessary for the conversion of bilirubin to a water-soluble form.
18. Describe the proper handling of serum or amniotic fluid specimens in order to protect the integrity of bilirubin in the sample.
19. Describe several methods used for quantitating serum bilirubin in newborns.
20. Describe how performing a spectral scan of amniotic fluid can contribute information about its bilirubin concentration.
21. Describe the urine screening procedures for metabolic disorders associated with the following compounds:
    A. glucose
    B. reducing substances
    C. keto acids
    D. sulfhydryl compounds

22. Describe the metabolic defect associated with phenylketonuria (PKU) and the clinical symptoms that occur when PKU is not detected at an early age.
23. Describe the ferric chloride and Guthrie assays used for detecting PKU.

INTRODUCTION

"And Adam . . . begat a son . . . and called his name Seth. . . ." (Genesis 5:3). Since the beginning of time, the process of pregnancy and birth has been one of fascination, mystery, and wonder. Although earlier generations had little useful science and technology at their disposal (but a great deal of folklore and practical knowledge), today we possess a variety of techniques for determining fertility, for detecting pregnancy, and for monitoring the development of the unborn child from the early weeks after conception through the time of birth and beyond.

The role of the laboratory in these processes is increasingly complex. Tests are available to help assess fertility or problems in conception. The burgeoning field of in vitro fertilization involves specialized laboratory studies and sophisticated techniques. Modern technology has provided us with an increasingly clearer picture of the developing child through all stages of pregnancy. Testing before birth permits life-saving medical intervention at earlier and earlier stages.

Our consideration in this chapter will be limited to some of the common laboratory contributions to the health of mother and child during the course of pregnancy. More sophisticated approaches, particularly to problems of fertility, are usually the domain of specialized laboratories and will not be discussed here.

17.1

BIOCHEMISTRY OF PREGNANCY

Beta HCG Measurements

The search for a rapid, inexpensive, and reliable means of determining pregnancy has been a long one. Since the days of the early Egyptians (if a woman vomited after drinking a mixture of pounded watermelon and breast milk from a woman who had a son, she was presumed to be pregnant), research has been carried out on the early detection of pregnancy. Current testing has focused on the measurement of **human chorionic gonadotropin (HCG)** and, more specifically, the beta subunit of this hormone **(beta-HCG).**

STRUCTURE OF THE MOLECULE

The protein hormone HCG is composed of two subunits—alpha and beta. These two polypeptides are linked together by a variety of noncovalent bonds, so they are easily dissociated under the right conditions. Each subunit is under separate genetic control and is synthesized independently of the other peptide. The alpha subunit appears always to be synthesized in amounts greater than those needed for full HCG formation, with the beta-subunit production being the limiting factor in determining the amount of intact hormone in the system. Each subunit contains several carbohydrate groups which are necessary for hormone activity, but which do not appear to be involved when HCG is assayed.

The alpha subunit of HCG is essentially identical in structure to the corresponding alpha subunits of **luteinizing hormone (LH), follicle-stimulating hormone (FSH),** and **thyroid-stimulating hormone (TSH).** The structure of the beta subunit is very similar to that of LH, but the beta subunit has an additional 30-amino acid portion which is responsible for the biochemical specificity of HCG. Since the alpha subunits of these four hormones are identical, assays for HCG must focus on measuring the beta subunit in order to avoid significant cross-reactivity.

PRODUCTION DURING PREGNANCY

Under normal circumstances, HCG is not detected in the urine or serum of a nonpregnant woman. When pregnancy occurs, the HCG level begins to be seen as early as 10 days after the preovulation LH surge (Fig. 17–1). Values rise rapidly during the early course of pregnancy, often doubling every other day. Depending on the specific test and its sensitivity, the increase in plasma HCG associated with pregnancy may be detected as early as 7 days after fertilization. Urine tests are less sensitive and do not show a positive response to the presence of HCG until 30–35 days after fertilization. The serum levels continue to rise markedly to a peak sometime between 60 and 80 days into the pregnancy, then decline significantly during the last trimester.

ASSAY TECHNIQUES

The clinical laboratory has come a long way from the classic **rabbit test** for pregnancy, in which urine

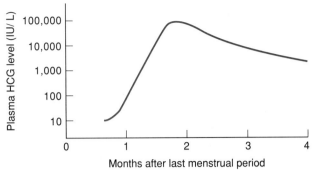

FIGURE 17–1 Changes in HCG concentration during pregnancy.

was injected into a rabbit and the presence of ovulation and formation of the corpus luteum was observed after the animal was sacrificed. Today a wide variety of immunological testing procedures are available for rapid detection of HCG. However, with these advances have also come problems in standardization and reproducibility.

Urine

The early chemical/immunological tests for pregnancy assessed the presence of HCG in the urine. Agglutination tests were used which allowed collection of a random urine sample (Fig. 17–2). In general, antiserum to HCG is added to the urine. If HCG is present in the sample, an antigen/antibody complex forms, binding to the anti-HCG antibody. A particle coated with HCG is then added (latex-coated particles are common, but erythrocytes are also frequently used). In a positive test, the absence of agglutination indicates that HCG is present in the patient sample. The added antibody attaches itself to the patient HCG and is not available to react with the HCG-carrier material added later. If agglutination occurs, it indicates that there is little or no patient HCG present, allowing the antibody to react with HCG on the carrier material.

The urine tests are simple to perform and are fairly rapid; an assay may require as little as 2 min or up to 2 h (mainly incubation time). These assays are normally used for a qualitative yes/no answer, although semiquantitative results can be obtained through the use of serial dilutions. False positive results may be seen if there is an increased amount of protein in the urine. A dilute urine may produce a false negative result, since the concentration of HCG might be lower than that expected in a more concentrated sample. Phenothiazines, barbiturates, chlorpromazine, methadone, and penicillin can produce false positive results.

Serum

The initial serum test for measurement of beta-HCG used a radioreceptor assay. Analogous to the familiar RIA approach, the test involved measurement of the competition between patient beta-HCG and HCG labeled with ^{125}I for binding to beta-HCG receptors in a tissue. The greater the amount of patient HCG, the less radiolabeled material bound, giving an inverse relationship between radioisotope binding and concentration of HCG. Although this technique allowed rapid measurement in serum and provided an earlier indication of a pregnancy (a positive test could be obtained within 7 to 10 days after conception), simpler and more sensitive immunoassay techniques soon supplanted this early methodology.

Currently, a wide variety of techniques have been applied to beta-HCG testing, for qualitative positive/negative results and quantitative determinations to monitor HCG levels in an ectopic pregnancy or for diagnosis of the presence of particular tumors. Radioimmunoassay, EIA techniques, and IRMA have all been employed for HCG measurement. Most of these assays assess the presence of the intact molecule, not simply the beta subunit.

Specificity of antibody is still somewhat of a problem in developing a truly accurate HCG assay. The molecule itself is rather heterogeneous, with reports of "big" HCG (a precursor form) and fragments frequently being detected. Cross-reactivity with LH has not been totally eliminated and occasionally presents difficulties if samples are collected midway through the menstrual period when LH levels can be quite high. If a patient is taking a medication which can produce increased serum LH values, some cross-reactivity with an HCG antibody may take place.

Problems in Standardization

Standards of hormones, unlike those for glucose or other molecules with relatively simple structures, cannot be prepared "off the shelf" with a highly purified reagent-grade standard material. The hormone must be isolated from a natural source and purified before use as a standard. Thus, the possibility exists that standards for a given hormone may vary from one another, depending on the purity of the material and the amounts of free subunits present.

The standardization crisis for HCG has come about due to the use of at least two different primary standards for assays. The initial standard was only

A   *P*-HCG + anti-HCG ⟶ *P*-HCG/anti-HCG complex

B   Anti-HCG + Carrier-HCG ⟶ Carrier-HCG/anti-HCG complex

FIGURE 17–2 Principle of urine hemagglutination HCG test. A. HCG present in patient urine. *Note.* Carrier-HCG cannot react to form an aggregate. B. No HCG in patient urine.

20% pure, leading to some specificity problems. The issue of cross-reactivity, particularly with LH, became critical as researchers attempted to raise antibodies specific only for the beta subunit of HCG. Years of clinical data were accumulated using this less desirable standard material. More recently, new standards have been developed that are purer and less contaminated by fragments. In addition, new antibodies are being used which are specific for the intact HCG molecule, not just the beta subunit. Data from these improved tests often do not correspond to numeric values provided by the earlier methodologies. The issues of standardization and assay agreement are still being resolved. Laboratory staff need to be very aware of the specifics of the HCG test they are using. The goal is to minimize confusion with regard to comparing sequential data on a patient and appropriate clinical interpretation.

HCG AND ECTOPIC PREGNANCY

After pregnancy begins, the HCG levels in serum double every few days. This rapid rise occurs for approximately 7 weeks and then begins to slow. During early pregnancy, however, the rate of HCG production in some patients may be very slow and levels may not reach the same high concentrations seen in a normal pregnancy. These data suggest the presence of an ectopic pregnancy. In this situation, the fertilized ovum does not implant in the uterus to begin normal development, but in some 95% of ectopic pregnancies, the ovum implants in the fallopian tube. Conditions are not suitable for fetal growth and a variety of biochemical and clinical abnormalities begin to develop. The production of HCG may continue, but at a much lower rate than expected. This decline in the rate of HCG synthesis is a warning sign of a possibly life-threatening situation.

HOME PREGNANCY TESTS

A variety of home pregnancy tests are available in pharmacies. These urine assays for HCG are usually based on some type of enzyme immunoassay procedure, which is usually fairly simple, with a color change indicating a positive reaction. In spite of reasonably clear written instructions, these tests have unacceptably high false positive and false negative rates due to problems in technique and failure to use an appropriate sample. Many pharmacists are not familiar with the tests and cannot provide good technical information. In addition, use of home pregnancy tests often delays entry into the healthcare delivery system if the woman is pregnant. A prevalent attitude is "What can the doctor tell me that I don't already know?" The tests are somewhat expensive and represent a false economy; a physician will need to have the test repeated by a laboratory when the patient is first seen for a prenatal visit. Laboratory personnel need to be familiar with these home tests, the techniques they employ, and their shortcomings. A trained technical staff can be of great help in providing reliable consumer information and in encouraging patients to obtain proper testing and care.

Biochemical Parameters and Alterations in Pregnancy

A wide variety of biochemical changes take place during pregnancy, often resulting in noticeable changes in reference ranges for many serum and urine constituents (Table 17–1). Serum protein levels shift, with a striking drop in albumin. This decrease probably results from the increased protein demand by the growing child. Immunoglobulin levels rise somewhat, perhaps in response to the need for increased immunological protection for both mother and child. Alkaline phosphatase values increase over the course of pregnancy, in part due to bone development but primarily as a result of elevations of placental alkaline phosphatase during the last trimester. Increases in most serum lipid fractions are seen as the body mobilizes lipids for nutrition for the child.

A variety of hormonal changes take place as pregnancy progresses. Both LH and FSH production diminish markedly since ovulation is no longer occur-

Table 17–1
BIOCHEMICAL CHANGES DURING PREGNANCY

Parameter	Observed Change
Albumin	Decrease
Immunoglobulins	Some increase
Alkaline phosphatase	Increase, especially during last trimester
Lipids	Increase
FSH and LH	Decrease
T3 and T4	Slight increases
Parathyroid hormone	Increase
Vitamin D	Increase
Estrogens	Increase, especially last trimester

ring. Slight increases in T3 and T4 values reflect the increased metabolic demands on the mother. Elevations of parathyroid hormone and vitamin D values provide some index of enhanced bone growth during this period. Steroid hormone concentrations rise, with changes in cortisol derivatives reflecting a variety of metabolic alterations in the mother. Estrogen increase, especially during the last trimester, is a reflection of the enhanced feto/maternal synthesis of these steroid hormones.

17.2

AMNIOTIC FLUID BIOCHEMISTRY

A pool of liquid surrounds the unborn child during its development in the mother. This fluid is a combination of filtrates from the maternal circulation and fetal metabolic and physiologic processes. At term, it has been estimated that the fetus swallows approximately 500 mL of amniotic fluid per day. Urine production accounts for some of the fluid volume and definitely affects the composition of this material.

Amniotic fluid plays several important roles during the child's prenatal development. The liquid serves as a cushion, providing some protection against the shocks and jolts which might be experienced during the pregnancy. The free flow of amniotic fluid lubricates the amniotic sac, minimizing irritation. There are undoubtedly other significant roles for amniotic fluid which have not yet been discovered.

The laboratory can use amniotic fluid as an indirect reflection of some aspects of prenatal biochemistry. Since the fluid is inhaled, internal lung secretions partially dissolve in the material and can be measured. Urine production and release by the fetus alter the osmolality and composition of the amniotic fluid, often providing valuable clues to fetal functioning. Measurement of hemoglobin or specific enzymes allows the detection of a variety of metabolic disorders, including sickle-cell anemia, Tay-Sachs disease, glycogen storage problems, and mucopolysaccharidoses.

Collection of amniotic fluid involves a certain level of risk to baby and mother, so is performed only by a physician. After the fetus is located by ultrasound, a needle is inserted into the mother's abdomen. Several milliliters (10–15 mL are usually adequate) of amniotic fluid are withdrawn and placed in a tube. The tube should be sealed and protected from light to avoid deterioration of any bilirubin present, and the sample should be refrigerated until assayed. Measurement of amniotic fluid constituents should be carried out as soon as possible to minimize any breakdown of the fluid constituents. For long-term storage, freezing is usually acceptable, depending on the tests to be performed.

17.3

FETAL MATURITY AND VIABILITY

Estrogen Production and Fetal Viability

Although the assessment of estrogen levels (or estrogen excretion) involves a urine or serum sample instead of amniotic fluid, measurement of this constituent can provide a fairly reliable picture of some aspects of fetal maturity. Because of the biochemical interactions between mother and unborn child leading to the increased production of estrogens during pregnancy, examination of the levels of these steroids can provide a rather unique look at the developing fetus.

SOURCE OF ESTROGENS DURING PREGNANCY

In the nonpregnant woman, estrogens are produced by the ovaries, operating at low levels of synthesis. Total estrogen excretion in the urine may range from 4 to 100 μg/day, depending on the phase of the menstrual cycle. However, during pregnancy, estrogen production rates rise markedly, reaching levels greater than 40 mg/24 h at term, close to a 500-fold increase over nonpregnant rates. This increased synthesis is caused by joint participation of mother and child in the biochemical production of estrogens.

The relative amounts of estrogens produced vary in the pregnant and nonpregnant individual. The nonpregnant woman synthesizes mainly estradiol (only in microgram amounts). During the 9 months of pregnancy, major production shifts over to estriol, with only small amounts of estradiol and other similar materials being synthesized. This shift takes place as the site of estrogen production moves from the ovaries to the placenta during pregnancy.

The major precursor of estrogens during pregnancy is a steroid known as **dehydroepiandrosterone sulfate (DHEA-S)**. This material is synthesized by both mother and child. The DHEA-S from the mother moves to the placenta for further conversion. The fetus either releases DHEA-S to the placenta or adds a hydroxyl group to the molecule before transporting this derivative to the placenta. A number of biochemical reactions take place in the placenta resulting in the synthesis of estriol and other derivatives. Both unconjugated and conjugated forms of these steroids are produced through the placental reactions. Only when the entire maternal/fetal system is intact does the enhanced production of estrogens take place.

ANALYTICAL CONSIDERATIONS

There is considerable debate about the best approach for assessment of estrogen production. The classic technique requires the collection of a 24-h urine specimen for measurement of urine total estrogens by the Kober technique, previously discussed in Chapter 15. Major drawbacks with this approach are the time needed for collection of the sample, problems in obtaining a proper specimen, and the lengthy analytical procedure. Analysis of plasma estriol represents an alternative which is becoming increasingly useful.

A 24-h urine specimen is required for most urine estrogen studies. If a repeat study is needed, a lapse of a day between the first collection and the initiation of the second collection is common (due to the time needed for analysis and reporting). Since decreases in estrogen output presage problems with the pregnancy, this delay is often not acceptable. Some recent studies have indicated that measurement of estriol and creatinine in a random urine specimen provide a fairly reliable index of estrogen excretion while decreasing the time required for sampling.

Since most specimens for estrogen analysis during pregnancy are collected on an out-patient basis, obtaining a proper 24-h sample is often difficult. Inadequate explanation of how to collect a specimen, and patient's failure to follow instructions properly or personal reluctance in handling a urine specimen all contribute to incomplete collection. Many laboratories routinely measure creatinine to assess whether or not a complete 24-h specimen was obtained.

The analytical methodology is often lengthy. Whether a lab employs a colorimetric approach or an RIA procedure, initial treatment of the specimen involves hydrolysis to destroy water-soluble conjugates so the entire estrogen mixture can be quantitatively extracted into an organic solvent. Enzymatic methods are lengthy (and sometimes incomplete), and acid hydrolysis approaches result in some destruction of material as well as adding time to the overall procedure.

If plasma samples are analyzed, some of the same hindrances still exist. To assay for plasma total estriol, an initial hydrolysis step (usually enzymatic) is required. Since production of estriol is somewhat irregular over the course of any given day, plasma values tend to fluctuate more than is the case when urine specimens are used. However, use of plasma eliminates the problems associated with collection of a timed urine specimen.

CLINICAL SIGNIFICANCE

During the first two trimesters of pregnancy, estrogen values rise slowly and are not of any great clinical utility. In the third trimester, a sharp and steady increase of estriol is seen (Table 17–2), which can be of great value in monitoring the integrity of the fetoplacental unit. A fairly steady rise indicates that development of the unborn child is proceeding

Table 17–2
URINE ESTRIOL OUTPUT DURING PREGNANCY

Gestational Age (weeks)	Total Estriol (mg/24 h)
24	4–12
28	6–20
32	8–29
34	9–34
36	10–36
38	11–39
40	12–40 or >

Adapted from Tietz, N. W. (ed.), *Textbook of Clinical Chemistry.* Philadelphia: W. B. Saunders Co., 1986, p. 1823.

well. Erratic changes in values or declines in estrogen production signify complications with the pregnancy. A drop in estrogen output of 30% or more indicates poor fetal development and warrants medical intervention.

Estrogen levels may stop rising or decline for several reasons. Patients with diabetes mellitus often have complicated pregnancies. Hypertension is frequently associated with developmental problems in the fetus. A pregnancy which extends beyond the usual 42-week period is often associated with sharp drops in estrogen production. Babies who are **small for date** (poor fetal development) may be detected by changes in estrogen levels along with other testing procedures. If previous pregnancies had complications, monitoring of estrogen output during subsequent pregnancies is usually warranted.

Fetal Lung Maturity

SURFACTANT DEVELOPMENT AND RESPIRATORY DISTRESS SYNDROME

A major concern for fetal viability is lung function: Will the child be able to breathe normally after birth or will there be significant respiratory problems? Factors affecting respiration in the newborn are of particular importance when deciding whether to induce labor earlier than usual or when delivery is premature.

The child's ability to breathe without assistance depends on many factors. One major consideration is the degree of development of materials on the inner surface of the lungs. These biochemicals are called **surfactants** because they reduce the surface tension of the pulmonary mucosa, forming air bubbles. They coat the inner lining of the lung. During respiration, as the air leaves the lung, the inner linings come in close contact with one another. Without the surfactant, the linings stick together. At the next inhalation, the lung does not expand appropriately and the baby cannot take in sufficient air. This problem, called **respiratory distress syndrome,** is of particular concern in a premature birth.

BIOCHEMISTRY OF LUNG SURFACTANTS

Surfactants are a group of compounds known as **phospholipids.** One portion of the phospholipid molecule can interact with water-soluble materials, while another part can attach to nonwater-soluble materials, thus enabling the regulation of surface tension in the lung.

The major surfactant is a compound called **lecithin,** which makes up some 70% of the total phospholipid content in the lungs. A variety of other related phospholipids have also been identified, including sphingomyelin, phosphatidylglycerol, phosphatidylserine, phosphatidylinositol, and lysolecithin. The characteristic structure of a phospholipid is illustrated by the lecithin molecule (Fig. 17-3). The glycerol backbone contains two long-chain fatty acids in ester linkage at carbons-1 and 2. These organic chains allow the molecule to interact with other lipid components in the system. The third glycerol carbon is linked to a phosphate group (providing some ionic charge to the molecule). This phosphate linkage is then attached to a choline group, containing a positively charged nitrogen atom. The presence of these two ionized groups permits water-solubility. The overall structure thus creates a "bridge" between aqueous and nonaqueous systems, the environment of the internal structure of the lung.

While the fetus is developing, synthesis of lecithin takes place at a low rate during the first 26 weeks or so. After this time, the rate of lecithin formation begins to increase, with sharp rises in concentration occurring between 34 and 38 weeks' gestation. Then, the amount of surfactant material in the lung is sufficient for normal respiration in the newborn child. If birth takes place before this rapid increase in lecithin synthesis, the immature child develops respiratory distress syndrome.

LABORATORY ASSESSMENT OF AMNIOTIC FLUID SURFACTANTS

Because of the technical problems associated with lecithin quantitation and the impracticality of tracking the base-line levels, measurement of lecithin is linked to the assessment of the amount of

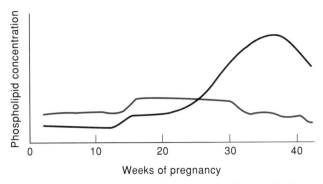

FIGURE 17-4   Lecithin/sphingomyelin changes during pregnancy. (Lecithin, black curve.)

sphingomyelin present (Fig. 17-4). This latter phospholipid is also formed during pregnancy but does not show the sharp rise in concentration during the last few weeks that lecithin does. **Sphingomyelin** serves as a base line for monitoring lecithin changes in the amniotic fluid. The usual parameter assessed is the **lecithin/sphingomyelin (L/S) ratio,** an index of the relative amounts of lecithin and sphingomyelin present in the amniotic fluid at any given time. Other types of studies have been carried out in attempts to monitor fetal maturity, but results obtained from these investigations are usually related to L/S ratios for comparison.

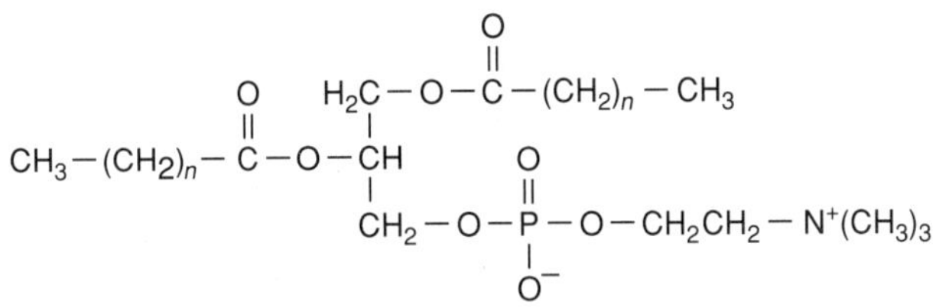

Sample Preparation

A key factor in correctly measuring the L/S ratio is the proper treatment of the sample. Most procedures call for an initial centrifugation to remove cellular debris before analysis is carried out. Unfortunately, a wide variety of times and centrifugation speeds have been published. In general, the longer the centrifugation and/or the higher the speed of the centrifuge, the more phospholipids precipitate and are lost for analysis. Lecithin is particularly susceptible to decreasing in concentration during centrifuging. Whatever the procedure, it must be carefully standardized to obtain reproducible results.

After the sample is spun and decanted to isolate the supernatant, this fluid must be gently and thoroughly mixed before analysis. Some layering of the phospholipid occurs during centrifugation which can alter the final results appreciably. It appears that refrigeration is preferable to storing of the sample at room temperature. The presence of blood or meconium can adversely affect the measurements. The effects of these contaminants are variable and cannot readily be predicted for any given sample.

Measurement of Lecithin/ Sphingomyelin Ratio

The phospholipids in the amniotic fluid sample are first extracted with an organic solvent, usually a 1:2 mixture of methanol and chloroform. Although

FIGURE 17-3   Structure of lecithin.

this method does not remove all lipids, lecithin and sphingomyelin seem to be adequately extracted. The extract is heated to concentrate the solvent, then is spotted on a thin-layer chromatography plate. Several solvent systems have been explored, but chloroform/methanol/water is most widely used. Lecithin and sphingomyelin separate during the chromatography. After drying, the plate is sprayed with a mixture of sulfuric acid and ammonium sulfate. The plate is then heated to 280°C until charred spots develop where the phospholipids are located. Scanning with a reflectance densitometer allows assessment of the relative amounts of lecithin and sphingomyelin present, permitting calculation of the L/S ratio.

Clinical Utility of Lecithin/ Sphingomyelin Ratio

The L/S ratio can provide a rough indication of fetal lung maturity. An L/S ratio greater than 2 suggests there is sufficient surfactant present for the baby to breathe normally. L/S ratios below 1 strongly indicate lung immaturity, and ratios between 1.5 and 2 are equivocal.

This test cannot be used with a high degree of confidence because of the many variables involved in the analysis. In babies born to mothers with diabetes mellitus, a variety of factors affect the L/S ratio. Values above 2 are not always clear indicators of fetal lung maturity in these cases.

Other Tests of Fetal Lung Maturity

In attempts to refine and improve the L/S ratio measurement, direct quantitation of phospholipids or measurement of other phospholipids has been explored. Use of two-dimensional thin-layer chromatography is common in the measurement of lipids other than lecithin and sphingomyelin. Increases in phosphatidylglycerol appear to indicate fetal lung maturity, even if the L/S ratio is below 2. Techniques for quantitation of this and other phospholipids need to improve before these tests can be more widely used.

The **foam stability test (shake test)** provides a simple, although not very reliable, method of assessing the amount of surfactant material in amniotic fluid. An equal volume of 95% ethanol is added to the amniotic fluid sample, the mixture is shaken to produce foam, and allowed to stand for 15 min. If a complete ring of bubbles remains on the top of the fluid, sufficient surfactant material is thought to be present for mature lung function. Incomplete rings of bubbles (or no ring) indicate inadequate phospholipid concentration. There appears to be a reasonably good correlation between the foam stability test and the L/S ratio. Variability of technique and interpretation decrease the reliability of the foam stability test, but it does serve as a useful initial screen until more reliable results are provided.

17.4

BILIRUBIN AND NEONATAL PROBLEMS

The measurement of serum bilirubin levels in adults and children involves few technical problems while providing useful information regarding liver function, but the changing metabolic dynamics of the newborn present unique challenges to the laboratory asked to quantitate neonatal bilirubin levels. The combination of small sample volumes and the need for quick turnaround times make demands not encountered when routine measurements of bilirubin levels are needed. Since there is a very real risk of irreversible brain damage to the newborn child if bilirubin levels exceed a certain concentration, the laboratory must provide rapid, accurate assessments of bilirubin status for these special patients.

Bilirubin Formation and Excretion in the Neonate

The major source of bilirubin is hemoglobin released by the breakdown of mature red blood cells at the end of their life span. Another 25% of the bilirubin in the circulation results from ineffective formation of erythrocytes with release of the heme proteins into the bloodstream. In both instances, the heme portions of the molecules are broken down, first to biliverdin and then to bilirubin. This rate of breakdown may be four to five times greater than that in an adult, resulting in higher concentrations of bilirubin than those seen in older patients. The concentration may also change more rapidly because of this enhanced metabolic conversion of heme to bilirubin.

After bilirubin is formed, it rapidly attaches to serum albumin, either as a complex (**alpha bilirubin**) or with a covalent bond (**delta bilirubin**). Further conversion takes place in the liver as the bilirubin is uncoupled from the albumin by specific proteins. The water-insoluble bilirubin is then transformed to a water-soluble form (conjugated bilirubin) by the enzyme **uridine diphosphate glucuronyltransferase (UDPGT),** which adds a glucuronic acid molecule to the bilirubin. In the newborn, this enzyme is not active during the first few days after birth. Therefore, little water-soluble bilirubin is formed and excreted. The net result is a transient increase in total bilirubin concentration until the bilirubin can be conjugated and eliminated in the urine.

Conjugated bilirubin can also be excreted into the intestine. Here it undergoes further metabolism by bacteria to produce urobilinogens, which are then eliminated from the body in the feces. Since the newborn has no intestinal flora, this excretory pathway is absent. The bilirubin present in the gastrointestinal tract is then reabsorbed through the enterohepatic circulation, once again entering the bloodstream as unconjugated bilirubin.

Neonatal Hyperbilirubinemia

The major concern for the newborn with elevated bilirubin is the possibility of **kernicterus** (German for "nuclear jaundice"). In this situation, the high amounts of unconjugated bilirubin are deposited in the brain tissue, causing neurologic damage. The unconjugated form of bilirubin is very lipid-soluble and is readily taken up by the highly lipid brain cells. On the other hand, conjugated bilirubin is not very lipid-soluble and does not produce kernicterus.

There is still some controversy about the exact levels the unconjugated bilirubin must reach in serum before kernicterus occurs. The usually accepted value is somewhere between 15–20 mg/dL for total bilirubin. If the serum bilirubin concentration exceeds this level, exchange transfusion is initiated to remove the bilirubin physically from the system before damage can occur. However, kernicterus can take place at much lower levels (sometimes as low as 10–12 mg/dL total bilirubin). Newborns with elevated bilirubin values may also be treated by **phototherapy**. Exposure to ultraviolet light produces photodegradation of the bilirubin in the circulation, promoting a decrease in the serum concentration.

A number of clinical situations result in elevated plasma bilirubin levels (Table 17–3). Newborns of low birth weight (usually due to premature birth) tend to have higher bilirubin concentrations than full-term, normal-weight children. Babies who are breast-fed demonstrate increases in bilirubin concentration after a few days. This increase has been attributed to inhibition of the glucuronyltransferase reaction by a pregnanediol derivative found in breast milk and by the high concentrations of unsaturated fatty acids released from the milk lipids. There are also some rare metabolic disorders associated with increases in bilirubin levels. The biochemical deficiencies may be in the conjugation step or in a decreased level of one or more of the proteins involved in the release of bilirubin from albumin. Patients with hemolytic anemia due to a deficiency of an erythrocyte enzyme (such as glucose-6-phosphate dehydrogenase or pyruvate kinase) also demonstrate increased serum bilirubin concentrations.

Table 17–3
CAUSES OF NEONATAL HYPERBILIRUBINEMIA

Low birth weight	Congenital deficiency of
Breast feeding	erythrocyte enzymes
Rh incompatibility	ABO incompatibility
Congenital defects in bilirubin	
metabolism or transport	

Perhaps the major causes of neonatal hyperbilirubinemia are ABO and Rh incompatibilities between mother and child. ABO incompatibility can occur if the baby inherits the father's blood type and that blood type is incompatible with the mother's. Antibodies from the mother promote significant fetal hemolysis, leading to bilirubin build-up in the bloodstream. Although this situation can give rise to problems, it is usually much milder and less severe than the Rh incompatibility. The antigen/antibody reaction in a case of Rh incompatibility is often severe, producing striking elevations in serum bilirubin concentrations. Early identification of mothers susceptible to this problem and treatment with Rh immune globulin have significantly decreased the incidence of this particular medical problem.

Assay for Bilirubin in the Neonate

Most assays for neonatal bilirubin employ the classic Jendrassik-Grof approach to bilirubin quantitation for both total and direct bilirubin, although some laboratories prefer the Evelyn-Malloy procedure. In either case, consideration must be given to a microchemistry procedure, since only a small sample volume is possible from newborns. Because high concentrations of total bilirubin may be expected, standard curves should be linear to at least 25 mg/dL. The use of an elevated bilirubin control in the 15–20 mg/dL range is strongly recommended to assure accuracy at higher concentrations. Since most newborns with problems have serial assays over several days, accuracy and reproducibility are important in the high range.

Many facilities employ a direct-reading spectrophotometric method for total bilirubin in which the absorbance of serum at 455 nm is measured. Other interfering substances (such as carotenes) are usually not present in sufficient quantities in newborns to be a problem. Quality control and agreement with chemical methods are extremely important if this method is employed, since data from both procedures are often compared during the course of patient treatment.

Measurements of total and conjugated bilirubin do not always give a reliable estimate of the reserve binding capacity of albumin in the serum of the newborn. A variety of procedures have been explored in attempts to determine directly how much more bilirubin could bind to albumin. This information allows for better assessment of bilirubin binding status and monitoring of treatment.

Many of the efforts involve fractionation on Sephadex columns after excess bilirubin has been added. Although useful clinical data on binding capacity can be obtained from these studies, relatively large samples are required and the procedure is lengthy. In addition, several assays need to be run in a single assessment, creating problems of cost, time, and total sample requirements.

More recently, a direct fluorometric assay has been reported and is under commercial development. Only bilirubin which is strongly bound to albumin fluoresces. Measurement of fluorescence can be made on a whole-blood sample before and after an excess of bilirubin is added to the blood. The difference in fluorescence is an indication of the amount of albumin which is not yet saturated by bilirubin from the patient. The method is rapid and reliable, requiring small blood samples. Total bilirubin can also be measured with this device. A detergent is added to the sample, forming a complex with all bilirubin present. The bilirubin/detergent complex demonstrates fluorescence proportional to the bilirubin concentration.

Amniotic Fluid Bilirubin

On occasion, the clinical history of the mother may indicate the possibility of Rh incompatibility prior to the birth of the baby. Since early assessment of this disorder is important for treatment, measurement of bilirubin in the amniotic fluid is sometimes employed to detect the medical problem.

The amniotic fluid sample is collected by the physician and should be brought to the laboratory as soon as possible. Since bilirubin is light-sensitive, the sample tube should be wrapped in aluminum foil or heavily taped to prevent light exposure. To enhance stability further, the fluid should be kept on ice until the analysis is performed. Under these conditions, bilirubin is stable for at least 1 month.

Bilirubin has an absorbance maximum at 455 nm (Fig. 17-5). To compensate for background absorbance, a scan must be made in the wavelength range of 350–600 nm. The background absorbance shows a steady rise as shorter wavelengths are approached. If bilirubin is present in measurable amounts, a broad peak centering at 455 nm is observed. Strong absorbance at approximately 410 nm indicates the presence of hemoglobin. The difference between the peak 455-nm absorbance and the baseline absorbance at the same wavelength indicates the concentration of bilirubin present. Values are plotted on a Liley graph against the assumed gestational age to determine the presence of a hemolytic problem in the fetus.

17.5

NEONATAL SCREENING

Concept of Metabolic Disorders

Although most diseases are acquired, for the newborn there are a number of medical complications which are inherited. The genetic problem leading to the specific disorder is a part of the child's DNA make-up from conception. These malfunctions in normal biochemistry are called **metabolic disorders,** or **inborn errors of metabolism.**

The foundation for our understanding of metabolic disorders was laid by a British physician, Archibald Garrod, in the very early years of this century. He postulated that various hereditary disorders might be produced by a deficiency of a specific enzyme involved in a particular metabolic pathway (Fig. 17-6). Since this enzyme is present in greatly decreased amounts, certain metabolites accumulate to high concentrations, unlike the low levels which occur if the metabolite is being converted to other products in the normal flow of metabolism. This proposal, considered rather radical at the time, has provided the direction for almost a century of research in this frustrating, but challenging, scientific arena.

Metabolic disorders would be little more than interesting scientific anomalies if the consequences were not so disastrous in many situations. When a metabolite normally present in minor concentration begins to accumulate, it quite often produces toxic effects. Usually the medical consequences of this increase in a minor biochemical component include mental retardation, brain damage, and poor physical development. If detected shortly after birth, the chances for healthy development are considerably enhanced through proper therapy (often involving a change in diet). Failure to identify the biochemical defect in time generally leads to the development of irreversible physical and mental impairment.

A wide variety of metabolic disorders exist; well over a thousand distinct disorders have been identified to date. Impairment of metabolic pathways runs the gamut of carbohydrates, lipids, amino acids, and hereditary lack of specific proteins. In time, we will probably see a defined metabolic disorder linked to each enzyme in every known metabolic pathway. Although the frequency of any par-

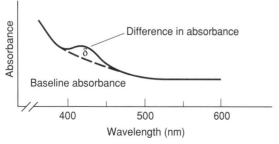

FIGURE 17-5 Spectrum of bilirubin in amniotic fluid.

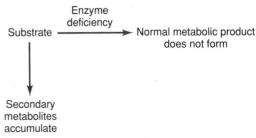

FIGURE 17-6 Concept of metabolic disorder.

ticular disorder is quite low, the combined effect of all the known metabolic problems can be very significant in its impact on the health of a group of people.

Urine Screening for Metabolic Disorders

It is obviously impractical to screen routinely for all known metabolic disorders. However, a few basic screening procedures are available which permit detection of many of the more common diseases (Table 17–4). These initial tests require a random urine sample and are quite simple to perform. Although a positive test does not pinpoint a particular disorder, it sends a warning signal which should initiate further testing to identify the medical problem.

In addition to testing which may be routinely carried out by the hospital, most (if not all) states have certain mandatory testing procedures for all newborns. The two most common state-mandated screening tests are for phenylketonuria (a disorder of amino acid metabolism) and for low thyroid hormone production. The incidences of these two disorders is sufficiently high that testing of all newborns is a cost-effective way of detecting the disorders. Treatment can then be initiated early enough to prevent or diminish the adverse mental and physical effects produced by the metabolic problems.

SAMPLING FOR METABOLIC SCREENING

For routine screening, a random urine sample is sufficient. When the sample is collected, information about food intake and medications should be recorded (particularly any intake prior to or at the time the sample is being taken). Dietary intake often affects the levels of specific biochemical intermediates, and some medications produce interferences in the testing procedures.

Both urine pH and bacterial contamination should be checked prior to carrying out any specific

Table 17–4
URINE SCREENING FOR METABOLIC DISORDERS

Class of Compound	Screening Procedure
Glucose	Glucose oxidase (dipstick)
Reducing substances	Copper reduction
Keto acids	Ferric chloride
	2,4-dinitrophenylhydrazine
Sulfhydryl compounds	Nitroprusside

chemical tests. An alkaline urine may indicate the presence of bacteria and often interferes with proper chemical reactions. Screening the sample with the nitrite portion of a commercial dipstick indicates the presence of nitrite-producing bacteria.

CARBOHYDRATE TESTING

Testing for glucose (using an enzyme reaction) and a nonspecific assessment of reducing substances in the urine may indicate the presence of high levels of some carbohydrate. The nonspecific assay should employ some variation of the copper reduction method so that carbohydrates other than glucose can be detected. Tablets are commercially available for this measurement. False positive results may occasionally be seen with vitamin C or some antibiotics.

AMINO ACID SCREENING

The screening procedure used to detect a disorder of amino acid metabolism or excretion tests not for amino acids, but rather for the keto acids formed through alternative metabolic pathways. When the conversion of a specific amino acid is blocked, a separate aminotransferase reaction often produces large quantities of a keto acid.

The presence of a keto acid can be observed using the ferric chloride test. A mixture of ferric chloride and ferrous ammonium sulfate in dilute HCl is added to the urine. In the presence of a keto compound, the iron forms a colored complex. This assay is particularly useful for the detection of phenylpyruvic acid, formed in phenylketonuria when phenylalanine conversion to tyrosine is blocked. The ferric chloride test gives a false positive result in the presence of salicylates, bilirubin, or acetoacetate (from diabetic ketoacidosis). The test is unreliable in the presence of alkaline pH or high phosphate concentration.

The use of 2,4-dinitrophenylhydrazine is also recommended for the detection of some keto compounds. An insoluble yellow precipitate is formed after the urine specimen is mixed with the phenylhydrazine derivative and allowed to react for 5 min. Samples containing the anticonvulsant valproic acid may produce a false positive test.

One specific metabolic disorder (cystinosis) involves the accumulation of sulfur-containing amino acids. The sample is treated with concentrated ammonium hydroxide and sodium cyanide to break down the disulfide groups and form the corresponding $-SH$ derivatives. Addition of sodium nitroprusside produces a magenta color in the presence of sulfhydryl groups. Increases in the concentration of cystine and homocystine produce positive results. If the urine specimen has a high osmolality, a false positive test may occur.

CONFIRMATORY TESTING

The preliminary screening tests outlined above are just that—preliminary. It is clear that no specific disorder can be identified conclusively on the basis of these measurements, although the presence of a problem can be indicated. For positive identification of the particular metabolic defect, a variety of sophisticated assays must be employed. These tests are usually carried out at special reference laboratories.

The first step involves the characterization of the metabolic disorder itself. The elevated metabolite must be identified for a proper diagnosis to be made. If an amino acid problem is suspected, a variety of chromatographic techniques are employed to identify the elevated compounds and the distorted pattern of amino acid excretion. Although paper chromatography still plays a role, more sophisticated techniques using HPLC or gas chromatography are increasingly employed for these studies. Both lipid and carbohydrate abnormalities can be detected using sophisticated chromatographic instrumentation.

Often it becomes necessary to measure enzyme activity to confirm the deficiency of a specific biochemical catalyst. Sampling may involve isolation of platelets or other blood components. In many instances, tissue sampling is necessary for proper elucidation of the medical problem. Detection of a low level of an enzyme (and quantitation of the extent of the decrease) plays an important role in both therapy and ongoing counseling of the patient and family.

Phenylketonuria (PKU)

Perhaps the best known and most thoroughly studied metabolic defect is **phenylketonuria (PKU)**. This disorder of amino acid metabolism occurs when the enzyme **phenylalanine hydroxylase (PAH)** is synthesized in low amounts. PAH catalyzes the hydroxylation of phenylalanine to form tyrosine, an important precursor for many hormones and neurotransmitters. The reaction sequence is illustrated below:

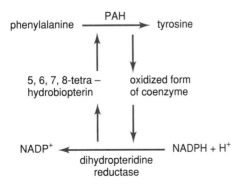

Although either enzyme in the sequence can be deficient, low levels of dihydropteridine reductase are rarely observed.

When phenylalanine hydroxylase activity is low, significant amounts of other derivatives are formed, including phenylpyruvic acid, phenylethylamine, and phenyllactic acid. The importance of this shift to an alternative pathway is not clearly understood at present. Phenylethylamine has been identified as a possible neurotransmitter in humans, but its role in the brain damage resulting from PKU is not understood.

If PKU is not detected at an early age, severe mental retardation invariably occurs, accompanied by abnormal electroencephalographic readings, hyperactivity, and aggressive behavior. Patients who go beyond 3 months without therapeutic intervention generally experience irreversible damage. Treatment for the disorder involves use of a special diet which contains little or no phenylalanine. In many instances, the child eventually outgrows the need for the diet and can gradually begin consuming regular food.

The initial screening test uses ferric chloride to detect the presence of high concentrations of phenylpyruvic acid. The major drawback to this test in PKU patients is the low initial levels of the ketoacid metabolite. The concentration of this metabolite may not increase until several days after birth as the baby begins to receive nourishment containing significant amounts of protein. The current trend toward early discharge after birth makes the detection of PKU in the newborn a rather difficult proposition. Many authorities do not recommend the ferric chloride test as a screen for the early detection of PKU. Most screening programs are effective in detecting the disorder only at 2–3 weeks after birth.

A more reliable screening procedure is the **Guthrie assay,** which examines the inhibition of microbial growth. In this procedure, the growth of *Bacillus subtilis* can be inhibited by a compound called beta-2-thienylalanine. If the microorganism is grown in the presence of a normal blood sample, growth is inhibited by the thienyl derivative. However, when high levels of phenylalanine are present, the amino acid cancels the effect of the inhibitor, allowing normal bacterial growth. Bacterial growth in the presence of the inhibitor is a positive test for the presence of elevated concentrations of phenylalanine (and for the presence of PKU).

The test is normally carried out in a reference laboratory, usually a state lab since most states require mandatory screening of newborns for PKU. A blood sample is obtained by heel-stick and blotted onto a piece of filter paper. This paper is then mailed to the lab for analysis. Using a hole punch, a disk is punched out of the paper and placed on a petri dish containing a nutrient medium, the bacteria, and the inhibitor. No ring of growth is seen around the disks containing blood from normal patients. The phenylalanine present in the patient with PKU overcomes the inhibitor, and a ring of bacterial growth is seen around those disks.

Hypothyroidism

The other major inherited disorder for which mandatory screening exists in most states is neonatal hypothyroidism. This disorder was discussed in considerable detail in Chapter 14. Since a whole-blood sample is required for this assay, the usual practice is to employ one heel-stick sample for both PKU and hypothyroid screening.

SUMMARY

Detection of pregnancy is accomplished by monitoring the amount of beta-HCG present in urine or serum. Although urine samples have an advantage in ease of collection, immunoassay methodology permits detection of HCG in plasma at an earlier stage of pregnancy. Assay reliability is complicated by problems in antibody specificity and in disagreement over standardization.

Assessment of fetal maturity can involve study of several parameters. Estrogen production (measured either in urine or plasma) is used to detect problems of fetal development and growth. Lung maturity in the unborn child can be predicted with some certainty through measurement of the L/S ratio in amniotic fluid. More specific studies of other phospholipids are also proving to be of value in this situation.

Since elevated bilirubin levels in the newborn can produce irreversible brain damage, accurate measurement of serum bilirubin is important. Standard methods for quantitation are reliable, but quality control must extend to measurements in the 20-mg/dL range. Spectrophotometric scanning for detection of bilirubin in the amniotic fluid can be of value in monitoring some high-risk pregnancies associated with Rh incompatibilities.

Our knowledge of inborn errors of metabolism has increased markedly in recent decades, as has our awareness of the wide variety of these diseases. In each case, the problem is associated with an extremely low level of a specific enzyme. As a consequence, necessary metabolites do not form and other toxic by-products are produced in levels much higher than normal. Screening procedures for detection of metabolic disorders include qualitative measurement of carbohydrates, keto compounds, and amino acids in urine. If the screening test is positive, confirmatory testing is necessary to identify the specific metabolic disorder. Procedures employing thin-layer chromatography, HPLC, or gas chromatography/mass spectrometry are commonly employed for confirmation.

Phenylketonuria is one common metabolic disorder whose screening is mandated by law in most states. This disorder is characterized by increased production of phenylpyruvic acid, formed as a by-product of the deficiency of phenylalanine hydroxylase, which impairs the conversion of phenylalanine to tyrosine. Mental retardation and a variety of neurologic problems are seen in this disorder. A positive test using ferric chloride suggests the presence of PKU. Confirmatory screening is carried out by examining the growth of *Bacillus subtilis* in the presence of blood from the newborn and an inhibitor of bacterial metabolism.

FOR REVIEW

Directions: For each question, choose the best response.

1. Which of the following polypeptide hormones may be described as having alpha chains that are biochemically identical but beta chains that are biochemically unique?
   A. FSH, TSH, ACTH, HPL
   B. TSH, LH, TRH, HCG
   C. LH, HPL, HCG, TRH
   D. HCG, FSH, TSH, LH

2. Identification of the following hormone is useful in determining the occurrence of pregnancy:
   A. beta HCG
   B. FSH
   C. HCG
   D. both A and B
   E. both A and C

3. When the chemical/immunological agglutination test is used to determine pregnancy, the absence of agglutination may
   A. be considered a positive test
   B. indicate the absence of HCG in the sample
   C. indicate the presence of HCG in the sample
   D. both A and B
   E. both A and C

4. Compared with a normal pregnancy, cases of ectopic pregnancy exhibit HCG serum levels that will
   A. rise more rapidly than normal
   B. not increase at the normal rate
   C. be the same as a normal pregnancy
   D. have no clinical significance associated to the disorder

5. All of the following biochemical changes are seen during pregnancy *except*
   A. decreased albumin
   B. increased alkaline phosphatase
   C. increased FSH and LH
   D. increased estrogens

6. Following the collection of amniotic fluid, the specimen should be
   A. centrifuged immediately
   B. protected from light
   C. treated with iodoacetate
   D. treated with EDTA

7. During pregnancy the primary site of estrogen production is the
   A. adrenal cortex
   B. adrenal medulla
   C. ovaries
   D. placenta

8. The integrity of the fetoplacental unit is in question when the estrogen production drops by _____ or more.
   A. 10%
   B. 15%
   C. 20%
   D. 30%

9. An assay performed to assess fetal lung maturity is the
   A. L/S ratio
   B. foam stability test
   C. total lipid analysis

D. both A and B

E. both A and C

10. All of the following statements apply to the function of uridine diphosphate glucuronyltransferase *except:*

    A. uridine diphosphate catalyzes the conversion of bilirubin to a water-soluble form

    B. uridine diphosphate catalyzes the conjugation of bilirubin

    C. uridine diphosphate facilitates the addition of glucuronic acid to the bilirubin molecule

    D. uridine diphosphate facilitates the addition of albumin to the bilirubin molecule

11. *Kernicterus* refers to the accumulation of bilirubin in the

    A. brain

    B. liver

    C. sclera

    D. skin

12. Phenylketonuria (PKU) is a metabolic defect caused by the absence or decreased production of

    A. phenylacetic acid

    B. phenylpyruvic acid

    C. phenylalanine hydroxylase

    D. phenylenediamine hydroxylase

13. The screening procedure useful in detecting PKU is

    A. copper reduction

    B. ferric chloride

    C. glucose oxidase

    D. nitroprusside

BIBLIOGRAPHY

Bakerman, S. et al., "Neonatal hyperbilirubinemia," *Lab. Manage.* March:23–29, 1985.

Bates, H. M., "Estriol assays for monitoring fetal/placental function," *Lab. Manage.* February:17–21, 1983.

Berry, H. K., "Screening newborns for genetic disease: The PKU model," *Diag. Med.* January:50–59, 1984.

Berry, H. K., "The spectrum of metabolic disorders," *Diag. Med.* March:39–53, 1984.

Broderssen, R., "Bilirubin transport in the newborn infant, reviewed with relation to kernicterus," *J. Pediat.* 96:349–356, 1980.

Brown, L. M., and Duck-Chong, C. G., "Methods of evaluating fetal lung maturity," *CRC Crit. Rev. Clin. Lab. Sci.* 16:85–159, 1982.

Chung, S. J., "Review of pregnancy tests," *South. Med. J.* 74:1387–1389, 1981.

Di Pietro, D. L., "Ectopic pregnancy: Interpreting hCG levels," *Lab. Manage.* November:43–51, 1981.

Freer, D. E., and Statland, B. E., "Measurement of amniotic fluid surfactant," *Clin. Chem.* 27:1629–1641, 1981.

Gabbe, S. G., and Hagerman, D. C., "Clinical application of estriol analysis," *Clin. Obstet. Gynecol.* 21:353–362, 1978.

Garry, P. J., and Anaokar, S. G., "Enzymatic assays for fetal lung maturity," *Lab. Manage.* May:41–46, 1980.

Hammond, K. B., and Wells, R., "Current approaches to evaluating the jaundiced neonate: A new look at bilirubin assays," *Lab. Med.* 14:239–245, 1983.

Hanley, W. B. et al., "Maternal phenylketonuria (PKU)—A review," *Clin. Biochem.* 20:149–156, 1987.

Hill, A. et al., "Difficulties and pitfalls in the interpretation of screening tests for the detection of inborn errors of metabolism," *Clin. Chim. Acta* 72:1–15, 1976.

Hussa, R. O., "Clinical utility of human chorionic gonadotropin and alpha-subunit measurements," *Obstet. Gynecol.* 60:1–12, 1982.

Kubasik, N. P., "Human chorionic gonadotropin," *Clin. Chem. News,* September, 1984.

Lamola, A. A., and Fanaroff, A. A., "Bilirubin fluorescence and prevention of kernicterus," *Diag. Med.* February:9–12, 1984.

Levy, H. L., and Mitchel, M. L, "The current status of newborn screening," *Hosp. Prac.* July:89–97, 1982.

Marcus, J. I., "Pregnancy testing—The real and the ideal," *Lab. Manage.* March:9–12, 1981.

Natelson, S., Scommegna, A., and Epstein, M. B., *Amniotic fluid: Physiology, biochemistry, and clinical chemistry,* New York: John Wiley and Sons, 1974.

Netzloff, M. L., "Neonatal screening for phenylketonuria," *Ann. Clin. Lab. Sci.* 12:368–371, 1982.

Painter, P. C., "Discordant hCG results in pregnancy: A method in crisis," *Diag. Clin. Testing.* July:20–24, 1989.

Pappas, A. A., Mullins, R. E., and Gadsden, R. H., "The role of amniotic fluid phospholipids in determining fetal lung maturity," *Ann. Clin. Lab. Sci.* 12:304–308, 1982.

Perelman, R. H. et al., "Developmental aspects of lung lipids," *Ann. Rev. Physiol.* 47:803–822, 1985.

Rooney, S. A., "The surfactant system and lung phospholipid biochemistry," *Amer. Rev. Respir. Dis.* 131:439–460, 1985.

Saxena, B. B., "New methods of pregnancy testing in adolescent girls," *Pediat. Clin. North Am.* 28:437–453, 1981.

Scriver, C. R., et al., "Mendelian hyperphenylalaninemia," *Ann. Rev. Genet.* 22:301,321, 1988.

Statland, B. E., and Freer, D. E., "Assessing fetal health and maturity," *Diag. Med.* November/December:73–86, 1979.

van Hell, H., and Helmich, J., "The evolution of pregnancy testing: Applications of monoclonal technology," *Amer. Clin. Prod. Rev.* March:38–45, 1984.

Warkentin, D. L., "From A to hCG in pregnancy testing," *Diag. Med.* April:35–43, 1984.

Chapter Eighteen

ELECTROLYTES AND WATER BALANCE

1. State the average amounts of water excreted daily through urine formation and other body processes.
2. List the seven principal minerals required by the mammalian system and six that are needed in trace amounts.
3. Identify the major role in the body of each of the following electrolytes.
   A. sodium
   B. potassium
   C. chloride
4. Name the primary extracellular and intracellular cations.
5. State the range at which blood pH should be maintained.
6. Define the following terms:
   A. buffer
   B. acidosis
   C. alkalosis
7. Name the two predominant buffer systems and two additional buffer systems that function in the blood to help maintain blood pH.
8. Describe how the bicarbonate and hemoglobin/oxyhemoglobin buffer systems contribute to the maintenance of blood pH.
9. Name two tissues which play important roles in maintaining hydrogen ion balance within the body.
10. Describe the process by which respiratory control influences the pH of blood.
11. Describe how each of the following contributes to renal control of blood pH:
    A. bicarbonate system
    B. phosphate system
    C. ammonia formation
12. Name the two major hormone systems that contribute to fluid regulation and electrolyte balance in the body.
13. State the principle of flame emission photometry and the composition of the ion-selective electrodes used to measure sodium and potassium in body fluids.
14. Describe the colorimetric and ion-selective electrode methods for quantitating chloride in body fluids.
15. Name the compound in the blood which constitutes the majority of dissolved CO_2.
16. Describe the colorimetric, ion-selective electrode, and enzymatic methods for quantitating carbon dioxide in the blood.
17. Identify the type of blood sample required for blood pH and blood gas analyses.
18. Describe the electrode systems used to measure blood pH, pCO_2, and pO_2.

19. Explain the alterations in acid/base balance that cause
    A. respiratory acidosis
    B. respiratory alkalosis
    C. metabolic acidosis
    D. metabolic alkalosis
20. Identify the hormonal disorder and its corresponding effect on fluid and electrolyte balance for each of the following hormones:
    A. antidiuretic hormone
    B. aldosterone
    C. renin
21. Identify the clinical symptoms and the abnormal laboratory test results associated with cystic fibrosis patients.
22. Describe methods used to elicit sweat production and methods employed to quantitate the chloride and sodium concentrations in sweat.

INTRODUCTION

Water has always been recognized as essential to life and has often been imbued with mystical powers. Electrolytes, particularly salt, also played important roles in the economies of many civilizations. Part of our current interest in health and fitness revolves around concern for the proper amount of sodium in the diet and its effect on blood pressure. We recognize the need to restore both electrolyte and fluid balance after strenuous exercise. The dual influences of water and electrolytes are intimately intertwined as important determinants of proper body function and good health.

18.1

WATER DISTRIBUTION IN THE BODY

Water Composition of the Body

Water plays a number of important, and often hidden, roles in biochemical processes in the body. Our perception that we are solid flesh is misleading. Approximately 60% of the total weight of the adult male is water. The adult female has a water composition of about 50%, and an infant is some 77% water. As weight increases, the fraction of total weight found as body water decreases (Table 18-1).

The water composition of tissues varies from moment to moment, and measurement is very inexact. However, we can easily follow the changing concentration in the circulation. Whole blood is approximately 65% water, which provides volume in the circulation and allows easy movement of materials from one place to another in the body. Water distri-

Table 18-1
BODY TYPE AND WATER COMPOSITION

Body Type	Percentage Total Weight as Water	
	Male	Female
Lean	70	60
Normal	60	50
Obese	50	42

bution between the inside and outside of the cell is approximately equal, with 55% of the water being intracellular and 45% being found outside the cell.

Sources of Body Water

Water either enters the body through consumption or is produced by metabolism (Table 18-2). Drinking is a major route of water intake. The average adult consumes about 1000 mL of water a day. Depending on the diet, another 1000-1200 mL of water enters the body from food. In addition, water is generated as part of several biochemical proc-

Table 18-2
SOURCES OF BODY WATER

Source	Amount/Day
Drink	1000 mL
Food	1000-1200 mL
Fat metabolism	100 mL/100 g
Protein metabolism	44 mL/100 g
Carbohydrate metabolism	60 mL/100 g

esses. This **water of oxidation** may amount to 300 mL or so each day. For every 100 g of fat metabolized, 100 g of water are produced. Oxidation of 100 g of carbohydrate yields 60 g of water, and 100 g of protein produces 44 g of water when oxidized.

Routes of Water Excretion

The average adult loses about 2500 mL of water each day through excretion. Approximately 1500 mL is lost in the urine. The other 1000 mL is excreted through a variety of processes including perspiration, exhalation of water vapor through the lungs, and water excreted from the intestine with fecal material.

18.2

ELECTROLYTE COMPOSITION OF BODY FLUIDS

At present, there are some 109 known elements, with others being synthesized through nuclear chemistry techniques. Of these 109 elements, fewer than 20 are of physiological importance in the normal human being. A number of other elements (such as lead) become significant only when their concentration produces toxic effects. The four most prevalent elements in the human body are carbon, hydrogen, nitrogen, and oxygen. The mammalian system requires seven principal minerals: sodium, potassium, chloride, sulfur, magnesium, calcium, and phosphorus. Several other minerals (such as iron, copper, iodine, manganese, cobalt, and zinc) are present (and required) in trace amounts. These trace elements are also essential to the body. Selenium, aluminum, and cadmium have been detected in the body, but the functions of these minerals are unknown. There may be both beneficial and toxic effects associated with this last group, depending on the concentration.

18.3

ROLES OF ELECTROLYTES IN BIOCHEMICAL PROCESSES

Sodium

The major role for sodium appears to be in the regulation of water balance in the body. As the primary extracellular cation, sodium contributes greatly to the osmotic pressure in compartments outside the cell. Water shifts from one compartment to another in response to changes in the sodium concentration. Movement of sodium into the neuron is a major factor in the process of nerve transmission. The intricate mechanism for maintaining blood pressure involves sodium as a key player along with enzymes and hormones to alter fluid balance. Sodium is also important in the enzyme Na^+, K^+-ATPase, which cleaves adenosine triphosphate to produce biochemical energy for transport processes. However, the particular role of sodium levels in regulating the enzyme is not clear at present. The sodium cation may have more influence on the movement of materials in and out of cells than we currently realize.

Potassium

As the primary intracellular cation, potassium is an integral part of the transmission of nerve impulses. Movement of potassium across the nerve tissue membrane permits the neural signal to move down the nerve fiber. Potassium also seems to be involved in synaptic processes, where the impulse "jumps" from one nerve fiber to another.

Chloride

This anion appears to have a fairly passive role in water balance, moving from compartment to compartment with sodium to provide ionic balance to the positively charged sodium ions. Both sodium and chloride seem to be transported together to maintain the charge balance, although some movement of chloride with potassium may also be seen. The chloride ion is the only known anion to serve as an enzyme activator, stimulating the starch hydrolysis reaction catalyzed by the enzyme amylase.

18.4

ACID/BASE BALANCE

Importance of Proper pH

The body requires very close pH regulation for proper metabolic functioning. Elaborate mechanisms are in place to maintain the blood pH at 7.35–7.45. Seemingly minor deviations of 0.1 pH units above or below these limits can profoundly alter metabolic processes. Enzyme activity, oxygen transport, cellular transport of materials—these and many other processes depend on precise regulation of pH for their proper functioning.

Sources of Protons

The body produces hydrogen ions in a variety of ways (Fig. 18-1). For every molecule of glucose metabolized, two molecules of lactic acid can be formed. At physiological pH, the lactic acid dissociates into the lactate anion and a proton. In a similar fashion, as triglycerides (or their component fatty acids) are broken down, acetoacetate and protons are produced. The end products of amino acid metabolism are urea, carbon dioxide, and hydrogen ions. Protons are moved in and out of the mitochondria during the complicated process of oxidative phosphorylation, which utilizes oxygen and generates ATP.

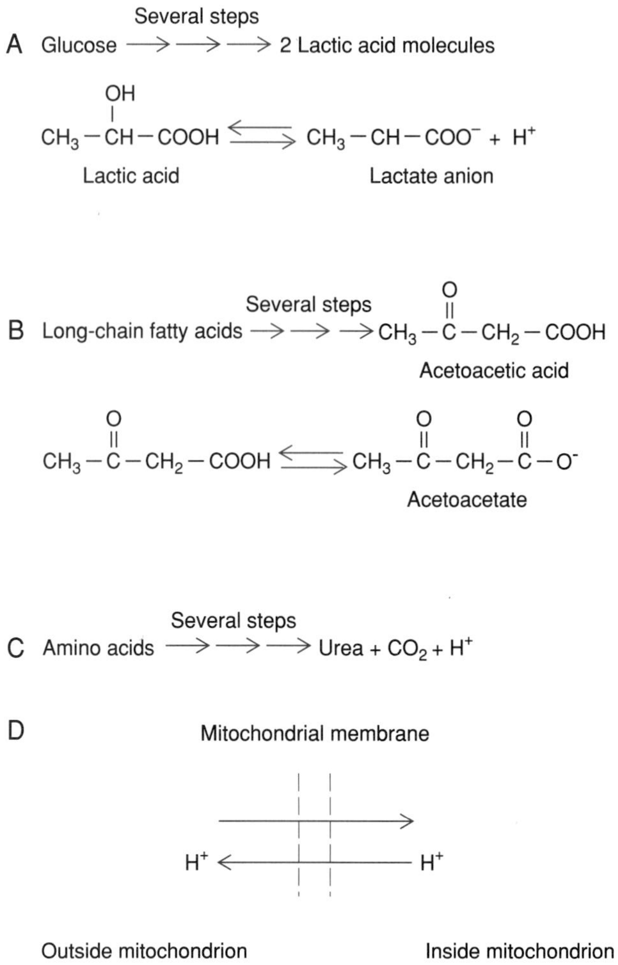

Buffer Systems in the Body

DEFINITION OF A BUFFER

To maintain pH at the proper level, buffer systems must be available. A **buffer** is a mixture of a weak acid and its salt with the capability of combining with protons or releasing protons in response to external shifts in pH. The purpose of a buffer is to furnish (or neutralize) protons in solution to maintain a defined pH. The buffer mixture consists of an equilibrium between a protonated species and its nonprotonated counterpart:

$$HA \rightleftharpoons H^+ + A^-$$

HA stands for the undissociated component of the buffer system. The proton which dissociates is designated by H^+ and the anion component of the buffer is represented by A^-. As protons are added to the system, the equilibrium shifts to form more HA, consuming the extra protons. If the proton level decreases (increase in pH), the equilibrium shifts to the right, releasing protons into solution. By careful selection of the concentration of the undissociated form of the buffer and the concentration of the anion, a specific pH range can be defined.

BUFFER SYSTEMS IN THE BODY

Since the measurement of intracellular pH is technically very difficult and not very informative from a clinical standpoint (at least, not at present), we will restrict our discussion to pH regulation in the bloodstream. The body has several buffer systems in the circulation to maintain blood pH. The predominant buffer systems are the plasma bicarbonate buffer and the erythrocyte hemoglobin/oxyhemoglobin buffer.

Other systems which contribute to pH stabilization include the organic and inorganic phosphate

FIGURE 18-1   Sources of protein in the body. A. Glucose conversion to lactic acid. B. Fatty acid breakdown. C. Amino acid metabolism. D. Mitochondrial release and uptake.

buffers and the plasma proteins. Phosphate acts as a buffer by way of the following equilibrium:

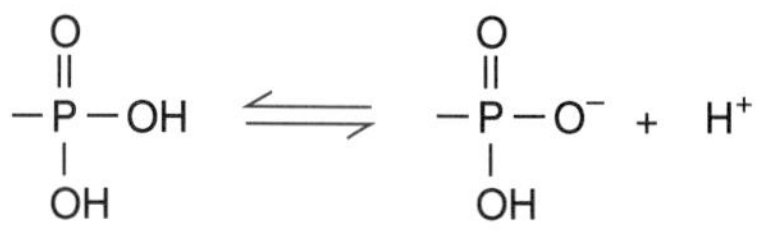

A second dissociation is possible, losing a proton from the remaining $-OH$ group. Proteins can buffer with both the $-NH_2$ groups and the $-COOH$ groups on amino acid side chains in the molecule. The amino group can reversibly capture a proton, forming $-NH_3^+$. Dissociation of the proton from the carboxyl group yields $-COO^-$ and H^+ in a reversible manner. Histidine groups can also provide some buffering through proton exchange with nitrogen atoms.

THE BICARBONATE BUFFER SYSTEM

The bicarbonate buffer system is comprised of water, carbon dioxide, carbonic acid, protons, and bicarbonate. The relationship among these components is given in the following equation:

$$H_2O + CO_2 \rightleftharpoons H_2CO_3 \rightleftharpoons H^+ + HCO_3^-$$

One of the driving forces in this system is carbon dioxide. As the concentration of this component increases (due to cellular metabolism), the equilibrium shifts to the right, producing more hydrogen ions and bicarbonate anions. The reaction can shift to the left under two different sets of circumstances:

1. an increase in protons
2. a decrease in CO_2 (through respiration)

The carbonic acid (H_2CO_3) is simply an intermediary in the equilibrium. It decomposes either to form water and carbon dioxide or protons and bicarbonate, depending on the concentrations of other chemicals in the system. The enzyme carbonic anhydrase speeds up the process, allowing for very rapid fine-tuning of this buffer system as it responds to shifts in concentrations of protons or CO_2.

THE HEMOGLOBIN/OXYHEMOGLOBIN BUFFER SYSTEM

The hemoglobin/oxyhemoglobin buffer system is intimately related to oxygen transport (Fig. 18-2). Protons and oxygen molecules interchange on hemoglobin as a result of pH changes in the cells and the circulation. When hemoglobin is protonated, it does not bind oxygen. As the oxygen content increases, a proton is released and an oxygen molecule attaches to the heme portion of the protein.

If we start in the lung, we see protonated hemoglobin entering the bloodstream. The high oxygen content in the lung (from respiration) adds oxygen

to hemoglobin with the loss of a proton. The oxyhemoglobin travels through the arterial circulation and reaches the cells. Cellular metabolism has generated protons, which attach to hemoglobin and cause oxygen to dissociate and enter the cell. The deoxyhemoglobin then carries the protons back to the lung, encounters oxygen, and the process begins again. A major role of hemoglobin in pH maintenance is to remove protons from cellular sites and carry them to the lung where they are utilized by the bicarbonate buffer system, as explained below.

Regulation of Acid/Base Parameters

To function properly, the body requires very tight pH control. Two major mechanisms exist to keep blood pH within the 7.35-7.45 normal physiological range. Both the lung and the kidney play important roles of intake and excretion in balancing the hydrogen ion concentration within the body.

RESPIRATORY CONTROL

The lung provides for pH control through transport of hydrogen ions by hemoglobin and the disposal of carbon dioxide (Fig. 18-3). Hemoglobin entering the lung contains protons which are displaced as oxygen binds to this protein. The oxyhemoglobin moves through the arterial circulation to the cell. Metabolism within the cell has produced carbon dioxide, which combines with water to form carbonic acid. Dissociation of carbonic acid yields hydrogen ions and bicarbonate anions. The hydrogen ions combine with oxyhemoglobin, displacing the oxygen

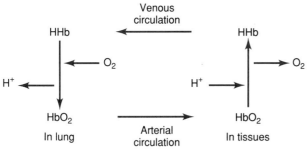

FIGURE 18-2   The hemoglobin/oxyhemoglobin buffer system. HHb stands for hemoglobin plus proton. HbO$_2$ stands for oxyhemoglobin.

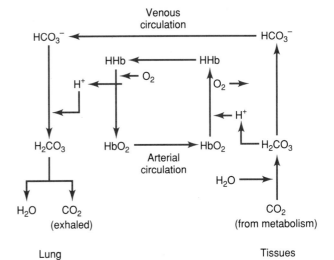

FIGURE 18-3   Respiratory control of pH.

and making it available to the cells for further metabolism. Both bicarbonate ions and protonated hemoglobin travel through the venous circulation back to the lung. In the lung, oxygen from respiration displaces protons, reforming oxyhemoglobin. The displaced protons combine with bicarbonate to form carbonic acid, which quickly dissociates to carbon dioxide and water. The carbon dioxide is then removed from the lung by exhalation. The previously formed protons are now part of water molecules.

RENAL CONTROL

The kidney plays an important and complex role in the maintenance of blood pH. Several interrelated processes take place at various sites in the proximal and distal tubule to trap and excrete hydrogen ions. In both the proximal and distal tubules, CO_2 and water combine to form carbonic acid, which then dissociates into H^+ and HCO_3^-. The protons are then excreted into the tubular filtrate when sodium ions are reabsorbed (Fig. 18-4). Charge balance is provided both by the exchange of hydrogen ions for sodium ions and by concurrent movement of bicarbonate ions back into the circulation with the reabsorbed sodium. The excreted protons can combine with bicarbonate in the tubular filtrate to produce water and carbon dioxide. A portion of the CO_2 is reabsorbed, and the water may be reabsorbed or excreted in the urine.

A second process to neutralize hydrogen ions takes place in the distal tubule. As protons are secreted into the tubular filtrate, they can combine with phosphate anions and be carried out in the urine:

$$H^+ + HPO_4^{2-} \rightarrow H_2PO_4^- \quad \text{or} \quad H_3PO_4$$

Hydrogen ion secretion is again balanced by sodium reabsorption and cotransport back into the circulation with bicarbonate.

A number of amino acids are deaminated in the distal tubule, creating ammonia as one of the prod-

ucts. When ammonia is secreted into the tubular filtrate, it combines with protons to form the ammonium ion, which is then excreted in the urine:

$$NH_3 + H^+ \rightleftharpoons NH_4^+ \text{ (excreted in urine)}$$

Again, hydrogen ions are trapped and removed from the system in combination with other materials.

18.5

REGULATION OF FLUID AND ELECTROLYTE LEVELS

Intake of Electrolytes

The various electrolytes enter the body as part of the regular diet. Sodium is so prevalent in our foods (either occurring naturally or as a supplement) that it is virtually impossible for sodium deficiency to exist. Instead, the major problem associated with sodium intake is how to reduce the amount of this electrolyte in the diet. Potassium is not as prevalent in foods but can be found in significant amounts in potatoes, bananas, raisins, watermelon, and cantaloupe. All the electrolytes are readily absorbed from the gastrointestinal tract into the circulation. No transport proteins are needed to carry these materials throughout the body.

Cellular Uptake and Excretion

Once in the body, the electrolytes distribute themselves inside and outside the cell. Enzymatic pumping systems which require ATP for energy maintain intracellular and extracellular electrolyte balance. Sodium is the chief extracellular electrolyte. The concentration of sodium in serum or plasma is approximately 135–145 mEq/L. Although the concentration of extracellular potassium in serum is only about 3.5–5.0 mEq/L, the intracellular potassium in the erythrocyte is in the 130–140 mEq/L range. Sodium values inside the red cell are in the same low range as potassium values outside the cell. Similar distributions of sodium and potassium are seen in other types of cells in the body.

Excretion by Kidney

WATER

Water excretion by the kidney begins at the glomerulus. This process was described in detail in Chapter 11 and will only be summarized here. Fil-

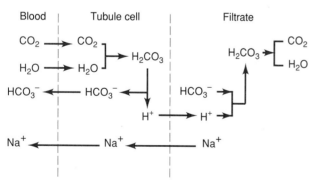

FIGURE 18-4 Renal Na^+ reabsorption and H^+ excretion.

tration of blood at this site produces a fluid which is nearly isotonic with plasma, since only materials above 60,000 mol wt are held back. As the fluid enters the proximal tubule, water reabsorption begins. By the time the material has moved to the loop of Henle, approximately 80–85% of the water has been reabsorbed. Movements of electrolytes and other small molecules affect the process, altering the extent of water reabsorption. In the distal tubule, secretion of water and other materials occurs, resulting in a fluid which is again almost isotonic with plasma. Some concentration may take place in the collecting ducts, resulting in a urine which has somewhat more solute content than does plasma.

ELECTROLYTES

All electrolytes are initially filtered at the glomerulus, since they are very small entities and readily pass through the membrane pores. Reabsorption processes restore the proper plasma electrolyte levels, balancing electrolyte intake from the diet and excretion to maintain appropriate concentrations.

Approximately 60–70% of the sodium filtered at the glomerulus is reabsorbed in the proximal tubule. Bicarbonate formed within the tubule cells moves back into the circulation with sodium, while protons are excreted. Another 25% or so of the filtered load is reabsorbed in the loop of Henle. Here chloride is the anion that transports with sodium as it returns to the bloodstream. In the distal tubule, hormonal regulation and sodium/potassium/proton interchanges determine the ultimate amount of sodium retained by the body.

The potassium filtered at the glomerulus is rapidly and almost completely reabsorbed in the proximal tubule. Any potassium eliminated from the body is lost by secretion in the distal tubules. This secretion is under hormonal control by the same system which regulates sodium retention.

Hydrogen ions, chloride, and bicarbonate are all more passive players in the retention/excretion process. The anions are transported with sodium to maintain ionic balance. Protons are frequently eliminated as sodium is retained. The extent of protonation of ammonia is determined by equilibrium conditions and is not actively regulated by processes other than those governing the rate of ammonia formation.

Hormonal Regulation

Two major hormone systems contribute to the regulation of fluid and electrolyte balance in the body. Antidiuretic hormone is responsible for the promotion of water retention, and the renin/angio-

tensin/aldosterone system controls sodium retention. Together, these two biochemical processes maintain the appropriate concentrations of electrolytes and the proper fluid level in the circulation.

Antidiuretic hormone (ADH) is a protein initially produced as a prohormone in the hypothalamus. This material is then transported to the pituitary, converted to the active hormone, and released into the bloodstream when the osmolality of the blood increases beyond a certain point. The major action of ADH is to increase tubule permeability to water so more is reabsorbed into the circulation.

Whereas ADH regulates total concentration of the blood through water balance, the **renin/angiotensin/aldosterone system** concerns itself more with electrolyte balance and fluid volume (Fig. 18-5). When sodium levels decline or total blood volume decreases, **renin** (a proteolytic enzyme) is formed from **prorenin** (the precursor). This process takes place in the kidney. Renin then cleaves the peptide angiotensinogen to form the 10-amino acid peptide angiotensin I. The physiologically inactive angiotensin I is further hydrolyzed by angiotensin-converting enzyme to angiotensin II, a vasopressor. **Angiotensin II** stimulates constriction of the blood vessels (to maintain proper blood pressure) and promotes the synthesis of **aldosterone,** a steroid hormone formed in the adrenal gland. The major action of aldosterone is to facilitate sodium retention by the kidney, also helping to increase blood volume. By conserving sodium and retaining water, blood volume and pressure are maintained at the proper levels.

18.6

MEASUREMENT OF ELECTROLYTES

Sodium and Potassium Assay by Flame Photometry

Flame emission photometry has long been a standard means of assaying for the concentrations of so-

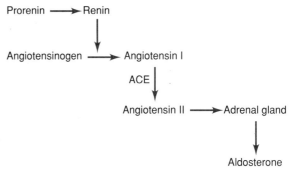

FIGURE 18-5   Renin/angiotensin/aldosterone system.

dium and potassium in body fluids. When the ions are heated sufficiently, electrons are excited to higher energy states within the atom. As these electrons drop back down to the ground state again, they emit light energy of a specific wavelength. The detector measures the amount of light given off, which is proportional to the number of ions present in the solution.

Sodium and potassium can be measured simultaneously using emission techniques and multiple detectors. A sample is aspirated and diluted (often 1:100 or 1:200) automatically. Lithium is included in the diluent to provide a strong background reference. This cation has its own specific emission lines. Comparison of the sodium (or potassium) emission to the signal given by lithium compensates for any variations in aspiration rate or fluctuation in the flame.

Sodium and Potassium Assay by Ion-Selective Electrodes

Since certain hazards and some operational difficulties are associated with flame emission equipment, most current automated systems employ ion-selective electrodes for electrolyte assay. The cation interacts with the membrane of the electrode (glass for Na^+, valinomycin for K^+), producing a potential difference relative to a reference electrode. This difference in potential is translated into a concentration by preparing a calibration curve using known standards. The principle is the same one employed in adjusting a pH meter.

The sample is either first diluted and then introduced into the electrode sample chamber or introduced directly without prior dilution. Interaction with the electrode membrane generates a potential difference proportional to the concentration of the specific electrolyte present. These electrode processes are widely used in automated systems, often employing flow-through electrodes for faster processing of samples.

Chloride

Chloride assays are carried out using either colorimetric or ion-selective electrode methods (Fig. 18-6). The most commonly employed colorimetric method involves a reaction with mercuric thiocyanate. Mercuric chloride and thiocyanate anion are the products of the first reaction. The thiocyanate anion formed then combines with iron (III) to form ferric thiocyanate, which absorbs strongly at 480 nm. The intensity of absorbance is proportional to the concentration of chloride present in the original

A $\quad 2\,Cl^- + Hg(SCN)_2 \longrightarrow HgCl_2 + 2\,SCN^-$

$Fe^{3+} + 3\,SCN^- \longrightarrow Fe(SCN)_3$
Strong absorbance at 480 nm

B $\quad Cl^- + Ag \longrightarrow AgCl + electron$

FIGURE 18-6 Chloride assay methodologies. A. Colorimetric assay. B. Ion-selective electrode method.

sample. An obvious problem with this widely used assay is disposing of the mercury salts.

In recent years, chloride ion-selective electrodes have become more widely used for the assay of chloride in body fluids. These electrodes are usually composed of a silver wire coated with silver chloride. In the presence of chloride anion, an oxidation/reduction reaction occurs. The silver metal forms silver (I) cation and an electron. The change in potential is detected electronically and reflects the chloride concentration in the sample. Serum or plasma can be assayed in a variety of automated systems using chloride electrodes, either directly or on a diluted sample.

Bicarbonate

Measurements of carbon dioxide and bicarbonate are usually considered interchangeable tests. Most of the dissolved CO_2 is in the form of bicarbonate in blood. Many automated systems assay for this parameter by acidifying the sample to release the carbon dioxide. This gas is then quantitated either by using a specific electrode or by diffusion into another chamber containing phenolphthalein. As the released gas redissolves in the indicator solution, the pH shifts, promoting a color change in the dye. The intensity of this change is determined colorimetrically, allowing calculation of the amount of CO_2 in the original sample.

An enzymatic assay (Fig. 18-7) has been employed by at least one automated system (DuPont aca). In this method, the sample is made alkaline. All the carbon dioxide is converted to bicarbonate. A series of indicator reactions are then enzymatically carried out, converting the bicarbonate to oxaloacetate. An oxidation/reduction reaction with NADH (catalyzed by malate dehydrogenase) forms malate and NAD^+. The decrease in absorbance at 340 nm as NADH is oxidized can be monitored to calculate the total CO_2 present.

A. $HCO_3 + H^+ \rightleftharpoons CO_2 + H_2O$

  1. Measure carbon dioxide with ion-selective electrode, or

  2. Diffuse carbon dioxide into solution containing phenolphthalein indicator. Reverse reaction occurs, regenerating protons; pH changes, with shift in color of indicator. Measure change in absorbance.

B. 1. Bicarbonate + Phosphoenolpyruvate $\xrightarrow{\text{Phosphoenolpyruvate carboxylase}}$ Oxaloacetate + Phosphate

  2. Oxaloacetate + NADH + H^+ $\xrightarrow{\text{Malate dehydrogenase}}$ Malate + NAD^+

FIGURE 18–7 Assays for bicarbonate (CO_2). A. Assay as carbon dioxide. B. Enzymatic assay for bicarbonate.

Blood pH and Blood Gases

Measurement of blood gases is a specialized technique requiring a great deal of care and skill. In most cases, an arterial sample is required. Since there is a significant risk to the patient associated with arterial punctures, personnel must be specially trained in this technique.

Most of the instrumentation available today for blood gas measurement is automated. Quantitation is performed using special electrodes to assess the concentration of each parameter. For pH assessment, a glass electrode is employed which is sensitive to hydrogen ions. The reference electrode is either Ag/AgCl or calomel. Electrodes for blood gases are covered with permeable membranes to allow penetration of the gas to the electrode surface. For pCO_2 measurements, the carbon dioxide in the sample interacts with a solution of $NaHCO_3$. The change in pH during this reaction is detected with an electrode. Oxygen measurement is made polarographically. The oxygen reacts with a platinum cathode, generating a current flow which is proportional to the oxygen content in the sample.

18.7

CLINICAL SITUATIONS INVOLVING ELECTROLYTE IMBALANCES

Acid/Base Disorders

A wide variety of disorders can result in a distortion of the acid/base balance in the body (Table 18–3). The underlying problem may be respiratory, producing difficulty in removing carbon dioxide. Alternatively, there may be a metabolic disturbance resulting in excess production of hydrogen ions or an impaired ability to eliminate them from the body. The term **acidosis** describes those situations resulting in a blood pH of less than 7.35, whereas **alkalosis** refers to situations in which the blood pH is greater than 7.45. Both acidosis and alkalosis may arise from a respiratory or a metabolic cause.

RESPIRATORY ACIDOSIS

Any impairment of respiration or decrease in the lungs' ability to exhale CO_2 results in **respiratory acidosis**. Accumulation of carbon dioxide in the blood shifts the bicarbonate buffer system in the direction of hydrogen ions and bicarbonate, leading to a drop in the pH. Patients with a wide variety of respiratory diseases (such as pneumonia, emphysema, asthma, chronic obstructive lung disease) can develop acidosis. Myasthenia gravis may also lead to respiratory acidosis as a secondary effect of impairment of the transmission of nerve impulses. Anesthesia or drug overdose (by morphine, for example) may depress the respiratory center and result in the impairment of CO_2 elimination by the lung.

RESPIRATORY ALKALOSIS

An excessive loss of carbon dioxide through hyperventilation is the primary cause of **respiratory alkalosis**. The increase in exhalation of CO_2 shifts the bicarbonate equilibrium toward the formation of

Table 18–3
ACID/BASE DISTURBANCES

Type	Cause
Respiratory acidosis	Excess CO_2 accumulation
Respiratory alkalosis	Excess CO_2 loss
Metabolic acidosis	Excess H^+ production
Metabolic alkalosis	Excess H^+ loss or excess alkali intake

carbon dioxide and water, decreasing the hydrogen ion concentration. The result is an increase in blood pH. Patients who are highly anxious may hyperventilate, exhaling increased levels of CO_2. Failure to maintain proper carbon dioxide levels is frequent among patients on mechanical ventilators. Some central nervous system disorders alter respiratory patterns, leading to an increase in blood pH. This disorder is seen in individuals with hepatic cirrhosis, hepatic coma, or in the early stages of aspirin overdose.

METABOLIC ACIDOSIS

A wide variety of situations can result in metabolic production of excess hydrogen ions, resulting in decreased blood pH and **metabolic acidosis.** In renal failure, some of the materials normally excreted in the urine are retained in the circulation. If these compounds carry hydrogen ions with them (such as ammonia and related nitrogen-containing derivatives), the hydrogen ions are also retained and accumulate in the circulation to higher than normal levels. Conversely, renal disease may result in excessive loss of other electrolytes (particularly sodium), with hydrogen ions being retained. In ketoacidosis (produced either by starvation or by diabetes mellitus), increased amounts of beta-hydroxybutyric acid are formed from lipid breakdown. This end product dissociates into excess hydrogen ions and beta-hydroxybutyrate anion, increasing loss of sodium and potassium as it is excreted in the urine. A number of toxic materials are metabolized to acids in the body, leading to increase in H^+. Among these materials are methanol (producing formic acid), ethanol (acetic acid), aspirin (salicylic acid), and ethylene glycol (oxalic acid). Accidental or intentional ingestion of large amounts of these materials can produce profound metabolic acidosis.

Lactic acidosis is a somewhat common cause of metabolic acidosis. Normally, little lactic acid is produced from glucose metabolism. Most of the material is converted to pyruvate and then passes through the Krebs cycle to produce biochemical energy. However, if oxygen is low, more lactic acid is formed because of impairment of pyruvate conversion to other products. Hypoxia caused by a wide variety of respiratory or circulatory problems enhances production of lactic acid and metabolic acidosis.

Bicarbonate loss produces metabolic acidosis as the equilibrium in the buffer system shifts to restore the proper bicarbonate anion concentration. Since the bicarbonate loss is selective (protons are not lost at the same time), regeneration of bicarbonate produces even more hydrogen ions, decreasing blood pH. Loss of bicarbonate may also occur through diarrhea or suction of the small intestine.

METABOLIC ALKALOSIS

Excessive intake of alkali or significant loss of hydrogen ions results in elevated levels of bicarbonate, leading to **metabolic alkalosis.** Patients being treated for peptic ulcer may ingest large amounts of antacids containing bicarbonate. Loss of hydrogen ions from the digestive tract can occur through prolonged vomiting or excessive removal of gastric contents. Diuretic therapy may enhance acid excretion by the kidney, leaving bicarbonate in the system. Some disorders of cortisol metabolism (such as Cushing's disease) or increases in aldosterone produce excessive potassium loss. The increase in potassium excretion is accompanied by a rise in excreted chloride. To balance the ionic equilibrium, more bicarbonate anion is formed, the bicarbonate buffer equilibrium shifts in the direction of proton utilization to form CO_2 and water, resulting in alkalosis.

Hormonal Causes

Hormonal regulation of fluid and electrolyte imbalance is complex and still not well understood. Several hormones may be involved with malfunctions of this complex process (Table 18-4). Antidiuretic hormone stimulates water retention by the kidney. The renin/angiotensin/aldosterone system is involved with a complicated process of fluid and electrolyte retention and excretion. Glucocorticoids alter fluid and electrolyte balance as part of other disease states.

Irregularities in antidiuretic hormone production are fairly rare but do have significant impact on water balance. ADH acts on the tubules and collecting ducts of the kidney, causing them to be more permeable to water. Decrease in ADH production can lead to **diabetes insipidus,** a condition characterized by extreme water loss (sometimes as much as 20 L of urine per day may be excreted). Excess

Table 18-4
HORMONAL CAUSES OF FLUID AND ELECTROLYTE IMBALANCE

Hormone	Effect
Antidiuretic hormone	
Increase	Fluid retention, low serum Na^+
Decrease	Fluid loss
Aldosterone	
Increase	Hypertension, low serum K^+
Decrease	Low serum Na^+, high serum K^+
Renin	
Increase	Hypertension

release of ADH from the pituitary promotes increased water retention accompanied by a decrease in sodium concentration (from dilution by the extra retained water).

Primary aldosteronism can be produced either by a specific tumor of the adrenal gland or by overdevelopment of the gland itself. The major clinical features are hypertension and low serum potassium. Decreases in aldosterone production are accompanied by a drop in serum sodium levels and a rise in serum potassium. Acid/base abnormalities may also develop.

Changes in renin production manifest themselves primarily through alteration of aldosterone output. In many cases of hypertension, monitoring renin levels does not provide much useful information. However, if the hypertension is due to blockage of the renal arteries or some other related type of kidney damage, the detection of an increase in renin production is diagnostically significant.

Renal Causes

A variety of kidney problems are associated with water and electrolyte loss. In general, problems with tubular reabsorption and secretion produce profound shifts in electrolyte excretion and retention. The changes in electrolyte concentrations may be the primary manifestation of the disease state or could be secondary to water imbalance. Some of these medical situations have a hormonal cause (such as inappropriate secretion of antidiuretic hormone). Others are produced by damage to the kidney, leading to destruction of tubules and impairment of normal processes. In some patients, excessive use of diuretics results in electrolyte loss which can create serious medical problems.

Cystic Fibrosis

Perhaps one of the most interesting and enigmatic diseases involving electrolyte imbalance is **cystic fibrosis.** This autosomal recessive disorder affects perhaps 1 in every 2000 Caucasian children (the incidence in African and Asian populations is much lower). Cystic fibrosis manifests itself clinically in respiratory difficulties, liver problems, and pancreatic abnormalities. Some months after birth, the lungs begin to accumulate mucus secretions, obstructing the air passages. Further blockage leads to lung scarring, inflammation, and infection. With proper treatment, many affected children may live to reach early adulthood.

The key laboratory diagnostic feature is an elevation of sweat electrolytes. Both Na^+ and Cl^- are increased due to the inability of the body sweat glands to reabsorb these materials before the perspiration is secreted. Although a large number of biochemical studies have been carried out to identify enzymatic and other abnormalities, the laboratory diagnosis hinges on the demonstration of increases in sweat electrolytes.

Generation of sufficient perspiration for analysis has always been a major problem in the assay for sweat electrolytes. Early methods involved wrapping the child's back in plastic and a blanket, a lengthy and often hazardous procedure with infants. At present, an electrical device is used to elicit sweat. Pilocarpine nitrate penetrates the skin, driven by an electric current. The process, referred to as **iontophoresis,** generates a flow of perspiration, which can either be collected on a weighed filter paper or assayed directly. Care must be taken to obtain at least 100 mg of sample.

There is some disagreement over the best method to assess electrolyte concentrations. One widely employed technique involves the use of a chloride ion-selective electrode. After the perspiration forms on the skin, the electrode is pressed firmly to the site of iontophoresis. Any chloride present interacts with the silver/silver chloride electrode and generates a potential change proportional to the chloride content of the sweat. The chloride level is then read from a meter previously calibrated with known standards. A number of problems occur, mainly due to the inexperience of technicians with the equipment and procedure. Many authorities recommend that electrolyte tests for cystic fibrosis be performed only at regional centers where a sufficient volume of assays is available for staff to maintain appropriate proficiency. It should be noted that the Cystic Fibrosis Foundation has expressed reservations about this electrode procedure.

In major reference centers, the sweat is assayed for both sodium and chloride. Flame photometry and titration with mercuric nitrate are the methods of choice. These processes tend to be more reliable, but are cumbersome. The sweat first must be weighed and then extracted from the collecting paper before assays are initiated. Since any center can expect only a low volume of tests, automation is not feasible.

SUMMARY

Water and electrolytes are in a delicate balance, with changes in the concentration of one parameter often producing shifts in other components of the system. Sodium is the primary extracellular electrolyte, with potassium serving as the major intracellular cation. Chloride seems to function in a passive role to maintain charge balance. Sodium significantly affects water balance, whereas potassium appears to play an important part in the transmission of nerve impulses.

The pH of body fluids is normally maintained between very close limits. The normal range for blood pH is 7.35–7.45. Several buffer systems control pH shifts in the bloodstream. Major buffers are the hemoglobin/oxyhemoglobin system and the bicarbonate buffer, which control pH through excretion of carbon dioxide by the lung and excretion of protons (either free or combined) by the kidney.

Sodium reabsorption by the kidney is important in maintaining the appropriate concentration of this cation in serum. Some potassium retention also occurs. The level of antidiuretic hormone affects water retention, with low levels of ADH leading to profound water loss. The renin/angiotensin/aldosterone system helps control blood pressure and sodium retention.

Sodium and potassium in serum or urine are measured either by flame photometry or by ion-selective electrodes. Most automated systems employ electrodes (often of the flow-through type) for these analyses. Chloride assays can be performed colorimetrically (using mercuric thiocyanate) or with a silver/silver chloride electrode. Bicarbonate measurements are carried out with phenolphthalein as indicator or by measuring evolved CO_2. Determination of blood gases requires special instrumentation.

Acid/base disturbances can be grouped into four categories: metabolic acidosis (excessive production of protons), metabolic alkalosis (retention of bicarbonate or loss of H^+), respiratory acidosis (retention of CO_2), and respiratory alkalosis (excessive CO_2 loss). Disturbances in antidiuretic hormone production (decrease in synthesis) lead to water loss. Increases in renin production can cause vasoconstriction, sodium retention, and high blood pressure. Defective kidney function results in electrolyte and water imbalances.

Cystic fibrosis is an inherited disease characterized by lung obstruction and increased excretion of sweat electrolytes. Laboratory diagnosis is often unreliable unless the personnel maintain proficiency in the testing process. Although chloride ion-selective electrode methods are available, most authorities prefer assay of sweat sodium and chloride by traditional methods.

FOR REVIEW

Directions: For each question, choose the best response.

1. What is the average volume of urine that an adult excretes daily?
   A. 500 mL
   B. 1000 mL
   C. 1500 mL
   D. 2500 mL

2. The primary extracellular cation is
   A. calcium
   B. chloride
   C. potassium
   D. sodium

3. The cation present in the greatest concentration in the cell is
   A. chloride
   B. phosphorus
   C. potassium
   D. sodium

4. Which electrolyte plays a major role in the regulation of water balance in the body?
   A. chloride
   B. phosphorus
   C. potassium
   D. sodium

5. Which electrolyte is significantly involved in the transmission of nerve impulses?
   A. iron
   B. phosphorus
   C. potassium
   D. sodium

6. The blood pH should be maintained within the range of
   A. 7.00–7.50
   B. 7.25–7.35
   C. 7.35–7.45
   D. 7.35–7.60

7. One of the major buffer systems employed by the body to maintain the blood pH level is
   A. ammonia/ammonium
   B. organic/inorganic phosphate
   C. plasma bicarbonate
   D. plasma proteins

8. Which body tissues play an important role in maintaining the blood pH level?
   A. liver and lung
   B. liver and kidney
   C. lung and kidney
   D. kidney and pancreas

9. The ultimate result of hydrogen ion transport by hemoglobin to the lung is the formation of
   A. ammonia
   B. bicarbonate
   C. carbonic acid
   D. carbon dioxide

10. The kidney exercises its metabolic control over blood pH by altering the retention or excretion of
    A. bicarbonate
    B. carbonic acid
    C. carbon dioxide
    D. sodium chloride

11. Due to the intracellular and extracellular concentration differences of this electrolyte, a hemolyzed serum specimen causes a false increase in
    A. chloride
    B. iron
    C. potassium
    D. sodium

12. Metabolic alkalosis is characterized by a(an)
    A. excess of bicarbonate
    B. deficit of bicarbonate
    C. excess of dissolved carbon dioxide
    D. deficit of dissolved carbon dioxide

13. Respiratory acidosis is characterized by a(an)
    A. excess of bicarbonate
    B. deficit of bicarbonate
    C. excess of dissolved carbon dioxide
    D. deficit of dissolved carbon dioxide

14. When measuring potassium using an ion-selective electrode, the membrane
    is composed of
    A. glass
    B. plastic
    C. silver
    D. valinomycin

15. The composition of the electrode used to measure blood pH is
    A. glass
    B. plastic
    C. platinum
    D. valinomycin

16. The laboratory method of choice for detecting cystic fibrosis is
    A. serum chloride
    B. amylase
    C. lipase
    D. sweat test

17. In cystic fibrosis which sweat electrolytes are elevated?
    A. calcium and chloride
    B. chloride and sodium
    B. sodium and potassium
    D. potassium and chloride

BIBLIOGRAPHY

Adrogue, H. J., and Madias, N. E., "Changes in plasma potassium concentration during acute acid-base disturbances," *Amer. J. Med.* 71:456–4667, 1981.

Batelle, D. C., and Arruda, J. A. L., "Renal tubular acidosis syndromes," *Miner. Electrolyte Metab.* 5:83–99, 1981.

Bia, M., and Thier, S. O., "Mixed acid/base disturbances: A clinical approach," *Med. Clin. North Am.* 65:347–361, 1981.

Chan, J. C. M., "Renal tubular acidosis," *J. Pediat.* 102:327–340, 1983.

Cox, M., "Potassium homeostasis," *Med. Clin. North Am.* 65:363–384, 1981.

Del Greco, F. et al., "The renin-angiotensin-aldosterone system in primary and secondary hypertension," *Ann. Clin. Lab. Sci.* 11:497–505, 1981.

Dobbins, J. W., and Binder, H. J., "Pathophysiology of diarrhoea: Alterations in fluid and electrolyte transport," *Clin. Gastroenterol.* 10:605–625, 1981.

Elms, J. J., "Potassium imbalance. Causes and prevention," *Postgrad. Med.* 72:165–171, 1982.

Feig, P. U., "Hypernatremia and hypertonic syndromes," *Med. Clin. North Am.* 65:271–290, 1981.

Ganguly, A. et al., "Primary aldosteronism: The etiologic spectrum of disorders and their clinical differentiation," *Arch. Intern. Med.* 142:813–815, 1982.

Goldberg, M., "Hyponatremia," *Med. Clin. North Am.* 65:251–269, 1981.

Goldsmith, B., "Aldosterone," *Clin. Chem. News.* January, 1984.

Gruskin, A. B. et al., "Serum sodium abnormalities in children," *Pediat. Clin. North Am.* 29:907–932, 1982.

Hammond, K., "Problems in diagnosing cystic fibrosis: Will new tests help?" *Diag. Med.* April:35–41, 1985.

Harrington, J. T., "Evaluation of serum and urinary electrolytes," *Hosp. Prac.* March:28–39, 1982.

Heeley, A. F., and Watson, D., "Cystic fibrosis—Its biochemical detection," *Clin. Chem.* 29:2011–2018, 1983.

Hilton, P. J., "Cellular sodium transport in essential hypertension," *New Eng. J. Med.* 314:222–229, 1986.

Hobbs, J., "Metabolic acidosis," *Am. Fam. Prac.* 23:220–227, 1981.

Hruska, K. et al., "Renal tubular acidosis," *Arch. Intern. Med.* 142:1909–1913, 1982.

Knochel, J. P., "Neuromuscular manifestations of electrolyte disorders," *Amer. J. Med.* 72:521–535, 1982.

LeGrys, V. A., "Cystic fibrosis: Recent diagnostic developments," *Lab Manage.* November:43–50, 1986.

Levin, M. L., "Renal control of sodium homeostasis," *Ann. Clin. Lab. Sci.* 11:322–326, 1981.

Masoro, E. J., "An overview of hydrogen ion regulation," *Arch. Intern. Med.* 142:1019–1023, 1982.

Michell, A. R., "Sums and assumptions about salt," *Persp. Biol. Med.* 27:221–233, 1984.

Mitch, W. E., and Wilcox, C. S., "Disorders of body fluids, sodium, and potassium in chronic renal failure," *Amer. J. Med.* 72:536–550, 1982.

Morgan, D. B., "Why plasma electrolytes?" *Ann. Clin. Biochem.* 18:275–280, 1981.

Morrison, G., and Murray, T. G., "Electrolyte, acid-base, and fluid homeostasis in chronic renal failure," *Med. Clin. North Am.* 65:429–447, 1981.

Narins, R. G., and Gardner, L. B., "Simple acid-base disturbances," *Med. Clin. North Am.* 65:321–346, 1981.

Narins, R. G. et al., "Diagnostic strategies in disorders of fluid, electrolyte, and acid-base homeostasis," *Amer. J. Med.* 72:496–520, 1982.

Ondetti, M. A., and Cushman, D. W., "Enzymes of the renin-angiotensin system and their inhibitors," *Ann. Rev. Biochem.* 51:283–308, 1982.

Re, R. N., "The renin-angiotensin system," *Med. Clin. North Am.* 71:877–895, 1987.

Schrier, R. W., "Treatment of hyponatremia," *New Eng. J. Med.* 312:1121–1123, 1985.

Sherman, R. A., and Eisinger, R. P., "The use (and misuse) of urinary sodium and chloride measurements," *J. Amer. Med. Assoc.* 247:3121–3124, 1982.

Sterns, R. H. et al., "Internal potassium balance and the control of the plasma potassium concentration," *Medicine.* 60:339–354, 1981.

Weiberger, M. H., "Sodium chloride and blood pressure," *New Eng. J. Med.* 317:1084–1086, 1987.

Weisberg, L. S., "Pseudohyponatremia: A reappraisal," *Amer. J. Med.* 86:315–318, 1989.

Winter, S. D., "Measurement of urine electrolytes: Clinical significance and methods," *CRC Crit. Rev. Clin. Lab. Sci.* 14:163–187, 1981.

Chapter Nineteen

MINERAL METABOLISM

1. State the biochemical roles played in body functions by each of the following minerals:
   A. calcium
   B. magnesium
   C. iron
   D. zinc
   E. copper
   F. manganese
   G. cobalt
   H. molybdenum
   I. chromium
   J. nickel
2. Identify the three forms of calcium as they exist in the blood, noting the form that is physiologically active.
3. Describe the regulatory effects of each of the following compounds on calcium:
   A. parathyroid hormone
   B. calcitonin
   C. vitamin D metabolites
4. Describe how a change in the serum ionized calcium level effects the release of parathyroid hormone.
5. Identify the tissues directly affected by the regulatory influence exerted by parathyroid hormone and calcitonin.
6. Describe the mechanisms by which parathyroid hormone and 1,25-dihydroxyvitamin D_3 stimulate an increase in and by which calcitonin causes a decrease in the serum calcium level.
7. Describe two ways by which the precursor to the active form of vitamin D can be obtained, noting the name of the active form of vitamin D.
8. Explain the effects of abnormal concentrations of serum protein on calcium concentrations in the blood.
9. Describe the methods used to quantitate total calcium levels and those employed to quantitate ionized (free) calcium.
10. Explain how each of the following may interfere with calcium analysis and how this interference may be avoided:
    A. protein
    B. magnesium
    C. bilirubin
    D. lipemia
    E. hemoglobin
11. Describe specimen collection and handling required for analysis of ionized calcium.
12. Describe the respective methods employed for measuring parathyroid hormone, calcitonin, and vitamin D metabolites.
13. Identify the two phosphate anions commonly referred to as phosphorus.
14. Describe the regulatory effect of vitamin D on phosphorus absorption in the small intestine.
15. Describe how phosphate and magnesium are handled by the kidney.
16. Describe the respective methods employed for measuring phosphate and magnesium.
17. Discuss the relationship between calcium and phosphate concentrations and their clinical significance.
18. List the primary components of bone tissue.

19. Discuss the clinical significance and laboratory findings associated with hypercalcemia and hypocalcemia.
20. List the three forms of magnesium as they exist in the blood.
21. Discuss the clinical significance and laboratory findings associated with abnormal magnesium levels.
22. Describe the absorption, transport, and storage of iron.
23. State the roles of ferritin and transferrin in iron metabolism.
24. Describe the methods employed for measuring serum total iron, total iron-binding capacity, transferrin, and ferritin.
25. Discuss the clinical significance and laboratory findings associated with iron deficiency and iron overload.
26. Describe the symptoms associated with Wilson's disease and the role of ceruloplasmin in copper metabolism.

INTRODUCTION

In addition to the important biochemical roles played by sodium and potassium, several other cations serve significant functions in the body. Calcium, magnesium, and iron are the major minerals in the circulation contributing to the proper operation of the living system. In each case, the metal exists in the cation form in the circulation. We will often refer to a specific metal as "calcium" or "iron" rather than its cation. In some instances, when different cation forms must be distinguished, as in the case of Fe(II) and Fe(III), we will indicate the oxidation state.

19.1

ESSENTIAL MINERALS AND THEIR ROLES

Calcium

Approximately 99% of the total calcium in the body is found in bone. The other 1% of the rough total of 1 kg of this mineral exists in the circulation and within the cell. In addition to its significant structural contribution to bone and tooth formation, calcium plays an important role in a number of biochemical processes. The calcium ion is an important activator of the coagulation system. Many anticoagulants employed in blood collection prevent blood clotting because they sequester calcium and remove it from the reaction sites. Calcium also serves as a necessary cation in a number of enzyme reactions. Nerve function and the transmission of the nerve impulses are regulated in part by calcium.

Magnesium

The average adult body contains about 20–28 g of magnesium, a little over 50% of which is in bone,

and another 25% or so in muscle. Like calcium, magnesium has an important role in a variety of enzyme reactions, especially those involving ATP formation and breakdown. These reactions are required for much of the biochemical energy of the body. Processes of muscle function, including muscle contraction, depend heavily on an appropriate availability of magnesium.

Iron

More than 2 g of iron are found in the healthy adult. Iron is a major constituent of the hemoglobin molecule and plays an important role in oxygen transport. About 30% of the iron in the body is stored iron, which is available when needed for hemoglobin synthesis. In addition to its contribution to oxygen utilization, iron is a component of several enzymes involved in the Krebs cycle and in the energy-producing reactions of mitochondria.

Other Metals

A wide variety of metals are present in very small amounts in the human body. Because of their low

Table 19-1
BIOCHEMICAL ROLES FOR MAJOR CATIONS

Cation	Functions
Ca	Coagulation Contributor to structure of bone and teeth Activator for enzymes Neurotransmission regulator
Mg	Contributor to bone structure Enzyme activator Muscle contraction
Fe	Oxygen binding in hemoglobin Cofactor for enzymes

concentrations, these minerals are frequently referred to as **trace metals.** Sometimes iron is included in the category of trace metals. Other important metals are copper, nickel, zinc, cobalt, chromium, manganese, and vanadium. In many instances, the metal serves as a necessary cofactor for a particular enzyme or enzyme class. In some instances, the role of the trace metal is not clearly understood at present. The roles of calcium, magnesium, and iron are summarized in Table 19–1.

19.2

CALCIUM

Forms of Calcium in Blood

Calcium exists in the circulation in three forms: bound, "ionized," and complexed. The bound form of calcium is made up of those ions attached to a transport protein, albumin. **Ionized calcium** is that portion of calcium which is not attached to protein and does not exist in a complex with other small molecules. This form circulates free in the bloodstream. The term *ionized* is a bit of a misnomer, since all three forms are calcium ions. The third form of calcium is a dissociable complex with anionic materials, particularly the carboxylate anion:

$$R-\overset{\overset{\text{O}}{\|}}{C}-O^- \text{————} Ca^{2+} \text{————} {}^-O-\overset{\overset{\text{O}}{\|}}{C}-R$$

The complexed form is not as available as the ionized form for participation in biochemical reactions. On occasion, the term *ultrafiltrable* calcium is used to describe a component that consists of both the ionized and the complexed calcium, since these two fractions can easily pass through a porous membrane, leaving the protein-bound calcium behind.

In the normal adult, approximately 50% of the calcium in the circulation exists in the ionized form (Table 19–2), with another 40% bound to albumin, and 10% in the complexed state. Urine calcium is some 85% ionized and 15% complexed. Since albumin does not normally filter through the glomerulus, essentially no protein-bound calcium is seen in the urine under normal circumstances.

Table 19–2
CALCIUM DISTRIBUTION IN BLOOD

Form of Calcium	Amount
Protein-bound	40%
Ionized	50%
Complexed	10%

Table 19–3
HORMONAL REGULATION OF CA

Hormone	Mode of Regulation
Parathyroid hormone	Enhance resorption from bone Stimulate vitamin D synthesis Enhance renal tubular reabsorption
Calcitonin	Stimulate calcium uptake by bone Decrease renal tubular reabsorption
Vitamin D metabolites	Enhance intestinal absorption Enhance resorption from bone Increase renal tubular reabsorption

Regulation of Calcium Levels

Calcium concentrations in the plasma are determined by a wide variety of dietary and hormonal factors in the healthy individual. Diet provides calcium for absorption into the system. Parathyroid hormone, calcitonin, and vitamin D all exert varying hormonal effects on the absorption and excretion of calcium, as well as on the phenomena of bone formation and resorption (Table 19–3). A complex interplay of biochemical processes provides a finely tuned mechanism for the control of calcium use by the body.

DIETARY

Calcium enters the body through the food we eat. A minimum daily requirement of 400–500 mg calcium intake for adults has been recommended by the World Health Organization, with increased intake for pregnant and lactating women. Milk and cheese are good dietary sources of this mineral. In developed countries, bread frequently contains reasonably high levels of calcium owing to supplementation with calcium sulfate, calcium phosphate, and organic calcium complexes used as preservatives and dough conditioners. However, bread consumption may provide only 2% or so of the daily required calcium intake.

Calcium is absorbed into the bloodstream from the duodenum and jejunum in the small intestine. Vitamin D metabolites play an important role in regulating calcium uptake by the intestine. Parathyroid hormone does not appear to be involved in this aspect of calcium metabolism. Some of the calcium is not absorbed across the gastrointestinal tract and is lost in the stool. In addition, over 200 mg of calcium per day is also secreted into the gut and eliminated from the body.

PARATHYROID HORMONE

Synthesis and Secretion

Parathyroid hormone (PTH) is a polypeptide composed of 84 amino acids. PTH is synthesized in the parathyroid glands, four small pieces of tissue imbedded in the thyroid glands in the neck (two parathyroid glands in each thyroid gland). As is the case with insulin and several other peptide hormones, PTH is synthesized as a longer chain precursor called pre-pro-parathyroid hormone. Upon synthesis, the initial 25-amino acid portion is enzymatically removed as the hormone moves to its storage site within the cell. The 90–amino acid prohormone is stored until needed. While in the storage granule, a significant amount of the prohormone is enzymatically cleaved to form the 84–amino acid active parathyroid hormone.

A decrease in the serum ionized calcium level triggers the release of parathyroid hormone. This release is very sensitive to the ionized Ca level, since a drop of only 0.1 mg/dL below the normal limit of ionized calcium is sufficient to trigger an increase of PTH in the circulation. The 84–amino acid chain is the primary component secreted into the bloodstream, although some smaller fragments (formed by proteolysis during storage) are also found. None of the parathyroid hormone precursors apparently enter the circulation. Magnesium is another essential ingredient for PTH release. If serum magnesium levels drop much below the normal limits, PTH secretion is noticeably inhibited.

Metabolism and Excretion

Once in the circulation, the serum concentration of parathyroid hormone decreases rapidly. The intact hormone has a half-life of about 5 min, as is the case with the *N*-terminal fragment of the hormone. The *C*-terminal segment has a half-life of roughly 1 h. Major sites of PTH breakdown are liver, kidney, and bone.

The liver is the primary location for fragmentation of the intact 84–amino acid PTH chain. From this hydrolysis comes the major amount of the *C*-terminal parathyroid hormone fragments found in the circulation (composed of amino acids 38–84 and portions of this chain). Neither the *C*-terminal nor the *N*-terminal (amino acids 1–34) segments appear to be further converted by the liver.

The kidney is important to the metabolism and excretion of PTH and its fragments. The small molecular weight of the intact molecule (under 10,000) means this molecule is filtered readily at the glomerulus. The fragments also undergo glomerular filtration since they are even smaller than the intact hormone. In addition, the intact form of the hormone and the *N*-terminal fragment undergo tubular secretion. A fraction of all three forms of the hormone are reabsorbed by the tubule.

Little metabolic uptake of intact hormone by bone occurs (10% or less of the total utilization by liver and kidney). The *N*-terminal fragment apparently has a higher bone uptake than do the other two forms of the hormone. Some inactivation of PTH by bone cells must occur after the hormone interacts

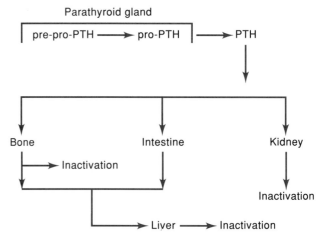

FIGURE 19–1 Synthesis and metabolism of PTH.

with a specific receptor to promote calcium resorption back into the circulation. The process of PTH synthesis and metabolism is summarized in Figure 19–1.

Biochemical Effects

Parathyroid hormone stimulates an increase in serum calcium through a variety of means (Table 19–4). The direct effect of PTH is on bone. Here the hormone causes calcium (and phosphate) to be released from bone back into the circulation. In addition, parathyroid hormone acts on the kidney, inhibiting the tubular reabsorption of phosphate while enhancing the reabsorption of calcium and magnesium. Further indirect effects of PTH at the kidney stimulate the formation of a vitamin D derivative which is also involved in calcium retention by the kidney and in the enhancement of calcium absorption across the intestinal tract.

Table 19–4
PARATHYROID EFFECTS ON SERUM CALCIUM LEVELS

Tissue Affected	Biochemical Effect Produced
Bone	Resorption of calcium and phosphate
Kidney	Direct enhancement of calcium and magnesium tubular reabsorption Direct blockage of phosphate tubular reabsorption Increased vitamin D metabolite formation
Intestine	Stimulation of vitamin D metabolite synthesis—leads to enhanced gastrointestinal uptake of calcium

CALCITONIN

Synthesis and Metabolism

Calcitonin is a peptide hormone produced by specific cells in the thyroid gland. This protein is synthesized in the cell as a 15,000-mol-wt precursor to the active hormone. Before release into the circulation, small peptides from each end of the precursor are removed by enzymes to yield the smaller, active calcitonin.

With a half-life of only about 10 min, calcitonin is rapidly cleared from the bloodstream once it is in the circulation. Although there appear to be materials in the plasma which can break down this peptide, the major means of calcitonin removal is the kidney. The peptide is filtered at the glomerulus, partially reabsorbed in the tubules, and apparently metabolized to small fragments (or amino acids) while still in the kidney. Little of the intact hormone is found in the urine of healthy individuals.

Biochemical Effects

The major effect of calcitonin on calcium metabolism is to lower blood levels of this mineral. When calcitonin interacts with bone, calcium is absorbed by bone and bone mass increases. Under the influence of increased calcitonin, the kidney decreases the renal tubular absorption of calcium and phosphate, as well as of magnesium, and more of these materials are excreted into the urine. Many details of the biochemical actions of calcitonin are still very unclear.

VITAMIN D AND METABOLITES

The role of vitamin D in the regulation of calcium metabolism is complex. Inactive precursors must first be converted to active compounds and then interact at a variety of sites to facilitate increase in serum calcium levels (Fig. 19–2). Proper functioning of the liver, kidney, and parathyroid gland is necessary for this intricate process to occur.

The precursor to the active form of vitamin D can be obtained in two ways. A portion of the necessary vitamin D_3 is consumed in the diet. A significant fraction of the D_3 is also formed in the skin through conversion of 7-dehydrocholesterol by ultraviolet radiation, usually from sunlight. Individuals in climates with little or no sunshine during a part of the year more frequently manifest vitamin D deficiency than do people in other areas of the world.

Table 19–5
EFFECTS OF VITAMIN D ON CALCIUM METABOLISM

1. Increased absorption from intestine
2. increased resorption from bone
3. increased reabsorption from kidney tubule

When vitamin D_3 enters the circulation, it binds to vitamin D transport protein (a globulin) to be carried through the bloodstream. The D_3 then travels to the liver, where it is converted to 25-hydroxyvitamin D_3. This component is still inactive and must be further hydroxylated to 1,25-dihydroxyvitamin D_3 by the kidney. The specific enzyme involved in this activation is itself controlled by parathyroid hormone. When serum calcium levels drop, PTH is released and the hydroxylase enzyme becomes active. The 1,25-dihydroxy derivative is then biochemically active, working in concert with parathyroid hormone to regulate calcium levels in the circulation. During pregnancy, the placenta has the capacity to carry out this conversion to a significant degree. An inactive 24,25-dihydroxy metabolite is also produced. The ratios of the 1,25 form and the 24,25 form are useful in assessing vitamin D status in the body.

There are three major actions of 1,25-dihydroxyvitamin D_3 (Table 19–5). In the intestine, this vitamin stimulates the synthesis of proteins which bind to calcium. More calcium is trapped and absorbed into the circulation through this process. In conjunction with parathyroid hormone, active vitamin D_3 increases the amount of calcium resorbed from bone and returns it to the bloodstream. The vitamin also acts on the kidney to enhance the reabsorption of calcium by the tubule, decreasing the amount of this mineral excreted in the urine. By stimulating an increase in intestinal absorption, an increase in bone resorption, and a decrease in urine excretion of calcium, 1,25-dihydroxyvitamin D_3 produces a rise in the serum total calcium concentration.

SERUM PROTEIN CONCENTRATION

The amount of protein in the circulation affects the calcium concentration by altering the bound-to-free calcium ratio. If the serum albumin level decreases, less protein is available to bind calcium. A transient increase in the amount of unbound calcium results. This rise in the ionized calcium fraction is detected by the parathyroid, which decreases

FIGURE 19–2
Vitamin D_3 metabolism and activation.

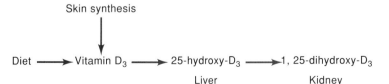

PTH production and lowers the total calcium concentration. Although the total calcium value may be low, the unbound fraction of calcium has been restored to a normal level.

Conversely, in some clinical situations (such as multiple myeloma) the concentrations of certain proteins increase. More calcium than usual binds to these proteins, causing a decrease in the amount of ionized calcium present. The body then adjusts parathyroid hormone production to restore the ionized calcium value to normal, but increases the total calcium concentration as a result.

Assays for Calcium and Hormones

TOTAL CALCIUM

Routine analysis for total calcium in serum is performed by colorimetric methods. A wide variety of organic dyes may be used to form complexes with Ca^{2+} and produce a colored derivative. The most commonly employed dye is cresolphthalein complexone; it produces a reddish complex with an absorbance maximum at 570 nm. Other dyes employed are alizarin, methylthymol blue, and arsenazo III (in the Kodak Ektachem).

Protein interference may be a problem, particularly if the calcium is complexed to protein (such as albumin) and cannot react fully with the dye. Adjustment of pH allows dissociation of calcium from protein. In some automated systems (such as Technicon), the large protein molecules can be separated from the calcium ions by dialysis before reaction with the organic dye.

Another prominent interferent is magnesium, which can also form complexes with many of the dyes. Addition of 8-hydroxyquinoline produces a magnesium/quinoline complex, removing this cation from reaction in the calcium assay.

High concentrations of bilirubin cause false elevations in the value for total calcium when using the cresolphthalein complexone method. Lipemia can also give rise to falsely increased values. Hemoglobin and elevated levels of plasma proteins tend to interfere slightly with the calcium/dye reaction and lower values somewhat.

Fluorometric assay methods for calcium have been widely explored, but are not frequently used in clinical settings. Fluorescence measurements are more sensitive, but also have several problems not seen in colorimetric assays. In addition, more laboratories have the facilities for colorimetric analysis than for fluorescence measurements.

The reference method for total calcium assay is atomic absorption spectroscopy. This approach is sensitive and specific, but does not lend itself well to production approaches for the delivery of large numbers of laboratory results. Atomic absorption is useful in establishing the values for standards and as a check on other methods.

IONIZED CALCIUM

Measurement of total calcium provides information about the amount of calcium available in the circulation, but it is the "free" calcium level which is of physiological and clinical importance. The calcium attached to protein (and the complexed calcium, to some extent) are not accessible to the cell. Measurement of ionized (free) calcium provides direct information about the amount of biochemically available calcium.

The most common instrument for quantitating ionized calcium in serum is the ion-selective electrode. In this method, the calcium ion interacts with a molecule on the electrode surface to produce a potential change which is proportional to the concentration of calcium ions in the sample. A number of commercially available electrodes can be purchased for this analysis.

Since pH plays an important role in the dissociation of calcium from proteins, the sample must be treated in such a way as to minimize pH shifts from the time of collection to the time of analysis. This control is best exerted by anaerobic collection of blood (using heparin as an anticoagulant), followed by assay of either whole blood or plasma. Whole blood is the preferred sample since less handling is required after collection.

A second analytical approach involves the determination of ultrafiltrable calcium—the sum of ionized calcium and the calcium complexed with anions. After the sample is collected and the red cells are removed, the plasma is passed through a membrane with pores small enough to block all proteins. Only the calcium not bound to albumin or other proteins is filtered and separated. After the ultrafiltration, a colorimetric or other method can be used for calcium determination. This approach allows a simpler analytical method and avoids the problems associated with the more complicated ion-selective electrode system.

PARATHYROID HORMONE

Radioimmunoassay (RIA) represents the primary means of quantitating parathyroid hormone, in spite of a wide variety of technical and clinical problems. At present, there are several different methods for measuring PTH, either as the intact molecule or as the C-terminal or N-terminal fragment. With the intact molecule, antibodies can be specifically directed to either the N-terminal or C-terminal portion of the hormone. In addition, there is a **middle molecule** assay available which has some clinical utility.

The major problems associated with specificity in the assay of parathyroid hormone center on the heterogeneity of serum proteins which react with a particular antibody. If an assay is employed for the *C*-terminal portion of the hormone molecule, the antibody detects both the intact PTH molecule and the *C*-terminal fragment produced by proteolysis. Similar problems occur with an *N*-terminal assay. Measurement of the midmolecule is ostensibly superior in detecting intact parathyroid hormone, but also quantitates any of the fragments with the amino acid sequence corresponding to the middle region of the intact hormone. Deciding which PTH assay to employ must be done in conjunction with information about the particular clinical situation, since the various assays differ in their usefulness in the several disease states in which calcium and PTH metabolism are altered.

CALCITONIN

The major assay for calcitonin at present is radioimmunoassay. With the advent of synthetic human calcitonin, it has been relatively easy to develop fairly specific antibodies to this hormone, but there are still some problems with specificity. Human calcitonin appears to be somewhat heterogeneous from an immunochemical standpoint. Antibodies form which seem specific to various regions of the peptide hormone. However, during the long incubation period (some assays require 1–3 days), the calcitonin may degrade and resulting multiple fragments may complicate the accurate measurement of intact protein.

Serum samples seem to yield higher calcitonin levels than do plasma samples. Hemolysis falsely elevates values in some RIA procedures. Sensitivity is a problem in most assays; it is frequently not possible to measure values accurately throughout the normal range. Although an upper limit for the reference range can be somewhat reliably defined, the lower limit is still unclear. Therefore, only elevated calcitonin values can be assessed with confidence, not decreased values. Variability in the purity of standards and the problems with stability of standards further complicate the assay for calcitonin.

VITAMIN D

Quantitation of vitamin D metabolites in plasma is a complex process, complicated by the low concentrations involved and the presence of a variety of materials with very similar structures. All current assay approaches involve some type of extraction to remove the vitamin from its transport protein. An organic solvent is employed to extract the vitamin since its structure (somewhat similar to the cholesterol molecule) gives it lipid characteristics.

Further purification is usually necessary to separate the various vitamin D components. Often, multiple column fractionations are employed to remove unwanted materials such as cholesterol and derivatives. In many assays, the different vitamin D metabolites are separated by high-performance liquid chromatography (HPLC), followed by some immunoassay method to quantitate the fractions of interest. Direct measurement by HPLC is usually not feasible owing to the low ultraviolet absorbance of the fractions. Development of new detector systems may alleviate this problem.

Immunoassay measurement can take several directions. Competitive protein-binding assays use the vitamin D transport protein to bind patient and labeled vitamin D. This method is particularly useful for measuring 25-hydroxyvitamin D_3—the metabolite with the longest half-life and the highest concentration in serum. Specific antibodies have been developed for measurement of the 1,25-dihydroxy metabolite, found in much lower concentrations. Improvements in developing appropriate antibodies are still needed. Radioreceptor assays (using vitamin D receptors isolated from thymus or from osteosarcoma cells) are reasonably accurate and sensitive in detecting and quantitating the 1,25-dihydroxy form of vitamin D_3.

Role of Phosphate in Calcium Metabolism

FORMS AND ROLE OF PHOSPHORUS

Phosphate is somewhat of an "orphan" in the scheme of calcium metabolism. This inorganic anion is frequently measured (owing to the wide use of multichannel analyzers), but not always appreciated. Referred to as "phosphorus," the entity actually measured is a mixture of two phosphate anions which exist in equilibrium:

$$H_2PO_4^- \rightleftharpoons HPO_4^{2-} + H^+$$

The ratio between the two forms varies markedly with pH; a small change in hydrogen ion concentration shifts the equilibrium strongly. Because of this, the parameter is reported as "phosphate" with no attempt to differentiate the various ionic species.

Most phosphorus present in the body does not exist in the free form. Approximately 80% is found in bone as part of calcium phosphate complexes. Another 15% is located in muscle. Phosphate exists as part of the structures of a wide variety of intracellular compounds. Most carbohydrates are phosphorylated before being metabolized further. A variety of nucleosides have phosphate attached to the carbohydrate portion. Enzyme removal and addition of phosphate to these nucleosides are important reactions for biochemical energy storage or release (ATP) or cell receptor function (guanosine triphosphate, GTP). The phosphate group is an important part of the phospholipids, whose roles include providing structural integrity for the cell membrane and furnishing surfactant to the lungs.

SOURCE OF PHOSPHATE

The average person consumes between 800 and 1200 mg phosphorus per day from a wide variety of foods. Almost 80% of this material is absorbed in the small intestine, regulated to a great extent by vitamin D. Significant reabsorption of phosphate occurs in the proximal tubule of the kidney under normal circumstances. This reabsorption process is markedly enhanced in cases of phosphate depletion. Approximately 30–40% of the daily intake is excreted in the stool. Excessive use of antacids has been shown to facilitate fecal loss of phosphate.

ASSAYS FOR PHOSPHATE

The classic (and still widely employed) assay for serum phosphate involves the reduction of this anion to form a molybdenum blue complex. An initial complex is formed with molybdate which can then be reduced to molybdenum blue by a variety of materials. Although stannous chloride is frequently used in both manual and automated systems, other agents also find application. In each instance, the goal is to generate a stable form of the molybdenum blue complex using a small sample and rapid reaction conditions.

More recently, enzymatic approaches have been developed which employ complex series of reactions. In most instances, the phosphate group is incorporated (using an enzyme reaction) into either a carbohydrate or a nucleoside, forming the corresponding phosphorylated derivative. This derivative can then participate in another enzyme reaction which allows meaurement of the phosphorylated material. The enzymatic systems often involve use of three enzymes, making them much more expensive than the colorimetric methods. The relative accuracy and cost-effectiveness of these new assays remain to be determined.

CALCIUM/PHOSPHATE RELATIONSHIPS

Interpretation of phosphate levels alone can be misleading and provides little useful information. In many instances, the primary pathologic condition may alter calcium levels, leading to secondary shifts in the phosphate values. Frequently, there is an inverse relationship between calcium and phosphorus values. When calcium levels increase (as in primary hyperparathyroidism or some cases of malignancy), the phosphate levels decrease. If the calcium in-

crease is due to renal failure, phosphate values also rise. Both calcium and phosphate values are elevated in cases of multiple myeloma and vitamin D overdose, as well as in cases of cancers which metastasize to bone.

Calcium, Hormones, and Bone Metabolism

COMPONENTS OF BONE

The processes of bone formation and resorption are complex and still somewhat poorly understood. Bone tissue is in a state of dynamic flux, with bone cells forming and breaking down constantly. This build-up and turnover of cells is regulated by the complicated interplay of diet, parathyroid hormone, calcitonin, and vitamin D.

The protein portion of bone is composed primarily of **collagen,** a long fibrous protein. These chains are assembled into the protein matrix of the bone after synthesis by the **osteoblast,** a specialized cell responsible for bone formation. The long fibers constitute the foundation upon which the calcium salts deposit to form the mineral component of bone.

Calcium and phosphate are the primary constituents of the mineralized portion of bone. Bone is formed from either calcium phosphate or the more complex salt hydroxyapatite, which has the formula $Ca_{10}(PO_4)_6(OH)_2$. A variety of binding proteins facilitate the incorporation of the mineral into the collagen matrix to form a strong protein/mineral material.

Another specialized cell participates in the breakdown of bone tissue. After the proper stimuli, **osteoclasts** produce and release enzymes responsible for collagen breakdown. When collagen is resorbed, the minerals are released back into the circulation. Calcium levels increase until hormonal and excretory control processes restore the proper balance.

HORMONAL REGULATION OF BONE FORMATION AND METABOLISM

The three hormones PTH, calcitonin, and vitamin D act in concert to maintain calcium homeostasis and appropriate bone development. This balance is accomplished by effects on the intestinal absorption of calcium, rate of calcium deposition into bone or loss from bone, and reabsorption of calcium by the kidney. Although each hormone has its own specific effects, there is also some interaction between hormones, especially parathyroid hormone and the vitamin D metabolites.

Changes in the serum ionized calcium concentration trigger the release of specific hormones to maintain the proper calcium balance in the system. A decrease in ionized calcium (for whatever reason) promotes an increase in the blood levels of parathyroid hormone and vitamin D. If ionized calcium levels rise, calcitonin concentrations increase to promote calcium uptake by the bone.

A decrease in ionized calcium triggers a complex series of events. Parathyroid hormone output is increased, which stimulates loss of calcium from bone and increased calcium resorption in the proximal tubule of the kidney. At the same time, PTH also acts on the kidney to produce larger amounts of the 1,25-dihydroxyvitamin D derivative necessary for enhanced absorption of calcium from the intestine. Vitamin D also plays a secondary role in the resorption of calcium from bone. During this time, the production of calcitonin is markedly diminished.

When ionized calcium levels are high, the opposite processes occur. Calcitonin levels rise to promote calcium uptake by the bone. PTH and 1,25-dihydroxyvitamin D concentrations decrease, allowing less calcium to be absorbed by the intestine and permitting more calcium excretion in the urine.

Clinical Situations Involving Altered Calcium Levels

HYPERCALCEMIA

Calcium levels above the established reference range of 8.5–10.5 mg/dL are common in a wide variety of diseases (Table 19–6). Patients demonstrate a number of neurologic symptoms, including fatigue, muscle weakness, and disorientation. In extreme cases stupor and coma can result. Impairment of renal function and the presence of kidney stones is quite common. Nausea, vomiting, and gastrointestinal distress occur frequently. In some instances, psychiatric abnormalities may be seen, including depression or psychotic behavior.

A high percentage of patients with cancer exhibit increased serum calcium values. If the tumor has metastasized to bone, the explanation is clear: bone

destruction by the cancer is occurring, producing increased release of bone mineral. When metastasis is absent, some increases can be explained by the biochemistry of the tumor. Many cancers produce parathyroid hormone within the cancer cell (ectopic production of PTH). This fraction of parathyroid hormone is not regulated by the normal feedback mechanisms. If calcium levels rise, there is no compensatory shutdown of ectopic hormone release. As a result, bone deterioration occurs and calcium loss is enhanced. In many instances, direct production of PTH has not been documented, but the presence of other peptides which appear to have some impact on bone resorption has been demonstrated.

Primary hyperparathyroidism is responsible for most cases of hypercalcemia in nonhospitalized patients. In the majority of instances, the excess release of parathyroid hormone is caused by a tumor on the parathyroid gland. In some 15% of situations, parathyroid hyperplasia exists (enlargement of the parathyroid gland). In either situation, there is a higher blood level of PTH and increased serum calcium values.

Thiazide diuretics increase levels of serum calcium, although the mechanism is unclear. A complex interplay may occur between increased bone resorption and inhibition of calcium excretion by the kidney.

Excessive intake of vitamin D promotes calcium increases through enhanced absorption of calcium by way of the gastrointestinal tract and increased bone resorption. Patients with sarcoidosis also demonstrate increased serum calcium levels, perhaps owing to supersensitivity to the effect of vitamin D.

In patients with multiple myeloma, elevated calcium levels could be attributed to two factors. The first problem is associated with tumors of the bone and increased calcium produced by this phenomenon. In addition, patients with multiple myeloma exhibit high concentrations of immunoglobulin fragments (Bence Jones proteins). This marked increase in protein concentration distorts the equilibrium between ionized and serum total calcium. More calcium is bound by the excess proteins, with a consequent lowering of the ionized calcium fraction. To compensate, the body releases more calcium from bone, giving rise to an elevated total calcium value.

HYPOCALCEMIA

A significant decrease in serum calcium (Table 19–7) can produce distinct clinical signs. Perhaps the most striking feature of this problem is **tetany**: muscle spasm, cramps, and irritability produced by lowered availability of calcium for the contraction/relaxation processes in muscle tissue. Dementia, mental retardation, and other neurologic problems are often associated with lowered serum calcium.

Table 19–6
FACTORS INVOLVED IN HYPERCALCEMIA

Cause	Reason for Elevated Ca
Cancer	Bone destruction Ectopic PTH synthesis
Hyperparathyroidism	Tumor on parathyroid gland Parathyroid hyperplasia
Diuretics	Complex effects on excretion
Excess vitamin D	Enhanced resorption
Multiple myeloma	Bone destruction Increased protein binding; lowers ionized Ca fraction

Table 19-7
DECREASES IN SERUM CALCIUM

Cause	Reason for Lowered Calcium
Low serum albumin	Increase in ionized Ca fraction
Hypoparathyroidism	Impaired production of PTH
Decreased vitamin D	Impaired intestinal absorption Impaired release from bone Impaired renal tubular reabsorption

Decrease in the serum albumin concentration frequently produces hypocalcemia. Since albumin serves as a major transport protein for calcium, any drop in the concentration of this protein lowers the amount of bound calcium and increases the level of ionized calcium. To restore the proper amount of ionized calcium, the body activates those mechanisms designed to decrease serum total concentrations of this mineral.

Hypoparathyroidism is a common cause of lowered serum calcium. The damage to this gland may be the result of surgery or produced as a secondary effect of other unrelated disease processes. In some instances, hypocalcemia occurs because of the manufacture of defective parathyroid hormone.

A variety of problems associated with decreases in vitamin D intake and metabolism are linked with lowered serum calcium levels. Inadequate diet deprives the body of sufficient calcium. Malabsorption may occur in the intestinal tract. In cases of liver disease, impaired conversion of precursor to active forms of the vitamin may be seen. Often, medications which induce enzymes associated with the metabolism of the 1,25-dihydroxy form of the vitamin contribute to a drop in the vitamin D concentration and a lowering of serum calcium. Anticonvulsant drugs are well known for this effect, which can also be seen in some people with excessive ethanol intake. The presence of renal disease, with a decrease in the conversion of the inactive form of the vitamin to the active 1,25-dihydroxy compound, often produces lowered serum calcium values. Partial loss of the capacity to reabsorb calcium in the proximal tubule also plays a role in this phenomenon.

19.3

MAGNESIUM

Absorption and Utilization of Magnesium

The only source of magnesium for the body is dietary intake. Cereal grains, green vegetables, and meat provide most of the magnesium in the diet. Water can be a significant contributor to magnesium consumption in areas where the water has a high mineral content.

Absorption of magnesium takes place in the small intestine. Of an average daily intake of some 300–350 mg, approximately 40% is absorbed through the gastrointestinal tract. As the level of magnesium in the intestine increases, the fraction of magnesium absorbed decreases. We do not understand most of the factors involved in the uptake of magnesium by the intestine.

Magnesium exists in the circulation in three forms: protein-bound (33%), free (55%), and complexed with anions (12%). Most of the protein-bound magnesium is complexed with albumin; a small fraction interacts with globulins. Only the free fraction is biochemically active; the complexed and protein-bound portions do not participate directly in biochemical processes.

Slightly over 50% of the magnesium in the body is found in bone, but it is not as tightly complexed in this tissue as is calcium. Some 25–28% of total body magnesium is located in muscle, signifying the important role this cation plays in the complex process of muscle contraction. Another 20% is found in soft tissues; the liver has a relatively high magnesium content. Only 0.3% of the total body magnesium is located in serum, with another 0.5% in erythrocytes.

Magnesium excretion is accomplished by the kidney. Filtration occurs at the glomerulus, with approximately 20–30% of the filtered magnesium being reabsorbed. Little or no tubular secretion of magnesium appears to take place in the kidney. Normal daily excretion of magnesium is roughly equivalent to the amount taken in with the diet in healthy individuals.

Unlike calcium, there is no known hormonal regulation of magnesium. Variations in parathyroid hormone release may have an indirect impact on serum and tissue magnesium levels (owing to changes in calcium metabolism), but no direct effect of hormone production on magnesium levels has been observed.

Assays for Serum Magnesium

In the routine clinical laboratory, magnesium determination is carried out using either colorimetric or fluorometric assays. For colorimetric analysis, various dyes are employed which absorb light at a given wavelength in the presence of Mg^{2+}. Calgamite (a naphthol sulfonic acid derivative) is widely used. In the presence of polyvinylpyrrolidone (used to minimize the effects of serum proteins), a violet complex forms which absorbs light at 532 nm. Other dyes employed include methylthymol blue (used on the DuPont aca) and Titan Yellow.

Fluorometric approaches include the use of 8-hydroxyquinoline or calcein. Problems common to other fluorometric assays are encountered, including variable blanks and difficulties with background stability.

Measurement of serum magnesium has been largely restricted to determinations of serum total content. With the advent of ion-selective electrodes, the free fraction can be determined independently. These analytical techniques are still in the experimental stage and have not found wide application in the clinical laboratory. The use of ultrafiltration followed by analysis of the sum of the free and complexed fractions is also undergoing study. Although atomic absorption spectrometry is highly reliable, specific, and sensitive, this method is not readily available outside of large, well-equipped laboratories.

Since erythrocytes contain almost twice as much magnesium as does serum, any hemolysis artificially raises the serum magnesium content.

Clinical Utility of Magnesium Measurements

The reference range for serum magnesium in adults is approximately 1.3–2.1 mEq/L, with children and adolescents having essentially the same values as adults. Diuretics induce severe urinary loss of magnesium, decreasing the serum value. Thiazide and loop diuretics cause the greatest excretion of magnesium. Diuretics which facilitate the conservation of potassium by the body frequently contribute to magnesium retention. Other drugs, such as antibiotics, some anticancer agents, and cardiac glycosides, enhance magnesium loss in the urine.

Increases in serum magnesium are usually associated with renal failure, some forms of leukemia, and situations involving diabetes mellitus. Patients with necrosis of the liver may also manifest an elevation of magnesium.

Magnesium-deficient patients experience a wide variety of neuromuscular problems, including tetany, convulsions, irritability, and significant cardiac arrhythmia. A major concern for any cardiac patient is his or her magnesium status, since this cation plays an important role in the contractile processes of the heart. Common causes for hypomagnesemia include impaired dietary intake and absorption or excessive loss in the urine or other body fluids. Magnesium depletion is frequently associated with loss of other electrolytes such as sodium, potassium, or calcium. In a wide variety of renal diseases (including glomerulonephritis, renal tubular acidosis, and after kidney transplants), excessive magnesium excretion is frequently seen.

19.4

IRON

Iron plays several important roles in proper biochemical functioning. As a component of the metal-loprotein hemoglobin, iron serves to bind oxygen for transport to cells. Over 65% of the iron in the adult male is found in hemoglobin. Several enzymes involved in the utilization of oxygen and the formation of biochemical energy have iron as part of their active site. These proteins contain only 1% of the total body iron. The body has mechanisms for the storage of iron to allow ready release for hemoglobin synthesis when needed. Some 30% or so of the iron in our bodies is in the stored forms. Iron is an essential element for our survival.

Dietary Intake of Iron

Iron enters the system through the diet. Major sources for iron are meat (particularly liver), beans, and soybean flour. Many bakery products are now supplemented with iron to provide additional intake. Some authorities question the rationale for supplementation for the general population. In most cases, the average daily iron requirement is about 1 mg, although menstruating or pregnant women require higher amounts. Since a good diet provides much more iron than necessary, iron supplementation for a large portion of the population is usually unnecessary.

Iron loss occurs primarily through breakdown of epithelial cells in the intestine and skin, resulting in a daily loss of approximately 1 mg. During menstruation, another milligram may be lost each day. Significant iron loss quite often points to some type of internal bleeding problem.

Absorption and Transport of Iron

The intake of iron from the intestine to the bloodstream is regulated to a great extent by the intestinal mucosal cells. Major factors controlling iron uptake are the amount of stored iron already present and the rate of red blood cell synthesis. Of the two, the level of stored iron has much more effect on the rate of iron uptake. As the amount of serum **ferritin** (an iron-storing protein) increases, the rate of intestinal absorption of iron declines. Since iron is not lost through the urine, the body must carefully regulate its uptake in order to avoid iron overload.

Storage and Utilization of Iron

When iron first enters the intestinal mucosal cell, it exists as the Fe(II) ion (Fig. 19–3). Inside the cell, the iron is oxidized to the +3 (ferric) state. It can then be bound as ferritin or further transported into the circulation. Once in the bloodstream, the Fe(III) attaches to **transferrin,** an iron-transport protein. The iron can then move to the **hepatocyte** (a specialized liver cell), where it is stored until needed. The hepatocyte is also responsible for breakdown of

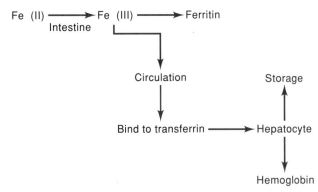

FIGURE 19-3 Iron uptake and transport.

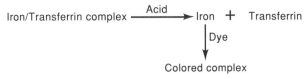

FIGURE 19-4 Assay for serum iron.

red cells, followed by recycling of the iron contained in these structures. Transferrin also delivers iron to reticulocytes and bone marrow so it can be employed directly for heme synthesis and incorporation into the hemoglobin molecule.

Two other proteins are involved in iron transport, but have different functions from that of transferrin. **Haptoglobin** binds hemoglobin iron and serves to facilitate disposal of this molecule. **Hemopexin** attaches to heme to aid in its removal from the circulation. Other proteins are concerned with intracellular iron transport, but little is definitely known at present about these molecules and their roles in iron metabolism.

Ferritin serves as the major storage form of iron. This intracellular molecule consists of a somewhat spherical protein shell inside of which the iron is stored. Some ferritin is released into the circulation and can be measured to assess iron deficiency. Ferritin has several **isoforms** (the transport protein equivalent of an isoenzyme). The various isoforms may have differing metabolic roles in iron utilization. Another molecule concerned with iron storage is **hemosiderin,** presumably an aggregate of several ferritin molecules.

Assessment of Iron Status

To obtain a reliable picture of iron metabolism and utilization, several parameters must be explored. The simplest (but perhaps least sensitive) avenue is to measure hemoglobin alone or in conjunction with hematocrit. This determination indicates the amount of iron used to form an adequate number of red blood cells. More specific approaches involve quantitation of serum iron and iron-binding capacity (to assess the level of serum transferrin). The amount of stored and metabolically available iron can be ascertained from determination of serum ferritin levels. Which test or tests to employ depends on the clinical picture and the severity of the problem.

TOTAL SERUM IRON

Before iron reacts with a reagent to form a measurable complex, it must first be dissociated from transferrin (Fig. 19-4). Treatment with acid solution results in fairly complete removal of iron from its transport protein. In some procedures, the protein is then precipitated and removed, or the iron is dialyzed out of the protein-containing material. In other systems, a direct measurement of iron is made without protein removal. A variety of organic chemicals are available which form colored complexes with iron, allowing colorimetric quantitation of serum iron.

Although slight hemolysis creates no significant problems for the accurate determination of serum iron, specimens with any major degree of hemolysis should be avoided. Standards should contain protein to approximate the conditions involved in the assay of patient samples. Care must be taken to use distilled, deionized water. Iron contamination of the water supply results in high blank values and erroneous increases in serum iron levels.

IRON-BINDING CAPACITY

The most common method of assessing transferrin levels is to measure the total iron-binding capacity (TIBC) (Fig. 19-5). Excess ferric salts are added to an aliquot of serum to saturate all the binding sites on transferrin. The unbound iron is then precipitated with solid magnesium carbonate. After centrifugation, the supernatant is analyzed for

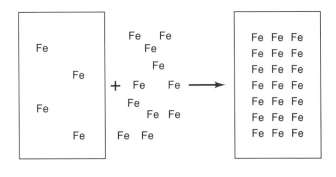

FIGURE 19-5 Assay for total iron-binding capacity.

iron content using the same method employed for serum iron determination. The amount of iron measured shows the total capacity for iron transport of the transferrin present in the patient serum. The unsaturated iron-binding capacity can then be calculated, if desired, by subtracting the serum iron concentration from the TIBC to get an estimate of the amount of transferrin available for further iron binding.

TRANSFERRIN

Immunochemical methods are available for the direct measurement of serum transferrin as a protein. The most commonly reported approach is nephelometric assay. There does not now seem to be any particular advantage to the direct assay of transferrin instead of TIBC measurement. Comparison of TIBC and serum iron values establishes a quick and direct relationship between serum iron content and the total transport capacity available.

FERRITIN

Ferritin assays are becoming increasingly important to assessing depletion of iron stores. A variety of immunoassay techniques (both IRMA and enzyme-linked systems) are commercially available. Particular attention must be paid to reliability at the low end of the standard curve in order to detect iron-deficient patients.

Iron and Clinical Situations

Reference ranges for serum iron are approximately 65–180 μg/dL, with women showing slightly lower concentrations than men. There is a strong diurnal pattern to serum iron levels, with values being some 10% higher in the morning than in the evening. Total iron-binding concentrations in adults are between 250 and 450 μg/dL. Ferritin values differ markedly between men (20–300 μg/L) and women (10–120 μg/L), depending on the specific assay employed.

IRON DEFICIENCY

Although a decrease in iron intake or an increase in iron loss produces anemia, a wide variety of clinical situations also lower hemoglobin concentrations without affecting iron stores. Liver disease, kidney disease, various types of cancer, thalassemia, abnormal hemoglobins (HbS and others), erythrocyte enzyme deficiencies, and chronic disease are some of the many situations which contribute to the development of an anemic state.

A decline in body iron may occur because of inadequate dietary intake, loss of blood (injury, menstruation, internal bleeding), or increased demand (pregnancy). There are three major stages in the development of overt iron-deficiency anemia. The first stage involves depletion of stored iron. Iron reserves decrease and serum ferritin drops below 12 μg/dL. Since iron is still available to fuel red cell formation, serum iron, TIBC, and hemoglobin levels are usually within the normal range. Iron-deficient erythropoiesis (the second stage of iron deficiency) is reached when the iron supply declines below normal (decrease in serum iron value), but circulating hemoglobin concentrations have not yet been affected. Overt iron-deficiency anemia is manifest by a drop in hemoglobin, indicating impairment of normal red cell synthesis. The stages of iron-deficiency anemia are summarized in Table 19–8.

One recent development in the battery of tests used to assess iron-deficiency anemia is the measurement of zinc protoporphyrin (ZPP). In the heme synthesis pathway, Fe(II) is added to protoporphyrin, followed by incorporation of the protoporphyrin/iron complex into the heme molecule. If iron is not present in adequate amounts, a zinc ion is inserted into the porphyrin ring in place of the iron. As iron stores (as assessed by ferritin measurements) decline, more ZPP forms. When ferritin levels drop to approximately 50 μg/L (still within normal limits), zinc protoporphyrin concentrations begin to increase. The rise in ZPP levels is parallel to the further decline in ferritin concentrations, allowing an earlier detection of iron depletion. ZPP can be measured fluorometrically with a specialized instrument.

IRON OVERLOAD

Iron overload occurs when there is excessive intake of iron. Under normal circumstances, the body has no way to excrete this excess iron. Accumulation occurs primarily in the liver, although other organs can be affected. Progressive hepatic damage develops, culminating in cirrhosis and hepatic failure.

Table 19–8
STAGES OF IRON DEFICIENCY

Stage	Laboratory Findings
Iron depletion	Low ferritin Normal serum Fe, TIBC, hemoglobin
Iron-deficient erythropoiesis	Low ferritin, serum Fe, and TIBC Normal hemoglobin
Iron deficiency	Low ferritin, Fe, TIBC, and hemoglobin

There are a variety of causes of iron overload. **Hemochromatosis** is a hereditary disease in which intestinal mucosal uptake of iron increases. As a result, frequent blood transfusions or injection of iron can cause iron accumulation. Iron overload can also be induced through excessive dietary intake over a long time.

The biochemical parameter which rises first in iron overload is the serum ferritin value. Serum iron values also increase, as does the percentage saturation of transferrin (often referred to as the **unsaturated iron-binding capacity**). TIBC may not be affected, since this parameter assesses only the total amount of iron able to bind to transferrin if the protein is fully saturated.

Iron overload is usually treated by therapeutic phlebotomy, since one 500-mL blood sample may contain up to 250 mg of iron. Administration of iron-chelating agents (such as desferrioxamine) allows the chelated iron to be excreted in the urine. This approach is effective, but much slower than phlebotomy in restoring normal iron concentrations.

19.5

CLINICAL SIGNIFICANCE OF OTHER MINERALS

A wide variety of other metals are considered essential for life (Table 19–9). Since these materials are usually present in very low amounts, they are often referred to as **trace metals.** Each is believed essential since it is found in all tissues of the body. Decrease in the concentration of an element below a certain level results in biochemical changes which produce specific clinical abnormalities. In most instances, the clinical problem can be reversed by dietary supplementation with the necessary trace metal.

Zinc

A required component of a large number of enzymes, zinc is involved with several metabolic processes. Synthesis of DNA and RNA, development of connective tissue, and regulation of several enzymes (including alcohol dehydrogenase, alkaline phosphatase, and carbonic anhydrase) are all affected by zinc concentrations in the body.

Copper

Copper is also a necessary metal cofactor in several important enzymes, including that involved in the oxidative phosphorylation pathway integral to ATP synthesis and oxygen utilization. Changes in copper levels result in significant alterations in the production of hemoglobin, mainly owing to interference with iron absorption and transport.

In the liver, copper is incorporated into the protein ceruloplasmin for transport throughout the body. A metabolic defect, known as **Wilson's disease,** causes a marked decrease in the amount of ceruloplasmin produced. Copper accumulates in the liver and is found in high amounts in the spinal fluid. Since copper excretion by way of the biliary tract is impaired, body stores of copper rise, producing the neurological and other problems associated with this disease.

Manganese

The active sites of enzymes such as pyruvate carboxylase and mitochondrial superoxide dismutase contain manganese. This metal helps regulate the synthesis of cholesterol and various mucopolysaccharides—gelatinous compounds that lubricate the joints and hold cells together. Manganese also plays an important role in oxidative phosphorylation. In animals, manganese deficiency impairs growth and produces central nervous system disorders. The impact of manganese deficiency in humans is still unclear.

Cobalt

This trace metal is an integral part of vitamin B_{12} (cobalamin), the coenzyme responsible for methyl transfer reactions. Deficiency of cobalt (or cobalamin) produces anemia because of B_{12}'s role in the formation of red blood cells. Assay for vitamin B_{12} is an important part of any work-up to identify the specific cause of anemia.

Table 19–9
OTHER ESSENTIAL TRACE METALS

Metal	Roles in Body
Zinc	Active site of enzymes Synthesis of DNA and RNA
Copper	Active site of enzymes Hemoglobin synthesis
Manganese	Active site of enzymes Cholesterol synthesis
Cobalt	Part of vitamin B_{12} Red cell synthesis
Molybdenum	Active site of enzymes Uric acid synthesis
Chromium	Part of "glucose tolerance factor"
Nickel	Stabilize nucleic acid structure

Molybdenum

The active sites of the enzymes involved in synthesis of xanthine and hypoxanthine (precursors of uric acid) contain molybdenum. Although changes in the concentration of this metal affect the activities of these enzymes, the effect of molybdenum deficiency in humans is not well defined at present.

Chromium

Chromium is an essential component of the **glucose tolerance factor.** This factor is thought to play a role in maintaining normal insulin response to a glucose load, perhaps by affecting membrane response to insulin. Diabetic individuals often show improvement in their condition when supplemental chromium is added to the diet.

Nickel

Although nickel has been shown to be an essential trace element, its role in growth and metabolism is not clear. Evidence suggests that nickel somehow stabilizes nucleic acid structure, allowing more reliable synthesis of protein. A role in membrane function has also been proposed for nickel. However, no clinical disorder related to nickel deficiency has been reported in humans.

Other Trace Metals

Many trace metals not mentioned above are detected in humans. Some of these materials have known biochemical functions, but their nutritional or clinical importance has not been well characterized. Other trace metals are important mainly because of their toxicity. Heavy-metal poisoning by lead or mercury sometimes becomes a major health concern. Aluminum toxicity is receiving more attention, since this material appears to be found in higher concentrations in patients with Alzheimer's disease. Environmental pollution brings with it a host of problems associated with trace-metal contaminants and related health problems, many of which are vaguely defined and poorly understood.

SUMMARY

Trace-metal utilization in the body is a complex process. Although calcium, magnesium, and iron are three major components of this system, a large number of other minerals are involved in biochemical processes.

Calcium exists in the circulation in three forms: ionized (unattached to any carrier material), complexed (attached loosely to carboxylic acids), and bound (attached to albumin and other proteins). Ionized calcium is the metabolically active form. Any change which alters the amount of binding proteins affects the concentration of ionized calcium.

Assay for total calcium is usually accomplished by colorimetric methods. After dissociation of calcium from albumin, a dye is added which changes color in the presence of calcium. Ionized calcium is best measured with ion-selective electrodes. Since pH alters the amount of binding to protein, this parameter must be carefully controlled.

Calcium absorption, circulating calcium concentrations, and bone formation and metabolism are hormone-dependent processes. Parathyroid hormone stimulates calcium resorption from bone and promotes reabsorption of calcium in the proximal tubule of the kidney, serving to increase calcium concentrations in the bloodstream. Calcitonin acts in the opposite manner to enhance bone deposition of calcium and lower serum calcium levels. Vitamin D metabolites facilitate calcium absorption from the small intestine and exert a secondary effect on bone resorption. Parathyroid hormone activates the production of the important 1,25-dihydroxyvitamin D metabolites necessary for these changes to take place.

Elevated serum calcium occurs in a variety of clinical conditions. In many cancers, metastasis to the bone promotes bone degradation, increasing serum total calcium. Even if metastases are not present, some tumors may secrete materials which facilitate breakdown of bone (in some instances, these materials are similar to parathyroid hormone). In patients with multiple myeloma, the increase in serum protein leads to more binding of calcium. The drop in the ionized calcium fraction stimulates further release of calcium from the bone and results in an elevated serum total calcium. Parathyroid malfunction or tumor increases production of PTH, promoting calcium release.

One result of a decline in serum albumin levels is lowered serum total calcium. The shift in binding-protein concentration leads to an increase in ionized calcium. Regulating mechanisms then adjust the ionized calcium to normal levels, producing a decrease in serum total calcium. Hypoparathyroidism is a common cause of lowered serum calcium. This condition can be produced by surgery or be secondary to some other medical problem. Decrease in the amount of available vitamin D often leads to lowered serum total calcium.

Magnesium represents a simple system with no known hormonal regulation. As with calcium, magnesium exists in three forms in the circulation, only the free (unbound) fraction of which is biochemically active. A major regulator of magnesium levels is the kidney through glomerular filtration and tubular reabsorption.

Assay for magnesium uses either colorimetry or fluorometry. Formation of a complex with a dye is followed by measurement of the color formed or the fluorescence produced. Some work has been done using ion-selective electrodes, but this approach is still in the research stage.

Increases in magnesium can be seen in diabetes, some forms of renal failure, and some cases of liver disease. Magnesium deficiency causes a wide variety of neuromuscular problems. This deficiency can be the result of inadequate diet, loss by the kidney, or be associated with losses of other cations.

Iron is an important constituent of hemoglobin and is involved in the transport of oxygen throughout the body. After iron is absorbed in the intestine, it attaches to transferrin (a transport protein) and moves to the liver. Some of the iron is then transported to other cells for use in hemoglobin synthesis or is stored as ferritin.

Assay for iron involves removal of the iron from transferrin, followed by colorimetric measurement of an iron complex containing a large organic molecule. Transferrin levels can be assessed indirectly by determining the total amount of iron the molecule could carry if fully saturated. Ferritin measurement is done using a variety of immunoassay techniques.

Iron deficiency is usually the result of inadequate iron intake or is due to blood loss. First, ferritin stores decline, followed by a decrease in serum iron. The final stages of iron depletion involve a decrease in circulating hemoglobin.

Iron overload can occur if excess iron is consumed in the diet. Some individuals have a hereditary disorder known as hemochromatosis, in which there appears to be excessive absorption of iron from the gastrointestinal tract. The initial sign of iron overload is a rise in the serum ferritin levels. Liver damage is a common result of iron overload.

A wide variety of trace metals exist in very small concentrations in the body. In some instances, the role of a specific metal is unclear, although a need for that mineral can be demonstrated. For many metals, a decrease results in neurologic problems.

FOR REVIEW

Directions: For each question, choose the best response.

1. The role of calcium in body metabolism is that of
   A. structural contribution to bone formation
   B. an activator of the coagulation system

C. facilitating transmission of nerve impulses

D. both A and C

E. A, B, and C

2. The physiologically active form of calcium is

A. complexed

B. ionized

C. lipid-bound

D. protein-bound

3. A compound whose function is to increase the intestinal absorption of calcium and phosphorus is

A. cAMP

B. calcitonin

C. parathyroid hormone

D. vitamin D

4. Which hormones play a significant role in calcium metabolism?

A. aldosterone and calcitonin

B. calcitonin and parathyroid hormone

C. thyroid hormone and calcitonin

D. aldosterone and parathyroid hormone

5. When the serum ionized calcium level decreases, the concentration of parathyroid hormone will

A. decrease

B. increase

C. not be affected

D. become variable

6. Calcitonin exerts its major effect on calcium metabolism by

A. lowering blood levels of calcium

B. stimulating the uptake of calcium by bone

C. increasing the renal tubular absorption of calcium

D. both A and B

E. A, B, and C

7. To avoid the dissociation of calcium from protein, it is important that blood drawn for ionized calcium analysis be

A. handled aerobically

B. handled anaerobically

C. treated with EDTA

D. treated with citrate

8. The formation of a molybdenum blue complex is associated with the quantitation of

A. calcium

B. iron

C. magnesium

D. phosphate

9. Which of the following blood changes characterize primary hyperparathyroidism?

A. increased calcium and phosphate

B. decreased calcium and phosphate

C. increased calcium and decreased phosphate

D. decreased calcium and increased phosphate

10. The primary constituents of the mineralized portion of bone are

A. calcium and iron

B. calcium and magnesium

C. phosphate and magnesium

D. phosphate and calcium

11. Which of the following clinical disorders is/are associated with magnesium deficiency?
    A. tetany
    B. convulsions
    C. abnormal cardiac rhythm
    D. both A and C
    E. all of the above

12. The iron-transport protein found in the blood is
    A. albumin
    B. ferritin
    C. gamma globulin
    D. transferrin

13. The total iron-binding capacity (TIBC) test is performed to assess the blood level of
    A. ferritin
    B. hemopexin
    C. iron
    D. transferrin

14. Which of the following characterize iron-deficiency anemia?
    A. decreased serum iron and decreased TIBC
    B. decreased serum iron and increased TIBC
    C. increased serum iron and decreased TIBC
    D. increased serum iron and increased TIBC

15. Which trace metal accumulates in Wilson's disease?
    A. cobalt
    B. copper
    C. nickel
    D. zinc

16. Which protein is produced in a decreased amount in Wilson's disease?
    A. albumin
    B. ceruloplasmin
    C. haptoglobin
    D. hemopexin

BIBLIOGRAPHY

Agus, Z. S. et al, "Disorders of calcium and magnesium homeostasis," *Amer. J. Med.* 72:473–488, 1982.

Aisen, P., "Current concepts in iron metabolism," *Clin. Haematol.* 11:241–257, 1982.

Armitage, E. K., "Parathyrin (parathyroid hormone): Metabolism and methods for assay," *Clin. Chem.* 32:418–424, 1986.

Arosio, P. et al, "Ferritin: Biochemistry and methods of determination," *Ligand Quart.* 4:45–51, 1981.

Austin, L. A., and Heath, III, H., "Calcitonin. Physiology and pathophysiology," *New Eng. J. Med.* 304:269–278, 1981.

Bakerman, S., and Khazanie, P., "Calcium metabolism and hypercalcemia," *Lab. Manage.* July:17–25, 1982.

Berner, Y. N., and Shike, M., "Consequences of phosphate imbalance," *Annu. Rev. Nutr.* 8:121–148, 1988.

Bouillon, R., and van Baelen, H., "Transport of vitamin D: Significance of free and total concentrations of the vitamin D metabolites," *Calcif. Tissue Int.* 33:451–453, 1981.

Brittenham, G. M. et al., "Assessment of bone marrow and body iron stores: Old techniques and new technologies," *Semin. Hematol.* 18:194–221, 1981.

Cali, J. P. et al. "A referee method for the determination of total calcium in serum," *Clin. Chem.* 19:1208–1213, 1973.

Charlton, R. W., and Bothwell, T. H., "Definition, prevalence, and prevention of iron deficiency, *Clin. Haematol.* 11:309–325, 1982.

Clemens, T. L., "Vitamin D: Recent advances in basic research and clinical assay methodology," *J. Clin. Immunoassay.* 9:183–192, 1986.

Cook, J. D., "Clinical evaluation of iron deficiency," *Semin. Hematol.* 19:6–18, 1982.

DeLuca, H. F., and Schnoes, H. K., "Vitamin D: Recent advances, *Annu. Rev. Biochem.* 52:411–439, 1983.

Elin, R. J., "Assessment of magnesium status," *Clin. Chem.* 33:1965–1970, 1987.

Faulkner, W. R., "The trace elements in laboratory medicine," *Lab. Manage.*, July:21–35, 1981.

Finch, C. A., and Huebers, H., "Perspectives in iron metabolism," *New Eng. J. Med.* 306:1520–1528, 1982.

Forman, D. T., and Parker, S. L., "The measurement and interpretation of serum ferritin," *Ann. Clin. Lab. Sci.* 10:345–350, 1980.

Gagel, R. F., "Calcitonin in thyroid carcinoma and other disease," *Lab. Manage.*, July:35–48, 1982.

Gollan, J. L., "Diagnosis of hemochromatosis," *Gastroenterol.* 84:418–421, 1983.

Halliday, J. W., and Powell, L. W., "Iron overload," *Semin. Hematol.* 19:42–53, 1982.

Holland, H. K., and Spivak, J. L., "Hemochromatosis," *Med. Clin. North Am.* 73:831–845, 1989.

International Committee for Standardization in Haematology, "The measurement of total and unsaturated iron-binding capacity in serum," *Br. J. Haematol.* 38:281–290, 1978.

International Committee for Standardization in Haematology, "Recommendations for measurement of serum iron in human blood," *Br. J. Haematol.* 38:291–294, 1978.

Juan, D., "Clinical review: The clinical importance of hypomagnesemia," *Surgery.* 91:510–517, 1982.

Knochel, J. P., "The clinical status of hypophosphatemia: An update," *New Eng. J. Med.* 313:447–449, 1985.

Labbe, R. F., and Rettmer, R. L., "Zinc protoporphyrin: A product of iron-deficient erythropoiesis," *Semin. Hematol.* 26:40–46, 1989.

Lum, G., "The differential diagnosis of hypercalcemic disease," *Lab. Manage.* March:29–40, 1983.

Morrell, G., and Bordens, R. W., "Recent advances in PTH assay: Suggested clinical approaches," *Lab. Manage.* April:27–30, 1980.

Norman, M. E., "Vitamin D in bone disease," *Pediat. Clin. North Am.* 29:947–971, 1982.

Offenbacher, E. G., and Pi-Sunyer, F. X., "Chromium in human nutrition," *Annu. Rev. Nutr.* 8:543–563, 1988.

Raisz, L., and Kream, B. E., "Regulation of bone formation," *New Eng. J. Med.* 309:29–35; 83–89, 1983.

Rajagopalan, K. V., "Molybdenum: An essential trace element in human nutrition," *Annu. Rev. Nutr.* 8:401–427, 1988.

Reichel, H. et al, "The role of the vitamin D endocrine system in health and disease," *New Eng. J. Med.* 320:980–991, 1989.

Robertson, W. G., and Marshall, R. W., "Calcium measurements in serum and plasma—total and ionized," *CRC Crit. Rev. Clin. Lab. Sci.* 11:271–304, 1979.

Robertson, W. G., and Marshall, R. W., "Ionized calcium in body fluids," *CRC Crit. Rev. Clin. Lab. Sci.* 15:85–125, 1981.

Sherwood, L. M., "Vitamin D, parathyroid hormone, and renal failure," *New Eng. J. Med.* 316:1601–1603, 1987.

Shils, M. E., "Magnesium in health and disease," *Annu. Rev. Nutr.* 8:429–460, 1988.

Skikne, B. S., "The regulation and control of iron absorption: Some recent findings," *Lab. Manage.* March:43–45, 1987.

Slatopolsky, E. et al., "Current concepts of the metabolism and radioimmunoassay of parathyroid hormone," *J. Lab. Clin. Med.* 99:309–316, 1982.

Stoff, J. S., "Phosphate homeostasis and hypophosphatemia," *Amer. J. Med.* 72:489–495, 1982.

Swales, J. D., "Magnesium deficiency and diuretics," *Br. Med. J.* 285:1377–1378, 1982.

Toffaletti, J., "Ionized calcium measurement: Analytical and clinical aspects," *Lab. Manage.,* July:31–35, 1983.

Toffaletti, J., "Ionized calcium: Parts I and II," *Clin. Chem. News.* July–August, 1989.

Ulmer, D. D., "Trace elements," *New Eng. J. Med.* 297:318–321, 1977.

Webb, A. R., and Holick, M. F., "The role of sunlight in the cutaneous production of vitamin D_3," *Annu. Rev. Nutr.* 8:375–399, 1988.

Wills, M. R., "Diagnostic utility of immunoassays for parathyrin in hyper- and hypocalcemic states," *Clin. Chem.* 34:1955–1956, 1988.

Worwood, M., "The clinical biochemistry of iron," *Semin. Hematol.* 14:3–30, 1977.

Zak, B. et al, "Modern iron ligands useful for the measurement of serum iron," *Ann. Clin. Lab. Sci.* 10:276–289, 1980.

Zaloga, G. P., and Chernow, B., "Hypocalcemia in critical illness," *J. Amer. Med. Assoc.* 256:1924–1929, 1986.

CHAPTER OUTLINE

404

Chapter Twenty

THERAPEUTIC DRUG MONITORING

UPON COMPLETION OF THIS
CHAPTER, THE STUDENT WILL BE
ABLE TO

1. State two reasons for measuring drug levels in body fluids.
2. List three major routes for drug administration.
3. Discuss the importance of knowing the time and specifics of drug administration for properly determining sampling times.
4. Identify carrier proteins that function to transport drugs in the blood.
5. Discuss the importance of free drug in relationship to cellular uptake and excretion by the kidney.
6. Describe the formation of metabolites from the parent drug.
7. Define the following terms:
   A. generic drug
   B. steady state
   C. subtherapeutic range
   D. therapeutic range
   E. toxic range
   F. half-life
8. State several advantages and disadvantages associated with the following techniques used to assess drug concentrations:
   A. spectrophotometric
   B. immunoassay
   C. chromatographic
9. List the physiological effects and examples of drugs for each of the following categories:
   A. aminoglycosides
   B. antiarrhythmics
   C. anticonvulsants
   D. antidepressants
10. Describe the therapeutic purposes for administering
    A. lithium carbonate
    B. theophylline
    C. salicylate
    D. acetaminophen
11. Describe the techniques for measuring salicylate.
12. Discuss the effects that age has on the metabolism and excretion of drugs.
13. Discuss the effects of pregnancy on drug metabolism and excretion.
14. Explain how a change in serum protein levels affects free drug concentration and tissue uptake.
15. Explain the effects of decreased renal output on drug levels.

INTRODUCTION

The role of the clinical chemistry laboratory is an ever-expanding one. Once limited to a few fairly simple testing procedures, the laboratory is now asked to provide information on the rapidly changing hormonal status of an individual, monitor enzyme shifts during an acute myocardial infarction, and quantitate metabolic products of bacterial or viral activity during an infection. Nowhere is the explosion in the demand for clinical chemistry services more apparent than in the field of drug monitoring. Advances in pharmacology and in our ability to analyze low levels of materials rapidly and accurately have led to increased and more complex demands on the laboratory in the realm of drug analysis.

Measurement of drug levels in body fluids provides information for two clinical purposes: determination of drug abuse or overdose and therapeutic drug monitoring. The issues related to substance abuse will be covered in the next chapter. The concept of therapeutic drug monitoring is a relatively recent one based on the realization that not all patients respond in the same manner to the same amount of medication, even when allowance is made for body mass. Measurement of drug levels in body fluids allows an individualized approach to patient treatment and provides a much higher level of safety than was possible in the past.

20.1

PRINCIPLES OF PHARMACOLOGY

To understand better the principles of therapeutic drug monitoring, we first need a brief introduction to the field of pharmacology, which deals with the study of drug action and drug conversion by the body. Our understanding of therapeutic drug-monitoring data generated by the laboratory is enhanced after we learn how drugs are introduced into the body, utilized by the cells, metabolized, and excreted.

Role of Drugs in Patient Treatment

Medications are used for many purposes in patient treatment. Some replace a natural biochemical the body no longer manufactures (or cannot make in adequate amounts). Thyroid hormones and estrogens are frequently prescribed for individuals who have deficiencies in these needed natural products. Antibiotics such as gentamicin and tobramycin interact with bacteria to block some metabolic process and destroy the microorganism. A large number of medications increase the ability and strength of heart muscle contraction and stabilize the cardiac

rhythm. Other compounds are used to decrease the convulsions associated with epilepsy and other neurologic disorders. There is increasing use of pharmacologic agents to treat depression (antidepressants), manic-depressive illness (lithium), and schizophrenia (haloperidol), as well as benzodiazepines and other compounds prescribed to deal with the tensions and stresses of everyday living. There are similar examples of medications for treating diabetes, hypertension, cancer, and a host of other disease states.

Our knowledge of how these drugs act is often very incomplete. We know a great deal about how the body absorbs, metabolizes, and eliminates acetylsalicylic acid (aspirin). We have rapid, accurate methods for analysis of this compound. But what we do not know at present is exactly how aspirin provides its analgesic effect. Although this information is not absolutely necessary, knowledge of the pharmacologic effects produced and the mechanism(s) of the processes involved enhances the laboratory staff's ability to provide reliable and useful information. In addition, awareness of pharmacologic issues permits better decision making about the timing of sample collection, interferences in assays, and unexpected problems associated with the use of a specific drug.

Administration

There are three major routes for the administration of medications: oral, intravenous, or intramuscular. Although other approaches (rectal and subcutaneous, for example) may be employed at times, we will not concern ourselves with these less typical situations but will look only at the main methods for introducing drugs into the body.

ORAL ADMINISTRATION

Because of its convenience, safety, and relative lack of discomfort, the oral route is the most prevalent technique for giving medications. Pharmaceuticals may be given in either liquid or solid form by mouth. Assuming no difficulty in swallowing or in passage of the medication through the esophagus, the drug reaches the stomach in a few seconds. If the stomach is empty, passage into the small intestine is also rapid. Some medications are to be given at mealtime, either to minimize acid breakdown of the drug in the stomach or to slow down absorption into the bloodstream, allowing the drug to be in the system for a longer time and lowering the possibility of a rapid (and possibly hazardous) increase in the blood level of the medication. Particulars of administration are the responsibility of the clinical staff, but laboratory personnel need to be somewhat knowledgeable about the specific requirements of common drugs to decide sampling times and to provide accurate data interpretation.

Once in the small intestine, the drug is transported across the intestinal wall and absorbed into the circulation. Prior to this absorption, some of the parent medication may be converted to another form. This metabolic change may be necessary for the body to take up the drug efficiently. A medication is often given in a form which creates fewer side effects, but the structure may ultimately be pharmacologically less effective. The gradual metabolism of a medication into another form also allows for slow absorption into the circulation, providing another means of controlling the blood level of the drug.

After absorption into the bloodstream, the drug then passes through the circulatory system of the liver. This **first-pass route** puts the drug in contact with the many enzymes contained in the various liver cells and creates both problems and opportunities. A number of liver enzymes convert the drug to other forms, both active and inactive. Some enzymes attach other molecules (such as glucuronic acid or sulfate) to the compound in order to make it more water-soluble. These modifications may help the drug be absorbed more readily by the target cells or they may permit the medication to be excreted more quickly by the kidney. Depending on the metabolic changes produced, the drug may be in a more active form, better able to produce the therapeutic effect. Conversely, the drug could be changed to a form that is less pharmacologically effective. These conversion processes are under active investigation by researchers and make up one of the more interesting and challenging areas of clinical pharmacology. Some knowledge of drug transformation is necessary for the laboratory person who wishes to accurately interpret data for the clinical team.

INTRAVENOUS ADMINISTRATION

Intravenous administration is a direct means of introducing a drug into the system. Problems of absorption by the gastrointestinal tract are eliminated, and conversion of the drug by enzyme systems in the small intestine is not a problem. The first-pass changes produced by the liver are minimized somewhat, although these metabolic alterations eventually do occur as the drug in the bloodstream circulates through the liver. Time of administration of intravenous drugs can be more carefully controlled and documented; both the time of initial infusion and the time the infusion was completed should be noted. Control of dosage through a long-term (hour or longer) administration is very easy to achieve if the medication is diluted and given as part of an intravenous saline or glucose regimen. Again, details about time and specifics of administration need to be carefully noted for proper interpretation of data.

INTRAMUSCULAR ADMINISTRATION

Injection of medication into muscle has both benefits and drawbacks. The route is not as direct as the intravenous method, since absorption from muscle tissue into the circulation takes place over time. Rate of absorption varies considerably depending on the amount of muscle mass available for the injection of the drug. Advantages of intramuscular administration include decreased early conversion of drug before it reaches the circulation and the somewhat controlled timed release of drug as it is absorbed from the muscle tissue into the bloodstream.

Transport

Once the drug enters the circulation, it must somehow reach the appropriate site of action (quite often inside a cell). Many drugs may be only partially water-soluble and are transported in the circulation by attaching to some carrier protein. Albumin is the major protein for circulatory transport of drugs to cells. Other proteins which bind specific classes of medications include the lipoproteins and globulins. Hemoglobin also makes a significant contribution to the transport of some drugs through the bloodstream.

The degree of binding (related to solubility and other factors) varies from one drug to another. Some drugs are quite water-soluble, with only a small fraction of the compound being attached to proteins. Other materials are relatively insoluble in water and the majority of the chemical is protein-bound. Knowledge of the degree of attachment to protein is important, since only the unbound portion of the medication is available for cellular uptake or for elimination by the kidney. At present, methods for the analysis of drugs measure the total amount of drug present in the body fluid and do not readily provide data on the bound-to-free ratio. A number of approaches are being developed which help in providing this additional information for selected medications.

Individual variations in the amount of transport proteins are greater than often realized. Even in healthy individuals there may be significant differences in the amount of albumin or other proteins needed to transport a given drug. Various disease states alter the levels of these proteins markedly. Proper interpretation of drug monitoring data must take into account the quantities of transport proteins available and their binding properties with the drug in question.

Cell Uptake and Metabolism

The processes of cellular uptake and metabolism of drugs are complicated. There is a great deal of variation from one pharmaceutical agent to another, although some general principles can be discussed. We will look briefly at these two processes in the following two sections, touching on specific details for some medications.

UPTAKE BY THE CELL

Our hypothetical medication is now somewhere in the circulation, being carried in part by some specific transport protein(s). As the blood passes through the capillary system, the drug can enter the extracellular fluid, again becoming at least partially bound to some specific protein or proteins. Another bound-to-free ratio has now been established, with only the free chemical being available to the cell.

One fate for the drug can be diffusion into tissues other than the site of action. Although the desired pharmacologic effect is not produced, the drug is stored and can later be released back into the circulation. Fat-soluble drugs are particularly likely to be taken up by adipose tissue in this manner. Little metabolism appears to take place in these tissues, although binding to specific intracellular proteins is common. The extent of this type of uptake and storage varies from one medication to another but does have a measurable effect on the amount of drug available to produce the desired therapeutic effect. It is difficult to predict how much this phenomenon contributes to drug distribution in the body. On occasion, problems arise when the administration of a drug is cut back or discontinued. Material then begins to escape from tissues, sometimes producing a paradoxical increase in the blood level of a medication even after its administration has been terminated.

The drug reaches its site of action by the same process as it enters other tissues. Keep in mind that we are discussing only the unbound portion of the medication participating in this diffusion process. The medication may either bind to the cell membrane at a specific site or enter the cell, depending on the drug's mechanism of action. Some drugs bring about their action by attaching to a receptor site on the outside of the cell membrane. This attachment may activate the membrane receptor to stimulate some specific biochemical changes which would take place inside the cell. Conversely, the drug may block the receptor so that other natural biochemicals cannot attach to it and produce their ordinary metabolic changes.

If the drug enters the cell before exerting its pharmacologic effect, it again distributes itself between a bound form and a free form. The amount of drug in each category depends on the presence and concentration of intracellular binding proteins available. The free drug is available to attach to specific receptors inside the cell, producing the desired biochemical changes.

The process of drug absorption and uptake is summarized in Figure 20-1.

DRUG METABOLISM

Conversion of the parent drug to one or more metabolites may take place anywhere in the system. Realistically, most of the transformation occurs in the liver. The enzyme systems in this organ usually do not metabolize specific drugs but produce changes in classes of drugs because of the structural similarities of the organic molecules. For example, there is a system responsible for the addition of hy-

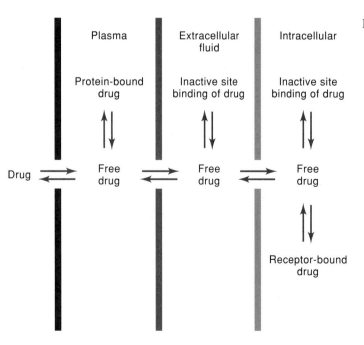

FIGURE 20-1 Drug absorption and uptake.

droxyl groups onto benzene rings. Any drug with a benzene ring in its structure can theoretically be converted by these enzymes. For the most part, drugs converted by these enzyme systems are pharmacologically less active and more water-soluble. Therefore, they are more readily excreted by the kidneys.

Patients vary markedly in their degree and rate of drug metabolism. Some people are *fast* metabolizers, meaning that their enzyme systems are very active and drug removal from the system is much more rapid than expected. In other instances, prolonged use of one or more drugs may stimulate the formation of more of these particular enzyme systems by the liver (a process known as **enzyme induction**). Over time, the ability of the system to convert and remove drugs increases, with the rate of metabolism being much higher later in the course of drug treatment. These situations cannot be readily predicted but can be recognized by proper monitoring.

Excretion

The main route of drug excretion is the kidney. As drugs are converted to more water-soluble forms, they are filtered at the glomerulus and excreted in the urine. Any clinical situation which results in diminished kidney function impairs drug excretion. As a result, the plasma level of a given medication decreases at a much lower rate. If drug dosage is not decreased, the medication can accumulate, often with very adverse effects.

Generic Drugs

When the patent on a specific pharmacologic agent expires, other companies are free to manufacture the drug and market it under their own name. In many instances, this process results in lower cost for the drug, since the new companies did not have to invest the millions of dollars necessary to develop a new medication. Occasionally some problems result if the "new" medication is not produced and compounded in exactly the same way as the "standard" form. Although the active ingredient may be the same and present in the same amount, changes in binders and other components of the tablet or capsule affect how rapidly the active ingredient is absorbed into the system. If the new mixture results in the medication being taken up by the system faster than expected (or more completely than before), the dosage which previously produced a predictable and desirable effect for a given patient may now result in the patient receiving an excessive amount of the drug. This problem can be avoided to a great extent through cooperation among the laboratory, pharmacy, and the clinical team in monitoring medication administration and promptly sharing information about adverse effects and problems.

20.2

FUNDAMENTAL CONCEPTS OF THERAPEUTIC DRUG MONITORING

Reasons for Therapeutic Drug Monitoring

The response of any given patient to drug treatment is a highly individual and variable one. Depending on age, physical condition, and genetic make-up, patients differ in their response to the same medication. Because the response to pharmacologic treatment varies among patients, the administration and management of drug therapy should be dealt with on an individual basis.

This concept of individualized treatment and management is relatively new. The approach has been made possible because of advances in our ability to measure blood levels of drugs and metabolites in a rapid and accurate manner. This individualized system is referred to as **therapeutic drug monitoring (TDM)** and deals with the measurement of drug concentrations during therapy with pharmaceutical agents. By following the patient on a regular basis, important information can be obtained which helps greatly in the overall management of the particular health-care problem.

Basic Principles of Therapeutic Drug Monitoring

In the past, administration of a given drug was carried out somewhat by trial and error. A rough estimate of dosage was made based on the patient's weight, the medication was given, and the patient was observed for a response. If there was no improvement in the clinical situation within a certain period, the dosage of the drug was increased and the process repeated. If the patient began to show an adverse reaction to the drug, the dosage was reduced. There was little in the way of organized principles to draw on for determining the proper medication or the appropriate dosage for a specific patient. Some of the hazards to the patient are obvious.

With the advent of sensitive and accurate testing procedures, knowledge about drug metabolism and excretion increased rapidly. One practical application of this basic research was the development of a data base for determining plasma (or other body fluid) levels of a particular medication in order to assess the appropriate level for the drug in treating an individual patient. Trial and error was replaced by sound scientific principles, definitely to the benefit of the patient.

Two major principles undergird the TDM approach: the idea of a steady-state drug level (complete with peaks and troughs) and the realization that a drug concentration in blood can be subtherapeutic (providing no benefit), therapeutic, or toxic.

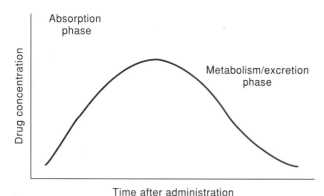

FIGURE 20-2 Single-dose drug kinetics (oral administration).

All TDM analyses develop and draw on data which focus on these two fundamental principles.

When a single dose of a drug is administered (assume oral administration for our example), the blood level changes markedly over time (Fig. 20-2). Initially the level rises as the drug is absorbed into the circulation from the intestine. Simultaneously, the material is being transported, taken into tissues, metabolized, and excreted. At some time, the concentration in the plasma reaches its highest point (this time is defined as the **peak** for sampling purposes) and then begins to drop off as metabolism and excretion processes predominate.

For single-dose administration, the rate of decline in concentration is expressed in terms of **half-life,** the time required for the concentration of the drug to decrease by 50%. This half-life is different for each drug and must be experimentally determined. The rates of metabolism and excretion affect the half-life of any given drug. Data on half-life are determined by administering a single dose of the medication to healthy volunteers, since clinical situations affect the half-life. Knowledge of half-life allows us to predict how the drug levels will change over time.

No clinical improvement can be expected after a single administration of a medication (except for the occasional headache), so we need to look at a more complex picture of drug metabolism. To do so, we must first understand some terminology common to TDM discussions. The goal of drug administration is to achieve the **therapeutic range,** that level of concentration in the bloodstream which provides the optimum amount of medication for treatment of the clinical disorder. The therapeutic range is specific for each drug and does not depend on the individual patient as long as other biochemical parameters (such as serum protein concentrations) are within normal limits. A blood level of medication below the therapeutic range is considered **subtherapeutic,** meaning it provides no clinical benefit. Many medications can also reach a level which is toxic to the patient. The goal of TDM is to achieve the therapeutic range rapidly, while ensuring that the patient is receiving sufficient medication to treat the problem, but not such an amount as to be toxic.

To reach this desired level of the drug, **multiple dosing** is employed. Simply put, the patient is given a set amount of the drug at regular intervals. The individual doses produce the expected rise and fall in the blood level, but the cumulative effect (Fig. 20-3) is a gradual increase in the concentration of the drug in the system. At **steady-state,** the rate of administration is equal to the rates of metabolism and excretion, so the level remains essentially constant. The small fluctuations in blood level of the drug between doses are important points to notice. The peak and trough values (highest and lowest points) are frequently measured to assure that drug administration is being carried out in the most efficient manner. These two values are also fairly constant for a given medication.

Applications of Therapeutic Drug Monitoring

Apart from using TDM to achieve the appropriate dosage of a medication for a given patient when beginning drug therapy, there are other practical applications for this approach to drug monitoring. The noncompliant patient can be identified readily. A large percentage of out-patients do not take their medications as directed, losing the therapeutic effectiveness. A low or undetectable level of drug in a patient who has shown no improvement (or has declined in health) suggests failure to follow the prescribed treatment plan. This information can often be used in counseling the patient to improve his following of directions. Changes in the disease state may necessitate changes in drug treatment. With proper monitoring of drug levels, any change in the physiologic situation due to a new or changing medical problem can be compensated for by altering medication dosage based on the laboratory data. Changes in the normal physiologic situation have profound implications when monitoring takes place over a span of years.

Pregnancy alters the metabolism and excretion of a large number of drugs, changing their therapeutic effectiveness. As a child matures through adolescence to adulthood or an adult passes through mid-

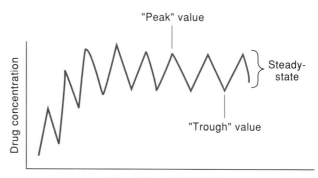

FIGURE 20-3 Multiple dosing to achieve therapeutic range.

dle age into the later years of life, the body's ability to utilize medications changes and drug treatment approaches need to vary accordingly. These normal stages in life can be accommodated through a good TDM program.

Precautions

As is true for any single piece of laboratory data, TDM information cannot be interpreted in a vacuum. Proper application of this data requires knowledge of the entire clinical picture and a solid background in the appropriate pharmacologic areas. Information about protein status, liver function, renal impairment (if any), and other parameters must be available and taken into account when the laboratory provides interpretive services for the clinical team. Simple reliance on a table of information or a package insert will prove very inadequate when consultation is required.

20.3

OVERVIEW OF ASSAY TECHNIQUES: GENERAL PRINCIPLES

Spectrophotometric: Colorimetric and Fluorometric

The earliest methods for measuring drugs in body fluids examined some spectral property of the analyte compound. The cardioactive agent quinidine can be easily quantitated after removal of protein simply by measuring the sample's fluorescence. Salicylate is assayed with a colorimetric reagent, the color being produced through a complex formed with iron. Although spectrophotometric methods are simple and do not require complex equipment, the approach has some severe limitations. Not all compounds form derivatives which absorb light or that fluoresce. A metabolite of the drug may also react, falsely elevating the value for the medication in question. In addition, the sensitivity of these methods does not always allow detection at the low levels encountered in serum or other fluid.

Immunoassays

The advent of immunoassay techniques opened the door to a new era in therapeutic drug monitoring. With the use of specific antibodies, only the derivative of interest (in most cases) would be measured. Detection limits were lowered dramatically, improving the sensitivity of the assays. On the other side, however, came increases in complexity and cost. More sophisticated equipment was required, which needed more expensive reagents. Techniques were often more complicated, thus requiring a higher level of technician proficiency. The issue of pharmacologically active metabolites was not dealt with to any extent other than marketing another assay kit to measure the metabolite separately.

Immunoassay approaches have proliferated in the last 15 years. Radioimmunoassay, enzyme immunoassay techniques, fluorescence immunoassay—the entire scope of these analytical schemes has been brought to bear on the problem of therapeutic drug monitoring. Application of these methods allows rapid, accurate measurement of drug levels at reasonable cost, whether there is a need for only a single measurement on an emergency basis or a batch run of several samples.

Chromatographic

The techniques of gas chromatography (GC) and high-performance liquid chromatography (HPLC) have wide application in the field of therapeutic drug monitoring. These approaches allow measurement of very low levels of drugs. Perhaps the major advantage of GC and HPLC is their ability to measure the amount of parent drug and metabolites simultaneously, allowing quantitation of all significant contributors to the therapeutic effect.

Both techniques demand considerable time and skill. Expensive equipment is required and a high degree of technical ability is called for. For GC, many compounds must first be converted to volatile derivatives that must still be stable at the temperature of separation. If HPLC is used, we must be able to detect the compound after it comes off the column. If the drug has an absorbance (usually in the UV region) or can be converted to a derivative which absorbs or fluoresces, detection is not a problem. Other medications have required very sophisticated detection systems, increasing the cost and the complexity of operating the equipment.

20.4

AMINOGLYCOSIDES

Properties

Gentamicin and **tobramycin** (Table 20-1) are **aminoglycosides,** composed of three carbohydrate rings (connected by way of oxygen atoms) with amino acids attached to the rings. These antibiotics are particularly effective in treating severe infections caused by a wide variety of gram-negative bacteria. They act by inhibiting bacterial protein synthesis. Amino acid incorporation into the growing protein chain is blocked by the drug, lowering the amount of protein manufactured and causing the bacteria to misread the genetic code, resulting in the production of "incorrect" proteins.

After treatment with gentamicin gained popularity, increasing concern was raised about its toxicity,

Table 20–1
PHARMACOLOGIC PROPERTIES OF AMINOGLYCOSIDES

Property	Gentamicin	Tobramycin
Administration	Intravenous, intramuscular	Intravenous, intramuscular
Percentage bound to protein	<10%	<10%
Time to reach peak plasma level	1–1.5 h (intramuscular)	Same
	30 m (intravenous)	Same
Plasma half-life	2 h	Same
Excretion in urine	90% parent drug	Same
Active metabolites	No	No
Therapeutic range (μg/mL)	4–8 (peak)	4–10 (peak)
Toxic range (μg/mL)	>12 (peak)	Same
	>2 (trough)	Same

particularly its effect on the kidney. Tobramycin was developed and shown to be less harmful to the kidney. Tobramycin has essentially the same characteristics as gentamicin with regard to absorption, protein binding, and excretion from the system unchanged.

Assay Techniques

A number of assay techniques have been developed for gentamicin and tobramycin over the years, some of which are primarily applicable for research purposes. The two most appropriate analyses for routine clinical use are either RIA or enzyme-linked immunoassay. A number of kits are commercially available for quantitative measurement of this antibiotic.

Sample collection requires careful attention to a number of parameters. Samples should not be collected using heparin as an anticoagulant because it provides undefined interference with the assay. Sample collection times should be closely coordinated with the clinical team so that actual times of administration (and end of infusion for intravenous situations) can be checked. This information should be written on the requisition by the individual doing the sample collection. Samples should be collected from the arm opposite the one where the drug was administered. Peak levels are obtained at the appropriate time determined by the administration protocol. Trough samples should be collected 30 min or less prior to the next dose. For most efficient monitoring of aminoglycoside blood levels, both peak and trough data should be obtained.

20.5

ANTIARRHYTHMICS AND OTHER CARDIOACTIVE DRUGS

Specific Compounds

The pharmacokinetic properties of several common **cardioactive drugs** are listed in Table 20–2.

Table 20–2
PHARMACOLOGIC PROPERTIES OF CARDIOACTIVE DRUGS

Property	Digoxin	Disopyramide	Procainamide	Quinidine
Administration	Oral	Oral	Oral	Oral
Percentage absorbed	60–75	70–95	65–100	100
Percentage bound to protein	20–40	65–90	15	80–90
Time to reach peak plasma level (hours)	1.5–5	1–2	1–2	1.5–2
Plasma half-life (hours)	20–60	4–10	2–4	4–7
Excretion in urine (% unchanged)	80–90	50	50–60	15–40
Active metabolites	No	No	35% N-acetylprocainamide	60–85% many active
Therapeutic range	1.0–2.0 ng/mL	2–5 μg/mL	4–10 μg/mL	2–5 μg/mL
Toxic range	>2.0 ng/mL	>7 μg/mL	>16 μg/mL 5–30 μg/mL*	>10 μg/mL

* For sum of procainamide and N-acetylprocainamide.

The most widely used are digoxin and procainamide, with quinidine and disopyramide being employed less frequently.

DIGOXIN

Since its discovery in 1785 as an active cardiac agent found in the foxglove plant, **digoxin** and other related compounds have been widely used to alter the excitability of cardiac muscle. The major effect seems to be on the Na^+, K^+-ATPase pump which is involved in the transport of these cations across the cell membrane. Digoxin inhibits the enzyme system and alters the strength of contraction in the heart muscle.

DISOPYRAMIDE

A fairly recent addition to the cardiac drug field, **disopyramide** has been in use since 1977. Its major effect appears to be as an **antiarrhythmic** drug, that is, one that stabilizes the heartbeat.

PROCAINAMIDE AND N-ACETYLPROCAINAMIDE

Procainamide has been known as an effective means of treating ventricular arrhythmias since the early 1950s. The drug acts directly on the heart muscle and stabilizes the rhythm within 30 min after administration. The active metabolite **N-acetylprocainamide** is also an effective pharmacologic agent.

QUINIDINE

Perhaps one of the earliest known antiarrhythmic agents known (the first recorded instance of quinidine use was in 1749), **quinidine** has been widely used in treatment of certain cardiac conditions but frequently has been overshadowed by other drugs. This medication is a myocardial depressant, acting to decrease the heart's excitability. The heart's ability to conduct an electric current is lowered and its contractions decrease. These effects allow the cardiac rhythm to stabilize.

Assay Techniques

Analysis for digoxin in serum is generally performed by RIA in most laboratories, although some new enzyme-linked immunoassay approaches are gaining popularity. Since digoxin is usually administered once a day and results are best interpreted when a trough sample is obtained, there is no real justification for a stat request for a digoxin level.

Several analytical approaches utilizing gas chromatography or HPLC have been reported for disopyramide. In addition, an enzyme immunoassay procedure is available. The advantage of the chromatographic methods is their ability to identify any metabolites which might otherwise cross-react in a less specific assay.

Although a variety of methods exist for the measurement of procainamide, probably the easiest and quickest analysis involves a commercially available enzyme immunoassay approach. In recent years, as the pharmacologic activity of N-acetylprocainamide has come to be more readily recognized, monitoring for procainamide involves the quantitation of both compounds in the sample.

There are several methods for quantitation of quinidine in serum or plasma. The least specific approach involves precipitation of protein using *meta*phosphoric acid followed by measurement of the fluorescence and comparison to a standard curve. This method is rapid, but it has the drawback of measuring all the forms of quinidine in the sample, both active and inactive. One modification involves a solvent extraction prior to measurement of fluorescence. The value obtained is often half that seen in the less specific assay, but it more reliably reflects active quinidine and metabolite concentrations. HPLC approaches have the advantage of being able to determine the amount of each metabolite separately as the mixture fractionates on a column. However, the method is much slower, more technically demanding, and requires expensive equipment.

Digoxin/Quinidine Interactions

When quinidine is administered to a patient who is taking digoxin, a marked increase in the digoxin concentration is noted. Conversely, after discontinuing quinidine, the digoxin concentration drops. The change is real, not an artefact produced by an interference with the digoxin assay. The most probable mechanisms for the digoxin increase appear to be quinidine's blockage of tissue uptake of digoxin and its interference in the excretion of digoxin by the kidney. The net results of these two actions is a rise in serum digoxin concentrations, sometimes as much as twofold.

Endogenous Digoxinlike Compounds

In recent years, endogenous substances in the blood have been discovered which cross-react with antibodies to digoxin. These substances seem to appear at times where there is expansion of the blood volume and an increase in blood pressure. These digoxinlike factors have been seen in acromegaly (excessive growth due to overproduction of growth hormone), in renal impairment, and in severe liver disease. During pregnancy, the level of this factor increases, then disappears from the circulation within 24 h or so after delivery. At present, we do not know the structure or function of the factor(s), but must keep in mind the possibility of some cross-reacting material being present in the samples from some patients. Detection of digoxin by RIA in a patient known not to be taking the medication might suggest the possible presence of the digoxinlike factor(s).

20.6

ANTICONVULSANTS

General Properties

The **anticonvulsants** (Table 20–3) are a group of medications with complex and sometimes poorly defined mechanisms of action. The overall function of this class of drugs (also referred to as **antiepileptics**) is to alter transmission of nerve impulses in some fashion to minimize the seizures which occur in epilepsy.

Analysis of Anticonvulsants

Many of the anticonvulsant drugs may be used either alone or in combination with other such medications. Assay methods are available which permit analysis of a serum or plasma level of a medication by itself or as a part of a separation and quantitation of several anticonvulsants in the same sample.

For individual assays, immunoassay techniques are the methods of choice. A number of companies market tests and equipment for analyzing the commonly used anticonvulsants. The assays take only a few minutes, and the techniques are fairly simple. Batch processing of large numbers of samples is possible using automated approaches, particularly with centrifugal analyzers.

For laboratories which receive a significant number of requests for multiple drug assays, systems involving gas chromatography or HPLC are more useful. A large number of published techniques describe the separation and quantitation of any or all of the commonly used anticonvulsants. These analytical approaches are especially valuable in situations where study of drug metabolites might be of value. A major metabolite of primidone is phenobarbital, an active anticonvulsant itself. More clinically useful information can be obtained in this case if levels of both the parent drug and the metabolite are available.

20.7

ANTIDEPRESSANTS

Properties

The **antidepressants** are a mixed group of pharmaceutical agents with poorly defined mechanisms of action. There is striking variation in the effectiveness of a single antidepressant from one patient to another. Antidepressants are divided into two major classes in terms of their pharmacologic effects: tricyclic antidepressants and monoamine oxidase inhibitors. The monoamine oxidase inhibitors are not widely used in the United States because of their profound effect on blood pressure. Our discussion,

Table 20–3
PHARMACOLOGIC PROPERTIES OF ANTICONVULSANTS

Property	Carbamazepine	Ethosuximide	Phenobarbital
Administration	Oral	Oral	Oral
Percentage absorbed	70	75–95	90
Percentage bound to protein	75–85	45–50	45–50
Time to reach peak plasma level (hours)	6–24	1–4	5–10
Plasma half-life (hours)	25–30	25–40	70–95
Excretion in urine (% unchanged)	<1	20	10–30
Active metabolites	Yes	No	No
Therapeutic range (μg/mL)	4–12	40–100	15–40
Toxic range (μg/mL)	>12	150	>50

	Phenytoin	Primidone	Valproic Acid
Administration	Oral, intravenous	Oral	Oral
Percentage absorbed	90–100	80–90	85–100
Percentage bound to protein	85–95	<10	90–95
Time to reach peak plasma level	3–12 h (oral)	2–4 h	30–90 m
Plasma half-life (hours)	variable	5–15	10–20
Excretion in urine (% unchanged)	1–5	45–50	3–7
Active metabolites	No	Yes, 15–25% to phenobarbital	No
Therapeutic range (μg/mL)	10–20	5–12	50–100
Toxic range (μg/ml)	>20	>15	>200

therefore, will focus predominantly on the tricyclic compounds.

It is believed that the **tricyclic antidepressants** act by blocking the uptake of norepinephrine and serotonin by the presynaptic nerve endings. This blockage prolongs the time of transmission of the nerve impulse. However, our understanding of the biochemistry of depression is not yet to the stage where we can relate this process reliably to either the cause or the treatment of this common psychiatric disorder. As a result, tricyclics are used slowly and cautiously and are frequently ineffective in the relief of depression.

We will not discuss to any great extent the behavior of individual tricyclic compounds but will consider the class as a whole. In practice, a given patient may well be receiving more than one of these medications. The pharmacologic and toxic effects observed are usually due to the sum of influences produced by individual drugs.

Pharmacokinetics

The primary route of administration is oral, with peak blood concentrations being achieved within 2 to 10 h after a single dose. Absorption of some of these compounds is erratic after intramuscular injection. First-pass metabolism by the liver occurs with between 30 and 70% of an individual dose being converted to other compounds after a single passage through the liver. Major routes of metabolism involve removal of methyl groups and addition of −OH followed by conjugation with glucuronic acid. Depending upon the specific drug, some of the metabolites possess significant pharmacologic activity of their own.

The fraction of drug bound to protein is high, up to 80%, depending on the specific drug. The major sites of binding are to albumin, a glycoprotein, and to lipoproteins. In addition, the uptake of drug by fatty tissues is quite high, allowing storage of the medication in the tissues. The net effect of the high protein binding and the uptake by tissues is a low level of free (and available) drug in the circulation.

Assay Techniques

Since any given patient may be receiving more than one tricyclic derivative and since many of the parent drugs have pharmacologically active metabolites, monitoring for levels of these drugs usually involves measurement of total amount of medication present in all its forms. At present, the most effective means of therapeutic drug monitoring for tricyclics involves the use of gas chromatography. The specific compounds can be separated and quantitated, and the total amount of active drug(s) can be reported. This technique permits identification of specific antidepressants if information on patient usage is not available (as in an overdose situation).

A flame ionization detector which is nitrogen-sensitive provides adequate means of identifying the components of the mixture. There are "rapid" chemical screening methods available, but their sensitivity is quite limited and they provide little useful clinical information in most cases. Although the GC approach is technically demanding, the information it yields can be life-saving.

Side Effects and Toxic Signs

The therapeutic levels for most tricyclics are poorly defined. There appears to be a range of blood concentrations for a given drug in which relief of depressive symptoms is seen in many patients. Concentrations of the drug below this level are ineffective, as are levels above the range. The problem is compounded by the fact that there appears to be several causes for depression, some of which are unresponsive to treatment with medications. A major area of research today involves the identification of patients who can be helped through antidepressant treatment.

A major concern in the use of tricyclics is the problem of overdose. A significant number of patients seen in emergency-room settings have attempted suicide with tricyclics or have accidentally ingested more than therapeutic amounts of their medication. The focus in the overdose situation is the effects of the drug on the cardiovascular system. The most common cause of death in this case is impairment of cardiac function, with increases in heart rate and irregularities in cardiac rhythm frequently observed. In addition, elevated amounts of the drug produce respiratory depression and seizures.

The utility of analyzing serum levels of tricyclic antidepressants in overdose situations has been strongly debated. An overdose patient who has a history of therapeutic use of these drugs should be assumed to have an excess amount of drug in his system. Quantitation of serum levels provides information only on the amount of circulating drug (both bound and free), but not the tissue load. However, autopsy studies have shown drug levels of 1100 ng/mL and higher in the blood of patients who have died of tricyclic overdose.

20.8

LITHIUM

Properties and Pharmacokinetics

The simple and inexpensive compound **lithium carbonate** has become the medication of choice for treatment of the manic phase of manic-depressive (bipolar) illness. This salt is quickly absorbed into the circulation after oral administration. Unlike most other medications, lithium does not bind to plasma proteins and is not metabolized. Equilibration in the brain and cerebrospinal fluid is complete within 24 h, suggesting its rapid distribution throughout the body.

Assay Techniques

Measurement of blood levels of lithium are best accomplished with a flame photometer. Most commercially available instruments can quickly be modified to measure lithium in addition to sodium or potassium or can be dedicated to lithium assays alone. Several companies are in the process of developing ion-specific electrodes for lithium measurement based on the same principles employed in the sodium- and potassium-specific electrodes.

Therapeutic Monitoring

The therapeutic range for lithium is approximately 1.0–1.5 mEq/L, with toxic symptoms being seen above 2 mEq/L. Some authorities recommend that assays be done daily during the initial stages of treatment. Samples should be collected 8–12 h after the previous dose and just prior to administration of the first morning dose. For out-patients receiving lithium, monitoring on a weekly or monthly basis should be sufficient, both to assess blood levels of lithium and to ensure compliance in taking the medication.

20.9

ANALGESICS

Salicylate

Aspirin in some form or another has been used as a medication for several hundred years. The active ingredient (**salicylate**) was first isolated from plant material but is now synthesized by the ton. Aspirin comes in a variety of preparations designed to control the rate of entry of active drug into the system and to lessen possible side effects. Aside from recent studies showing the ability of aspirin to decrease blood clotting and the mechanism of this process, we have no clear picture of how aspirin works as a pain-reliever and an antiinflammatory agent.

PROPERTIES

When purchased, aspirin is a mixture of **acetylsalicylic acid** (the precursor to the active ingredient) and various buffering agents. Since aspirin can produce gastrointestinal distress and bleeding in some individuals, the other materials in the pill or capsule are included to minimize these side effects and to influence the rate of absorption in the intes-

tine. A number of medications may contain acetylsalicylic acid and not clearly indicate its presence.

PHARMACOKINETICS

When acetylsalicylic acid is taken orally, hydrolysis to salicylic acid begins once the material reaches the intestine. The breakdown probably is initiated by the acid contents in the stomach but is also carried out to a great extent by enzymes released in the small intestine. After entering the bloodstream, first-pass hydrolysis also occurs as the material passes through the liver. The half-life for this initial metabolic processing is only about 15–20 min in a normal individual, assuming an empty stomach and no hindrance to absorption of the aspirin. Buffering agents or coatings on the tablet slow the absorption rate, altering the half-life of the hydrolysis step appreciably.

The removal of the acetate group produces salicylic acid (or salicylate if the pH of the system is greater than 7). The two terms are generally used interchangeably and refer to the same basic structure. Further metabolism converts the salicylate to the glucuronic acid and glycine conjugates as well as the oxidation product gentisic acid. Although these metabolites do not produce much in the way of a pharmacologic effect, they are forms which can be readily excreted in the urine, allowing a rapid clearance of the drug from the plasma.

Salicylate in the circulation is primarily protein-bound. Approximately 80–90% of the total salicylate in the system is bound, mainly to albumin. The free fraction is somewhat variable and can be as high as 50% when the total salicylate level in blood is quite elevated. Patients with low serum albumin levels are at greater risk for salicylate toxicity since the free fraction is higher than expected, even if the total amount of measured salicylate is still below the presumed toxic level.

Excretion of salicylate and metabolites takes place in the urine and is very sensitive to pH. As the urine becomes more alkaline, the excretion rate of salicylate increases markedly. At alkaline pH, the molecule exists primarily in the carboxylate anion form, which is more water-soluble and less likely to be reabsorbed by the proximal tubule back into the circulation. A 1-unit increase in urine pH may produce a doubling of the rate of excretion of salicylate and metabolites. Alkalinization of urine is one strategy often utilized for emergency treatment of salicylate toxicity.

ASSAY TECHNIQUES

A variety of techniques have been reported for the analysis of salicylates in body fluids. Many of the methods employ gas chromatography or HPLC, permitting separation of the specific metabolites and providing a high degree of specificity. However, these approaches are lengthy and do not lend themselves well to emergency screening. A rapid colorimetric method uses **Trinder's reagent,** composed of mercuric chloride, hydrochloric acid, and ferric nitrate. Proteins present in the sample are precipitated by the mercuric salt in acid medium. The ferric ion forms a colored complex (absorbance maximum at 540 nm) with salicylate derivatives in the supernatant. Although the method is not specific for salicylic acid alone, the results are reliable enough to be used in toxicology services for screening purposes.

Because of the variability of timing in the administration of aspirin, there is no set protocol for timing of sample collections. Consideration must be given to the purpose of the collection and assay (aspirin overdose or therapeutic monitoring) and the administration pattern for the individual patient. Specific details should be worked out in collaboration with the clinical team.

THERAPEUTIC MONITORING

Therapeutic levels for salicylate are approximately 15–30 mg/dL. Toxic symptoms also begin to appear in some patients at the upper limits of these stated levels, so interpretation must (to some extent) be made on an individual basis. Signs of toxicity include tinnitus (ringing in the ears) and possible deafness, liver damage (as seen in elevations of aminotransferases), gastrointestinal bleeding, some prolonging of clotting time, and impaired renal function.

Acetaminophen

PROPERTIES

Acetaminophen is a pain-reliever with a structure somewhat similar to aspirin. Both medications are also useful in treating a fever. Unlike aspirin, however, acetaminophen does not possess any anti-inflammatory properties, which renders it less useful for treatment of arthritis and related conditions. Acetaminophen is of particular value in treating patients who cannot tolerate aspirin for whatever reason. The amount of gastrointestinal bleeding and associated distress is greatly minimized by substituting acetaminophen for aspirin. A large number of preparations (both prescription and over-the-counter) contain acetaminophen, often in combination with other medications (including aspirin in some instances). As is the case with aspirin, we do not have a good understanding of the mechanism of action of acetaminophen.

PHARMACOKINETICS

Acetaminophen is fairly rapidly and completely absorbed in the small intestine. The drug does not undergo metabolism in the gastrointestinal tract but is absorbed intact. Peak serum levels may appear 30–40 min after administration under normal therapeutic usage. As the molecule passes through the liver, it is converted to a number of inactive metabolites. The drug is excreted in the urine mainly in the form of metabolites, with approximately 5% of the total amount of acetaminophen showing up in the untransformed state.

Unlike salicylate, acetaminophen is weakly bound to plasma proteins. Only about 10–25% of the material is transported by albumin and other proteins, with the bulk of the drug existing in the free state. Therefore, a higher fraction is available to be distributed throughout the body.

The major route for elimination of acetaminophen and metabolites is by way of the kidney. The plasma half-life of the drug after therapeutic dosing is about 2 h. Urine pH does not significantly affect the rate of excretion of materials in this situation. Renal impairment affects the rate of acetaminophen elimination, but the impact is lessened since most of the excretion products are inactive forms of the drug.

ASSAY TECHNIQUES

Several spectrophotometric methods have been developed for the quantitation of acetaminophen, but they are not used extensively in clinical situations because they are cumbersome and lengthy. Gas chromatography and HPLC are preferred because they are more reliable and selective. An enzyme immunoassay procedure is commercially available which couples specificity with rapid turn-around for emergency situations.

THERAPEUTIC MONITORING

As with aspirin, there are no clear-cut guidelines for sample timing or monitoring of acetaminophen utilization in clinical situations. Blood levels of this drug are ordinarily monitored to assess the possibility of acetaminophen overdose (usually a suicide attempt), indicated by a blood level of greater than 150 mg/L. Samples collected soon after drug ingestion may not reflect the actual level, since several hours are required for peak serum concentrations after an overdose. Serial monitoring of blood levels is necessary to obtain the most useful clinical data. Data indicate that blood levels of acetaminophen of 150–200 mg/L or higher between 4 and 12 h after ingestion are invariably accompanied by severe hepatic damage. If the plasma acetaminophen level is below 150 mg/L in this period, damage to the liver is much less likely.

Although the major metabolites of acetaminophen are pharmacologically inactive, some "minor" metabolites form which are extremely toxic. The mechanism of action of these components is somewhat complex, but the end result is the destruction of liver cells. Monitoring liver enzymes and bilirubin levels for several days after an overdose is usually necessary to establish the existence and extent of damage.

20.10

THEOPHYLLINE

Properties

Like its chemical relative caffeine, **theophylline** has been known for years to be an effective treatment for many types of respiratory problems. This medication functions primarily as a **bronchodilator**, relaxing bronchial smooth muscle and blood vessels. In addition, theophylline stimulates respiration and strengthens the action of cardiac muscle. The drug serves as a central nervous system stimulant, which further enhances respiration and decreases drowsiness.

Theophylline is administered primarily orally or intravenously. A number of commercial preparations are available, including slow-release capsules to allow finer control over the absorption of the drug. Small beads of theopylline are also available which can be given with food to babies and small children who may have difficulty in swallowing a tablet.

Pharmacokinetics

After oral administration, theophylline is rapidly and completely cleared from the gastrointestinal tract in the form of the parent drug. Peak blood levels after a single dose appear some 30 min–2 h after administration in adults. The absorption rate in children appears to be somewhat faster.

A fairly high proportion of theophylline is transported through the circulation bound to plasma proteins. In adults, some 53–60% of the drug is protein-bound. The degree of protein binding in children is considerably less, with approximately 35% of the drug attached to proteins in the blood. In patients with hepatic cirrhosis, the extent of protein binding is lowered, with some 30–35% of theophylline in the circulation existing in the bound state.

Metabolism of theophylline is extensive and carried out primarily by the liver. Several metabolites with structures similar to uric acid are formed. These conversion products are much more water-soluble and can be readily excreted in the urine. For the most part, the metabolites have little or no pharmacologic activity. Only one compound, 3-methylxanthine, has an appreciable theophyllinelike effect (approximately 70–80% of the parent drug), but it is present in very low amounts and probably does not contribute extensively to the therapeutic effects. Less than 10% of the theophylline itself is found excreted in the urine after administration.

The rate of elimination of theophylline varies markedly with age. Newborns (especially premature infants, who have a higher incidence of respiratory difficulties) have very slow rates of elimination, with half-lives of 24–30 h being routinely seen. As the infant ages, the rate of elimination increases enormously. Children between the age of 1 and 9 years show elimination half-lives of less than 4 h. Adults who are nonsmokers have slower rates of elimination (half-life approximately 8–9 h), whereas cigarette and marijuana smokers have elimination rates roughly twice as fast.

Since theophylline is processed through metabolism in the liver, any patient with liver disease or congestive heart failure has a slower rate of conversion to inactive metabolites. In addition, patients who are also receiving other medications may have altered clearance of theophylline. Medications such as phenytoin, carbamazepine, and related compounds stimulate the formation of high levels of drug-metabolizing enzymes in the liver. In these situations, the conversion of theophylline to inactive derivatives is enhanced. The effective level of theophylline in the circulation decreases much more rapidly, and dosage must be increased to compensate for this situation. Some antibiotics, such as erythromycin, interfere with the elimination of theophylline. The process by which this interference is carried out is not clear at present.

Assay Techniques

A number of HPLC methods have been developed and refined which are simple, rapid, and subject to little interference. The main advantage to HPLC is that caffeine can be quantitated simultaneously with theophylline. Since caffeine is a significant metabolite of the drug in newborns and small children (and produces some of the same pharmacologic effects), monitoring for the presence and amount of this metabolite can be useful when providing laboratory information for many pediatric patients. Enzyme immunoassay procedures are now commercially available which are most suitable for either batch or single-sample assays. These methods are also rapid, accurate, and essentially interference-free.

Therapeutic Monitoring

Because of the variability in practice of administration of theophylline, sampling times must be coordinated with the clinical team for maximum effectiveness in therapeutic drug monitoring. For adults, the therapeutic range for theophylline is 10–20 μg/mL, although toxic signs have been occasionally observed at serum concentrations as low as 15 μg/mL. Plasma levels for newborns should be lower, in the 5–10 μg/mL range.

As the theophylline level increases above the therapeutic range, a number of toxic symptoms are seen.

Nausea, vomiting, headaches, and anxiety may be manifested at concentrations between 15 and 20 $\mu g/$ mL. Increase in heart rate and alterations of cardiac rhythm begin to appear at concentrations between 20 and 40 $\mu g/mL$. Above this level, seizures and cardiac arrest are common.

20.11

SPECIAL FACTORS FOR INFANTS AND CHILDREN

The ability of a patient to metabolize and excrete drugs varies greatly with age. There appear to be three periods where major shifts occur in the capability of the body to convert and dispose of medications. The newborn has a diminished ability both to metabolize and eliminate these foreign compounds. From infancy to preadolescence, the body increases its capability markedly. Finally, during puberty there is a rather abrupt change from the childhood to the adult pattern of drug metabolism.

Prior to birth, fetal drug metabolism and excretion are handled by the mother. Although materials can readily pass across the placenta, the processes for drug disposal do not yet exist to any degree in the unborn child. After birth, the liver enzymes necessary for drug metabolism and conjugation begin to develop and the infant's capability for conversion and excretion of pharmacologic substances improves noticeably.

Children in the early school years (roughly 6–10 years) have a high capacity for drug metabolism. These patients utilize medications at approximately twice the level of adults, requiring appropriate adjustments in dosages to achieve the desired therapeutic response.

As the child reaches adolescence, profound changes take place in all body systems. One such change in a marked decrease in the rate of drug metabolism (elimination does not appear to be affected greatly). It is thought that the striking rise in the production of steroid hormones during puberty affects the process of drug metabolism because of competition with the appropriate enzymes in the liver. The shift in metabolic rate is seen somewhat earlier in females than in males, consistent with other changes seen during puberty. Within a few months, the shift has been made from the child's rate of metabolism to a lower rate more consistent with the drug utilization pattern of adults.

20.12

EFFECTS OF PREGNANCY ON DRUG METABOLISM

A number of physiological changes take place during pregnancy which measurably affect drug metabolism and elimination. These changes are compounded by the fact that many medications should be either discontinued or the dosage lowered during the course of the pregnancy. There is little experimental data on drug metabolism during pregnancy because ethical considerations proscribe their use. Drug administration during pregnancy can adversely affect the unborn child.

Plasma volume and total body water increase during the course of pregnancy. These increases help determine the distribution and concentration of the drug within the system. A large fraction of the total body water is thought to be in the extracellular portion of the tissues, affecting the amount of drug actually reaching the target cells.

Changes in plasma protein levels are quite striking during this time. Albumin levels decrease markedly, but lipoprotein concentrations may increase. These shifts in quantities of specific proteins alter the amount and degree of binding of drug to protein. Variations in binding change the amount of drug available to the specific effector site.

Levels of various steroids increase during pregnancy and compete to some extent with drugs for the sites in the liver where metabolism and conjugation take place. One result of this competition is a possible decrease in the rates of inactivation and elimination for a specific pharmacologic agent. On the other hand, the overall ability of the liver to metabolize compounds is also enhanced to some extent during this period. In addition, blood flow to the liver increases during pregnancy, increasing the transport of drug to this major site of metabolism. The overall effect differs depending on the medication in question, but the basic changes should be kept in mind when interpreting data.

Renal function is often affected by pregnancy and varies according to the stage. During the early months, blood flow to the kidney and the rate of filtration of materials from the blood increases greatly. These changes increase the mother's body's ability to remove a given drug by excretion. During the last month or so of pregnancy, there is a gradual decline in kidney function under normal circumstances and drug excretion usually slows down. In some instances, a pathologic situation severely impairs kidney function; drug elimination is sharply curtailed in these circumstances.

20.13

CONSIDERATIONS IN THE ELDERLY

Geriatric patients represent some unique problems for therapeutic drug monitoring. This age group has an increased incidence of adverse drug reactions relative to younger patients. Among the causes of these problems are differences in the body's ability to metabolize and excrete the medication.

In elderly patients there is a decrease in the level of serum proteins, particularly albumin. Consequently, drugs are less able to bind to protein. Therefore, the same dose leads to a higher free drug concentration, even though the total concentration (sum of bound and free drug) remains unchanged. Interpretation of the clinical effects of administration of any medication must include data regarding the serum protein levels. Increasingly, laboratories measure free drug concentrations so that this factor can be better assessed.

A decrease in the liver's ability to metabolize drugs is a major reason for the slower elimination of drugs from the bloodstream in elderly patients. Because of the diminished rate of metabolism, the drug remains at a higher level in plasma and stays for a longer time. By monitoring drug levels, we can adjust the dosage appropriately to compensate for this decline in metabolism.

Elderly patients also suffer from diminished renal output. Kidney function is decreased compared with younger individuals. Since most drugs are eliminated from the body by excretion in the urine, this diminished renal capability leads to slower excretion of the drug and prolonged higher levels in the circulation. This phenomenon can be compensated for if information regarding creatinine clearance or other data reflecting kidney function are available when interpreting drug levels.

20.14

MONITORING DRUGS IN SALIVA

In our earlier discussion, emphasis was placed on the fact that only the free drug was available to the tissues. Whatever medication was bound to protein was essentially unavailable until a new equilibrium was established. In many clinical situations, the amount attached to carrier protein varies depending on the protein level and the presence of other materials which compete for binding sites on that protein. Current techniques for monitoring drug levels do not allow us to measure this important parameter and provide information only on the total amount of medication present, not that portion actually available to the cell.

In recent years a number of approaches have been developed (with varying degrees of success) to obtain this needed information. At least one company has marketed a filtration system which allows a separation of bound and free drug. Unfortunately, the amount of sample needed is rather high and the preparation time is sometimes long. An intriguing alternative is to measure the drug level in saliva. A number of studies have shown that only free drug appears in saliva; material bound to protein does not pass through the membranes. Fairly good correlations have been reported between saliva levels of a particular compound and the level of unbound material in serum or plasma. This technique also has the advantage of being noninvasive: no venipuncture is needed to obtain a sample. One draw-

back is the quite variable rate of saliva production, which affects results to some extent. Further improvement in technique and further correlation studies may make the measurement of saliva levels of drugs a valuable addition to the field of therapeutic drug monitoring.

20.15

LABORATORY DELIVERY OF THERAPEUTIC DRUG-MONITORING SERVICES

The establishment of a successful and useful therapeutic drug-monitoring service involves careful consideration of many details. Decisions are necessary about which drugs need to be monitored. In most laboratories, assays for the various pharmacologic agents described in this chapter should be available. In addition, the clinicians in a specific setting may have unique needs which would justify the inclusion of other assays. Close interaction between the laboratory and the clinical staff is essential.

Decisions need to be made about the extent of services and the timing of assays. In some instances, a batch procedure is appropriate since results are not needed immediately. Samples for digoxin or aminoglycoside analysis might be run once a day, and the schedule set to provide results in adequate time for the clinical staff to adjust dosages before the next administration period. Other medications, such as theophylline, require a more flexible analysis schedule so that emergency situations can be handled promptly. TDM services in an out-patient setting require consideration of the patient load and the manner in which the information is to be used. Close collaboration with the clinical staff minimizes delays and misunderstandings.

Equipment is a major consideration both in selection of assay methodology and determination of tests to be offered. In very few cases should a laboratory attempt to offer comprehensive services in house. Acquisition of equipment should be decided in part on the basis of its versatility. A system capable of performing a wide variety of EMIT and colorimetric assays is much more useful than a specialized instrument dedicated to only one or two assays. Careful cost analysis is necessary before a decision is made to implement a specific assay. Use of a nearby reference laboratory or other facility may be a better alternative to doing an unusual test in house.

Careful consideration needs to be given to all aspects of the delivery of services, from sample collection to reporting results. Information on time and method of administration, as well as the rate of drug infusion (if pertinent) should be recorded on the laboratory requisition along with the time of sample collection. Proper reporting is crucial: stat requests should be performed promptly and telephoned to the physician requesting the assay. If a particular value exceeds the therapeutic limits for that medication and is in the toxic range, the report should be treated as a stat and called, even though the request was routine. The laboratory should have data available on therapeutic and toxic levels for all drugs assayed.

A therapeutic drug-monitoring service is an integral part of any laboratory. The extent of services available within the institution varies and should be discussed thoroughly with the clinical staff. Careful consideration needs to be given to what type of services are offered, when they are offered, and how results are communicated. By following appropriate laboratory practices, a TDM program can contribute greatly to the welfare of the patients.

SUMMARY

Medications are usually administered orally, intravenously, or intramuscularly. Each mode of administration affects the rate of absorption into the circulation. After entering the bloodstream, the drug may bind to plasma proteins. The extent of protein binding determines, in part, the rate of uptake by the target tissue. Only free (unbound) medication enters the cell. Metabolism of the parent pharmaceutical agent may take place, primarily in the liver. Both pharmacologically active and inactive products may be formed. Conjugation with glucuronic acid in the liver is a major means of forming more water-soluble components. Excretion of drug and metabolites in the urine eliminates these agents from the body.

The concept of therapeutic drug monitoring was developed to furnish accurate information regarding the level of a specific drug in a particular patient's system and to relate that concentration to therapeutic effectiveness. A subtherapeutic level provides no benefit, and a toxic level represents a hazard to the patient. By measuring the serum or plasma concentration at indicated times after administration (either peak level, trough level, or both, depending on the particular need), useful information can be generated allowing the clinical staff to adjust dosage and increase benefit and safety for the patient.

Each drug has its own rate of absorption, time of peak concentration after administration, degree of binding to plasma proteins, formation of active and/or inactive metabolites, and rate of excretion. The level required for therapeutic and toxic effects varies from one medication to another. Knowledge of specific properties is necessary for proper interpretation of data.

A number of measuring techniques are available for quantitation of drug levels in body fluids. Although spectrophotometric approaches are usually fairly simple, they often lack the speed and sensitivity of other analytical approaches. For the measurement of a specific compound, immunoassay techniques are widely used because of their specificity and speed. A single assay can be performed in a few minutes with a high degree of reliability. For more complex situations (measurement of a mixture of drugs or a drug and its active metabolites), gas chromatography or high-performance liquid chromatography are the methods of choice.

Various age groups manifest differences in drug disposition. Children usually have higher rates of metabolism and elimination. Elderly patients may metabolize materials more slowly. In addition, impaired renal function may be present in a geriatric patient, greatly affecting the rate of excretion of a medication. Pregnant women have altered rates of metabolism and excretion as well as special requirements to avoid harming their unborn children.

Providing a therapeutic drug-monitoring service requires careful planning and communication. Realistic expectations must be established, both about the range of services and the timing of analyses. Some medications need to be monitored on a daily basis and can be handled in a batch mode. Other requests warrant a stat response or at least a reasonably quick turnaround time. All analytical data should be accompanied by information regarding time of drug administration, rate of infusion (if appropriate), and time of sample collection. Reports to the clinical team should be done in a timely manner. A properly operated therapeutic drug-monitoring program can contribute significantly to improved patient health.

FOR REVIEW

Directions: For each question, choose the best response.

1. All of the following are considered major routes for drug administration *except*
   A. rectal
   B. oral
   C. intravenous
   D. intramuscular

2. The major carrier protein of drugs in the circulation is
   A. albumin
   B. hemoglobin
   C. myoglobin
   D. transferrin

3. For a drug to enter the tissue and elicit the desired effect, it must be
   A. bound to albumin
   B. bound to globulin
   C. in a free state
   D. injected intravenously

4. Transformation of the parent drug to its metabolites primarily occurs in the
   A. gallbladder
   B. liver
   C. pancreas
   D. spleen

5. Drug excretion is primarily accomplished by the
   A. intestines
   B. kidneys
   C. lungs
   D. skin

6. The therapeutic range of a drug may be defined as the blood level of the drug that
   A. elicits minimal side effects
   B. elicits no side effects
   C. cures the disorder
   D. provides optimum concentration of medication for treatment

7. For a drug to exhibit steady-state levels, the rate of drug administration must be _____ the rates of drug metabolism and excretion.
   A. equal to
   B. greater than
   C. less than

8. The analytical system that allows the simultaneous differential measurement of parent drug and metabolites is
   A. fluorometry
   B. spectrophotometry
   C. immunoassay
   D. high-performance liquid chromatography

9. Digoxin, procainamide, and quinidine are drugs that may be classified as
   A. aminoglycosides
   B. anticonvulsants

C. antidepressants

D. cardioactive

10. The medication of choice for treatment of manic-depression is
    A. carbamazepine
    B. lithium carbonate
    C. phenobarbital
    D. phenytoin

11. Which of the following drugs may be quantitated colorimetrically using Trinder's reagent (mercuric chloride, ferric nitrate, and hydrochloric acid)?
    A. acetaminophen
    B. phenobarbital
    C. salicylate
    D. theophylline

12. Which of the following details must be known in order to properly interpret drug concentration data received from a blood analysis?
    A. time and route of drug administration
    B. collection time of blood specimen
    C. serum protein concentration
    D. liver and renal function
    E. all of the above

13. The therapeutic serum drug level for a patient with cirrhosis should be maintained _____ that for a noncirrhotic individual.
    A. higher than
    B. lower than
    C. the same as
    D. cannot be determined

14. A person suffering from kidney disease is receiving a generally accepted therapeutic dose of a drug; you would expect the blood concentration of the drug to
    A. be higher than the therapeutic range
    B. be lower than the therapeutic range
    C. be within the therapeutic range
    D. fluctuate higher and lower than the therapeutic range

BIBLIOGRAPHY

Aziz, K., "Therapeutic drug monitoring," *Am. Lab.* April:78–81, 1988.

Bailey, R. R., "The aminoglycosides," *Drugs* 22:321–327, 1981.

Duffy, J. P., and Rumack, B. H., "Effects and treatment of acetaminophen overdosage," *Syva Monitor.* 1:1–7, 1983.

Fincham, R. W., "Clinical expectations of laboratories performing therapeutic drug analysis," *J. Clin. Immunoassay.* 7:266–267, 1984.

Follath, F. et al., "Reliability of antiarrhythmic drug plasma concentration monitoring," *Clin. Pharmacokinet.* 8:63–82, 1983.

Frade, P. D., and Miceli, J. N., "Antiarrhythmic drugs, part 2: Monitoring," *Lab. Manage.* February:22–27, 1988.

Friedman, H., and Greenblatt, D. J., "Rational therapeutic drug monitoring," *J. Amer. Med. Assoc.* 256:2227–2233, 1986.

Frommer, D. A. et al., "Tricyclic antidepressant overdose—A review," *J. Amer. Med. Assoc.* 257:521–526, 1987.

Graves, S. W., and Williams, G. H., "Endogenous digitalis-like natriuretic factors," *Annu. Rev. Med.* 38:433–444, 1987.

Henry, J. B., ed., "Therapeutic drug monitoring," *Clin. Lab. Med.* 4, no. 3, 1981.

Hollister, L. E., "Plasma concentrations of tricyclic antidepressants in clinical practice," *J. Clin. Psychiatry.* 43:66–69, 1982.

Leal, K. W., and Troupin, A. S., "Clinical pharmacology of antiepileptic drugs: A summary of current information," *Clin. Chem.* 23:1964–1968, 1977.

Lovell, M. A., and MacMillan III, R. H., "Aspirin overdose," *Clin. Chem. News.* July, 1985.

Mandelli, M., and Tognoni, G., "Monitoring plasma concentrations of salicylate," *Clin. Pharmacokinet.* 4:424–440, 1980.

Moyer, T. P., and Boeckx, R. L., eds., *Applied Therapeutic Drug Monitoring* (Vol. I) *Fundamentals.* Washington, D.C.: American Association for Clinical Chemistry, 1982.

Ogilvie, R. J., "Clinical pharmacokinetics of theophylline," *Clin. Pharmacokinet.* 3:267–293, 1978.

Pechere, J.-C., and Dugal, R., "Clinical pharmacokinetics of aminoglycoside antibiotics," *Clin. Pharmacokinet.* 4:170–199, 1979.

Pippenger, C. E., "Therapeutic drug monitoring: Pharmacologic principles," *Diag. Med.* June:28–36, 1983.

Preskorn, S. H., and Irwin, H. A., "Toxicity of tricyclic antidepressants—kinetics, mechanism, intervention: A review," *J. Clin. Psychiatry.* 43:151–156, 1982.

Riegelman, S. et al., "Factors affecting the pharmacokinetics of theophylline," *Eur. J. Respir. Dis.* 61(suppl. 109):67–82, 1980.

Schottelius, D. D., "Therapeutic drug monitoring: Fundamental pharmacological concepts," *J. Clin. Immunoassay.* 7:247–253, 1984.

Scoggins, B. A. et al., "Measurement of tricyclic antidepressants. part II. Applications of methodology," *Clin. Chem.* 26:805–815, 1980.

Shipe, Jr., J. R., and Herold, D. A., "Drug monitoring in the neonate," *Ann. Clin. Lab. Sci.* 12:296–303, 1982.

Smith, T. W., "Digitalis—mechanism of action and clinical use," *New Eng. J. Med.* 318:358–365, 1988.

Walker, R. J., and Duggin, G. G., "Drug nephrotoxicity," *Annu. Rev. Pharmacol. Toxicol.* 28:331–345, 1988.

Chapter Twenty-one

SCREENING FOR SUBSTANCE ABUSE

UPON COMPLETION OF THIS CHAPTER, THE STUDENT WILL BE ABLE TO

1. State the physiological effects of each of the following substances:
   A. ethanol
   B. methanol
   C. isopropanol
   D. amphetamines
   E. cocaine
   F. heroin
   G. marijuana
   H. phencyclidine
   I. anabolic steroids
2. Explain the process by which ethanol is metabolized by the liver.
3. Describe the proper blood collection procedure for alcohol analysis.
4. Describe the gas chromatographic and enzymatic methods used for quantitating ethanol.
5. Name the cocaine and heroin metabolites that are generally sought during drug quantitation.
6. Identify the major active ingredient of marijuana.
7. Discuss the meaning of the term *designer drugs*.
8. Explain the principle of thin-layer chromatography.
9. Describe the steps involved in performing thin-layer chromatography.
10. Calculate an R_f value when given the required information.
11. Discuss the use of the following methods for identifying and/or quantitating drugs of abuse:
    A. immunoassays
    B. HPLC
    C. GC/mass spectrometry
12. Discuss the importance of properly maintaining the chain of evidence when handling specimens for drug analysis.

INTRODUCTION

Substance abuse has become one of the most important and controversial issues of today. At one time, illicit drug use was confined to a small subculture of the population. Now, the problems associated with the use of stimulants, hallucinogens, cocaine, marijuana, and ethanol touch every aspect of our society. Hundreds of millions of dollars are spent yearly on law enforcement and treatment for dealing with the physical, psychological, and social damage produced by substance abuse. The evidence is now overwhelming that the vast majority of crimes are committed by individuals under the influence of drugs, often in order to obtain the means of purchasing more drugs.

21.1

ETHANOL ABUSE

Abuse of ethanol ("alcohol") ranks with cigarette smoking as one of the major substance abuse problems in this country. Approximately 12% of all those who consume ethanol are alcoholics. Very few fit the "skid row" stereotype and frequently do not recognize or admit their addictive behavior. Rarely have other substances had the impact ethanol has on the moral, social, and legal aspects of our lives. The clinical chemistry of ethanol is intimately entwined with a variety of legal and medical issues for the laboratory technician. Concerns for reliability of analysis, confidentiality of results, and the effects of ethanol on other biochemical systems all must be addressed as we survey the issue of ethanol abuse.

Pharmacology

ETHANOL CONTENT OF BEVERAGES

Ethanol is consumed in a wide variety of beverages (Table 21-1). Alcohol content is often ex-

Table 21-1
APPROXIMATE ETHANOL CONTENT OF
ALCOHOLIC BEVERAGES

Beverage	Ethanol Content (%)
Beer	3–6
Ciders	4–5
Wines	8–15
Sherry, madeira, port	18–20
Whiskey, gin	40–45
Vodka	40–50
Brandy	45–50
Rum	50–70

pressed as *proof,* twice the percentage of ethanol in the beverage. A 90-proof vodka would contain 45% ethanol. When considering the effects of drinking, we must be concerned with the total amount of ethanol intake in a given period, not how many "drinks" a person has consumed.

In addition to the ethanol content, most beverages contain **congeners** (small amounts of ethers, aldehydes, ketones, and essential oils that add aroma and flavor to alcoholic beverages). Methanol may also be present in small amounts. These congeners (over 100 in all) play an important role in the effects of consuming the beverage. They are also responsible for many of the unpleasant aftereffects of drinking. Some of the toxic effects of drinking ethanol are caused by the congeners, although their mechanisms of action are not clear at this time.

PHYSIOLOGICAL AND PSYCHOLOGICAL EFFECTS OF ETHANOL

Pharmacologically, ethanol functions as a depressant of the central nervous system, not a stimulant. The presumed stimulation is actually due to the blockage of inhibitory centers, leading to a perception of being stimulated. Risk-taking behaviors often increase after alcohol consumption: for instance, the level of aggressive humor may rise.

A major initial effect of ethanol consumption is an increase in heart rate and blood pressure. Simultaneously, enhanced vasodilation of peripheral blood vessels occurs, accompanied by a feeling of warmth. This expansion in blood vessels actually increases the rate of heat loss. Conversely, the arteries in the vicinity of the heart constrict, raising blood pressure.

The performance of any task which requires skill, thought, or attention diminishes after even light ethanol intake. The person's ability to hear and see decreases. Time perception is distorted, with time appearing to pass more rapidly. The speed of moving objects is underestimated, often with tragic consequences.

METABOLISM OF ETHANOL

When consumed orally, ethanol is rapidly absorbed from the gastrointestinal tract with no conversion to other products. Under most circumstances, approximately 60% of the ethanol consumed is in the bloodstream within 1 h. By 1.5 h, the amount absorbed has increased to over 90%. Absorption is slowed when ethanol is consumed with food.

Ethanol can also be passively absorbed by inhalation of vapors. This substance has a low molecular weight and is easily diffusible through pores, entering the circulatory system by way of the capillaries.

Some cases of fatal intoxication in distilleries have been due to this passive absorption phenomenon.

After entering the circulation, ethanol is metabolized by the liver. The enzyme, **alcohol dehydrogenase** (of which there are several isoenzymes), oxidizes the ethanol to form acetaldehyde:

$$CH_3CH_2OH + NAD^+ \xrightarrow[CH_3CHO + NADH + H^+]{\text{alcohol dehydrogenase}}$$

The acetaldehyde formed is then excreted or converted to acetate, which can undergo further metabolic transformation. The maximum rate of ethanol conversion in the normal adult is approximately 7 g of ethanol per hour. This load is equivalent to approximately 6–10 mL of ethanol (roughly one beer, one glass of wine, or one shot of whiskey).

In recent years, two other systems have been implicated in the metabolism of ethanol. A microsomal ethanol-oxidizing system also generates acetaldehyde by reaction with enzymes concerned with drug metabolism in general. Catalase is also suspected of being a part of the ethanol-metabolizing scheme, but there is insufficient evidence at present to discuss its exact role.

Formation of excess H^+ from ethanol creates some metabolic problems. A decrease in pH inhibits the excretion of uric acid by the kidneys, leading to hyperuricemia and kidney stones. In addition, the conversion of pyruvate (formed from glucose metabolism) is shifted from acetyl CoA to the nonbeneficial lactic acid. Prolonged accumulation of lactic acid leads to disturbances in acid/base balance and lactic acidosis. Furthermore, gluconeogenesis is blocked by excess lactic acid in the system. Many intoxicated persons are found to have low blood glucose levels, especially if there is a long history of ethanol abuse. Since chronic drinkers also have poor dietary habits, malnutrition becomes a major problem.

The major route of ethanol disposal is by way of metabolism; approximately 90–98% of the ethanol consumed is converted by the liver. The remainder is excreted, primarily by the kidney, with the rate of elimination in urine being approximately one half the rate of absorption by the stomach. A small fraction is lost through perspiration and respiration, with the lung disposing of approximately 0.5–1.5% of the total ethanol load.

Analysis

Our discussion of methods for analyzing ethanol will be restricted to those techniques which would find application in the average clinical chemistry laboratory. We will not consider forensic techniques such as breath analysis. Legal aspects (consent, sample custody in the chain of evidence) will be considered later in this chapter.

SAMPLING

Blood obtained by venipuncture is the sample of choice. Cleansing the site on the arm should not be done with an alcohol swab, because although the possibility of contamination is very slight, unnecessary questions can be raised about the validity of results. Instead, soap and water or some nonalcohol-containing detergent can be used to prepare the site.

Eight to 10 mL of whole blood in a sterile container containing oxalate and sodium fluoride is the preferred specimen. Plasma should be separated from cells as soon as possible after sampling. Care must be taken to prevent microbial formation of ethanol through metabolic processes which would falsely elevate the value. Since ethanol readily evaporates, the sample must be tightly stoppered at all times. Also, because an increase in temperature enhances the rate of evaporation, temperature should be controlled.

ANALYTICAL APPROACHES

The two common techniques for quantitative determination of ethanol levels are gas chromatography and enzymatic analysis. The colorimetric determination of volatile materials using a Conway diffusion plate is qualitative only and subject to many interferences. This technique is, therefore, not used much anymore. Measurement of sample osmolality has been described by a number of authors, but is cumbersome. In addition to the osmolality assay, values for electrolytes, protein, and other parameters must be incorporated into the calculations. The osmolality approach is too inaccurate for routine use.

Gas Chromatography

In terms of selectivity and specificity, gas chromatographic analysis of ethanol provides the most useful information. A serum sample is injected directly onto the column after being mixed with an internal standard. Separation occurs within a few minutes, depending on the column and the specific fractionation conditions. Identification of the alcohols methanol, ethanol, and isopropanol can easily be made, since they separate well from each other (Fig. 21–1). The technique employed should also be able to identify acetone, present in many diabetics with ketoacidosis. An internal standard (*n*-propanol is frequently used) allows for reliable quantitation.

A variation on this analytical theme employs **headspace analysis,** the measurement of ethanol in a vapor. The sample is warmed in a closed container

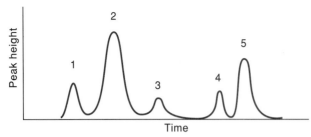

FIGURE 21–1 Alcohol fractionation by gas chromatography. 1: methanol, 2: ethanol, 3: acetone, 4: isopropanol, and 5: *n*-propanol (internal standard).

and the vapor which evolves is collected for analysis. Headspace analysis greatly decreases any problems of column contamination produced by direct injection of serum, but has some technical complexities which make it a more complicated process.

Enzymatic Analysis

Although gas chromatography is rapid and allows separation of several components of interest, it requires committed instrumentation and a high degree of skill on the part of the operator. A desirable alternative approach for the quantitation of blood ethanol is the enzymatic analysis. The reaction employs alcohol dehydrogenase with NAD^+ as coenzyme. Acetaldehyde is formed as the ethanol is oxidized, with the generation of NADH used as a marker for the reaction. Direct measurement of the increase in absorbance at 340 nm allows quantitation of the ethanol present. In some systems, a dye reaction is coupled to the formation of NADH to produce a color. These procedures also produce some nonspecific color formation and are not as reliable as the 340 nm measurement of NADH absorbance.

Since further oxidation of acetaldehyde is possible (catalyzed by the enzyme aldehyde dehydrogenase), a trapping agent is included in the assay system. Semicarbazide is employed to react with any acetaldehyde formed, blocking further oxidation and production of more NADH than would be generated by the ethanol reaction alone.

The major drawback of the enzymatic method is its lack of specificity relative to gas chromatography. Although GC measurement separates methanol and isopropanol (both possible abused substances) from ethanol, the enzyme system reacts with all three alcohols to produce NADH as a product. This is not a serious problem with methanol, since it is converted at a relatively slow rate by the enzyme. However, isopropanol may produce erroneous results in some systems. It is worthwhile to check the response of a given method with known dilutions of isopropanol to ascertain the exact extent of interference with this material. Another quick check can be

made by recording the change in absorbance over a period to determine the rate of absorbance change per unit time. If ethanol is present, the initial rate is quite high, leveling off after several minutes. Isopropanol, on the other hand, produces a very different rate pattern by showing a slow (but steady) rise over the same period.

UNITS OF REPORTING

How blood ethanol values are reported depends on local practice and the use to which the information is put. Two common modes are available: percentage ethanol or concentration in milligrams per deciliter. Reporting in micromoles per liter or similar units is as yet little more than an interesting exercise in unit manipulation. The stating of ethanol values in percentages (weight per volume) derives from the need to have results which are compatible in format with legislative statutes. The early work on the relationship between blood ethanol levels and impairment of performance (particularly as it relates to driving) reported values as percentage of ethanol; however, more current literature usually states results as milligrams per deciliter. A value of 0.1% ethanol is equivalent to 100 mg/dL.

The blood level of ethanol is roughly correlated to physiological symptoms and impairment. Table 21–2 summarizes some of the relationships between ethanol concentration and physiological response.

Physiological and Medical Effects of Long-Term Abuse

Long-term ethanol abuse can significantly impair most organ systems in the body. The primary tissue affected is the liver. Actual hepatic destruction may take place, leading to the formation of scar tissue

Table 21–2
STAGES OF IMPAIRMENT BY ETHANOL

Blood Alcohol (%, w/v)	Signs and Symptoms
0.01–0.05	No obvious impairment, some changes observable on performance testing
0.03–0.12	Mild euphoria, decreased inhibitions, some impairment of motor skills
0.09–0.25	Decreased inhibitions, loss of critical judgment, memory impairment, diminished reaction time
0.18–0.30	Mental confusion, dizziness, strongly impaired motor skills (staggering, slurred speech)
0.27–0.40	Unable to stand or walk, vomiting, impaired consciousness
0.35–0.50	Coma and possible death

Adapted from Dubowski, K., "Alcohol analysis: Clinical aspects. Part II," Lab. Med., April:35, 1982.

and the development of cirrhosis. Prior to the development of cirrhosis, alcoholic hepatitis is common. Lipids frequently accumulate, leading to the condition known as **fatty liver.** This situation is often subclinical, but can produce death on occasion.

A significant number of biochemical alterations take place in the liver as a result of alcohol abuse. The processes for amino acid utilization, carbohydrate and lipid metabolism, and drug conversion and detoxification are all changed as a result of ethanol consumption. These alterations have far-reaching effects on other systems in the body. One major impact is the decrease in the ability of blood to clot, opening the door to severe bleeding problems associated with the tissue damage also seen in alcoholics.

Alcohol ingestion has been widely associated with various lesions of the gastrointestinal tract. Ulcers, gastrointestinal bleeding, and digestive disturbances are common. Chronic pancreatitis is frequently seen, with severe pancreatic damage and impaired secretion of digestive enzymes. Damage to both cardiac and skeletal muscle occurs in many alcoholics. There is a marked increase in the incidence of cancer of the liver, esophagus, and mouth for in-patients with a history of heavy alcohol consumption, particularly for those who smoke.

Production of several hormones is altered by long-term ethanol intake. Cortisol levels are usually increased in alcoholics. There is frequently an increase in production of insulin. Synthesis of testosterone is impaired by a complex mechanism, and production of estrogen is enhanced. Thyroid hormone production does not appear to be affected by alcohol consumption.

Depression of the immune system is commonly seen in many cases of alcohol abuse. The lowering of antibody production results in an increase in the incidence of infectious diseases, particularly respiratory illnesses. Pneumonia and tuberculosis are much more prevalent among alcoholics than in the general population.

One major consequence of alcoholism is damage to the nervous system, produced in part by vitamin B_{12} and folate deficiencies. Peripheral neuropathy is common, with muscle weakness and diminished pain response. Alteration of the brain tissue and irreversible destruction of brain cells are frequent manifestations of neurologic damage produced by ethanol abuse. Depression is often seen, either as a primary disorder or as a result of severe alcohol intake.

Fetal Alcohol Syndrome

Within the last 10 years, **fetal alcohol syndrome** has become an issue of major concern. If the mother drinks during pregnancy, the odds are significantly greater that the child will be mentally retarded and have some sort of birth defect. Growth is reduced in both the pre- and postnatal period. Decrease in brain size in the affected child is quite common. In addition, there is a higher incidence of spontaneous abortions, stillbirths, and early infant deaths for the children of drinking mothers.

Ethanol and Other Drugs

There is a large body of literature which documents the adverse effects of using ethanol in combination with other drugs. In most instances, the effect is additive, meaning that the combination is more powerful than the effects produced by either drug used alone. In part, these phenomena are produced after ethanol has induced microsomal enzymes in the liver which are involved in the metabolism of drugs in general. At other times, the drug accentuates an adverse situation initially produced by long-term ethanol consumption.

Ethanol combined with tranquilizers or barbiturates can be lethal. The effects produced by each drug separately are enhanced by the presence of the other substance. Numerous overdoses of prescription drugs result from concurrent intake of alcoholic beverages. Many of these incidents are accidental, but a large proportion are attempted suicides.

Use of antihistamines with ethanol markedly increases the sedative effect of the antihistamine. If an alcoholic has a low prothrombin level (common with long-term abuse), ingestion of aspirin can produce massive gastrointestinal bleeding. The use of ethanol is common in people who abuse illicit drugs such as marijuana, cocaine, heroin, and various hallucinogens. The combination can have extremely adverse effects far beyond those expected if only one drug were being used.

Legal/Medical Issues

The request for a blood alcohol analysis triggers questions not usually asked when other laboratory tests are ordered. The purpose of the test should be ascertained carefully to determine whether the need is strictly medical or if some legal action may be forthcoming which would involve use of the test results. Documentation of sampling, consent, and laboratory data is handled somewhat differently for the two situations.

Even if the request is for medical purposes, many physicians are familiar with the physiological effects of ethanol at various concentrations shown in Table 12-2. In most states, a 0.10% or greater alcohol concentration in a driver's blood is presumptive evidence of driving under the influence of alcohol. Evidence from the American Medical Association suggests this level should be lowered to 0.08% or less.

Precautions should be taken in the assay procedure to determine the lower limit of detection for the method employed. Most assays have a sensitivity of 0.02–0.03% detection limit. If this limit has not been experimentally established, the laboratory could report a result that implies some ethanol intake when there was none. There is a difference between reporting "0.03% ethanol" and a result of "less than lower limit of detection." The first value implies that some ethanol was present in the sample. The second result merely indicates that the laboratory cannot measure below 0.03% and the sample gave a result that was below the limit of detection; in this case, there is no implication that any ethanol was present. For some legal situations, the presence or absence of ethanol in the sample may be as significant to the case as the actual level determined. Laboratory staff should be very careful about implying that ethanol was present when none was there.

Methanol and Isopropanol Abuse

The addictive nature of alcohol leads some alcoholics to drink whatever they can obtain. On occasion, the laboratory may be confronted with a case of methanol or isopropanol ingestion. Methanol is a major component of a number of paints and varnishes, as well as being found in paint removers. Isopropanol can be easily purchased in 70% solution as rubbing alcohol. Both materials are extremely toxic. Ingestion of methanol leads to formation of formaldehyde and formic acid. Alterations in acid/base balance, destruction of the pancreas, and blindness occur frequently in this situation. Metabolism of isopropanol produces acetone, also creating metabolic acidosis.

The ideal assay for these possibilities is gas chromatography. Increases in the amounts of either methanol or isopropanol or the presence of acetone is evidence for the possible presence of these materials. Alternatively, if isopropanol poisoning is suspected, a serum or urine screen for acetone can be performed. All data must be interpreted with caution, making sure that other medical possibilities have been excluded.

21.2

ILLICIT SUBSTANCE ABUSE

Amphetamines

At one time, amphetamines (Fig. 21–2) were widely used to treat obesity by encouraging weight loss. These stimulants suppress the appetite, but were later shown to be little more effective than

FIGURE 21-2   Structures of common amphetamines.

treatment through diet restriction alone. In the early 1970s, awareness of the addictive and psychologically damaging hazards of amphetamines led to a severe cutback in their manufacture and use as pharmaceutical agents. These drugs are still used in the treatment of narcolepsy (uncontrolled sleeping) and have some value in the management of hyperactive children.

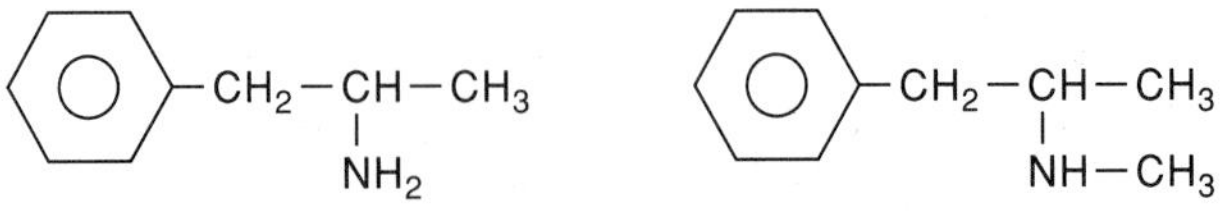

Amphetamines produce alertness, foster increase in mood, and reduce feelings of fatigue. To some degree they enhance the desire to work when therapeutic doses are employed. Amphetamines are frequently used by people who need to stay awake and alert for long periods (such as long-haul truck drivers and students). Athletes often abuse amphetamines to enhance their performance and diminish their feeling of tiredness.

Amphetamine abuse manifests itself in many ways. Dryness of the mouth, dizziness, blurred vision, and headaches are common at lower levels of usage. Sleeplessness and anxiety are frequently produced by small doses. As the intake increases, mood swings, delusions, and hallucinations become common. Agitation, skin disorders, and vitamin deficiencies are frequently observed. Panic states and psychoses are common in extreme abusers.

Amphetamines are considered to produce a state of physical dependency, a common situation for addictive drugs. Withdrawal results in marked depression, long periods of sleep, and extreme appetite. The loss of the sense of power and well-being felt while on amphetamines frequently drives an abuser back to using these substances, indicating that they are both physically and psychologically addictive.

Amphetamines are usually administered orally or by intravenous injection. A portion of the drug is converted by the liver, rendering it pharmacologically inactive. Significant amounts of the material are excreted unchanged in the urine within a few hours after administration, allowing easy detection of amphetamine usage by urine drug-screening techniques.

The source of the majority of illicit amphetamines is synthesis in an illegal laboratory. The precursors are fairly readily obtained and the synthetic techniques are reasonably simple. Many "underground" publications can be obtained which describe the synthetic process in detail. The major hazards involved are the flammable diethyl ether used as a solvent in the process and the toxic amine precursors.

Cocaine

Perhaps the major illicit drug of abuse is cocaine, which sprang from obscurity 20 years ago to increasing prominence today. A number of myths have arisen about cocaine usage, addiction, and analysis. Because of its frequent association with an affluent lifestyle, cocaine has a glamour not present with other drugs. However, as the price declines and the problems with usage become more apparent, we are recognizing the cocaine epidemic as one of the most destructive forces in our society.

Formally known as **benzoylmethylecgonine, cocaine** is an alkaloid extracted from specific plants. It can be employed in two ways. The hydrochloride salt can be ingested, administered intravenously, or inhaled. This compound is water-soluble and unstable when heated. The free base ("crack") is formed by neutralization of the hydrochloric acid present in the salt. The material then becomes soluble in organic solvents such as ethanol and ether. Since the free base form is volatile and stable at high temperatures, it can be smoked as well as be administered by other routes. The most popular route for cocaine intake appears to be intranasal, while most abusers prefer to smoke crack.

When taken nasally, absorption is fairly slow through the mucous membranes, although the effects may last an hour or more, compared with the approximately 20-min euphoria experienced when smoking cocaine. Once in the circulation, the drug is rapidly metabolized by plasma cholinesterases to form benzoylecgonine and ecgonine methyl ester as primary metabolites (Fig. 21–3). In addition to the enzymatic hydrolysis, slow conversion to these two metabolites can take place spontaneously through splitting of the ester linkages. The metabolites are excreted in the urine and can be readily detected by a variety of chemical testing procedures. Although

the half-life of cocaine in the circulation is only about 30 min to 2 h, benzoylecgonine accumulates, since its half-life is 6–7 h.

Laboratory testing usually does not focus on identification of cocaine itself, since it is fairly rapidly converted to other materials. Lowering of pH and temperature can delay this conversion process somewhat. The major component analyzed for is benzoylecgonine. Assays are commercially available to detect the presence of this material by RIA, enzyme immunoassay, or gas chromatography/mass spectrometry. Care must be taken when using any type of immunoassay to ascertain the degree of cross-reactivity. Since a metabolite is being determined, presence of benzoylecgonine may not correlate with the actual amount of cocaine presently in the bloodstream or with clinical symptoms at the time of sampling and analysis.

Cocaine induces a strong sense of euphoria and mental alertness, while diminishing feelings of fatigue. The drug produces both vasoconstriction and elevated heart rate, leading to significant increase in blood pressure. Restlessness and increased motor activity are seen, particularly as dosage is increased. Hallucinations and a sense of paranoia may be experienced. As the cocaine is metabolized and the pharmacological effects begin to decrease, a strong feeling of depression sets in, frequently leading to further usage. Cocaine has been shown to be both physically and psychologically addictive.

In recent years there have been a number of well-documented cases of death due to cocaine use. Myocardial infarct is one of the common causes of death, probably assisted by the increase in heart rate, vasoconstriction, and elevation of blood pressure that accompany cocaine use. Another effect of the drug is to destabilize the cardiac rhythm, which contributes to a heart attack. A significant number of cases have also been reported in which cocaine use appears to have led to a stroke.

Perhaps one of the more tragic consequences of cocaine use is seen in pregnancy. There is an increased incidence of spontaneous abortion and stillbirth among pregnant women on cocaine. A greater number of infants born to cocaine users have birth defects. Many of these babies are born addicted to cocaine and can have potential neurologic damage.

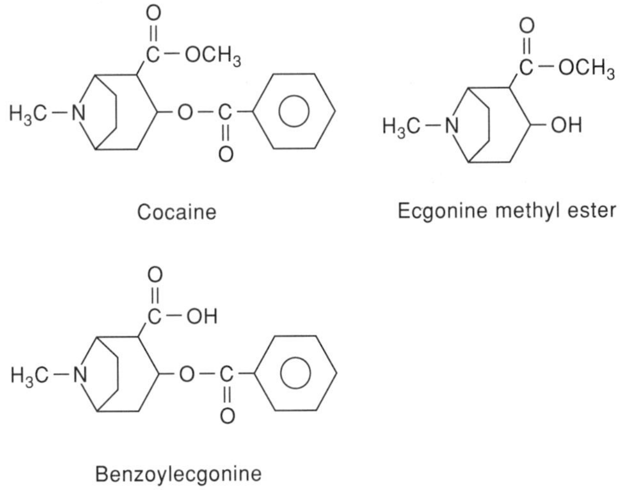

FIGURE 21–3 Cocaine and its metabolites.

Heroin

Heroin abuse has risen, fallen, and risen again during the 20th century. Once considered a major drug of abuse, its popularity declined in the 1960s and 1970s as other abused substances became more common. In recent years, the availability of "black tar," an inexpensive form of heroin, has led to some resurgence of use.

Although heroin can also be ingested or inhaled,

the major route of administration is intravenous. Commonly, the drug is diluted with other materials (sugar, quinine, starch) before injection. Once in the circulation, heroin (with a half-life of only 6–8 min) is rapidly metabolized to morphine derivatives, and little heroin is ever detected in routine analysis. The initial metabolic conversion is the loss of an acetyl group from heroin (diacetylmorphine) to form 6-monoacetylmorphine (Fig. 21–4). This metabolite has a half-life in the bloodstream of approximately 40 min, metabolizing further to morphine. The liver then conjugates the morphine, forming morphine glucuronide, which is then excreted in the urine.

Since heroin is metabolized so rapidly, the drug analyzed for in cases of suspected heroin usage is morphine. In some assays, the conjugated form is hydrolyzed first, particularly if gas chromatography is to be used for separation and identification. Using GC, all derivatives can be easily identified and quantitated. There is usually no value in determining the exact amount of morphine and derivatives present. A simple yes-or-no answer is generally sufficient to demonstrate the presence of the drug.

Use of heroin initially generates a feeling of extreme well-being. The psychological state is one in which tension is reduced and there is a slowing down of both physical activity and perception. Aggressive tendencies are markedly diminished, as is the sex drive. The relaxed, tranquil state of mind may last for several hours, accompanied by daydreams. When withdrawal begins, tremors, muscle cramps, nausea, vomiting, and diarrhea set in if another administration of heroin is not forthcoming.

A number of medical complications accompany prolonged heroin usage. The drug is physically addictive, and withdrawal can be extremely painful (but rarely fatal). Since heroin is administered by injection and needles are often shared, contracting hepatitis and AIDS are major health concerns. Constant damage to the veins is also a problem. Heroin use during pregnancy leads to an increased incidence of spontaneous abortions and stillbirths. Babies born to mothers using heroin are often addicted at birth, requiring the infant to go through withdrawal. Quite often the baby dies during this time due to cardiac stress.

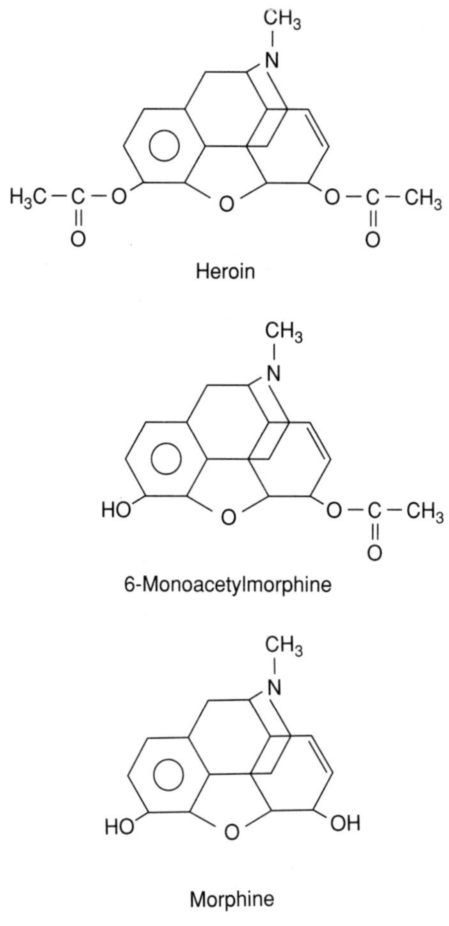

Marijuana

The use of marijuana has been a controversial subject for decades. At one time, the drug was considered highly dangerous and many erroneous concepts about its effects were prevalent. The pendulum of opinion swung drastically during the 1960s to a position which held that marijuana was not even as harmful as alcohol. Current research is beginning to show that there are definite hazards in the use of marijuana, even though many pieces of the puzzle are not yet in place.

Marijuana is prepared from the leaves and stems of the plant *Cannabis sativa,* which for centuries was grown as a source of hemp. The physiological and psychological effects of using marijuana are due to a group of compounds known as **cannabinoids,** the major active ingredient of which is **delta-9-tetrahydrocannabinol (THC)** (Fig. 21–5). When marijuana is smoked, this material is rapidly absorbed from the lungs into the bloodstream. The THC reaches the brain in about 30 s. Storage in fatty tissues is one reason for the prolonged retention of THC and metabolites in the body. Over one quarter of inhaled material from a single use can be detected in the body a week later. Some studies indicate that a month or more is required to clear all the active ingredients from the body after one marijuana dose.

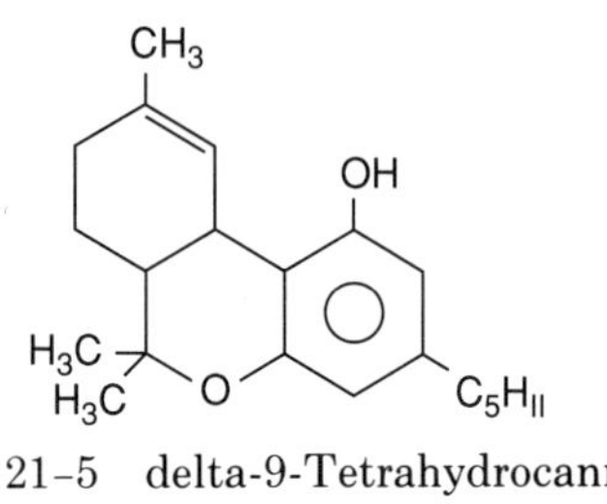

FIGURE 21–4 Heroin and its metabolites.

FIGURE 21–5 delta-9-Tetrahydrocannabinol.

Both THC and its metabolites are rapidly excreted in the urine, making urine a good sample for testing purposes. A positive test could be seen as soon as 1 h after smoking marijuana, with the THC concentration peaking at about 5 h. It is possible in some situations to obtain a positive test up to 3 days after use.

A variety of techniques are available for THC testing, including RIA and enzyme immunoassay (EIA). The EIA approach has the advantage of speed and lower expense, although the RIA method is somewhat more sensitive. Confirmation can be accomplished by gas chromatography/mass spectrometry.

Reports of the effects of marijuana on humans are contradictory, due in part to differences in potency of the material used and in the experience of the user. Definite impairment of motor skills has been amply documented. The psychological response differs from one individual to another. Some users report a sense of euphoria and relaxation, whereas others become more hyperactive. Time perception is altered and reaction time appears to be impaired. Oral communication skills diminish as the user is unable to focus on a sustained task. Some individuals experience panic reactions and states of high anxiety on occasion. In addition to its psychological addiction, there is strong evidence that marijuana is physically addictive, producing withdrawal symptoms similar to those of mild heroin addiction. The addictive aspects of marijuana use are still under active investigation.

Long-term effects, seen even after marijuana use is discontinued, are becoming more apparent. A decrease in motivation accompanies marijuana use and is sometimes seen even when use of the drug is stopped. Flashbacks similar to those experienced with LSD (lysergic acid diethylamide) or PCP (phencyclidine) use are infrequent, but do occur in some individuals. Evidence suggests some long-term alteration of brain structure, although more research needs to be done in this area. Damage to the respiratory system is well documented and similar to that caused by tobacco smoke, leading to alterations in lung tissue structure.

As with tobacco smoke, a major concern is passive inhalation. It has been clearly shown that an individual in the presence of someone smoking marijuana inhales sufficient fumes to experience some of the signs of marijuana use. In addition, this person tests positive for THC in the urine. Although the health hazards have not been completely studied, concern is growing for this aspect of marijuana abuse.

Phencyclidine

The use of **phencyclidine** (Fig. 21-6) has had a short but violent history in this country. First intro-

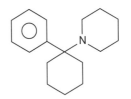

Phencyclidine

FIGURE 21-6 Phencyclidine.

duced as an anesthetic in the late 1950s, the drug 1-(1-phenylcyclohexyl) piperidine hydrochloride (PCP) was withdrawn for human use by 1965 because of its adverse effects. Illicit use became popular during the late 1960s, but declined after a number of reports of very unpleasant and hazardous side effects. After a period of relative unpopularity, PCP again began to be considered a major problem in the late 1970s and continues to present serious difficulties today.

PCP can be eaten, injected, inhaled, or smoked. The major mode of administration currently appears to be by smoking, either alone or in combination with marijuana. The pharmacologic aspects of PCP absorption and excretion are not well known in humans. Plasma half-life is some 11–12 hours. Both unchanged PCP and its metabolites are excreted in the urine. The drug is absorbed well by fatty tissues and is retained for a long time after administration, possibly contributing to the flashback phenomenon frequently experienced with this drug.

The major effects of PCP are a sense of dissociation, euphoria, and numbness when taken in small doses. At higher levels, there is a loss of feeling in the limbs (the drug was originally used as an anesthetic). Disorganized thoughts, bizarre and violent behavior, and disconnection from reality become apparent. Heart rate and blood pressure increase while the central nervous system is depressed. Death from overdose is frequent.

Analytical techniques for PCP include thin-layer chromatography, enzyme immunoassay, and gas chromatography. The method of choice depends on a number of factors including number of assays expected, turnaround time needed, and equipment available. More and more laboratories are exploring the implementation of some sort of PCP assay as this drug of abuse becomes more prevalent.

Anabolic Steroids

It is estimated that over one million athletes in the United States are using anabolic steroids as part of their personal physical development program. The widespread abuse of these materials warrants a discussion of their properties and means for detection. Steroid abuse has a number of medical and legal implications. It is quite possible that the clinical chemistry laboratory will become much more involved in the analysis of anabolic steroids in the future.

Unlike the other categories of abused substances discussed in this chapter, the use of **anabolic steroids** is not illegal. However, athletes in competition are penalized by their sports organizations if these materials are detected in urine samples or if other proof of steroid use is demonstrated. At present, the Food and Drug Administration has not listed anabolic steroids as controlled substances, making it more difficult to enforce bans on usage.

There are over a hundred different anabolic steroids presently known. Some have legitimate therapeutic value, either for human treatment or in veterinary practice. Most of the steroids (Fig. 21-7) abused by athletes are derivatives of testosterone. Some steroids have been voluntarily taken off the market by their manufacturers in the United States, but are still available through European or black-market sources.

Testing for the presence of anabolic steroids in urine is carried out primarily by gas chromatography/mass spectrometry. Some RIA procedures have been privately developed, but problems of specificity and sensitivity have plagued these methods. Some synthetic steroids have little affinity for the antibody employed, rendering them undetectable by the system. Other materials, such as oral contraceptives, produce a positive test since the antibody is not highly specific.

Analysis of steroids by GC/mass spectrometry involves an extraction procedure to isolate the steroids from urine, usually followed by hydrolysis (chemical or enzymatic) to remove glucuronide conjugates. Derivatization with a variety of methylsilyl compounds renders the steroids volatile so they can be separated by gas chromatography. Detection of the entire

mass spectrum of each compound is rarely carried out because it is so extensive. Instead, a few characteristic peaks are monitored to allow more rapid processing of samples.

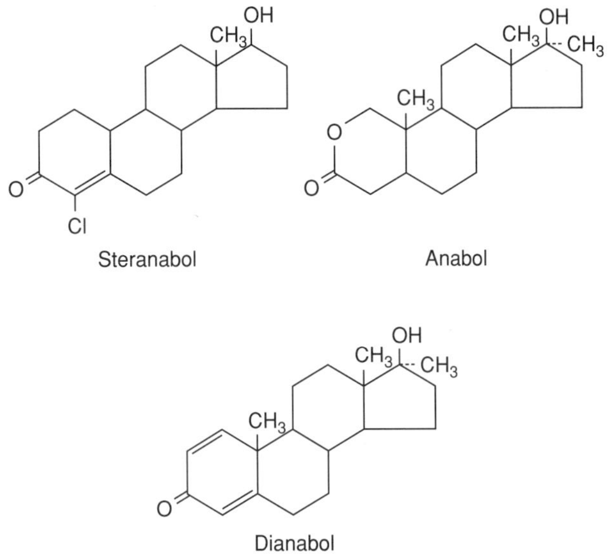

To date a major problem in characterizing these compounds has been obtaining reliable information about metabolites. Interconversions of steroids is a complicated process at best with naturally occurring compounds. Many of the metabolic pathways for the synthetic steroids have not been well characterized in humans, so specific information about their components is lacking in many cases.

Prolonged administration of anabolic steroids leads to a number of serious medical and psychological problems. Dosages up to 100 times the therapeutic level are frequently employed, with complex patterns of use of several steroids over time. Accompanying the striking increase in muscle mass are a variety of problems associated with the liver. Prolonged use results in hepatic damage, including cholestatic hepatitis (which is associated with blockages of the bile duct) and jaundice. Decrease in sperm formation and increase in breast size are frequently seen in males who abuse these compounds. Increases in low-density lipoproteins and diminished production of the HDL fraction are observed, with the total cholesterol unchanged. This phenomenon may explain some of the cardiovascular difficulties experienced by users. There are also increased incidences of liver cancer and kidney damage as a result of anabolic steroid use.

Significant alterations of hormone output are seen in athletes using these materials. Since many of these steroids have a structure which resembles testosterone, the normal synthesis of testosterone by the body is greatly diminished. Detection of decreased amounts of this hormone in urine of male athletes is often used to substantiate other data involving steroid abuse. Increases in estrogen production have been observed in some males. In women, the steroids induce changes resulting in a more masculine appearance. Frequently, these changes are not readily reversible.

One of the major psychological problems associated with steroid abuse is the striking increase in aggressiveness seen after heavy usage. Cases of violence believed to be associated with steroids are becoming more widely reported. Some users exhibit psychotic tendencies and manifest personality traits associated with drug addiction.

FIGURE 21-7   Representative anabolic steroids.

"Designer Drugs"

Perhaps one of the more perplexing issues for those combating substance abuse has to do with a class of compounds known as **"designer drugs."** These materials are analogs of either prescription pharmaceuticals or an illicitly abused drug. The modifications in structure, some of which are quite minor, are enough to create a new compound, which is not specifically banned by current legislation. Synthetic derivatives of heroin and amphetamines are currently available through illicit sources.

A fairly new law may make it easier for enforcement agencies to deal with this class of substances. Without spelling out what specific structures are illegal (as was the case with previous legislation), the new statute prohibits the manufacture and sale of any compound with a structure similar to a known controlled, psychoactive material. Similarity of structure implies similarity of effect and is presumed to be sufficient evidence for making the material illegal. The routine clinical laboratory probably does not need to be too concerned about this area of drug abuse except to be aware that these compounds exist and could be complicating factors in a laboratory drug-screening procedure.

21.3

ABUSE OF PRESCRIPTION DRUGS

Not too surprisingly, another major problem in the area of substance abuse is the abuse of prescription drugs. A significant percentage of drug overdose problems seen in hospital emergency rooms arise from excessive use of prescribed barbiturates, tranquilizers, antidepressants, and other legal mind-altering medications. This situation is further complicated when these materials are used in conjunction with ethanol. Often, an additive, or synergistic, effect results when medications are accompanied by consumption of alcoholic beverages, producing a physiological impact far in excess of what might be expected from the medication alone.

In many instances, the drug may be legal, but the individual using it has not obtained the material through legal means. Theft of prescription drugs is common, as is the manufacture of drug "look-alikes," materials of similar composition manufactured by less-than-reputable sources to appear like known prescription medications. There is a large black-market business in the illicit sale of prescription medications. Any drug-screening program needs to be prepared to handle analyses for these types of drugs in addition to the commonly abused illicit substances previously described.

21.4

LABORATORY IDENTIFICATION OF ABUSED SUBSTANCES

A variety of analytical approaches are available for screening samples for the presence of drug abuse. The techniques range from simple colorimetric screens (which are not very specific or sensitive) to sophisticated analytical schemes involving the latest in high technology. Application to a specific laboratory must be made on the basis of work load; types and frequency of requests; and the investment of time, equipment, and personnel the laboratory wishes to devote to drug-screening services.

Analysis for the presence of abused substances has focused primarily on the use of urine as the test sample of choice. Instead of observing the transient passage of a drug through the circulation, the accumulation of any parent drug present and metabolites of that material can readily be studied. The urine specimen represents the net load of the drug over a long period, whereas the blood sample provides only a quick picture of the drug level at a specific time.

Thin-Layer Chromatography

Perhaps one of the most versatile methods for initial screening for drugs of abuse is **thin-layer chromatography (TLC).** This procedure operates in much the same way as column chromatography, using a solid-phase support medium and a liquid mobile-phase separation system. Drugs are separated on the basis of their ability to dissolve in the solvent system and the strength of their interaction with the support phase. Color reactions are then used to locate and identify specific substances.

The urine sample has its pH adjusted to the slightly basic range (pH 8.5 is common). Drugs are extracted in one of two ways, depending on the specific methodology. The urine may be passed through an ion-exchange column, with the drugs adhering to the column. After a wash to remove interfering materials, all the drugs are eluted with specific solvents. Conversely, the urine sample may be extracted with an organic solvent system which is not water-soluble. The organic layer and the aqueous layer can be easily separated. In either case, the solvent system is evaporated to concentrate the drugs in the original sample, making the assay more sensitive.

The concentrated sample is then spotted near the bottom of a thin-layer chromatography plate coated with a silica gel preparation. One commercial method has the sample concentrated onto a sample disc which is then placed on the plate, eliminating the tedious process of spotting and trying to keep the spot small. Care must be taken to spot the samples above the location of the initial solvent level in the chromatography tank to avoid washing the sample off into the solvent.

The prepared plate is placed in the separation container, with the sample-containing end partially immersed in the solvent system (usually a mixture of organic solvents, weak organic acids, and water). As the solvent travels up the plate by capillary action, it partially dissolves the drugs in the sample.

Each material migrates upward at a rate which is affected by the solubility of that drug in the solvent system and the degree of interaction between drug and the silica solid phase. As a result, various compounds can be separated from one another on the basis of their relative mobilities in that specific system (Fig. 21-8).

When the solvent front gets near the top of the plate, the separation system is removed from the tank. The solvent front is marked for later measurements of relative mobilities and the plate is dried in a warm oven. A variety of color tests are available for detection of specific compounds. The plate is either sprayed with certain reagents or dipped into a solution which reacts with a particular drug or class of drugs. Exposure to UV light allows detection of some materials which fluoresce. The relative mobility of each located spot can be calculated by dividing the distance the analyte travels from the origin by the distance traveled by the solvent front.

$$R_f = \frac{\text{Distance from origin of sample spot}}{\text{Distance from origin of solvent front}}$$

The measurement of distance is usually made from the front of the spot (the part which has traveled the farthest on the plate) to the line marking the place where the sample was initially applied. This retardation factor (R_f) allows comparison with standards and some reasonably firm identification of specific drugs, especially when this information is combined with data regarding color development after different chemical treatment.

A major drawback to the use of TLC in drug identification is some lack of reproducibility from one chromatography run to the next. Among other factors, slight differences in the concentrations of solvents, the degree of equilibration with vapor phase in the separation chamber, and temperature render the R_f value somewhat variable from one analysis to another. The safest way to compensate for these variables is to perform separations of known mixtures of drugs on the same plate at the same time the unknown sample is analyzed. Any changes in conditions are reflected equally in the unknown and the standards alike. Often, a visual comparison with standards is all that is needed to suggest the presence or absence of a particular abused substance.

Identification of drugs by thin-layer chromatography cannot be considered conclusive proof of the use of a specific substance. If possible, all positive results should be confirmed by more specific methods, particularly if there are medical/legal issues involved.

Immunoassay Techniques

For the rapid screening of large numbers of samples, the enzyme immunoassay approach has considerable merit. Within a few minutes, aided by automated instrumentation, a preliminary profile of possible classes of abused substances can readily be determined. But this ease of operation also points out the major drawback to EIA analysis. Only classes of compounds are usually determined, not individual drugs. Urine screens are available for amphetamines, barbiturates, benzodiazepines, and cannabinoids. More specific analyses for the presence of cocaine (actually benzoylecgonine), methadone, heroin (as morphine), and phencyclidine are also available.

Detection by enzyme immunoassay is as sensitive as thin-layer chromatography and gas chromatography. Other methods (RIA and GC/mass spectrometry) have much lower limits of detection. Since the EIA tests are usually for groups of compounds, these data cannot be used as conclusive evidence, but must be followed by confirmation testing when a positive result is obtained. The false positive and false negative rates are both low (frequently less than 5%), making this technique suitable for rapid processing in laboratories with a high volume of drug-screen requests.

High-Performance Liquid Chromatography

The use of high-performance liquid chromatography for drug-screening purposes is in its infancy. The technique has value in being able to separate a variety of different materials fairly rapidly. The main drawback is specificity. There is no conclusive way to demonstrate that a particular peak eluted from the column is, in fact, a certain drug of abuse. Although the use of derivatization reactions (especially postcolumn techniques) might enhance spec-

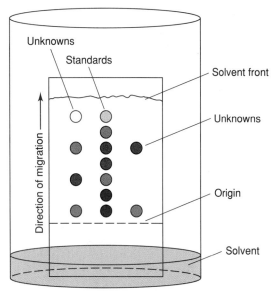

FIGURE 21-8 Separation of drugs by thin-layer chromatography.

ificity somewhat, the current legal climate requires that uncertainty be decreased to as low a level as possible. This uncertainty in being able to confirm conclusively the presence or absence of a particular drug dictates against HPLC as a standard method of analysis. Future developments, such as adaptation of HPLC to mass spectrometry, may change or eliminate this restriction.

Gas Chromatography/Mass Spectrometry

The most specific and sensitive method for drug screening is the coupling of gas chromatography to mass spectrometry. The selectivity and power of separation of GC is combined with the ability to identify each separated peak in terms of molecular weight and structural characteristics by mass spectrometry. The combination gives rise to a highly specific method for the elucidation of the presence and structural identity of abused substances.

Urine is the usual sample of choice, since both parent drug and metabolites might be detected. The sample is extracted (either with a column or with solvent) and the extract concentrated to maximize the ability to detect substances. Injection on the gas chromatograph and determination of the fractionation pattern was described in Chapter 2. Often a preliminary identification of specific materials can be made on the basis of the GC trace, although confirmation is required.

After the gas chromatograph run to determine the presence of compounds, a second injection of the sample is made into the chromatograph. The same separation on the column takes place, but the eluted fractions are then routed through a mass spectrometer (Fig. 21–9). Fragmentation of the molecule(s) in each peak occurs (by electron bombardment or other means), and the charged particles pass through the magnetic field. Separation of fragments is a function of the charge and mass of each piece of the original molecule. Current instruments are capable of separations of masses which differ from one another by one atomic mass unit (the mass of a hydrogen atom).

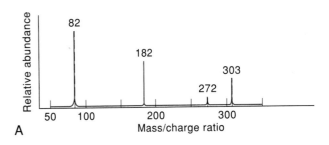

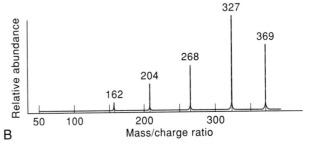

FIGURE 21–10   A. Mass spectra for cocaine. B. Mass spectra for heroin.

The data obtained from the fragmentation analysis is stored in a computer and processed for maximum information. Each GC peak has a mass spectrum determined, so a correlation between elution time from the column and structure can be easily made. The mass spectrum consists of a series of lines located across the x axis (giving the mass of each fragment). The height of the line corresponds to the relative amount of the fragment in the mixture. In most instances, only a few major peaks are examined to confirm the presence of a specific compound (Fig. 21–10).

The analysis of mass spectroscopy data is a complicated task. Many laboratories use a data base from the National Bureau of Standards which contains spectral data of over 15,000 compounds. Special libraries can be constructed for different classes of compounds of interest. Undue reliance should not be placed on computer interpretations, but these data bases can be extremely helpful in dealing with structural identification.

FIGURE 21–9
Diagram of a mass spectrometer (From *Organic Chemistry*, 4th ed., by Ralph Fresenden and Joan Fresenden. Copyright 1990 by Wadsworth, Inc. Reprinted with permission of Brooks/Cole Publishing Company, Pacific Grove, CA 93950.)

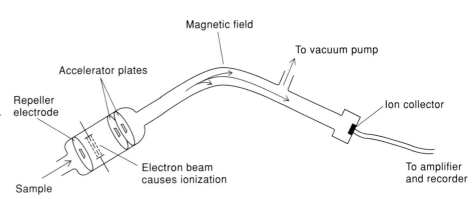

Reliability of Laboratory Analysis

An area of vital concern to both laboratories and those who are being screened for drug use is the reliability of the testing procedures. Reports in the news media as well as questions raised in both the scientific and legal communities have generated a number of issues about testing procedures.

One obvious issue concerns the qualifications of those performing the testing. Laboratory staff must be well trained, familiar with the methods employed, and conscientious. Several studies have pointed out the problems of using nonlaboratory personnel or those with minimal training in laboratory testing. A much higher level of error occurs in this situation. Failure to follow procedures, even by highly trained staff, can negate any training or quality assurance program.

Research on the testing procedures is constantly leading to improvements in methodology. Of the commercially available materials, significant changes have taken place even within the last few years to lower the false positive and false negative rates. If immunoassay techniques are employed, some other materials may cross-react. For example, a variety of prescription and nonprescription medications may produce false positive results when screens for amphetamines, barbiturates, and opiates are performed. In some instances, the specificity of antibody has been improved (resulting in less cross-reactivity). Most manufacturers now claim accuracies in excess of 97%.

The setting of cut-off points for assays plays an important, but often unrecognized, role in assessing whether a given sample is positive or negative for the presence of a drug. Testing procedures are established to minimize the number of false positive results. To do this, the sensitivity of detection for a compound is adjusted so that only drug concentrations above a certain level are detected. If a specific substance exists in a sample, but is there at a concentration below the cut-off point (the lowest concentration detectable by the assay), the sample would give a negative result, even though some of the analyte drug was present. Knowledge of cut-off points for assays is important for the interpretation and reporting of results (Table 21-3).

21.5

LEGAL/MEDICAL ISSUES

Consent Versus Unlawful Search

Obtaining a sample for drug screening can present as many problems as the analysis itself. An issue which cannot be ignored is consent. For any routine laboratory test, consent is required. We usually do not bother with documentation of that consent; the cooperation of the patient in allowing a venipuncture or in providing a urine sample is considered sufficient. Drug screening, however, is often being done in association with some legal issue (whether it be a violation of the law or an employer/employee dispute). Voluntary consent must be documented in such cases to avoid later problems.

In some situations, an "implied consent" statute may exist. Most states have such a law with regard to breath alcohol testing. When an individual obtains a driver's license, he or she also consents to be tested for alcohol level under some stated circumstances. Refusal to undergo this testing can result in automatic assumption of guilt and revocation of the license for a stated period of time.

The process of obtaining consent can be complicated and should be established in consultation with knowledgeable legal counsel. The procedure and documentation should be clearly spelled out and strictly adhered to by all staff. Consent must be obtained in writing before any sample collection can be carried out.

Chain of Evidence

Because of the possibility of sample tampering, a chain of evidence (or chain of custody) must be established for the sample. The chain begins when the sample is collected and terminates with the person who performs the actual analysis. All individuals who handle the sample must document the time and date when they had the sample in their control, as well as the particular process they carried out (sample collection, centrifugation, or other step). If a sample is to be stored until a later time, it must be sealed and stored in a secure site. Many laboratories have special locked boxes for these kinds of samples. Frequently, the sample is wrapped with tape and the tape is signed by the person preparing the specimen. Tampering can then be easily observed if the integrity of the wrapping and writing are disturbed. A written document is kept on the process, signed by each person who handles the sample in any way.

Table 21-3
LOWER LIMITS OF DETECTION FOR
VARIOUS SCREENING METHODS

Method	Approximate Lower Limit of Detection (mg/mL)
TLC	0.2-5
EIA	0.025-5
HPLC	0.02-10
GC alone	0.01-10
RIA	0.001-10
GC/mass spec	0.001-5

Limit of detection depends on drug being analyzed. Adapted from *J. Amer. Med. Assoc.* 257:3110-3114, 1987.

This is a legal document and should be designed in collaboration with an attorney or other qualified legal counsel.

Confidentiality and Reporting of Results

Reporting the results of drug screens presents some unique challenges. If the test is run in response to a strictly medical request, results can be reported in the usual manner to the physician. Even in this instance, however, results should not be given over the telephone. There have been instances in which law enforcement personnel have telephoned the laboratory and misrepresented themselves in order to obtain information on drug use by an individual. Taking the number and calling back later may help to avoid some of these situations.

When a drug screen is performed in relation to a legal situation (actual or potential), extreme care must be taken. Results are not given out by telephone and are not discussed with anyone except the person whose sample was analyzed. Information should be given in writing and only after proof of identification. If the case goes to court, the test results can be subpoenaed and entered as evidence in open court, where they are subject to cross-examination. Improper release of drug-screen information can result in legal action against the hospital, the laboratory, and the individual who wrongly provided the results.

SUMMARY

Substance abuse has reached epidemic proportions in the United States and elsewhere. Although ethanol is one of the most abused materials (along with cigarettes), illicit drugs constitute an important component of this problem. Cocaine, marijuana, heroin, and other controlled materials are significantly abused. There is a growing recognition of the problem of abuse of prescription and over-the-counter medications. Anabolic steroid misuse has become a major concern in athletics.

Ethanol is rapidly absorbed after ingestion, with blood levels peaking within 1 hour. The major routes of conversion and elimination are loss through respiration and perspiration, excretion in the urine, and conversion to acetaldehyde by enzyme systems in the liver. Primary means of ethanol quantitation are by gas chromatography and enzymatic oxidation to form acetaldehyde and NADH.

The primary abused illicit drugs all have some characteristics in common. Each drug has been shown to generate both physical and psychological addiction. In some instances, prolonged use of the material can lead to psychosis. Analysis involves detection of either the parent drug, a metabolite, or both. The preferred sample for a drug screen is urine. Heroin use is confirmed by detection of morphine in the urine. The major metabolite of cocaine is benzoylecgonine. Marijuana use is suggested by identification of various tetrahydrocannabinol derivatives. Passive absorption of marijuana is now a well-documented phenomenon.

A number of screening and confirmatory procedures are available to identify the presence of drugs of abuse. Thin-layer chromatography provides versatility and flexibility, but requires confirmatory tests. Enzyme immunoassay procedures screen mainly for classes of compounds and also need confirmatory testing. Although HPLC has some potential in drug screening, the method of choice for identification of specific unknown materials is gas chromatography/mass spectrometry.

Sampling, patient consent, chain of evidence, and confidentiality in reporting are areas which must receive particular attention in a drug-screening facility. Since the outcome of testing has significant impact on jobs, punitive measures by the courts, and other legal and social issues, special precautions must be taken to protect individual rights and provide accuracy in reporting.

FOR REVIEW

Directions: For each question, choose the best response.

1. All of the following are physiological effects of ethanol consumption *except*
    A. depresses central nervous system
    B. increases heart rate
    C. increases vasodilation
    D. decreases blood pressure

2. All of the following may be used to cleanse the skin when drawing blood for ethanol analysis *except*
    A. alcohol swab
    B. merthiolate
    C. soap and water
    D. zephiran

3. In the determination of ethanol by the alcohol dehydrogenase method, the product measured is
    A. acetaldehyde
    B. semicarbazide
    C. NAD^+
    D. NADH

4. Benzoylecgonine is the major metabolite of
    A. cocaine
    B. heroin
    C. marijuana
    D. phencyclidine

5. Morphine is the major metabolite of
    A. cocaine
    B. heroin
    C. marijuana
    D. phencyclidine

6. The major active component of marijuana is
    A. delta-9-tetrahydrocannabinol
    B. ecgonine methyl ester
    C. quinine
    D. phencyclidine

7. Substances with modified structures that are analogs of prescription pharmaceuticals or abused drugs are known as
    A. designer drugs
    B. generic drugs
    C. trade drugs
    D. toxic drugs

8. A chromatography system that requires a solid-phase support medium and a liquid mobile-phase separation system best describes:
    A. GC
    B. GLC
    C. HPLC
    D. TLC

9. The system of choice for drug analysis because of its specificity and sensitivity is
    A. GC
    B. HPLC
    C. TLC
    D. GC/mass spectrometry

10. In a medicolegal case it is important that the specimen be handled properly to avoid
    A. any tampering with the specimen
    B. any discrepancy in the chain of evidence

C. any contamination of the specimen

D. loss of the specimen

E. all of the above

BIBLIOGRAPHY

Ackerman, S., "Drug testing: The state of the art," *Am. Scientist.* 77:19–23, 1989.

Aziz, K. et al., "Drug abuse testing," *Amer. Clin. Prod. Rev.* October:44–46, 1987.

Carroll, C. R., *Drugs in Modern Society,* 2nd ed. Dubuque: Wm. C. Brown, 1989.

Council on Scientific Affairs, American Medical Association, "Scientific issues in drug testing," *J. Amer. Med. Assoc.* 257:3110–3114, 1987.

Cregler, L. L., and Mark, H., "Medical complications of cocaine abuse," *New Eng. J. Med.* 315:1495–1500, 1986.

Davidson, P. L., "Start-up considerations for a drug testing laboratory," *Lab. Manage.* November:47–53, 1987.

Dubowski, K., "Alcohol determination in the clinical laboratory," *Am. J. Clin. Pathol.* 74:747–750, 1980.

Dubowski, K., "Alcohol analysis: Clinical aspects. Parts I and II, *Lab. Med.,* March:43–54 and April:27–36, 1982.

Eckardt, M. J. et al., "Health hazards associated with alcohol consumption," *J. Amer. Med. Assoc.* 246:648–666, 1981.

Edmonson, H. A., "Pathology of alcoholism," *Am. J. Clin. Pathol.* 74:725–742, 1980.

Hume, D. N., and Fitzgerald, E. F., "Chemical tests for intoxication: What do the numbers really mean?" *Anal. Chem.* 57:876A–886A, 1985.

Kaye, S., "The collection and handling of the blood alcohol specimen," *Am. J. Clin. Pathol.* 74:743–746, 1980.

Poklis, A., "Drugs of abuse: An integrated approach to urine screening." *Lab. Manage.* July:41–49, 1986.

Spiehler, V., "Cocaine and metabolites," *Clin. Chem. News.* July, 1986.

Warner, M., "Jumping to conclusions," *Anal. Chem.* 59:521A–522A, 1987.

Willette, R. E., "Cannabinoids," *Clin. Chem. News.* December, 1983.

INDEX

■■■■■■■■

Note: Page numbers in *italics* refer to illustrations; page numbers followed by t refer to tables.